Technical Manual No 3-34.64/Marine
Corps Reference Publication 3-17.7G

Headquarters
Department of the Army
Washington, D.C., 25 September 2012

Military Soils Engineering

I0030420

Contents

DISTRIBUTION RESTRICTION: Approved for public release; distribution is unlimited.

* This manual supersedes FM 5-410, 23 December 1992.

Figures

Contents

Tables

Preface

SCOPE

Construction in the theater of operations is normally limited to roads, airfields, and structures necessary for military operations. This manual emphasizes the soils engineering aspects of road and airfield construction. The references give detailed information on other soils engineering topics that are discussed in general terms. This manual provides a discussion of the formation and characteristics of soil and the system used by the United States (US) Army to classify soils. It also gives an overview of classification systems used by other agencies. It describes the compaction of soils and quality control, settlement and shearing resistance of soils, the movement of water through soils, frost action, and the bearing capacity of soils that serve as foundations, slopes, embankments, dikes, dams, and earth-retaining structures. This manual also describes the geologic factors that affect the properties and occurrences of natural mineral/soil construction materials used to build dams, tunnels, roads, airfields, and bridges. Theater-of-operations construction methods are emphasized throughout the manual.

PURPOSE

This manual supplies engineer officers and noncommissioned officers with doctrinal tenets and technical facts concerning the use and management of soils during military construction. It also provides guidance in evaluating soil conditions, predicting soil behavior under varying conditions, and solving soil problems related to military operations. Military commanders should incorporate geologic information with other pertinent data when planning military operations, to include standing operating procedures.

The proponent of this publication is the US Army Engineer School. Submit changes for improving this publication on DA Form 2028 and forward it to: Commandant, US Army Engineer School, ATTN: ATSE-TD-D, Fort Leonard Wood, MO 65473-6650.

Unless otherwise stated, masculine nouns and pronouns do not refer exclusively to men.

Chapter 1

Rocks and Minerals

The crust of the earth is made up of rock; rock, in turn, is composed of minerals. The geologist classifies rocks by determining their modes of formation and their mineral content in addition to examining certain chemical and physical properties. Military engineers use a simpler diagnostic method that is discussed below. Rock classification is necessary because particular rock types have been recognized as having certain properties or as behaving in somewhat predictable ways. The rock type implies information on many properties that serves as a guide in determining the geological and engineering characteristics of a site. This implied information includes—

- A range of rock strength.
- Possible or expected fracture systems.
- The probability of encountering bedding planes.
- Weak zones.
- Other discontinuities.
- Ease or difficulty of rock excavation.
- Permeability.
- Value as a construction material.
- Trafficability.

This chapter describes procedures for field identification and classification of rocks and minerals. It also explains some of the processes by which rocks are formed. The primary objective of identifying rock materials and evaluating their physical properties is to be able to recommend the most appropriate aggregate type for a given military construction mission.

SECTION I - MINERALS

PHYSICAL PROPERTIES

1-1. Rocks are aggregates of minerals. To understand the physical properties of rocks, it is necessary to understand what minerals are. A mineral is a naturally occurring, inorganically formed substance having an ordered internal arrangement of atoms. It is a compound and can be expressed by a chemical formula. If the mineral's internal framework of atoms is expressed externally, it forms a crystal. A mineral's characteristic physical properties are controlled by its composition and atomic structure, and these properties are valuable aids in rapid field identification. Properties that can be determined by simple field tests are introduced here to aid in the identification of minerals and indirectly in the identification of rocks. These properties are—

- Hardness.
- Crystal form.
- Cleavage.
- Fracture.
- Luster and color.

- Streak.
- Specific gravity.

HARDNESS

1-2.   The hardness of a mineral is a measure of its ability to resist abrasion or scratching by other minerals or by an object of known hardness. A simple scale based on empirical tests has been developed and is called the Mohs Hardness Scale. The scale consists of 10 minerals arranged in increasing hardness with 1 being the softest. The 10 minerals selected to form the scale of comparison are listed in table 1-1. Hardness kits containing most of the reference minerals are available, but equivalent objects can be substituted for expediency. Objects with higher values on Mohs' scale are capable of scratching objects with lower values. For example, a rock specimen that can be scratched by a copper coin but not by the fingernail is said to have a hardness of about 3. Military engineers describe a rock as either hard or soft. A rock specimen with a hardness of 5 or more is considered hard. The hardness test should be performed on a fresh (unweathered) surface of the specimen.

Table 1-1.The Mohs hardness scale

Mineral	Relative Hardness	Equivalent Objects
Diamond	10	
Corundum	9	
Topaz	8	
Quartz	7	Porcelain (7)
Feldspar	6	Steel file (6.5)
Apatite	5	Window glass (5.5) Knife blade or nail (5)
Fluorite	4	
Calcite	3	Copper coin (3.5)
Gypsum	2	Fingernail (2)
Talc	1	

CRYSTAL FORM

1-3.   Most, but not all, minerals form crystals. The form, or habit, of the crystals can be diagnostic of the mineral and can help to identify it. The minerals galena (a lead ore) and halite (rock salt) commonly crystallize as cubes. Crystals of garnet (a silicate mineral) commonly have 12 or 24 equidimensional faces. Some minerals typically display long needle-like crystals. Minerals showing no crystal form are said to be amorphous. Figure 1-1 illustrates two of the many crystal forms.

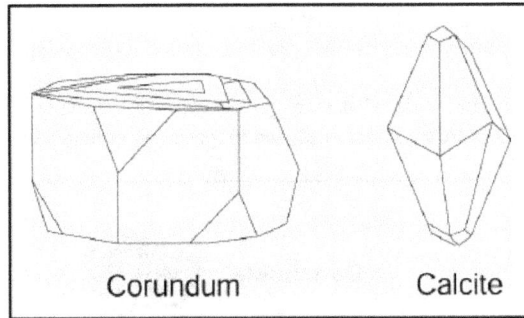

Figure 1-1. Crystal forms

CLEAVAGE

1-4. Cleavage is the tendency of a mineral to split or separate along preferred planes when broken. It is fairly consistent from sample to sample for a given mineral and is a valuable aid in the mineral's identification. Cleavage is described by noting the direction, the degree of perfection, and (for two or more cleavage directions) the angle of intersection of cleavage planes. Some minerals have one cleavage direction; others have two or more directions with varying degrees of perfection. Figure 1-2 illustrates a mineral with one cleavage direction (mica) and one with three directions (calcite). Some minerals, such as quartz, form crystals but do not cleave.

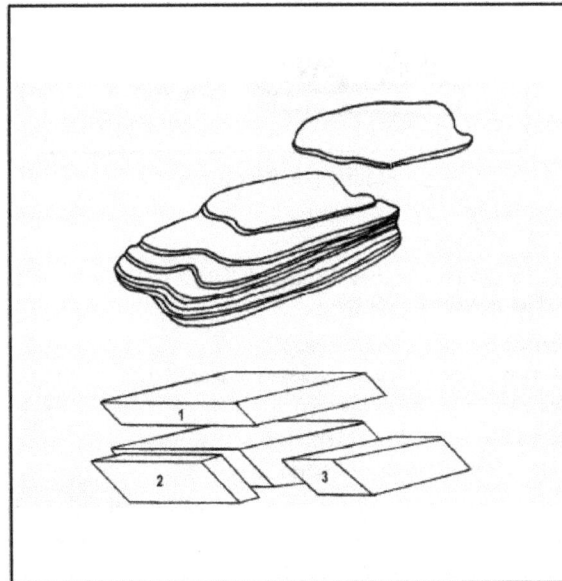

Figure 1-2. Cleavage

FRACTURE

1-5. Fracture is the way in which a mineral breaks when it does not cleave along cleavage planes. It can be helpful in field identification. Figure 1-3, page 1-4, illustrates the common kinds of fracture. They are—

- Conchoidal. This fracture surface exhibits concentric, bowl-shaped structures like the inside of a clam shell (for example, chert or obsidian).
- Fibrous or splintery. This fracture surface shows fibers or splinters (for example, some serpentines).
- Hackly. This fracture surface has sharp, jagged edges (for example, shist).
- Uneven. This fracture surface is rough and irregular (for example, basalt).

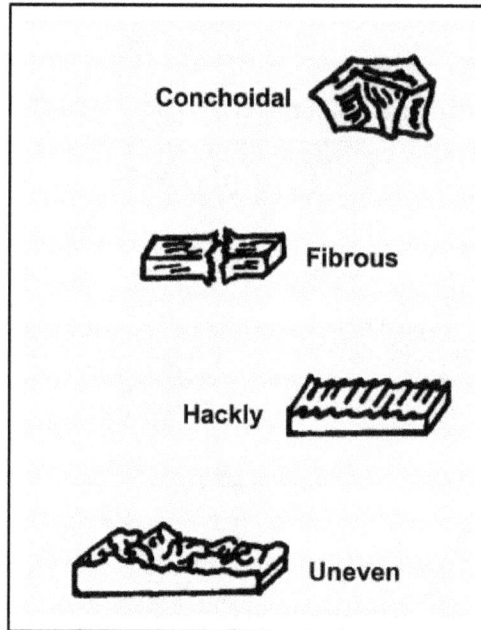

Figure 1-3. Fractures

LUSTER AND COLOR

1-6.   The appearance of a mineral specimen in reflected light is called its luster. Luster is either metallic or nonmetallic. Common non-metallic lusters are—

- Vitreous (having the appearance of glass).
- Adamantine (having the brilliant appearance of diamonds).
- Pearly (having the iridescence of pearls).
- Silky (having a fibrous, silklike luster).
- Resinous (having the appearance of resin).

1-7.   For some minerals, especially the metallic minerals, color is diagnostic. Galena (lead sulphide) is steel gray, pyrite (iron sulphide) is brass yellow, and magnetite (an iron ore) is black. However, many nonmetallic minerals display a variety of colors. The use of color in mineral identification must be made cautiously since it is a subjective determination.

STREAK

1-8.   The color of a powdered or a crushed mineral is called the streak. The streak is obtained by rubbing the mineral on a piece of unglazed porcelain, called a streak plate. The streak is much more consistent in a mineral than the color of the intact specimen. For example, an intact specimen of the mineral hematite (an iron ore) may appear black, brown, or red, but the streak will always be dark red. The streak is most useful for the identification of dark-colored minerals such as metallic sulfides and oxides. Minerals with hardness 6.5 will not exhibit a streak, because they are harder than a piece of unglazed porcelain.

SPECIFIC GRAVITY

1-9.   The specific gravity of a substance is the ratio of its weight (or mass) to the weight (or mass) of an equal volume of water. In field identification of minerals, the heft, or apparent weight, of the specimen is an aid to its identification. Specific gravity and heft are controlled by the kinds of atoms making up the mineral and the packing density of the atoms. For example, ores of lead always have relatively high specific gravity and feel heavy.

COMMON ROCK-FORMING MINERALS

1-10.   There are approximately 2,000 known varieties of minerals. Only about 200 are common enough to be of geologic and economic importance. Some of the more important minerals to military engineers are—

- Quartz.
- Feldspars.
- Micas.
- Amphiboles.
- Pyroxenes.
- Olivine.
- Chlorite.
- Calcite.
- Dolomite.
- Limonite.
- Clay.
- Quartz.

QUARTZ

1-11.   Quartz (silicon dioxide) is an extremely hard, transparent to translucent mineral with a glassy or waxy luster. Colorless to white or smoky-gray varieties are most common, but impurities may produce many other colors. Like man-made glass, quartz has a conchoidal (shell-like) fracture, often imperfectly developed. It forms pointed, six-sided prismatic crystals on occasion but occurs most often as irregular grains intergrown with other minerals in igneous and metamorphic rocks; as rounded or angular grains in sedimentary rocks (particularly sandstones); and as a microcrystalline sedimentary rock or cementing agent. Veins of milky white quartz, often quite large, fill cracks in many igneous and metamorphic rocks. Unlike nearly all other minerals, quartz is virtually unaffected by chemical weathering.

FELDSPARS

1-12.   Feldspars form very hard, blocky, opaque crystals with a pearly or porcelainlike luster and a nearly rectangular cross section. Crystals tend to cleave in two directions along flat, shiny, nearly perpendicular surfaces. Plagioclase varieties often have fine parallel grooves (striations) on one cleavage surface. Orthoclase varieties are usually pink, reddish, ivory, or pale gray. Where more than one variety is present, color differences are normally distinct. Crystalline feldspars are major components of most igneous rocks, gneisses, and schists. In the presence of air and water, the feldspars weather to clay minerals, soluble salts, and colloidal silica.

MICAS

1-13.   Micas form soft, extremely thin, transparent to translucent, elastic sheets and flakes with a bright glassy or pearly luster. "Books" of easily separated sheets frequently occur. The biotite variety is usually brown or black, while muscovite is yellowish, white, or silvery gray. Micas are very common in granitic rocks, gneisses, and schists. Micas weather slowly to clay minerals.

AMPHIBOLES

1-14. Amphiboles (chiefly hornblende) are hard, dense, glassy to silky minerals found chiefly in intermediate to dark igneous rocks and gneisses and schists. They generally occur as short to long prismatic crystals with a nearly diamond-shaped cross section. Dark green to black varieties are most common, although light gray or greenish types occur in some marbles and schists. Amphiboles weather rapidly to form chlorite and, ultimately, clay minerals, iron oxides, and soluble carbonates.

PYROXENES

1-15. Pyroxenes (chiefly augite) are hard, dense, glassy to resinous minerals found chiefly in dark igneous rocks and, less often, in dark gneisses and schists. They usually occur as well-formed, short, stout, columnar crystals that appear almost square in cross section. Granular crystals are common in some very dark gabbroic rocks. Masses of nearly pure pyroxene form a rock called pyroxenite. Colors of green to black or brown are most common, but pale green or gray varieties sometimes occur in marbles or schists. Pyroxenes weather much like the amphiboles.

OLIVINE

1-16. Olivine is a very hard, dense mineral that forms yellowish-green to dark olive-green or brown, glassy grains or granular masses in very dark, iron-rich rocks, particularly gabbro and basalt. Masses of almost pure olivine form a rare rock called peridotite. Olivine weathers rapidly to iron oxides and soluble silica.

CHLORITE

1-17. Chlorite is a very soft, grayish-green to dark green mineral with a pearly luster. It occurs most often as crusts, masses, or thin sheets or flakes in metamorphic rocks, particularly schists and greenstone. Chlorite forms from amphiboles and pyroxenes by weathering or metamorphism and, in turn, weathers to clay minerals and iron oxides.

CALCITE

1-18. Calcite is a soft, usually colorless to white mineral distinguished by a rapid bubbling or fizzing reaction when it comes in contact with dilute hydrochloric acid (HC1). Calcite is the major component of sea shells and coral skeletons and often occurs as well-formed, glassy to dull, blocky crystals. As a rock-forming mineral, it usually occurs as fine to coarse crystals in marble, loose to compacted granules in ordinary limestone, and as a cementing agent in many sedimentary rocks.

1-19. Calcite veins, or crack fillings, are common in igneous and other rocks. Calcite weathers chiefly by solution in acidic waters or water containing dissolved carbon dioxide.

DOLOMITE

1-20. Dolomite is similar to calcite in appearance and occurrence but is slightly harder and more resistant to solutioning. It is distinguished by a slow bubbling or fizzing reaction when it comes in contact with dilute HCl. Usually the reaction can be observed only if the mineral is first powdered (as by scraping it with a knife). Coarse dolomite crystals often have curved sides and a pinkish color. Calcite and dolomite frequently occur together, often in intimate mixtures.

LIMONITE

1-21. Limonite occurs most often as soft, yellowish-brown to reddish-brown, fine-grained, earthy masses or compact lumps or pellets. It is a common and durable cementing agent in sedimentary rocks and the major component of laterite. Most weathered rocks contain some limonite as a result of the decomposition of iron-bearing minerals.

CLAY

1-22. Clay minerals form soft microscopic flakes that are usually mixed with impurities of various types (particularly quartz, limonite, and calcite). When barely moistened, as by the breath or tongue, clays give off a characteristic somewhat musty "clay" odor. Clays form a major part of most soils and of such rocks as shale and slate. They are a common impurity in all types of sedimentary rocks.

SECTION II - ROCKS

FORMATION PROCESSES

1-23. A rock may be made of many kinds of minerals (for example, granite contains quartz, mica, feldspar, and usually hornblende) or may consist essentially of one mineral (such as a limestone, which is composed of the mineral calcite). To the engineer, rock is a firm, hardened substance that, in contrast to soil, cannot be excavated by standard earthmoving equipment. In reality, there is a transitional zone separating rock and soil so that not all "rock" deposits require blasting. Some "rock" can be broken using powerful and properly designed ripping equipment. The geologist places less restriction on the definition of rock.

1-24. Rocks can be grouped into three broad classes, depending on their origin. They are—

- Igneous.
- Sedimentary.
- Metamorphic.

1-25. Igneous rocks are solidified products of molten material from within the earth's mantle. The term igneous is from a word meaning "formed by fire." Igneous rocks underlie all other types of rock in the earth's crust and may be said to form the basement of the continents on which sedimentary rocks are laid down. Most sedimentary rocks are formed by the deposition of particles of older rocks that have been broken down and transported from their original positions by the agents of wind, water, ice, or gravity. Metamorphic rocks are rocks that have been altered in appearance and physical properties by heat, pressure, or permeation by gases or fluids. All classes of rocks (sedimentary, igneous, and metamorphic) can be metamorphosed. Igneous, sedimentary, and metamorphic rocks often occur in close association in mountainous areas, in areas once occupied by mountains but which have since eroded, and in broad, flat continental regions known as shields. Flat-lying sedimentary rocks form much of the plains of the continents and may occupy broad valleys overlain by recent or active deposits of sediments. The sediments being deposited in today's oceans, lakes, streams, floodplains, and deserts will be the sedimentary rocks of tomorrow. The rock-forming processes continually interact in a scheme called the rock cycle, illustrated in figure 1-4.

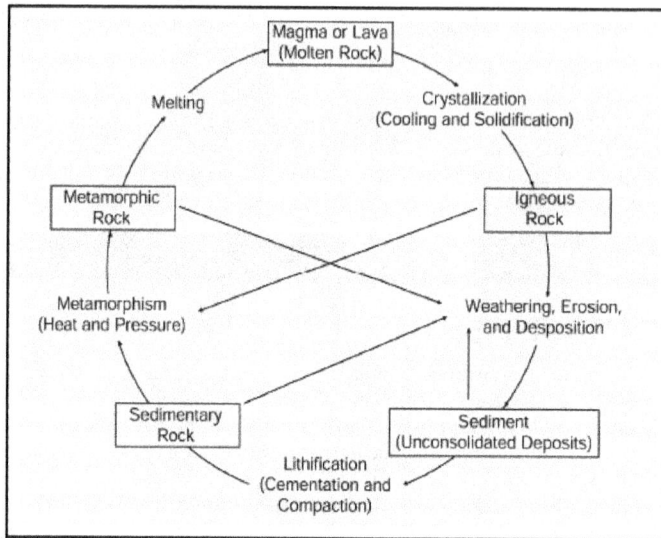

Figure 1-4. The rock cycle

IGNEOUS

1-26. Igneous rocks are solidified from hot molten rock material that originated deep within the earth. This occurred either from magma in the subsurface or from lava extruded onto the earth's surface during volcanic eruptions. Igneous rocks owe their variations in physical and chemical characteristics to differences in chemical composition of the original magma and to the physical conditions under which the lava solidified.

1-27. The groups forming the subdivisions from which all igneous rocks are classified are—

- Intrusive igneous rocks (cooled from magma beneath the earth's surface).
- Extrusive igneous rocks (cooled from magma on the earth's surface).

1-28. Figure 1-5 is a block diagram illustrating the major kinds of intrusive and extrusive rock bodies formed from the crystallization of igneous rocks. Dikes and sills are tabular igneous intrusions that are thin relative to their lengths and widths. Dikes are discordant; they cut across the bedding of the strata penetrated. Sills are concordant; they intrude parallel to and usually along bedding planes or contacts of the surrounding strata. Dikes and sills may be of any geologic age and may intrude young and old sediments. Batholiths are large, irregular masses of intrusive igneous rock of at least 40 square miles in area. A stock is similar to a batholith but covers less than 40 square miles in outcrop. Stocks and batholiths generally increase in volume (spread out) with depth and originate so deep that their base usually cannot be detected. Magma that reaches the earth's surface while still molten is ejected onto the ground or into the air (or sea) to form extrusive igneous rock. The molten rock may be ejected as a viscous liquid that flows out of a volcanic vent or from fissures along the flanks of the volcano. The flowing viscous mass is called lava and the lava flow may extend many miles from the crater vent. Lava that is charged with gases and ejected violently into the air forms pyroclastic debris consisting of broken and pulverized rock and molten material. The pyroclastics solidify and settle to the ground where they form deposits of ash and larger-sized material that harden into layered rock (tuff). Igneous rocks are usually durable and resistant and form ridges, caps, hills, and mountains while surrounding rock material is worn away.

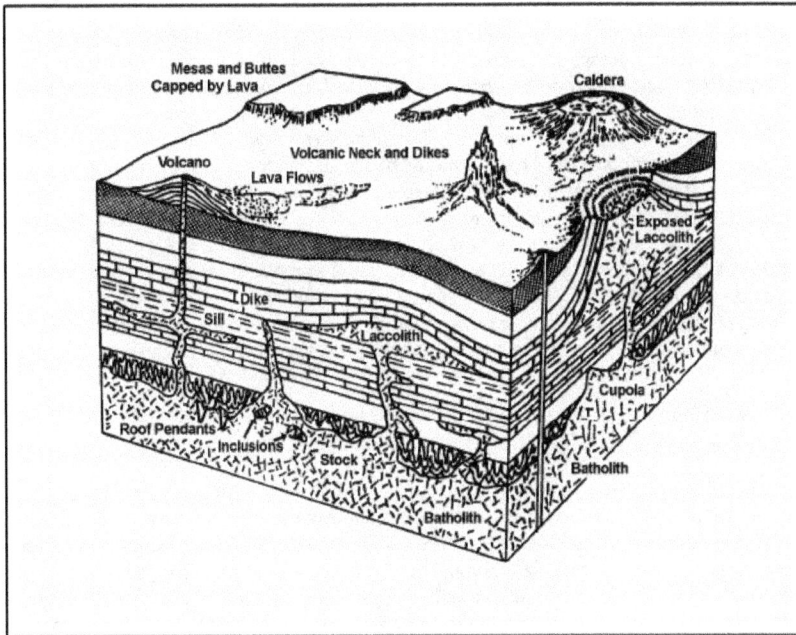

Figure 1-5. Intrusive and extrusive rock bodies

1-29. The chemical composition and thus the mineral content of intrusive and extrusive igneous rocks can be similar. The differences in appearance between intrusive and extrusive rocks are largely due to the size and arrangement of the mineral grains or crystals. As molten material cools, minerals crystallize and separate from solution. Silica-rich magma or lava solidifies into rocks high in silicon dioxide (quartz) and forms the generally light-colored igneous rocks. Molten material rich in ferromagnesian (iron-magnesium) compounds form the darker-colored igneous rocks, which are deficient in the mineral quartz. If the magma cools slowly, large crystals have time to grow. If the magma (or lava) cools quickly, large crystals do not have the chance to develop. Intrusive rocks are normally coarse-grained and extrusive igneous rocks are fine-grained for this reason. If lava cools too quickly for crystals to grow at all, then natural glass forms. Figure 1-6 illustrates the difference between intrusive and extrusive igneous rock crystals.

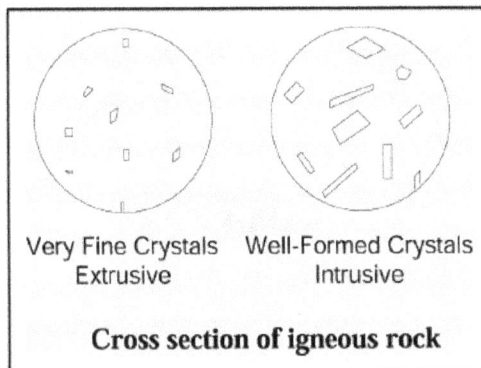

Figure 1-6. Cross section of igneous rock

SEDIMENTARY

1-30. Sedimentary rocks, also called stratified rocks, are composed of chemical precipitates, biological accumulations, or clastic particles. Chemical precipitates are derived from the decomposition of existing igneous, sedimentary, and metamorphic rock masses. Dissolved salts are then transported from the original position and eventually become insoluble, forming "precipitates"; or, through evaporation of the water medium, they become deposits of "evaporites." A relatively small proportion of the sedimentary rock mass is organic sediment contributed by the activities of plants and animals. Clastic sediments are derived from the disintegration of existing rock masses. The disintegrated rock is transported from its original position as solid particles. Rock particles dropped from suspension in air, water, or ice produce deposits of "clastic" sediments. Volcanically ejected material that is transported by wind or water and then deposited forms another class of layered rocks called "pyroclastics." Most pyroclastic deposits occur in the vicinity of a volcanic region, but fine particles can be transported by the wind and deposited thousands of miles from the source. Inorganic clastic sediments constitute about three-fourths of the sedimentary rocks of the earth's crust. Loose sediments are converted to rock by several processes collectively known as lithification. These are—

- Compaction.
- Cementation.
- Recrystallization.

1-31. The weight of overlying sediments that have accumulated over a long time produces great pressure in the underlying sediments. The pressure expels the water in the sediments by the process of compaction and forces the rock particles closer together. Compaction by the weight of overlying sediments is most effective in fine-grained sediments like clay and silt and in organic sediments like peat. Cementation occurs when precipitates of mineral-rich waters, circulating through the pores of sediments, fill the pores and bind the grains together. The most common cementing materials are quartz, calcite (calcium carbonate), and iron oxides (limonite and hematite). Recrystallization and crystal growth of calcium carbonate dissolved in saturated lime sediments develop rocks (crystalline limestone and dolomite) with interlocking, crystalline fabrics.

1-32. Sedimentary rocks are normally deposited in distinct parallel layers separated by abrupt, fairly even contact surfaces called bedding planes. Each layer represents a successive deposit of material. Bedding planes are of great significance as they are planes of structural weakness. Masses of sedimentary rock can move along bedding planes during rock slides. Figure 1-7 represents the layer-cake appearance of sedimentary rock beds. Sedimentary rocks cover about 75 percent of the earth's surface. Over 95 percent of the total volume of sediments consists of a variety of shales, sandstones, and limestones.

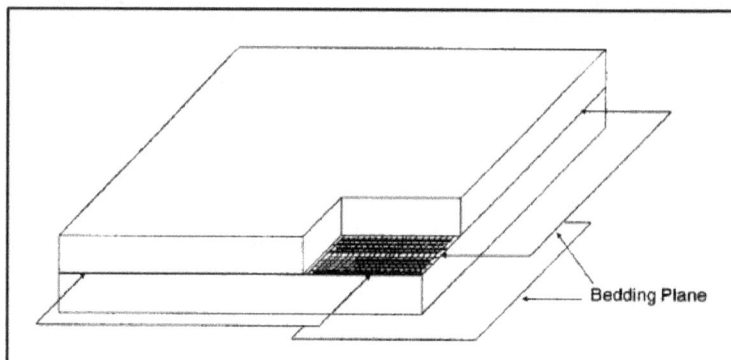

Figure 1-7. Bedding planes

METAMORPHIC

1-33. The alteration of existing rocks to metamorphic rocks may involve the formation within the rock of new structures, textures, and minerals. The major agents in metamorphism are—

- Temperature.
- Pressure.
- Chemically active fluids and gases.

1-34. Heat increases the solvent action of fluids and helps to dissociate and alter chemical compounds. Temperatures high enough to alter rocks commonly result from the intrusion of magma into the parent rock in the form of dikes, sills, and stocks. The zone of altered rock formed near the intrusion is called the contact metamorphism zone (see figure 1-8). The alteration zone may be inches to miles in width or length and may grade laterally from the unaltered parent rock to the highly metamorphosed derivative rock. Pressures accompanying the compressive forces responsible for mountain building in the upper earth's crust produce regional metamorphic rocks characterized by flattened, elongated, and aligned grains or crystals that give the rock a distinctive texture or appearance called foliation (see figure 1-9). Hot fluids, especially water and gases, are powerful metamorphic agents. Water under heat and pressure acts as a solvent, promotes recrystallization, and enters into the chemical composition of some of the altered minerals.

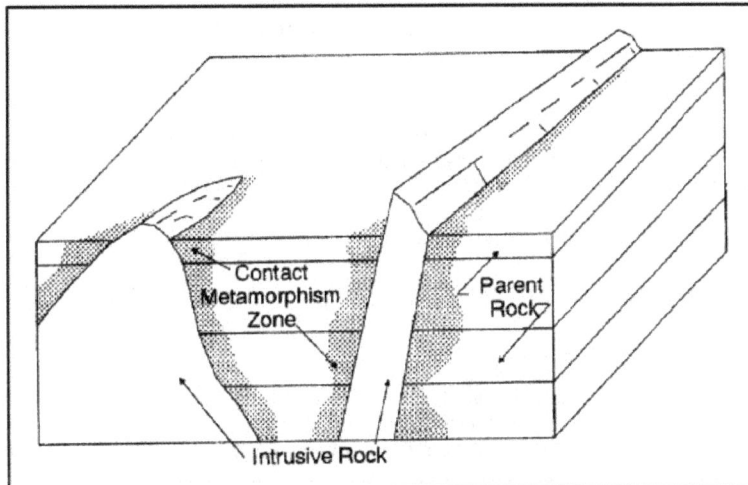

Figure 1-8. Contact metamorphism zone

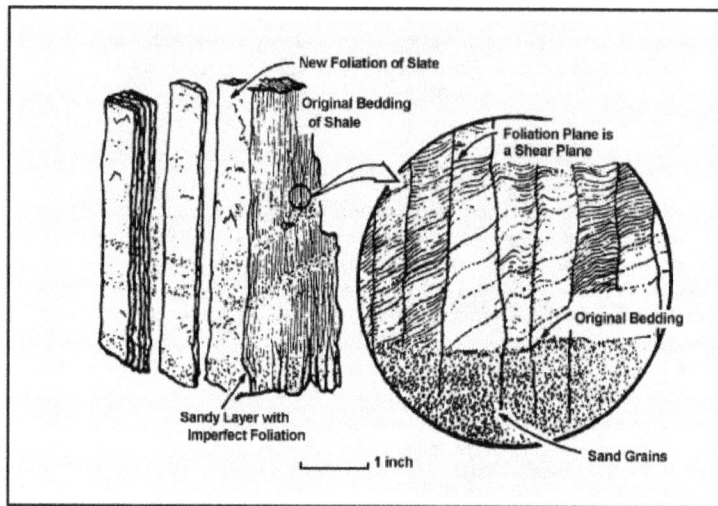

Figure 1-9. Metamorphic foliation

CLASSIFICATION

1-35. Igneous, metamorphic, and sedimentary rocks require different identification and classification procedures. The fabric, texture, and bonding strength imparted to a rock by its formation process determine the procedures that must be used to classify it.

IGNEOUS

1-36. Igneous rocks are classified primarily on the basis of—

- Texture.
- Color (or mineral content).

1-37. Texture is the relative size and arrangement of the mineral grains making up the rock. It is influenced by the rate of cooling of the molten material as it solidifies into rock. Intruded magmas cool relatively slowly and form large crystals if the intrusion is deep and smaller crystals if the intrusion is shallow. Extrusive lava is exposed abruptly to the air or to water and cools quickly, forming small crystals or no crystals at all. Therefore, referring to table 1-2, igneous rocks may have textures that are coarse-grained (mineral grains and crystals that can be differentiated by the unaided eye), very fine-grained (mineral crystals too small to be differentiated by the unaided eye), or of contrasting grain sizes (large crystals, or "phenocrysts," in a fine-grained "ground mass"). The intrusive igneous rocks generally have a distinctive texture of coarse interlocking crystals of different minerals. Under certain conditions, deep-seated intrusions form "pegmatites" (rocks with very large crystals). The extrusive (volcanic) igneous rocks, however, show great variation in texture. Very fine-grained rocks may be classified as having stony, glassy, scoriaceous, or fragmental texture. A rock with a stony texture consists of granular particles. Fine-grained rocks with a shiny smooth texture showing a conchoidal fracture are said to be "glassy." An example is obsidian, a black volcanic glass. Gases trapped in the extrusive lava may escape upon cooling, forming bubble cavities, or vesicles, in the rock. The result is a rock with scoriaceous texture. Fragmental rocks are those composed of lithified, pyroclastic material. The pyroclastic (volcanoclastic) deposits are made up of volcanic rock particles of various sizes that have drifted and accumulated by the action of wind and water after ejection from the volcanic vent. Fine-grained (smaller than 32 millimeters (mm)) ejecta (called lapilli, or ash) form deposits that become volcanic tuff. Lava flowing out of the volcanic vent or fissure forms a flow with a ropelike texture if the lava is very fluid. More viscous lava forms a blocky flow. Upon cooling, basaltic lava flows, sills, and volcanic necks sometimes crack. They often acquire columnar jointing characterized by near-vertical columns with hexagonal cross sections (see figure 1-10).

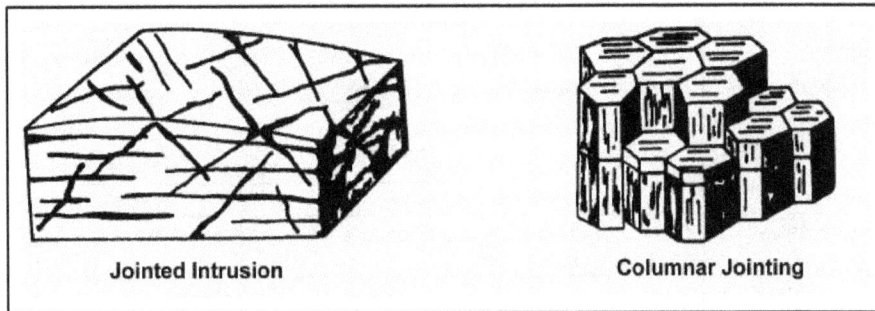

Jointed Intrusion   Columnar Jointing

Figure 1-10. Jointing in igneous rocks

1-38. Igneous rocks are further grouped by their overall color, which is generally a result of their mineral content (see table 1-2). The light-colored igneous rocks are silica-rich, and the dark-colored igneous rocks are silica-poor, with high ferromagnesian content. The intermediate rocks show gradations from light to dark, reflecting their mixed or gradational mineral content. An example illustrating the use of table 1-2 is as follows: lava and other ejecta charged with gases may form scoria, a dark-colored, highly vesicular basaltic lava or pumice, a frothy, light-colored felsite lava so porous that it floats on water.

1-39. The common igneous rocks are—

- Granite.
- Felsite.
- Gabbro and diorite.
- Basalt.
- Obsidian.
- Pumice.
- Scoria.

Table 1-2. Classification of igneous rocks

Origin	Dominant Texture*	Typical Mineral		
		Color		
		Light		Dark
Intrusive	Coarse-grained (distinguishable grains	Granite		Gabbro-diorite
Extrusive	Very fine (indistinguishable)	Stony	Felsite	Basalt
		Glassy	Obsidian	
		Scoriaceous	Pumice	Scoria
		Fragmental	Volcanic ash, cinder, bombs, and blocks	
Intrusive/ extrusive	Contrasting grain size	Porphyritic rocks		

* Rocks containing many scattered larger crystals are called "porphyritic," such as porphyri ic granite and porphyritic basalt

Granite

1-40. This is a coarsely crystalline, hard, massive, light-colored rock composed mainly of potassium feldspar and quartz, usually with mica and/or hornblende. Common colors include white, gray, and shades of pink to brownish red. Granite makes up most of the large intrusive masses of igneous rock and is frequently associated with (and may grade into) gneisses and schists. In general, it is a reasonably hard, tough, and durable rock that provides good foundations, building stones, and aggregates for all types of construction. Relatively fine-grained varieties are normally much tougher and more durable than coarse-grained types, many of which disintegrate rather rapidly under temperature extremes or frost action. Very coarse-grained and quartz-rich granites often bond poorly with cementing materials, particularly asphaltic cements. Antistripping agents should be employed when granite is used in bituminous pavements.

Felsite

1-41. This is a very fine-grained, usually extrusive equivalent of granite. Colors commonly range from light or medium gray to pink, red, buff, purplish, or light brownish gray. Felsites often contain scattered large crystals of quartz or feldspar. Isolated gas bubbles and streaklike flow structures are common in felsitic lavas. As a rule, felsites are about as hard and dense as granites, but they are generally tougher and tend to splinter and flake when crushed. Most felsites contain a form of silica, which produces alkali-aggregate reactions with portland cements. Barring these considerations, felsites can provide good general-purpose aggregates for construction.

Gabbro and Diorite

1-42. They form a series of dense, coarsely crystalline, hard, dark-colored intrusive rocks composed mainly of one or more dark minerals along with plagioclase feldspar. Since gabbro and diorite have similar properties and may be difficult to distinguish in the field, they are often grouped under the name gabbro-diorite. They are gray, green, brown, or black. Gabbro-diorites are common in smaller intrusive masses, particularly dikes and sills. As a group, they make strong foundations and excellent aggregates for all types of construction. However, their great toughness and high density make excavation and crushing costs very high, particularly in finer-grained varieties.

Basalt

1-43. This is a very fine-grained, hard, dense, dark-colored extrusive rock that occurs widely in lava flows around the world. Colors are usually dark gray to black, greenish black, or brown. Scattered coarse crystals of olivine, augite, or plagioclase are common, as are gas bubbles that may or may not be mineral-filled. With increasing grain size, basalt often grades into diabase, an extremely tough variety of gabbro. Both basalt and diabase make aggregates of the highest quality despite a tendency to crush into chips or flakes in sizes smaller than 2 to 4 centimeters.

Obsidian

1-44. This is a hard, shiny, usually black, brown, or reddish volcanic glass that may contain scattered gas bubbles or visible crystals. Like man-made glass, it breaks readily into sharp-edged flakes. Obsidian is chemically unstable, weak, and valueless as a construction material of any type.

Pumice

1-45. This is a very frothy or foamy, light-colored rock that forms over glassy or felsitic lava flows and in blocks blown from erupting volcanoes. Innumerable closely spaced gas bubbles make pumice light enough to float on water and also impart good insulating properties. Although highly abrasive, pumice is very weak and can usually be excavated with ordinary hand tools. It is used in the manufacture of low-strength, lightweight concrete and concrete blocks. Most varieties are chemically unstable and require the use of low-alkali portland cements.

Scoria

1-46. It looks very much like a coarse, somewhat cindery slag. In addition to its frothy texture, scoria may also exhibit stony or glassy textures or a combination of both. The color of scoria ranges from reddish brown to dark gray or black. Scoria is somewhat denser and tougher than pumice, and the gas bubbles that give it its spongy or frothy appearance are generally larger and more widely spaced than those in pumice. Scoria is very common in volcanic regions and generally forms over basaltic lava flows. It is widely used as a lightweight aggregate in concrete and concrete blocks. Like pumice, it may require the use of special low-alkali cements.

SEDIMENTARY

1-47. Sedimentary rocks are classified primarily by—

- Grain size.
- Composition.

1-48. They can be described as either clastic or nonclastic (see table 1-3). The clastic rocks are composed of discrete particles, or grains. The nonclastic rocks are composed of interlocking crystals or are in earthy masses. Clastic sedimentary rocks are further classified as coarse-grained or fine-grained. The coarse-grained rocks have individual grains visible to the naked eye and include sandstones, conglomerates (rounded grains), and breccias (angular grains). These are the rock equivalents of sands and gravels. The fine-grained rocks have individual grains that can only be seen with the aid of a hand lens or microscope and include siltstones, shales, clay stones, and mudstones.

Table 1-3. Classification of sedimentary rocks

Group		Dominant Composition		Rock Type
Clastic	Coarse-grained	Rock fragments larger than 2mm	Rounded	Conglomerate
			Angular	Breccia
		Mineral grains (chiefly quartz) 1/16 mm to 2 mm)		Sandstone
	Fine grained	Clay and silt-sized particles (smaller than 1/16 mm)		Shale
Nonclastic	Inorganic	Dolomite		Dolomite
		Microcrystalline silica		Chert
	Organic/inorganic	Calcite		Limestone
	Organic	Carbonaceous plant debris		Coal

1-49. Shales, claystones, and mudstones are composed of similar minerals and may be similar in overall appearance; however, a shale is visibly laminated (composed of thin tabular layers) and often exhibits "fissility," that is, it can be split easily into thin sheets. Clay stone and mudstone are not fissile. Mudstone is primarily a field term used to temporarily identify fine-grained sedimentary rocks of unknown mineral content.

1-50. The nonclastic sedimentary rocks can be further described as inorganic (or chemical) or organic. Dolomite is an inorganic calcium-magnesium carbonate. Chert, a widespread, hard, durable sedimentary rock, composed of microcrystalline quartz, precipitates from silica-rich waters and is often found in or with limestones. Limestone is a calcium carbonate that can be precipitated both organically and inorganically. A diagnostic feature of limestone is its effervescence in dilute HCl. Coal is an accumulation and conversion of the organic remains of plants and animals under certain environments.

1-51. Important features of the exposed or sampled portion of a deposit include stratification, the thickness of strata, the uniformity or nonuniformity of strata laterally, and the attitude (strike and dip) of the bedding planes. Special sedimentary bedding features are—

- Cross bedding (individual layers within a bed lie at an angle to the layers of adjacent beds (see figure 1-11, page 1-16), typical of sand dune and delta front deposits).
- Mud cracks (polygonal cracks in the surface of dried-out mud flats).
- Ripple marks (parallel ridges in some sediments that indicate the direction of wind or water movement during deposition).

1-52. Some typical sedimentary rocks are—

- Conglomerate and breccia.
- Sandstone.
- Shale.
- Tuff.
- Limestone.
- Dolomite.
- Chert.

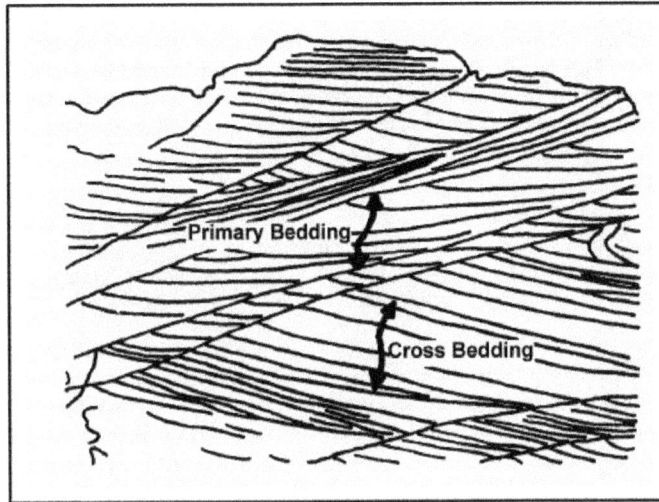

Figure 1-11. Cross bedding in sandstone

Conglomerate and Breccia

1-53. They resemble man-made concrete in that they consist of gravel-sized or larger rock fragments in a finer-grained matrix. Different varieties are generally distinguished by the size or composition of the rock fragments (such as limestone breccia, boulder conglomerate, or quartz pebble conglomerate). Wide variations in composition, degree of cementation, and degree of weathering of component particles make these rocks highly unpredictable, even within the same deposit. Generally, they exhibit poor engineering properties and are avoided in construction. Some very weakly cemented types may be crushed for use as fill or subbase material in road or airfield construction.

SANDSTONE

1-54. This is a medium- to coarse-grained, hard, gritty, clastic rock composed mainly of sand-sized (1/16 mm to 2 mm) quartz grains, often with feldspar, calcite, or clay. Sandstone varies widely in properties according to composition and cementation. Clean, compact, quartz-rich varieties, well-cemented by silica or iron oxides, generally provide good material for construction of all types. Low-density, poorly cemented, and clayey varieties lack toughness and durability and should be avoided as sources of construction material; however, some clay-free types may be finely crushed to provide sand.

Shale

1-55. This is a soft to moderately hard sedimentary rock composed of very fine-grained silt and clay-sized particles as well as clay minerals. Silica, iron oxide, or calcite cements may be present, but many shales lack cement and readily disintegrate or slake when soaked in water. Characteristically, shales form in very thin layers, break into thin platy pieces or flakes, and give off a musty odor when barely moistened. Occasionally, massive shales (called mudstones) occur, which break into bulky fragments. Shales are frequently interbedded with sandstones and limestones and, with increasing amounts of sand or calcite, may grade into these rock types. Most shales can be excavated without the need for blasting. Because of their weakness and lack of durability, shales make very poor construction material.

Tuff

1-56. This is a low-density, soft to moderately hard pyroclastic rock composed mainly of fine-grained volcanic ash. Colors range from white through yellow, gray, pink, and light brown to a rather dark grayish brown. When barely moistened, some tuffs give off a weak "clay" odor. Very compact varieties often resemble felsite but can usually be distinguished by their softness and the presence of glass or pumice fragments. Loose, chalky types usually feel rough and produce a gritty dust, unlike the smooth particles of a true chalk or clay. Tuff is a weak, easily excavated rock of low durability. When finely ground, it has weak cementing properties. It is often used as an "extender" for portland cement and as a pozzolan to improve workability and neutralize alkali-aggregate reactions. It can also be used as a fill and base course material.

Limestone

1-57. This is a soft to moderately hard rock composed mainly of calcite in the form of shells, crystals, grains, or cementing material. All varieties are distinguished by a rapid bubbling or fizzing reaction when they come in contact with dilute HCl. Colors normally range from white through various shades of gray to black; other colors may result from impurities. Ordinary limestone is a compact, moderately tough, very fine-grained or coarsely crystalline rock that makes a quality material for all construction needs. Hardness, toughness, and durability will normally increase with greater amounts of silica cement. However, more than about 30 percent silica may produce bonding problems or alkali-aggregate reactions. Clayey varieties usually lack durability and toughness and should be avoided. Weak, low-density limestones (including limerock and coral) are weakly recemented when crushed, wetted, and compacted. They are widely used as fill and base course material. In mild climates, some may prove suitable for use in low-strength portland cement concrete.

Dolomite

1-58. It is similar to limestone except that the mineral dolomite occurs in lieu of calcite. It is distinguished by a slow bubbling or fizzing reaction when it comes into contact with dilute HCl. Often the reaction cannot be seen unless the rock is first powdered (as by scraping with a knife). Limestone and dolomite exhibit similar properties, and often one grades into the other within a single deposit.

Chert

1-59. This is a very hard, very fine-grained rock composed of microcrystalline silica precipitated from seawater or groundwater. It occurs mainly as irregular layers or nodules in limestones and dolomites and as pebbles in gravel deposits or conglomerates. Most cherts are white to shades of gray. Very dark, often black, cherts are called flint, while reddish-brown varieties are called jasper. Pure, unweathered cherts break along smooth, conchoidal (shell-like) surfaces with a waxy luster; weathered or impure forms may seem dull and chalky-looking. Although cherts are very hard and tough, they vary widely in chemical stability and durability. Many produce alkali-aggregate reactions with portland cement, and most require the use of antistripping agents with bituminous cements. Low-density cherts may swell slightly when soaked and break up readily under frost action. Despite these problems, cherts are used in road construction in many areas where better materials are not available.

METAMORPHIC

1-60. Metamorphic rocks are classified primarily by—

- Mineral content.
- Fabric imparted by the agents of metamorphism.

1-61. They are readily divided into two descriptive groups (see table 1-4, page 1-18) known as—

- Foliated.
- Nonfoliated.

Table 1-4. Classification of metamorphic rocks

Structure	Characteristics	Rock Type
Foliated	Very fine-grained; cleaves readily into thin sheets or plates	slate
	Fine- to coarse-grained; thin semiparallel layers of platy minerals; splits into flakes between layers	Schist
	Fine- to coarse-grained; streaks or bands of differing mineralogic composition; breaks into bulky pieces	Gneiss
Nonfoliated	Mostly fused quartz grains	Quartzite
	Mostly calcite or dolomite	Marble

1-62. Foliated metamorphic rocks display a pronounced banded structure as a result of the deformational pressures to which they have been subjected. The nonfoliated, or massive, metamorphic rocks exhibit no directional structural features.

1-63. The common foliated or banded metamorphic rocks include—

- Slate.
- Schist.
- Gneiss.

Slate

1-64. This is a very fine-grained, compact metamorphic rock that forms from shale. Unlike shale, slates have no "clay" odor. They split into thin, parallel, sharp-edged sheets (or plates) usually at some angle to any observable bedding. Colors are normally dark red, green, purple, or gray to black. Slates are widely used as tiles and flagstones, but their poor crushed shape and low resistance to splitting makes them unsuitable for aggregates or building stones.

Schist

1-65. This is a fine- to coarse-grained, well-foliated rock composed of discontinuous, thin layers of parallel mica, chlorite, hornblende, or other crystals. Schists split readily along the structural layers into thin slabs or flakes. This characteristic makes schists undesirable for construction use and hazardous to excavate. However, varieties intermediate to gneiss may be suitable for fills, base courses, or portland cement concrete.

Gneiss

1-66. (Pronounced "nice") This is a roughly foliated, medium- to coarse-grained rock that consists of alternating streaks or bands. The banding is caused by segregation of light-colored layers of quartz and feldspar alternating with dark layers of ferromagnesian minerals. These streaks may be straight, wavy, or crumpled and of uniform or variable thickness. Gneisses normally break into irregular, bulky pieces and resemble the granitic rocks in properties and uses. With increasing amounts of mica or more perfect layering, gneisses grade into schists.

1-67. The common nonfoliated metamorphic rocks include—

- Quartzite.
- Marble.

Quartzite

1-68. This is an extremely hard, fine- to coarse-grained, massive rock that forms from sandstone. It is distinguished from sandstone by differences in fracture. Quartzite fractures through its component grains rather than around them as in sandstone, because the cement and sand grains of quartzite have been fused or welded together during metamorphism. Therefore, broken surfaces are not gritty and often have a splintery or sugary appearance like that of a broken sugar cube or hard candy. Quartzite is one of the hardest, toughest, and most durable rocks known. It makes excellent construction material, but excavation and crushing costs are usually very high. Because of its high quartz content, antistripping agents are normally required with bituminous cements. Even so, bonds may be poor with very fine-grained types.

Marble

1-69. This is a soft, fine- to coarsely crystalline, massive metamorphic rock that forms from limestone or dolomite. It is distinguished by its softness, acid reaction, lack of fossils, and sugary appearance on freshly broken surfaces. Marble is similar to ordinary compact or crystalline limestones in its engineering properties and uses. However, because of its softness, it is usually avoided as an aggregate for pavements on highways and airfields. White calcite or pinkish dolomite veins and subtle swirls or blotches of trace impurities give marble its typical veined or "marbled" appearance.

1-70. Metamorphic rocks have been derived from existing sedimentary, igneous, or metamorphic rocks as depicted in figure 1-4, page 1-7, and figure 1-12.

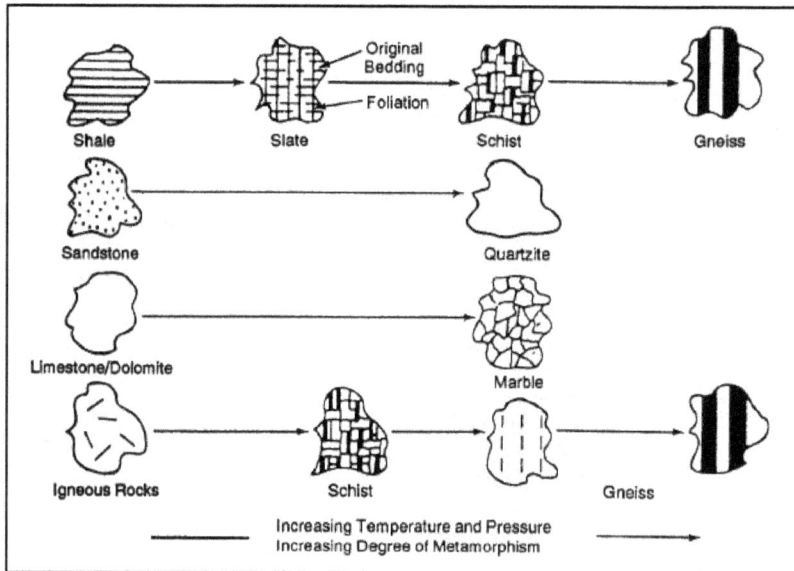

Figure 1-12. Metamorphism of existing rocks

IDENTIFICATION

1-71. Military engineers must frequently select the best rock for use in different types of construction and evaluate foundation or excavation conditions.

FIELD IDENTIFICATION

1-72. A simple method of identification of rock types that can be applied in the field will assist in identifying most rocks likely to be encountered during military construction. This method is presented in simple terms for the benefit of the field engineer who is not familiar with expressions normally used in technical rock descriptions.

1-73. The identification method is based on a combination of simple physical and chemical determinations. In some cases, the grains composing a rock may be seen, and the rock may be identified from a knowledge of its components. In other cases, the rock may be so fine-grained that the identification must be based on its general appearance and the results of a few easy tests.

1-74. The equipment required consists only of a good steel knife blade or a nail and a bottle of dilute (10 percent solution) HCl, preferably with a dropper. A small 6- to 10-power magnifying glass may also be helpful. HCl (muriatic acid) is available at most hardware stores and through the military supply system. Military hospitals are a typical source for HCl.

1-75. Samples for identification should be clean, freshly broken, and large enough to clearly show the structure of the rock. In a small sample, key characteristics (such as any alignment of the minerals composing the rock) may not be observed as readily as in a larger one. Pieces about 3 inches by 4 inches by 2 inches thick are usually suitable.

GENERAL CATEGORIES

1-76. The identification system of geologic materials is given in flow chart form (see table 1-5). In this method, all considerations are based on the appearance or character of a clean, freshly broken, unweathered rock surface. Weathered surfaces may exhibit properties that may not be true indicators of the actual rock type.

1-77. For most rocks the identification process is direct and uncomplicated. If a positive identification cannot be made, the more detailed rock descriptions (in the preceding paragraphs) should be consulted after first using the flow chart to eliminate all clearly inappropriate rock types. By the use of descriptive adjectives, a basic rock classification can be modified to build up a "word picture" of the rock (for example, "a pale brown, fine-grained, thin-bedded, compact, clayey, silica-cemented sandstone").

1-78. To use the flow chart, the sample must be placed into one of three generalized groups based on physical appearance. These groups are—

- Foliated.
- Very fine-grained.
- Coarse-grained.

Foliated

1-79. Foliated rocks are those metamorphic rocks that exhibit planar orientation of their mineral components. They may exhibit slaty cleavage (like slate) expressed by closely spaced fractures that cause the sample to split along thin plates. If the sample exhibits a parallel arrangement of platy minerals in thin layers and has a silky or metallic reflection, the sample has schistosity and is called a schist. If the sample exhibits alternating streaks or bands of light and dark minerals of differing composition, then the sample has gneissic layering and is called agneiss.

Very Fine-Grained

1-80. If the sample appears pitted or spongy, it is called frothy. The pits are called vesicles and are the result of hot gases escaping from magma at the top of a lava pool. If the sample is light enough to float on water and is light-colored, it is called pumice. If the frothy rock sample is dark-colored and appears cindery, it is called scoria.

1-81. If the sample has the appearance of broken glass, it is called either obsidian or quartz. Obsidian is a dark-colored natural volcanic glass that cooled too fast for any crystals to develop. Quartz is not a glass but is identified on the flow chart as light-colored and glassy. Quartz is a mineral and not one of the aggregates classifed for use in military construction. It is often found in its crystalline form with six-sided crystals. It is common as veins in both igneous and metamorphic rock bodies.

1-82. A hardness test is conducted to determine whether a stony sample is hard or soft. If the sample can be scratched with a knife or nail, then it is said to be soft. If it cannot be scratched, then it is said to be hard.

1-83. Fine-grained rocks may be glassy, frothy, or stony. The term "stony" is used to differentiate them from "glassy" and "frothy" rocks.

Table 1-5. Identification of geologic materials

Foliated		Very fine-grained; splits along thin planes			Slate
		Metallic reflection; splits into slabs and flakes or thin semitransparent sheets			Schist
		Contains streaks or bands of light and dark minerals; breaks to bulky, angular fragments			Gneiss
Very Fine Grained	**Frothy**	Light Colored; lightweight; easily crushed			Pumice
		Dark colored; cindery			Scoria
	Glassy	Light colored; massive; extremely hard			Quartz
		Dark colored; may have some gas bubbles			Obsidian
	Stony	**Soft**	No acid reaction	Earthy; clay odor; platy	Shale
				May have small pieces of glass; low density	Tuff
			Acid reaction	Sugary appearance	Marble
				Dull and massive	Limestone
		Hard	Waxy; very hard; weathers to soft white		Chert
			Dull; may contain some gas bubbles or visible crystals	Light colored / Dark colored	Felsite / Basalt
			Sandy; mostly one mineral (quartz)	Gritty sandpaper feel / Sugary; not gritty	Sandstone / Quartzite
Coarse-Grained	**Hard**		Sandy; mostly one mineral (quartz)	Gritty sandpaper feel / Sugary; not gritty	Sandstone / Quartzite
			Mixed minerals; salt-and-pepper appearance	Light Colored / Dark colored	Granite / Gabbro-diorite
			Fragmental; appearance of broken concrete		Conglomerate
	Soft		Fragmental; may contain small pieces of glass	Low density	Tuff
			Acid reaction	Sugary appearance	Marble
				Shell fragments	Limestone

1-84. If the sample is soft, a chemical determination must be made using dilute HCl, HCl tests determine the presence or absence of calcite (calcium carbonate) which comprises limestone/dolomite and marble. Very fine-grained rocks that are soft and stony in appearance and have an acid reaction are composed of calcium carbonate. If the sample is dull and massive, it is called a limestone. Dolomites are also dull and massive and are not separated from limestones on this chart, but they do not readily react to acid. Their surface must be powdered first by scratching the sample with a nail. The dolomite powder then readily reacts to the acid. If the sample exhibits a sugary appearance, then it is the metamorphic rock, marble. The sugary appearance is due to the partial melting or fusing of calcite crystals by heat and pressure. It is similar to the appearance of the sugar coating on a breakfast cereal. These rocks are susceptible to chemical attack by acidic solutions that form when carbon dioxide (CO_2) is absorbed in groundwater. This produces a mild carbonic acid that dissolves the calcite in the rock. This is the process that produces caves and sinkholes. If there is not an acid reaction (no effervescence) and the sample has a platy structure, it is a shale. A shale may be any color. It is normally dull and separates into soft, thin plates. When freshly broken, it may have a musty odor similar to clay. Shales are derived from lithification of clay particles and fine muds. Shale is a sedimentary rock and should not be confused with its metamorphic equivalent, slate. Shales often occur interbedded within layers of limestone and dolomite. Because they are often found with carbonate rocks, they may exhibit an acid reaction due to contamination by calcium carbonate.

1-85. If the sample has no acid reaction and has a relatively low density, with small pieces of glass in its fine-grained matrix, the rock is called a tuff. Tuff is a volcanic sedimentary rock comprised mostly of ash particles that have been solidified. By the flow chart, it is characterized as soft; however, it may be hard if it has been welded by hot gases during the eruption. If it is hard, tuff may exhibit physical properties that make it suitable for some construction applications.

1-86. Fine-grained, stony samples that are hard may be waxy, dull, or sandy. Waxy samples resemble candle wax and have a conchoidal fracture and sharp edges. These rocks are called chert. They often weather into soft, white materials. Dull samples are either light- or dark-colored igneous extrusive rocks. If the sample is light-colored, it is a felsite. Felsites are normally massive but may appear banded or layered. Unlike gneiss, the layers are not made up of alternating light and dark minerals. Felsites may also have cavities filled with other lighter-colored minerals. They represent a variety of lava rocks that are high in silica. Because of their great variety, they may be hard to identify properly. The dark counterpart to the felsites is basalt. Basalt is normally black and very tough. It is also a lava rock and often exhibits columnar jointing. It often occurs as dikes or sills in other rock bodies.

1-87. Hard, sandy, very fine-grained samples composed mostly of quartz may be either sandstone or quartzite. If the sample feels gritty like sandpaper, it is called sandstone. Sandstones are sedimentary rocks composed of sand grains that have been cemented together. If the sample appears sugary and does not feel like sandpaper, the rock is called quartzite. Quartzite is the metamorphic equivalent of sandstone. Like marble, it has a sugary appearance, which is due to the partial melting and fusing of the crystal grains.

Coarse-Grained

1-88. Coarse-grained rocks refer to those that have either crystals or cemented particles that are large enough to be readily seen with the unaided eye. Samples may be hard or soft. Hard samples may appear sandy, mixed, or fragmental. A sandy sample would be a sandstone or a quartzite. Because coarse sands grade into fine sands, there are coarse sandstones and fine sandstones. The sugary-appearing metamorphic version of a coarse sandstone is called a quartzite.

1-89. Hard, coarse-grained rocks that are comprised of mixed interlocking crystals have a salt-and-pepper appearance due to their light and dark minerals. If the rock is predominantly light in color, it is a granite. If it is predominantly dark in color, it is called a gabbro-diorite. Both of these rocks are igneous intrusive rocks that cooled slowly, allowing the growth of large interlocking crystals. If the sample appears to be made of round, cemented rock fragments similar to the appearance of concrete, it is called a conglomerate. If the rock fragments are angular instead of round, the rock is called a breccia.

1-90. If a coarse-grained sample is soft, it may either be fragmental or have an acid reaction. If it appears fragmental and has a low density and small pieces of glass, it is a tuff. Tuff has already been described as a very fine-grained, stony, soft rock material. This entry on the chart is for the coarse-grained version of the same material. The coarse-grained rock samples that are soft and have an acid reaction are coarse-grained limestones and marbles, as already described.

1-91. The system of identification of rock types used by military engineers serves to identify most common rock types. This method requires the user to approach rock identification in a consistent and systematic manner; otherwise, important rock characteristics may go unnoticed. Many important rock features may not appear in small hand specimens. To enable better identification and evaluation, personnel who are in charge of geologic exploration should maintain careful notes on such rock features as bedding, foliation, gradational changes in composition or properties, and on the associations of rocks in the field. During preliminary reconnaissance work, geologic maps or map substitutes should be used to make preliminary engineering estimates based on the typical rock properties.

ENGINEERING PROPERTIES

1-92. The following engineering properties and tests are provided to help make engineering judgments concerning the use of rock materials as construction aggregate:

- Toughness.
- Hardness.
- Durability.
- Crushed shape.
- Chemical stability.
- Surface character.
- Density.

1-93. Preliminary engineering estimates can be made based on the typical rock properties ascertained from these tests. If a rock sample cannot be identified using the flow chart, a decision can be made as to its suitability for use as a construction material by these tests.

Toughness

1-94. This is mechanical strength, or resistance to crushing or breaking. In the field, this property may be estimated by attempting to break the rock with a hammer or by measuring its resistance to penetration by impact drills.

Hardness

1-95. This is resistance to scratching or abrasion. In the field, this may be estimated by attempting to scratch the rock with a steel knife blade. Soft materials are readily scratched with a knife, while hard materials are difficult or impossible to scratch (see table 1-6).

Table 1-6. Field-estimating rock hardness

Hardness	Characteristics
Very hard	Not scratched by a steel file
Hard	Scratched by a steel file but difficult or impossible to scratch with a steel knife blade (or nail)
Moderately hard	Scratched by a knife but not by a copper coin
Soft	Scratched by a copper coin
Very soft	Scratched by a fingernail

Durability

1-96. This is resistance to slaking or disintegration due to alternating cycles of wetting and drying or freezing and thawing. Generally, this may be estimated in the field by observing the effects of weathering on natural exposures of the rock.

Crushed Shape

1-97. Rocks that break into irregular, bulky fragments provide the best aggregates for construction because the particles compact well, interlock to resist displacement and distribute loads, and are of nearly equal strength in all directions. Rocks that break into elongated pieces or thin slabs, sheets, or flakes are weak in their thin dimensions and do not compact, interlock, or distribute loads as effectively (see figure 1-13, page 1-24).

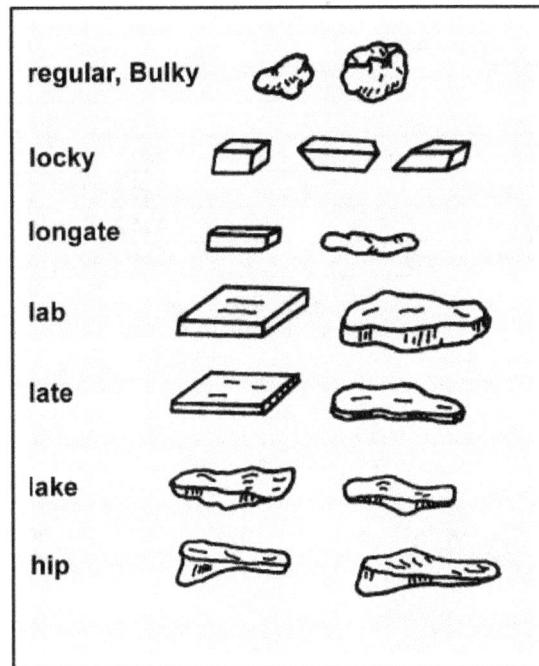

Figure 1-13. Crushed shape

Chemical Stability

1-98. This is resistance to reaction with alkali materials in portland cements. Several rock types contain impure forms of silica that react with alkalies in cement to form a gel. The gel absorbs water and expands to crack or disintegrate the hardened concrete. This potential alkali-aggregate reaction may be estimated in the field only by identifying the rock and comparing it to known reactive types or by investigating structures in which the aggregate has previously been used.

Surface Character

1-99. This refers to the bonding characteristics of the broken rock surface. Excessively smooth, slick, nonabsorbent aggregate surfaces bond poorly with cementing materials and shift readily under loads. Excessively rough, jagged, or absorbent surfaces are likewise undisturbed because they resist compaction or placement and require excessive amounts of cementing material.

Density

1-100. This is weight per unit volume. In the field, this may be estimated by "hefting" a rock sample (see table 1-7). Density reflects on excavation and hauling costs and may influence the selection of rocks for special requirements (such as riprap, jetty stone, or lightweight aggregate). Among rocks of the same type, density is often a good indicator of the toughness and durability to be expected. Table 1-8, page 1-26, lists the general ratings of rock properties for each of the 18 typical military construction aggregates. These ratings serve only as a guide; each individual rock body must be sampled and evaluated separately. Table 1-9, page 1-27, provides general guidance to determine the suitability of an unidentified rock sample for general military construction missions based on the evaluation of its physical properties.

Table 1-7. Field-estimating rock density

Description	Density (g/cm³)
Very dense	over 3.0
Dense	2.8 to 3.0
Moderately dense	2.6 to 2.8
Low density	2.4 to 2.6
Low density	2.4 to 2.6
Very low density	Below 2.4

1-101.   Table 1-10, page 1-27, provides a rating for selected geologic materials concerning their suitability as an aggregate for concrete or asphalt or for their use as a base course material. Rock materials that typically exhibit chemical instability in concrete are marked with an asterisk. These materials may cause a concrete-alkali reaction due to their high silica content. In general, felsites, chert, and obsidian should not be used as concrete aggregate. The end result of these reactions is the weakening, and in extreme cases the failure, of the concrete design. Apparently, silica is drawn out of the aggregate to make a gel that creates weaknesses within the mix. The gel may expand with temperature changes and prevent the proper bonding of the cement and aggregate.

1-102.   Rock types with two asterisks may not bond readily to bituminous materials. They require special antistripping agents to ensure that they do not "strip," or separate, from the pavement mix. Stripping severely reduces pavement performance.

Table 1-8. Engineering properties of rocks

	Rock Type	Toughness	Hardness	Durability	Chemical Stability	Crushed Shape	Surface Character	Density (g/cm)
Igneous	Granite	Good – very good	Good	Good	Excellent	Excellent	Good	2.65
	Gabbro-Diorite	Excellent	Excellent	Excellent	Excellent	Good	Good	2.96
	Basalt	Excellent	Excellent	Excellent	Excellent	Fair-good	Poor	2.86
	Felsite	Good	Good	Good	Questionable	Fair-good	Fair-good	2.66
	Obsidian	Poor	Good	Good	Questionable	Very poor	Very poor	2.3 – 2.4
	Pumice	Very poor	Very poor	Poor	Questionable	Good	Poor	<1.0
	Scoria	Poor	Poor	Poor	Good	Good	Poor	Variable
Sedimentary	Tuff	Poor	Poor	Poor	Questionable	Good	Good	Variable
	Congolmerate (Breccia)	Poor - fair	Poor - fair	Poor - fair	Good	Fair	Good	2.68
	Sandstone	Variable	Variable	Variable	Excellent	Good	Excellent	2.54
	Shale	Poor	Poor	Poor	Questionable	Poor	Good	1.8 – 2.5
	Limestone or Dolomite	Good	Fair-good	Good	Good	Good	Good	2.66 – 2.70
	Chert	Poor	Excellent	Variable	Questionable	Poor -fair	Fair	2.50
Metamorphic	Gneiss	Good	Good	Good	Excellent	Good-fair	Good	2.74
	Schist	Good	Good	Good	Excellent	Poor - fair	Poor - fair	2.85
	Slate	Poor	Poor	Poor	Excellent	Poor	Good	2.72
	Quartzite	Excellent	Excellent	Excellent	Excellent	Fair	Good - fair	2.69
	Marble	Good	Poor	Good	Good	Excellent	Excellent	2.63

Table 1-9. Aggregate suitability based on physical properties

Aggregate Used In	Toughness	Hardness	Durability	Chemical Stability	Crushed Shape	Surface Character
Portland Cement	Good	Fair	Good	Good	Good*	Good
Bituminous Surfaces	Good	Good	Good	Any/All	Good	Good
Base Coarse	Good	Good	Good	Any/All	Good	Good
* The preferred shape for portland cement aggregate is an irregular, bulky shape.						

Table 1-10. Use of aggregates for military construction missions

	Rock type	Use as Aggregates		Use as a Base Course or Subbase
		Concrete	Asphalt	
Igneous	Granite	Fair-good	Fair-good**	Good
	Gabbro-Diorite	Excellent	Excellent	Excellent
	Basalt	Excellent	Excellent	Excellent
	Felsite	Poor*	Fair	Fair-good
Sedimentary	Congolmerate	Poor	Poor	Poor
	Sandstone	Poor-fair	Poor-fair	Fair-good
	Shale	Poor	Poor	Poor
	Limestone	Fair-good	Good	Good
	Dolomite	Good		
	Chert	Poor*	Poor**	Poor-fair
Metamorphic	Gneiss	Good	Good	Good
	Schist	Poor-fair	Poor-fair	Poor-fair
	Slate	Poor	Poor	Poor
	Quartzite	Good	Fair-good**	Fair-good
	Marble	Fair	Fair	Fair
* Reacts (alkali-aggregate).				
** Antistripping agents should be used.				

Chapter 2

Structural Geology

Structural geology describes the form, pattern, origin, and internal structure of rock and soil masses. Tectonics, a closely related field, deals with structural features on a larger regional, continental, or global scale. Figure 2-1, page 2-2, shows the major plates of the earth's crust. These plates continually undergo movement as shown by the arrows. Figure 2-2, page 2-3, is a more detailed representation of plate tectonic theory. Molten material rises to the earth's surface at midoceanic ridges, forcing the oceanic plates to diverge. These plates, in turn, collide with adjacent plates, which may or may not be of similar density. If the two colliding plates are of approximately equal density, the plates will crumple, forming a mountain range along the convergent zone. If, on the other hand, one of the plates is more dense than the other, it will be subducted, or forced below, the lighter plate, creating an oceanic trench along the convergent zone. Active volcanism and seismic activity can be expected in the vicinity of plate boundaries. In addition, military engineers must also deal with geologic features that exist on a smaller scale than that of plate tectonics but which are directly related to the deformational processes resulting from the force and movements of plate tectonics.

The determination of geologic structure is often made by careful study of the stratigraphy and sedimentation characteristics of layered rocks. The primary structure or original form and arrangement of rock bodies in the earth's crust is often altered by secondary structural features. These secondary features include folds, faults, joints, and schistosity. These features can be identified and mapped in the field through site investigation and from remote imagery.

SECTION I - STRUCTURAL FEATURES IN SEDIMENTARY ROCKS

BEDDING PLANES

2-1. Structural features are most readily recognized in the sedimentary rocks. They are normally deposited in more or less regular horizontal layers that accumulate on top of each other in an orderly sequence. Individual deposits within the sequence are separated by planar contact surfaces called bedding planes (see figure 1-7, page 1-10). Bedding planes are of great importance to military engineers. They are planes of structural weakness in sedimentary rocks, and masses of rock can move along them causing rock slides. Since over 75 percent of the earth's surface is made up of sedimentary rocks, military engineers can expect to frequently encounter these rocks during construction.

2-2. Undisturbed sedimentary rocks may be relatively uniform, continuous, and predictable across a site. These types of rocks offer certain advantages to military engineers in completing horizontal and vertical construction missions. They are relatively stable rock bodies that allow for ease of rock excavation, as they will normally support steep rock faces. Sedimentary rocks are frequently oriented at angles to the earth's "horizontal" surface; therefore, movements in the earth's crust may tilt, fold, or break sedimentary layers. Structurally deformed rocks add complexity to the site geology and may adversely affect military construction projects by contributing to rock excavation and slope stability problems.

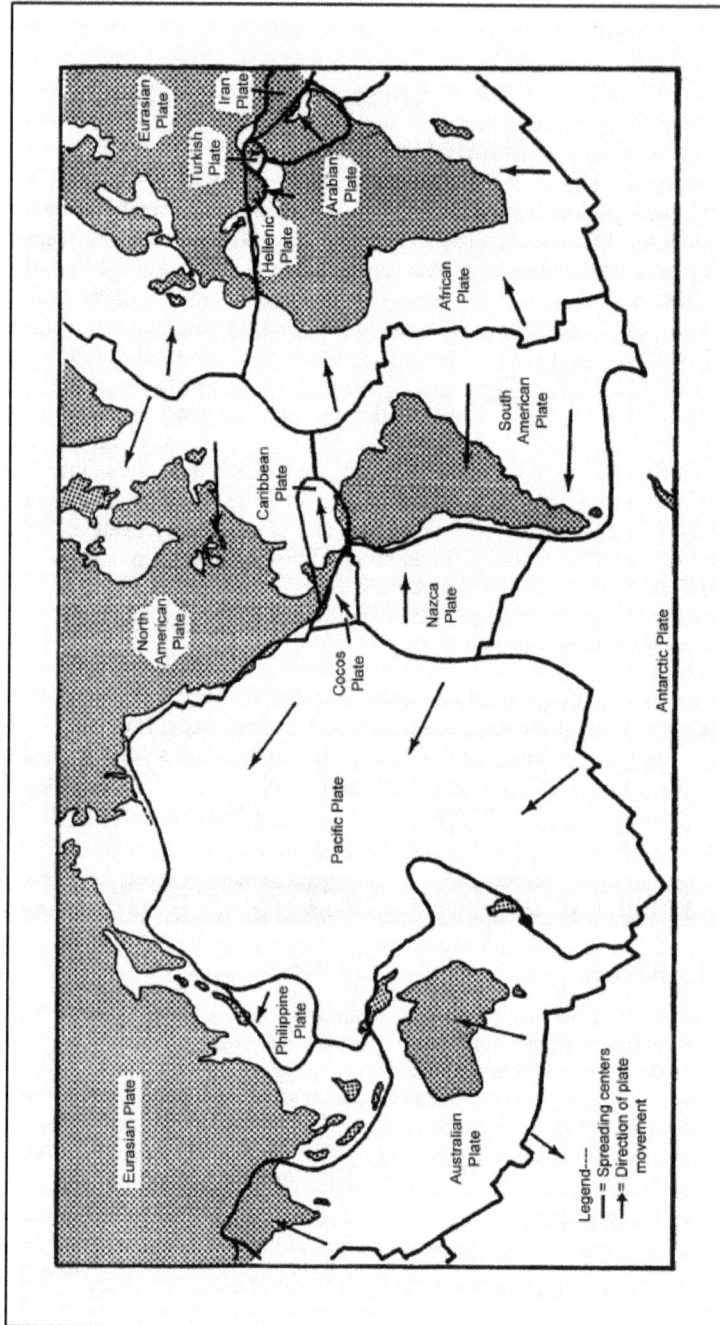

Figure 2-1. The major plates of the earth's crust

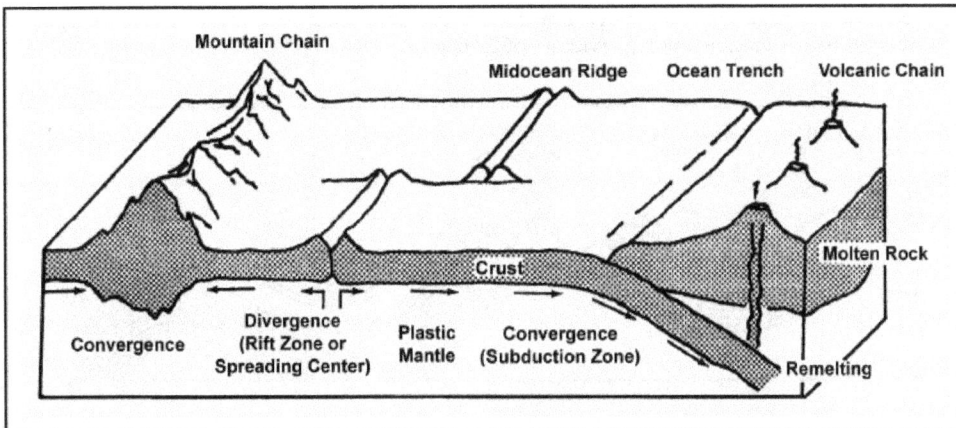

Figure 2-2. Major features of the plate tectonic theory

2-3.   Vegetation and overlying soil conceal most rock bodies and their structural features. Outcrops are the part of a rock formation exposed at the earth's surface. Such exposures, or outcrops, commonly occur along hilltops, steep slopes, streams, and existing road cuts where ground cover has been excavated or eroded away (see figure 2-3). Expensive delays and/or failures may result when military engineers do not determine the subsurface conditions before committing resources to construction projects. Therefore, where outcrops are scarce, deliberate excavations may be required to determine the type and structure of subsurface materials. To determine the type of rock at an outcrop, the procedures discussed in chapter 1 must be followed. To interpret the structure of the bedrock, the military engineer must measure and define the trend of the rock on the earth's surface.

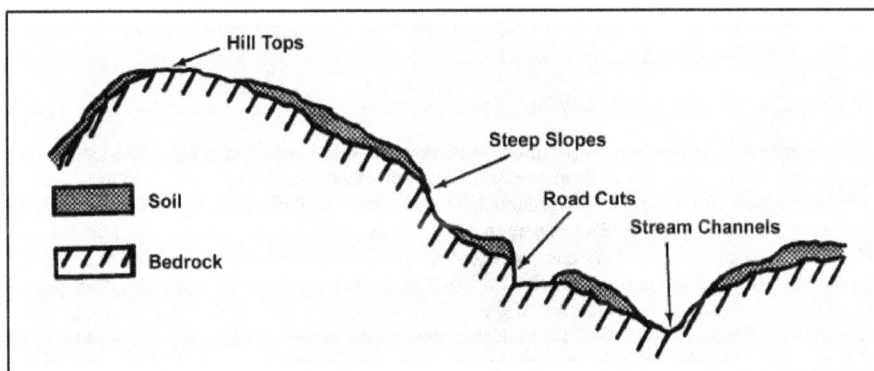

Figure 2-3. Location of rock outcrops

FOLDS

2-4.   Rock strata react to vertical and horizontal forces by bending and crumpling. Folds are undulating expressions of these forces. They are the most common type of deformation. Folds are most noticeable in layered rocks but rarely occur on a scale small enough to be observed in a single exposure. Their size varies considerably. Some folds are miles across, while others may be less than an inch. Folds are of significant importance to military engineers due to the change in attitude, or position, of bedding planes within the rock bodies (see figure 2-4, page 2-4). These can lead to rock excavation problems and slope instability. Folds are common in sedimentary rocks in mountainous areas where their occurrence may be inferred from ridges of durable rock strata that are tilted at opposite angles in nearby rock outcrops. They may also be recognized by topographic and geologic map patterns and from aerial photographs. The presence of tilted rock layers within a region is usually evidence of folding.

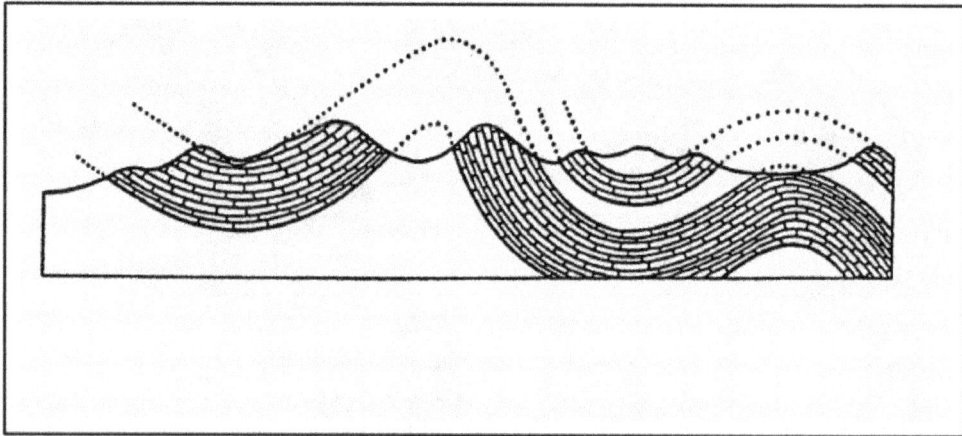

Figure 2-4. Folding of sedimentary rock layers

TYPES

There are several basic types of folds. They are—

- Homocline.
- Monocline.
- Anticline.
- Syncline.
- Plunging.
- Dome.
- Basin.

2-5. A rock body that dips uniformly in one direction (at least locally) is called a homocline (see figure 2-5a). A rock body that exhibits local steplike slopes in otherwise flat or gently inclined rock layers is called a monocline (see figure 2-5, b). Monoclines are common in plateau areas where beds may locally assume dips up to 90 degrees. The elevation of the beds on opposite sides of the fold may differ by hundreds or thousands of feet. Anticlines are upfolds, and synclines are downfolds (see figure 2-5, c and d, respectively). They are the most common of all fold types and are typically found together in a series of fold undulations. Differential weathering of the rocks composing synclines and anticlines tends to produce linear valleys and ridges. Folds that dip back into the ground at one or both ends are said to be plunging (see figure 2-6). Plunging anticline and plunging syncline folds are common. Upfolds that plunge in all directions are called domes. Folds that are bowed toward their centers are called basins. Domes and basins normally exhibit roughly circular outcrop patterns on geologic maps.

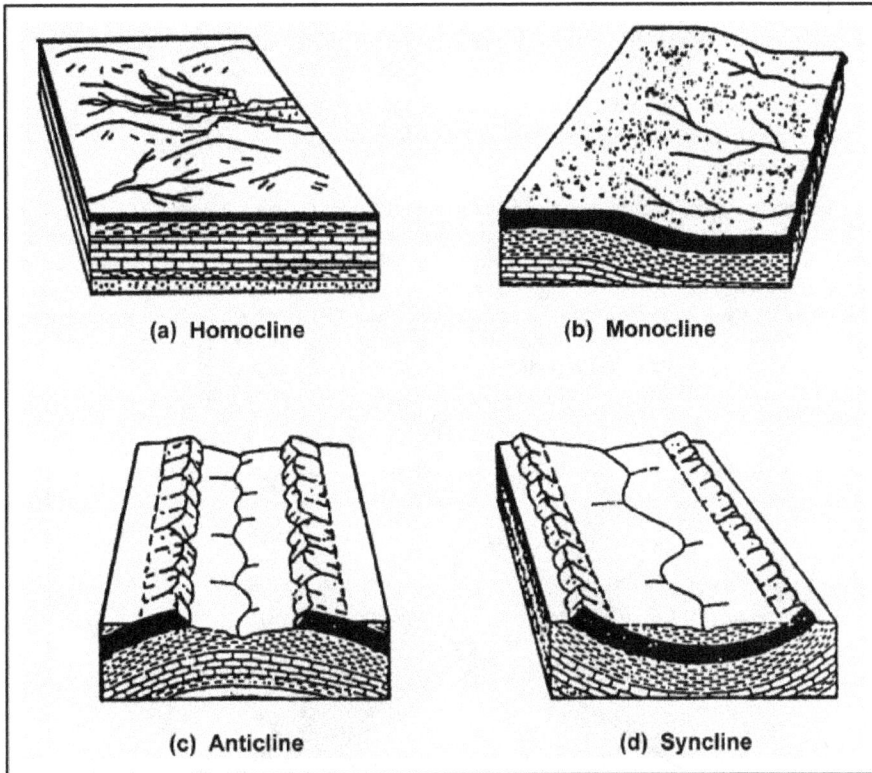

(a) Homocline   (b) Monocline

(c) Anticline   (d) Syncline

Figure 2-5. Common types of folds

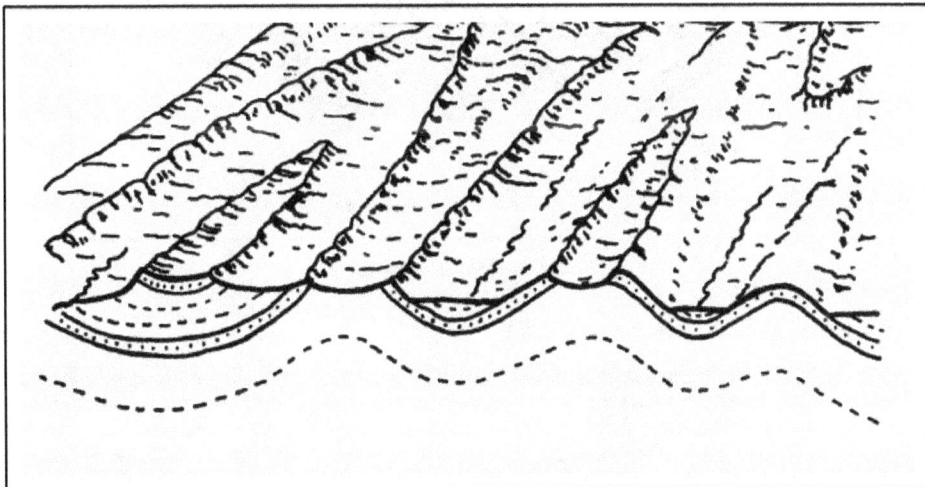

Figure 2-6. Topographic expression of plunging folds

SYMMETRY

2-6. Folds are further classified by their symmetry. Examples are—

- Asymmetrical (inclined).
- Symmetrical (vertical).
- Overturned (greatly inclined).
- Recumbent (horizontal).

2-7. The axial plane of a fold is the plane that bisects the fold as symmetrically as possible. The sides of the fold as divided by the axial plane are called the limbs. In some folds, the plane is vertical or near vertical, and the fold is said to be symmetrical. In others, the axial plane is inclined, indicating an asymmetrical fold. If the axial plane is greatly inclined so that the opposite limbs dip in the same direction, the fold is overturned. A recumbent fold has an axial plane that has been inclined to the point that it is horizontal. Figure 2-7 shows the components of an idealized fold. An axial line or fold axis is the intersection of the axial plane and a particular bed. The crest of a fold is the axis line along the highest point on an anticline. The trough denotes the line along the lowest part of the fold. It is a term associated with synclines.

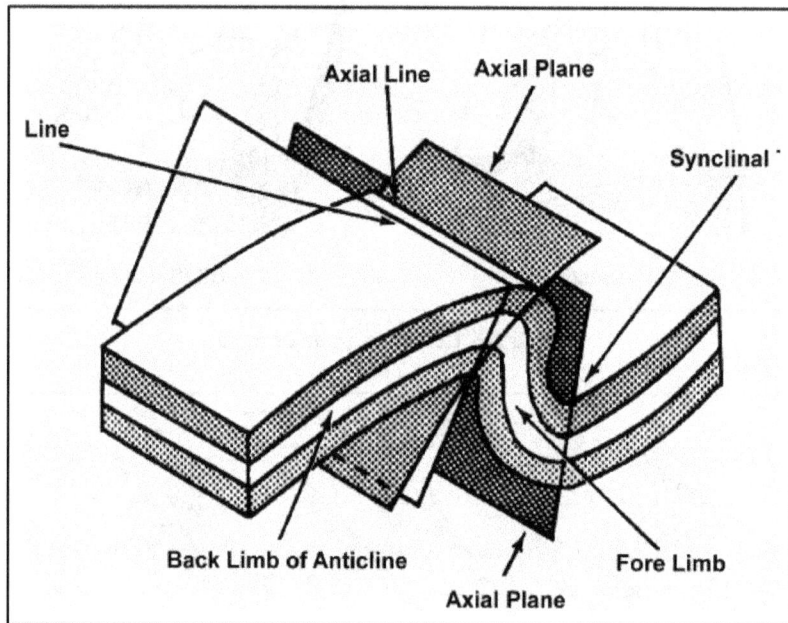

Figure 2-7. Fold symmetry

CLEAVAGE AND SCHISTOSITY

2-8. Foliation is the general term describing the tendency of rocks to break along parallel surfaces. Cleavage and schistosity are foliation terms applied to metamorphic rocks. Metamorphic rocks have been altered by heat and/or pressure due to mountain building or other crustal movements. They may have a pronounced cleavage, such as the metamorphic rock slate that was at one time the sedimentary rock shale. Certain igneous rocks may be deformed into schists or igneisses with alignment of minerals to produce schistosity or gneissic foliation. The attitudes of planes of cleavages and schistosity can be mapped to help determine the structure of a rock mass. Their attitudes can complicate rock excavation and, if unfavorable, lead to slope stability problems.

FAULTS

2-9. Faults are fractures along which there is displacement of the rock parallel to the fracture plane; once-continuous rock bodies have been displaced by movement in the earth's crust (see figure 2-8). The magnitude of the displacement may be inches, feet, or even miles along the fault plane. Overall fault displacement often occurs along a series of small faults. A zone of crushed and broken rock may be produced as the walls are dragged past each other. This zone is called a "fault zone" (see figure 2-9). It often contains crushed and altered rock, or "gouge," and angular fragments of broken rock called "breccia." Fault zones may consist of materials that have been altered (reduced in strength) by both fault movement and accelerated weathering by water introduced along the fault surface. Alteration of fault gouge to clay lowers the resistance of the faulted rock mass to sliding. Recognition of faults is extremely important to military engineers, as they represent potential weakness in the rock mass. Faults that cut very young sediments may be active and create seismic (earthquake) damage.

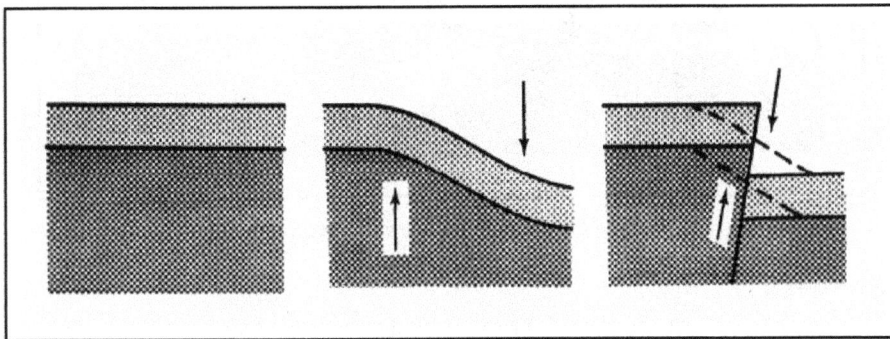

Figure 2-8. Faulting

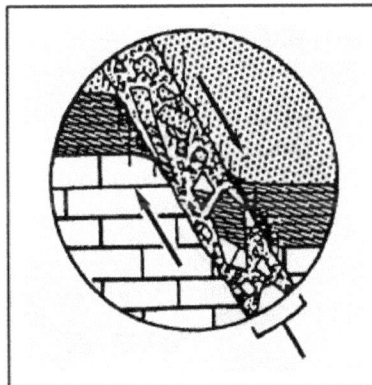

Figure 2-9. Fault zone

RECOGNITION

2-10. Faults are commonly recognized on rock outcrop surfaces by the relative displacement of strata on opposite sides of the fault plane and the presence of gouge or breccia. Slickensides, which are polished and striated surfaces that result from movement along the fault plane, may develop on the broken rock faces in a direction parallel to the direction of movement. Faulting may cause a discontinuity of structure that may be observed at rock outcrops where one rock layer suddenly ends against a completely different layer. This is often observed in road cuts, cliff faces, and streambeds. Although discontinuity of rock beds often indicates faulting, it may also be caused by igneous intrusions and unconformities in deposition. Faults that are not visibly identifiable can be inferred by sudden changes in the characteristics of rock strata in an outcrop or borehole, by missing or repeated strata in a stratigraphic sequence, or (on a larger scale) by the presence of long straight mountain fronts thrust up along the fault. Rock strata may show evidence of dragging along the fault. Drag is the folding of rock beds adjacent to the fault (see figure 2-8, page 2-7 and figure 2-10). Faults are identifiable on aerial photographs by long linear traces (lineations) on the ground surface and by the offset of linear features such as strata, streams, fences, and roads. Straight fault traces often indicate near-vertical fault planes since traces are not distorted by topographic contours.

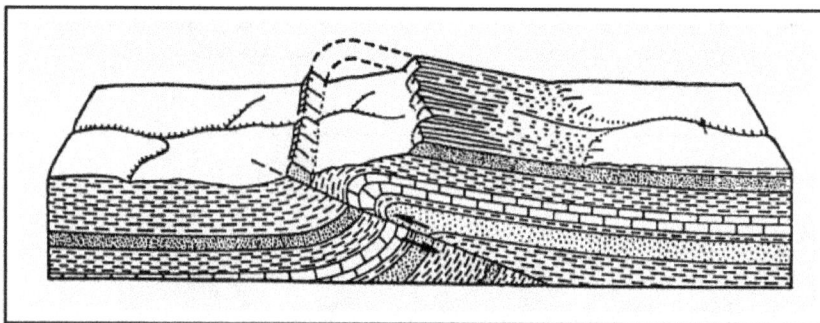

Figure 2-10. Thrust fault with drag folds

TERMINOLOGY

2-11. The strike and dip of a fault plane is measured in the same manner as it is for a layer of rock. (This procedure will be described later.) The fault plane intersection with the surface is called the fault line. The fault line is drawn on geologic maps. The block above the fault plane is called the hanging wall; the block below the fault plane is called the footwall. In the case of a vertical fault, there would be neither a hanging wall nor a footwall. The vertical displacement along a fault is called the throw. The horizontal displacement is the heave. The slope on the surface produced by movement along a fault is called the fault scarp. It may vary in height from a few feet to thousands of feet or may be eroded away (see figure 2-11).

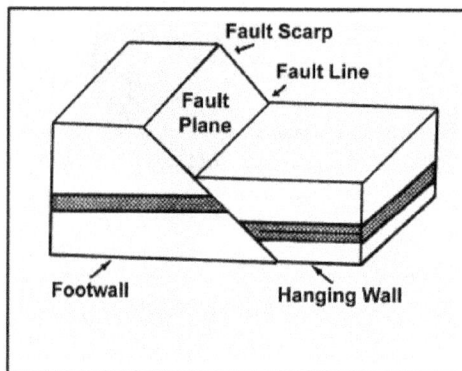

Figure 2-11. Fault terminology

TYPES

2-12. Faults are classified by the relative direction of movement of the rock on opposite sides of the fault. The major type of movement determines their name. These types are—

- Normal (gravity).
- Reverse (thrust).
- Strike-slip.

2-13. Normal faults are faults along which the hanging wall has been displaced downward relative to the footwall (see figure 2-12, a). They are common where the earth's surface is under tensional stress so that the rock bodies are pulled apart. Normal faults are also called gravity faults and usually are characterized by high-angle (near-vertical) fault planes. In a reverse fault, the hanging wall has been displaced upward relative to the footwall (see figure 2-12, b). Reverse faults are frequently associated with compressional forces that accompany folding. Low-angle (near-horizontal) reverse faults are called overthrust faults. Thrust faulting is common in many mountainous regions, and overthrusting rock sheets may be displaced many kilometers over the underlying rocks (see figure 2-10). Strike-slip faults are characterized by one block being displaced laterally with respect to the other; there is little or no vertical displacement (see figure 2-12, c). Many faults exhibit both vertical and lateral displacement. Some faults show rotational movement, with one block rotated in the fault plane relative to the opposite block. A block that is downthrown between two faults to form a depression is called a "graben" (see figure 2-13, a). An upthrown block between two faults produces a "horst" (see figure 2-13, b). Horsts and grabens are common in the Basin and Range Province located in the western continental United States. The grabens comprise the valleys or basins between horst mountains.

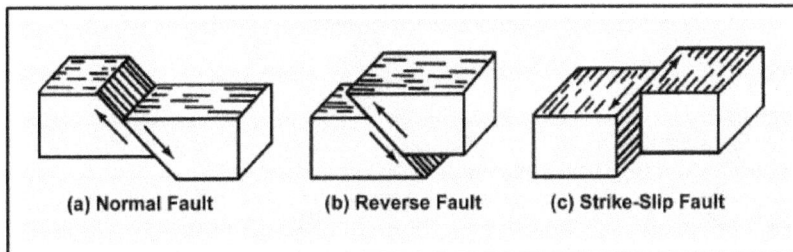

Figure 2-12. Types of faults

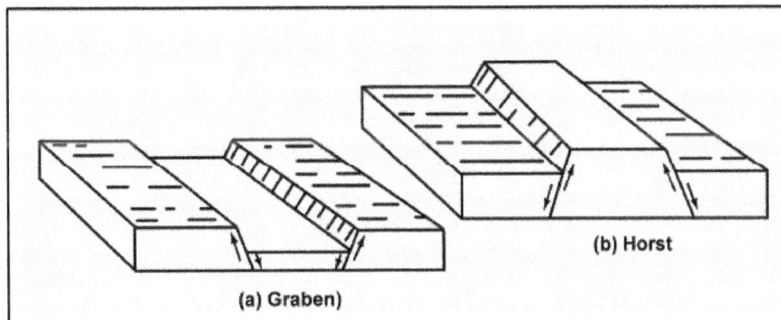

Figure 2-13. Graben and horst faulting

JOINTS

2-14. Rock masses that fracture in such a way that there is little or no displacement parallel to the fractured surface are said to be jointed, and the fractures are called joints (see figure 2-14, page 2-10). Joints influence the way the rock mass behaves when subjected to the stresses of construction. Joints characteristically form planar surfaces. They may have any attitude; some are vertical, others are horizontal, and many are inclined at various angles. Strike and dip are used to measure the attitude of joints. Some joints may occur as curved surfaces. Joints vary greatly in magnitude, from a few feet to thousands of feet long. They commonly occur in more or less parallel fractures called joint sets. Joint systems are two or more related joint sets or any group of joints with a characteristic pattern, such as a radiating or concentric pattern.

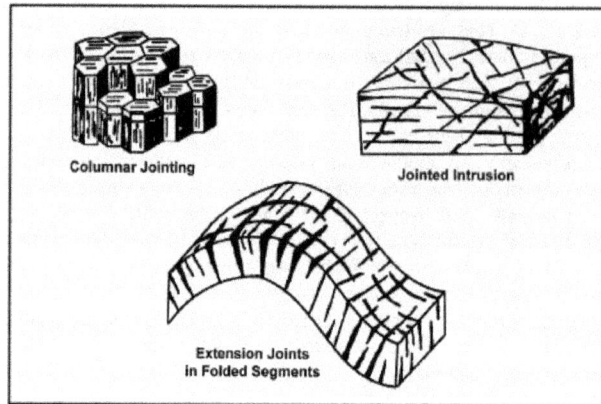

Figure 2-14. Jointing in sedimentary and igneous rock

FORMATION

2-15. Joints in rock masses may result from a number of processes, including deformation, expansion, and contraction. In sedimentary rocks, deformation during lithification or folding may cause the formation of joints. Igneous rocks may contain joints formed as lava cooled and contracted. In dense, extrusive igneous rocks, like basalt, a form of prismatic fracturing known as columnar jointing often develops as the rock cools rapidly and shrinks. Jointing may also occur when overlying rock is removed by erosion, causing a rock mass to expand. This is known as exfoliation. The outer layers of the rock peel, similar to the way that an onion does.

SIGNIFICANCE

2-16. Because of their almost universal presence, joints are of considerable engineering importance, especially in excavation operations. It is desirable for joints to be spaced close enough to minimize secondary plugging and blasting requirements without impairing the stability of excavation slopes or increasing the overbreakage in tunnels. The spacing of the joints can control the size of the material removed and can also affect drilling and blasting. The ideal condition is seldom encountered. In quarry operations, jointing can lead to several problems. Joints oriented approximately at right angles to the working face present the most unfavorable condition. Joints oriented approximately parallel to the working face greatly facilitate blasting operations and ensure a fairly even and smooth break, parallel to the face (see figure 2-14). Joints offer channels for groundwater circulation. In excavations below the groundwater table, they may greatly increase water problems. They also may exert an important influence on weathering.

STRIKE AND DIP

2-17. The orientation of planar features is determined by the attitude of the rock. The attitude is described in terms of the strike and dip of the planar feature. The most common planar feature encountered is a sedimentary bed. Strike is defined as the trend of the line of intersection formed between a horizontal plane and the bedding plane being measured (see figure 2-15). The strike line direction is given as a compass bearing that is always in reference to true north. Typical strikes would thereby fall between north 0 to 90 degrees east or north 0 to 90 degrees west. They are never expressed as being to the southeast or southwest.

Azimuths may be readily converted to bearings (for example, an azimuth of 350 degrees would be converted to a bearing of north 10 degrees west).

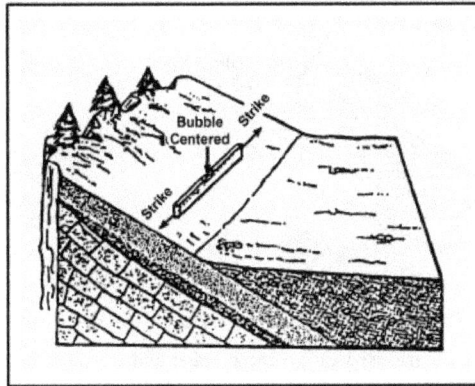

Figure 2-15. Strike

2-18. The dip is the inclination of the bedding plane. It is the acute angle between the bedding plane and a horizontal plane (see figure 2-16). It is a vertical angle measured at right angles from the strike line. The dip direction is defined as the quadrant of the compass the bed is dipping into (northeast, northwest, southeast, or southwest). By convention, the dip angle is given in degrees followed by the dip direction quadrant (for example, 30 degrees northeast).

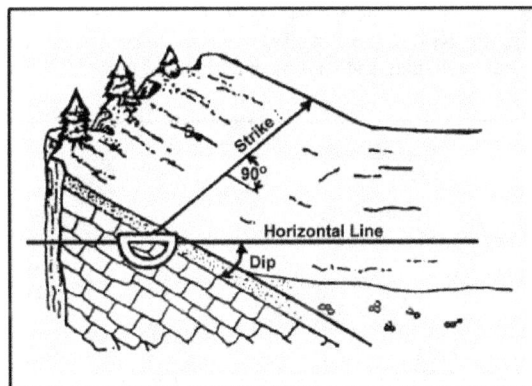

Figure 2-16. Dip

2-19. The strike and dip measurements are taken in the field on rock outcrops with a standard Brunton compass. The Brunton compass is graduated in degrees and has a bull's-eye level for determining the horizontal plane when measuring the strike direction. The strike is determined by aligning the compass along the strike direction and reading the value directly from the compass. Included with the Brunton compass is a clinometer to measure the dip angle. This angle is measured by placing the edge of the compass on the dipping surface at right angles to the strike direction and reading the acute angle indicated by the clinometer (see figure 2-17, page 2-12).

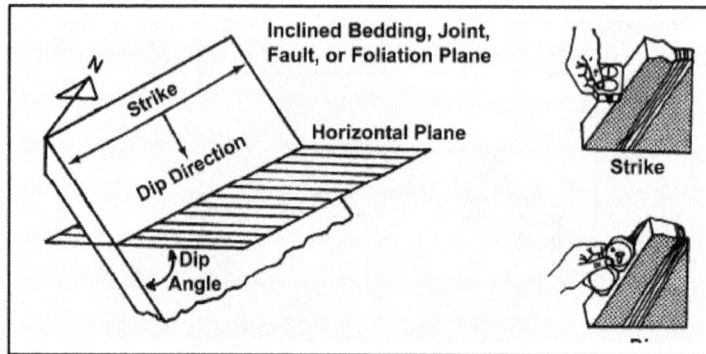

Figure 2-17. Measuring strike and dip with a Brunton compass

2-20. Strike and dip symbols are used on geologic maps and overlays to convey structural orientation. Basic symbols include those for inclined, vertical, and horizontal beds (see figure 2-18). For inclined beds, the direction of strike is designated as a long line that is oriented in reference to the map grid lines in exactly the same compass direction as it was measured. The direction of the dip is represented by a short line that is always drawn perpendicular to the strike line and in the direction of the dip. The angle of the dip is written next to the symbol (see figure 2-18, a). For vertical beds, the direction of the strike is designated as it is for inclined beds. The direction of the dip is a short line crossing the strike line at a right angle extending on both sides of the strike line (see figure 2-18, b). For horizontal beds, the direction of strike is represented by crossed lines which indicate that the rock strikes in every direction. The dip is represented by a circle encompassing the crossed lines. The circle implies that there is no dip direction and the dip angle is zero (see figure 2-18, c). These basic symbols are commonly used to convey attitudes of sedimentary rocks (see figure 2-19). Similar symbols are used to convey attitudes of other types of planar features, such as folds, faults, foliation, and jointing in other rock bodies.

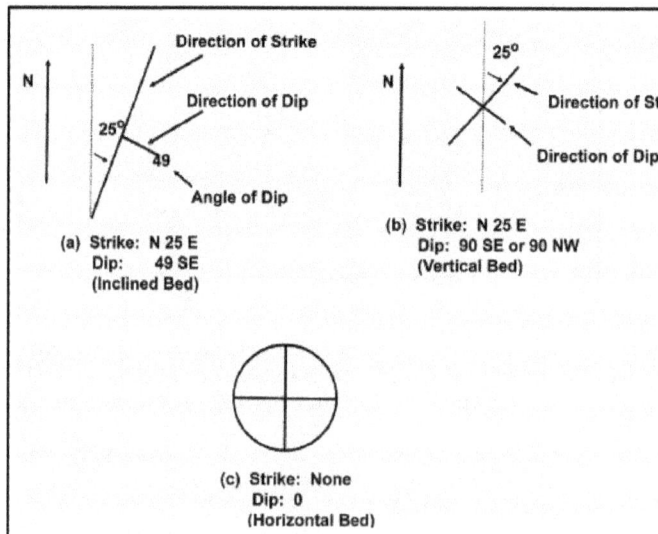

Figure 2-18. Strike and dip symbols

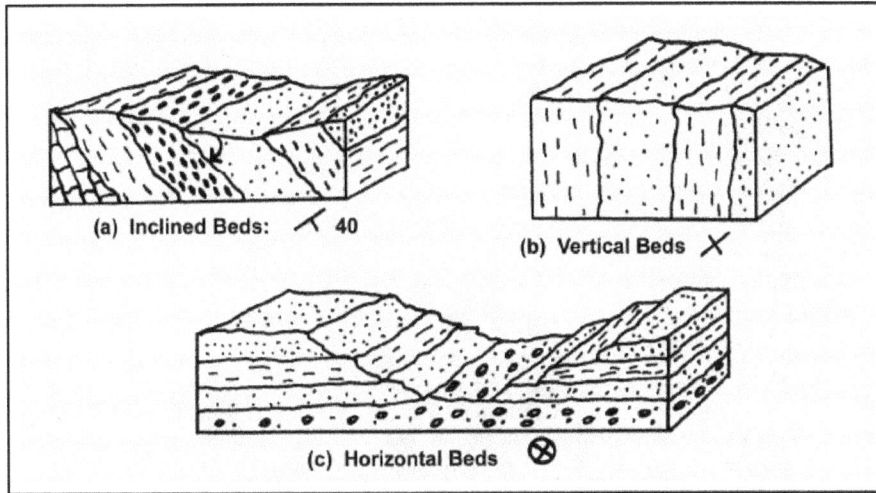

Figure 2-19. Strike and dip symbols of sedimentary rocks

SECTION II - GEOLOGIC MAPS

TYPES

2-21. Geologic maps show the distribution of geologic features and materials at the earth's surface. Most are prepared over topographic base maps using aerial photography and field survey data. From a knowledge of geologic processes, the user of a geologic map can draw many inferences as to the geologic relationships beneath the surface and also much of the geologic history of an area. In engineering practice, geologic maps are important guides to the location of construction materials and the evaluation of foundation, excavation, and groundwater conditions.

2-22. The following geologic maps are used for military planning and operations:

- Bedrock or aerial maps.
- Surficial maps.
- Special purpose maps.

BEDROCK OR AERIAL MAPS

2-23. These maps show the distribution of rock units as they would appear at the earth's surface if all unconsolidated materials were removed. Symbols on such maps usually show the age of the rock unit as well as major structural details, such as faults, fold axes, and the attitudes of planar rock units or features. Thick deposits of alluvium (material deposited by running water) may also be shown.

SURFICIAL MAPS

2-24. These maps show the distribution of unconsolidated surface materials and exposed bedrock. Surface materials are usually differentiated according to their physical and/or chemical characteristics. To increase their usefulness as an engineering tool, most surficial maps show the distribution of materials at some shallow mapping depth (often one meter) so that minor residual soils and deposits do not mask the essential features of engineering concern.

SPECIAL PURPOSE MAPS

2-25. These maps show selected aspects of the geology of a region to more effectively present information of special geologic, military, or engineering interest. Special purpose maps are often prepared to show the distribution of—

- Engineering hazards.
- Construction materials.
- Foundation conditions.
- Excavation conditions.
- Groundwater conditions.
- Trafficability conditions.
- Agricultural soils.
- Surface-water conditions.

2-26. Very detailed, large-scale geologic maps may show individual rock bodies, but the smallest unit normally mapped is the formation. A formation is a reasonably extensive, distinctive series of rocks deposited during a particular portion of geologic time (see table 2-1). A formation may consist of a single rock type or a continuous series of related rocks. Generally, formations are named after the locality where they were first defined. Formations may be grouped by age, structure, or lithology for mapping purposes.

Table 2-1. Geologic time scale

Era	Period or System		Symbol	Epoch or Series	Important Physical Events and Flora and Fauna	Million Years Ago
Cenozoic	Quaternary	Neogene	Q	Recent	Glaciers melted; milder climates; many mammals disappeared.	1
	Quaternary		Q	Pleistocene	Glaciation; fluctuating cold to mild climates; most invertebrate living species; dominance of large mammals; man.	
	Tertiary	Neogene		Pliocene	Continued uplift and mountain building; climate cooler, mammals reach peak in size and abundance.	10
	Tertiary			Miocene	Uplifts of Sierras and Rockies; moderate climates; rise of grazing mammals.	25
	Tertiary	Paleogene	T	Oligocene	Lands generally low, rise of Alps and Himalayas began; volcanoes in Rockies area; first saber-toothed cats.	40
	Tertiary			Eocene	Mountains eroded; many lakes in western North America; climates mild to very tropical; all modern mammals present (first horses).	
	Tertiary			Paleocene	Mountains high; climates mild to cool; primitive mammals; modern birds; new invertebrates.	60
Mesozoic	Cretaceous		K		Lands low and extensive; mild climates; last widespread oceans; flowering plants expand rapidly; giant reptiles become extinct; modern insects; ammonites die out; foraminifera; period closed with Laramide Revolution (Sierra uplifted).	70
	Jurassic		J		Continents low; large areas of Europe covered by seas; climates mild; mountains from Alaska to Mexico rise; eruptions and intrusions in the northwest; dinosaurs; marine reptiles; ammonites and belemnites abundant; ginkgoes; conifers; cycads.	135
	Triassic		T_R		Continents mountainous; large areas arid; eruptions in eastern North America and New Zealand; first dinosaurs and marine reptiles; first hexacorals, last conodonts.	180
Paleozoic	Permian		P		First mammal-like reptiles, other reptiles diversified; many marine invertebrates become extinct; last trilobites; period ended with Appalachian Revolution.	220
	Carboniferous	Pennsylvanian	PP		Lands low, covered by shallow seas or extensive coal swamps; climates warm; amphibians and reptiles reach large size; large insects; scorpions; cockroaches; fusuline foraminifera abundant.	270
	Carboniferous	Mississippian	M		Widespread seas retreated as result of mountain building in eastern United States, Texas, Colorado; climates warm; crinoids dominate; amphibians, sharks and bony fishes spread; insects develop wings.	
	Devonian		D		North America low and flat, but mountains and volcanoes present in eastern United States and Canada; Europe mountainous with arid basins; fishes dominant; first amphibians; many brachiopods corals, bryozoans and blastoids; first ammonites.	350
	Silurian		S		Continents relatively flat; mountain building in Europe; climates mild; much salt deposited; eurypterids and corals abundant; first air breathers; lycopod plants.	400
	Ordovician		O		Continents low with shallow seas; mountains rose at close in Europe and North America; abundant graptolites, trilobites, nautiloids, cystoids; first ostracods and conodonts; seaweeds and algae; climates uniformly mild.	440
	Cambrian		€		Extensive seas in major synclines on all continents; climates mild; marine invertebrates and algae abundant; trilobites dominant; all animal phyla probably existed.	500
Precambrian	Proterozoic (Algonkian)		P€		Shallow seas in geosynclines; climates warm and moist to dry and cold; glaciation in eastern Canada; bacteria; marine algae; worm burrows; sponge spicules; probably most phyla lived but left no record; few fossils; iron area of Lake Superior formed.	600 / 1000
	Archaeozoic (Primitive Life)				Extensive mountain building with intrusions and eruptions; iron deposits formed; earliest known life; blue-green algae; fungi; graphite and carbonaceous shales in Australia and Canada; carbon in Rhodesian rock.	3000
	Azoic (Without Life)				Formation of the Earth's crust; no rocks have been found, therefore cannot be dated by any known method.	6000

SYMBOLS

2-27. Symbols are used to identify various features on a geologic map. Some of those features are—

- Formations (see figure 2-21).
- Contacts (see figure 2-21).
- Attitudes (see figure 2-21).
- Fault lines and fold axes (see figure 2-21).
- Cross sections (see figure 2-23, page 2-18).

FORMATIONS

2-28. Letters, colors, or symbolic patterns are used to distinguish formations or rock units on a geologic map. These designators should be defined in a legend on the map. Letter symbols usually consist of a capital letter indicating the period of deposition of the formation with subsequent letters (usually lower case) that stand for the formal name of the unit (see table 2-1, page 2-15). Maps prepared by the US Geological Survey and many other agencies use tints of yellow and orange for Cenozoic rocks, tints of green for Mesozoic rocks, tints of blue and purple for Paleozoic rocks, and tints of red for Precambrian rocks. Symbolic patterns for various rock types are given in figure 2-20.

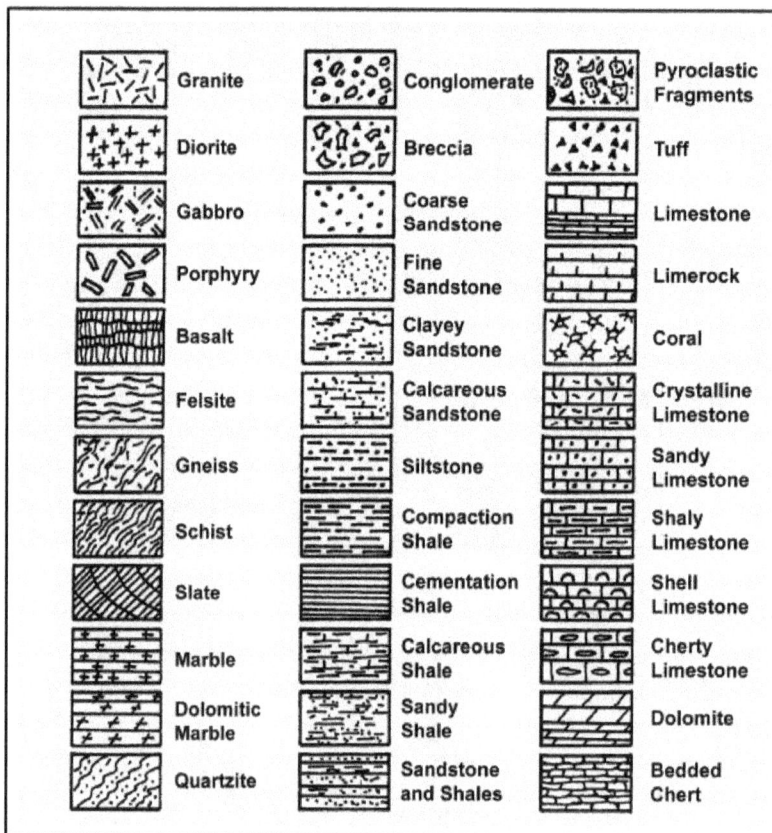

Figure 2-20. Symbolic patterns for rock types

Contacts

| Accurately Located | Approximately Located | Covered or Concealed |

Strike and Dip Symbols

	Inclined	Vertical	Horizontal	Overturned
For Strata:	40	or		40
For Joints:	40			
For Foliation:	40			

Axes of Folds

| Anticline | Syncline showing plunge of axes. | Overturned anticline and syncline showing dip of limbs. |

Fault Traces

65	U D	Ts
Fault showing dip of fault plane.	Fault showing relative movement (U, upthrown side, D, downthrown side).	Thrust fault (Ts or hachures on upper plate).

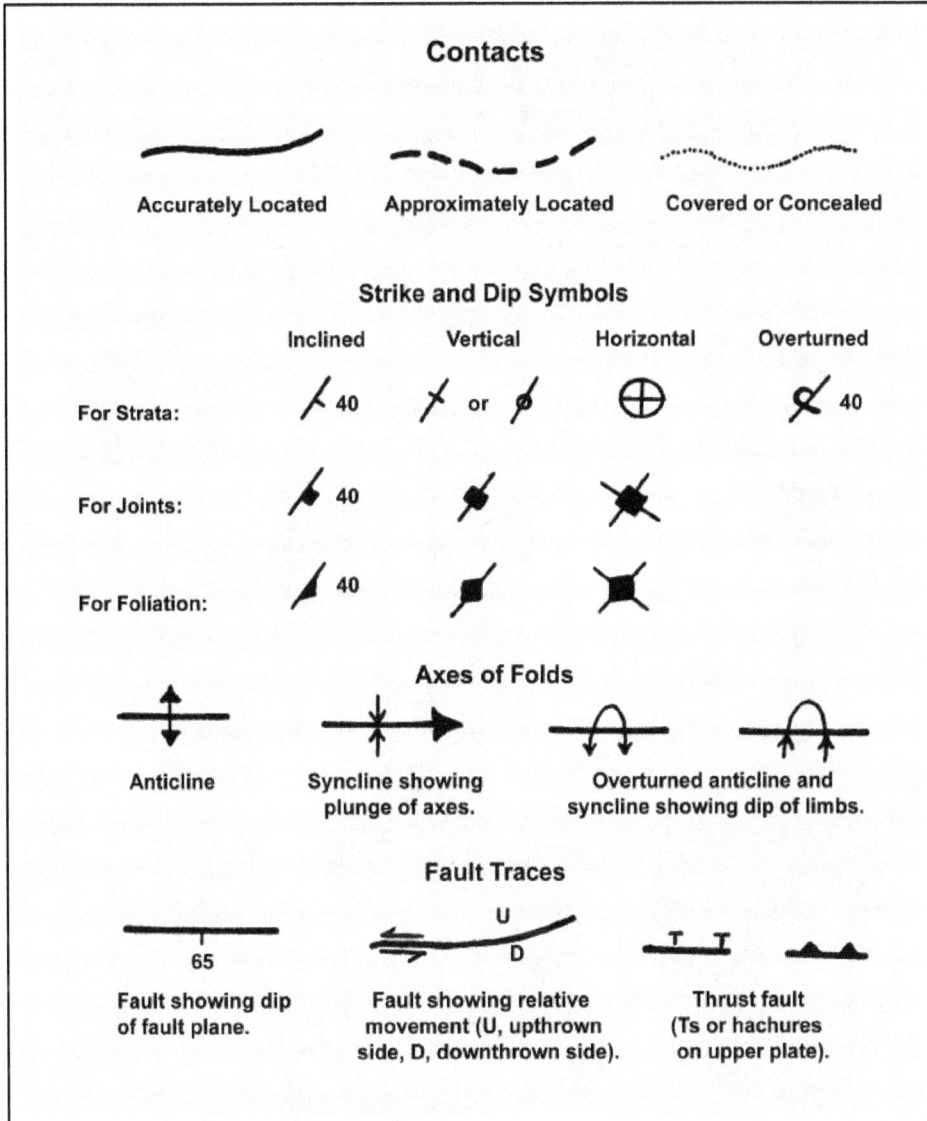

Figure 2-21. Geologic map symbols

CONTACTS

2-29. A thin, solid line shows contacts or boundaries between rock units if the boundaries are accurately located. A dashed line is used for an approximate location and a dotted line if the contact is covered or concealed. Questionable or gradational contacts are shown by a dashed or dotted line with question marks.

ATTITUDES

2-30. Strike and dip symbols describe planes of stratification, faulting, and jointing. These symbols consist of a strike line long enough so that its bearing can be determined from the map, a dip mark to indicate the dip direction of the plane being represented, and a number to show the value (in degrees) of the dip angle. The number is omitted on representations of both horizontal and vertical beds, because the values of the dips are automatically acknowledged to be 0 and 90 degrees, respectively. Figure 2-22 shows the placement of strike and dip symbols on a geologic map with respect to the location and orientation of a sedimentary rockbed.

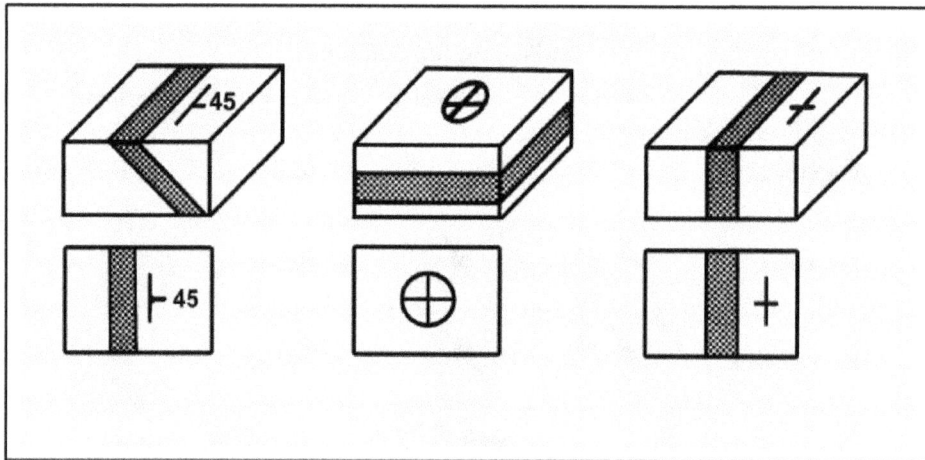

Figure 2-22. Placement of strike and dip symbols on a geologic map

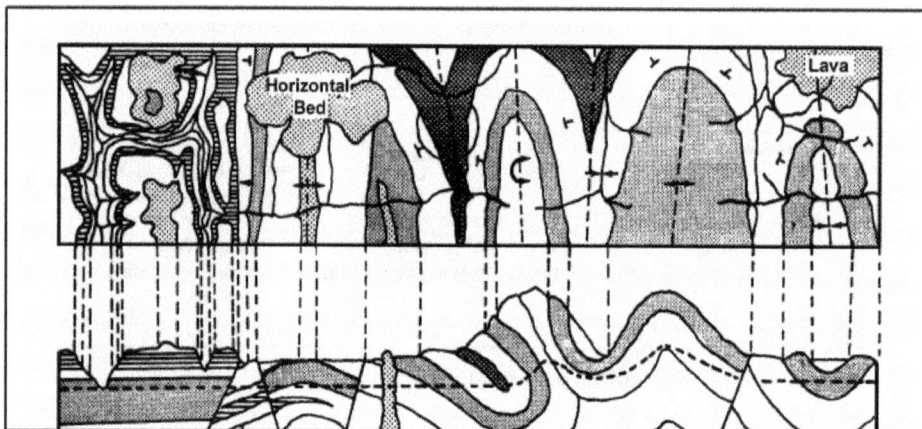

Figure 2-23. Geologic map and cross section

FAULT LINES AND FOLD AXES

2-31. Heavy black lines, which may be solid, dashed, or dotted (as described for contacts), show fault lines and fold axes. The direction of movement along faults is shown by arrows or by the use of symbols to indicate up thrown and down thrown sides. The arrows accompanying fold axes indicate the dip direction of the limbs and/or the plunge direction of the fold.

CROSS SECTIONS

2-32. Cross sections show the distribution of geologic features and materials in a vertical plane along a line on a map. Cross sections are prepared in much the same way as topographic profiles using map, field, and borehole data. Geologic sections accompany many geologic maps to clarify subsurface relationships. Like geologic maps, geologic sections are often highly interpretive, especially where data is limited and structures are complex or concealed by overburden. Maps and sections use similar symbols and conventions. Because of the wealth of data that can be shown, geologic maps and sections are the two most important means of recording and communicating geologic information.

OUTCROP PATTERNS

2-33. An outcrop is that part of a rock formation that is exposed at the earth's surface. Outcrops are located where there is no existing soil cover or where the soil has been removed, leaving the rock beneath it exposed. Outcrops may indicate both the type and the structure of the local bedrock. Major types of structural features can be easily recognized on geologic maps because of the distinctive patterns they produce. Figure 2-24 and figures 2-25 through 2-31, pages 2-20 through 2-23, show basic examples of common structural patterns.

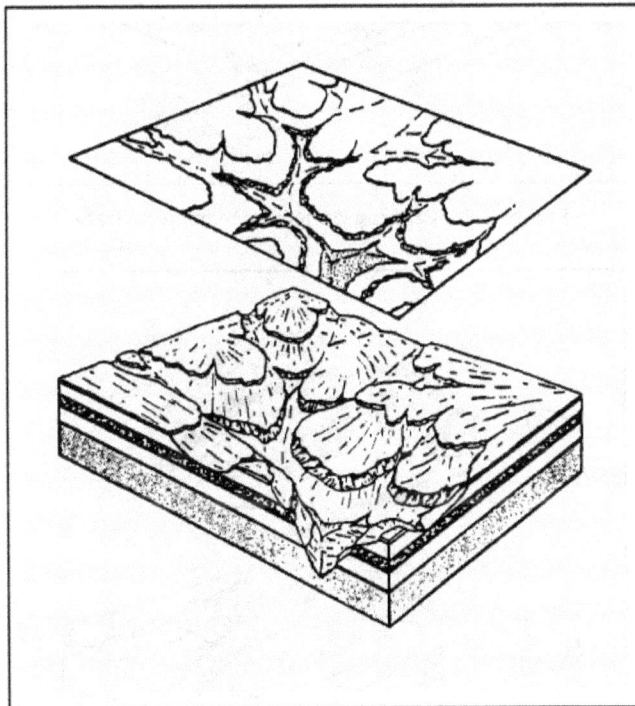

Figure 2-24. Outcrop patterns of horizontal strata

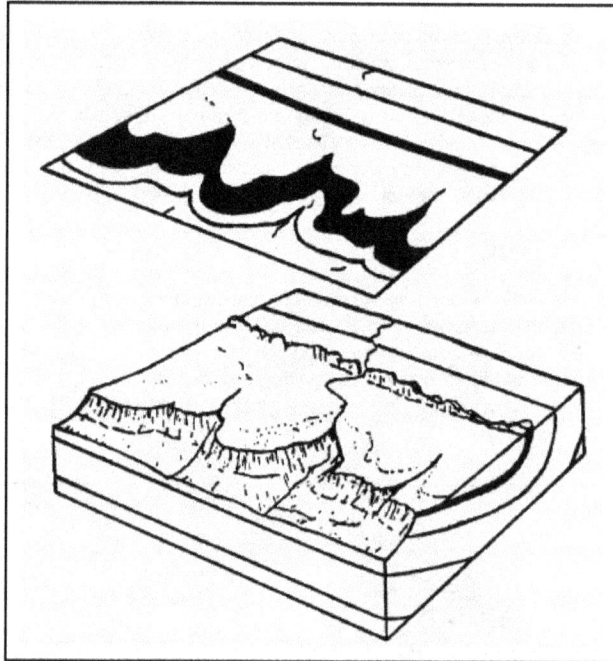

Figure 2-25. Outcrop patterns of inclined strata

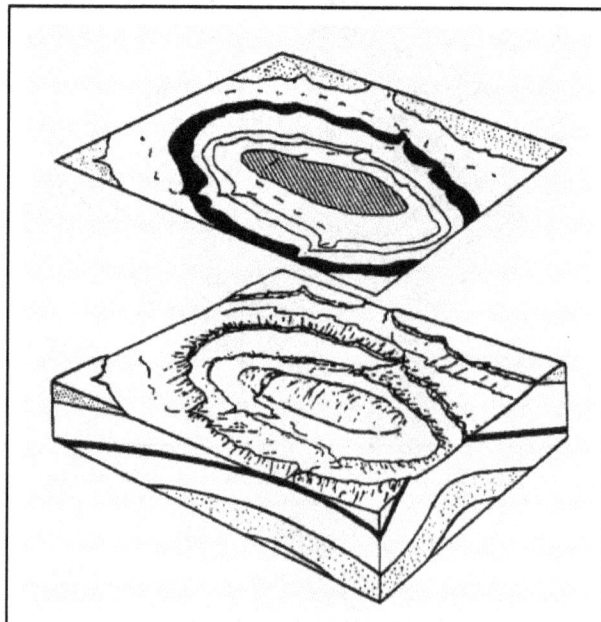

Figure 2-26. Outcrop patterns of an eroded dome

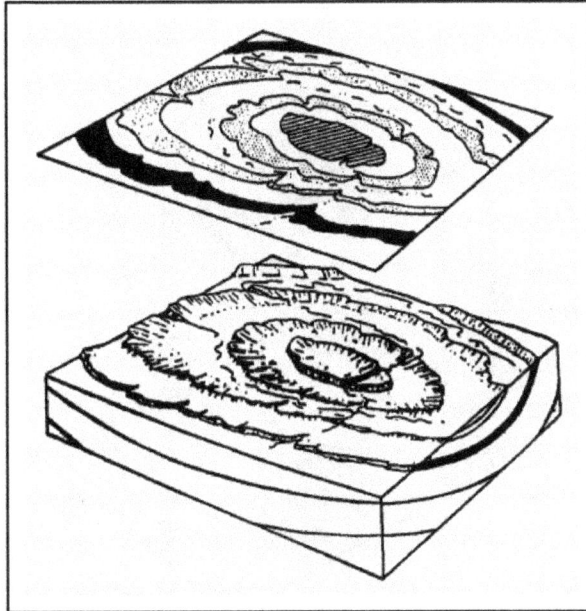

Figure 2-27. Outcrop patterns of and eroded basin

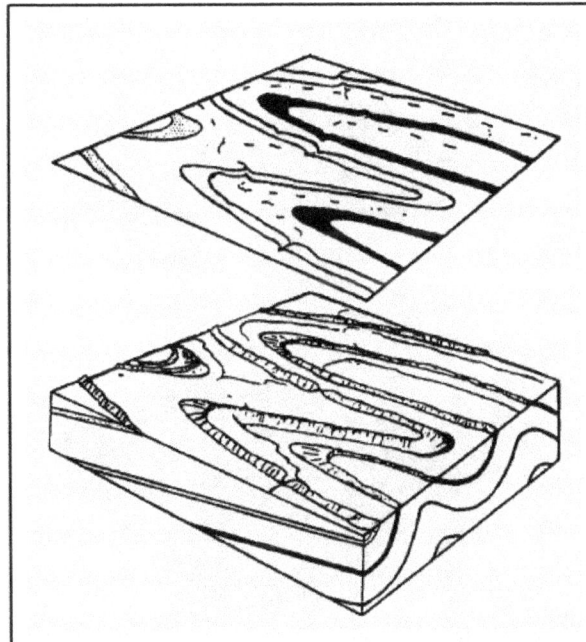

Figure 2-28. Outcrop patterns of plunging folds

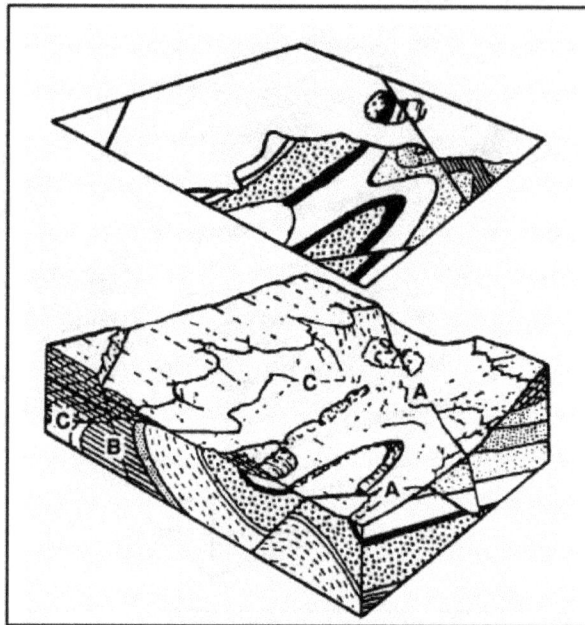

Figure 2-29. Outcrop patterns produced by faulting

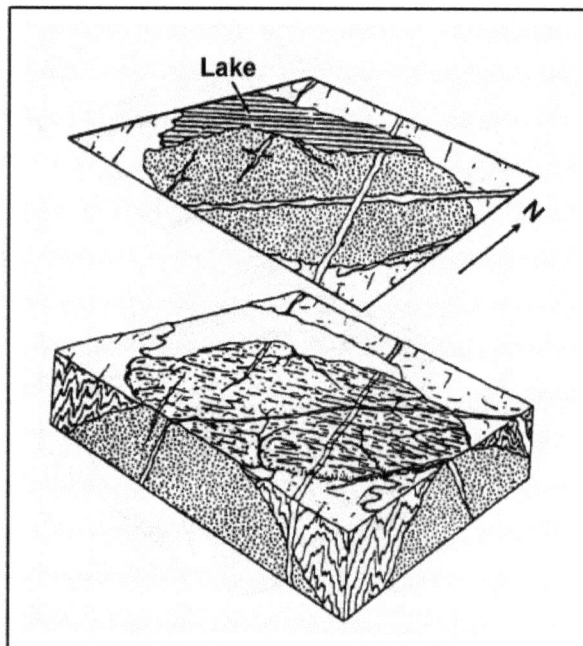

Figure 2-30. Outcrop patterns of intrusive rocks

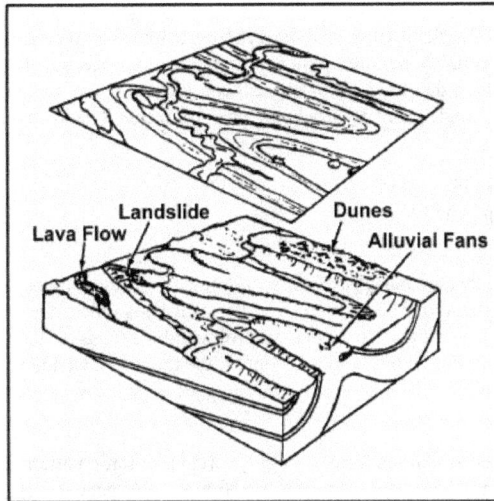

Figure 2-31. Outcrop patterns of surficial deposits

2-34. Each illustration contains a block diagram showing a particular structural feature along with its topographic expression. The outcrop pattern of each rock unit shown on the block diagram is projected to a horizontal plane, resulting in the production of a geologic map that is also shown. This allows the reader to readily relate the structure shown on the block diagram to the map pattern. Structural details can be added to basic maps using the symbols in figure 2-21, page 2-17. The illustrations include some of the following structural features:

- Horizontal strata.
- Inclined strata.
- Domes.
- Basins.
- Plunging folds.
- Faults.
- Intrusive rocks.
- Surficial deposits.

HORIZONTAL STRATA

2-35. Dendritic (branching or treelike) drainage patterns typically develop on horizontal strata and cut canyons or valleys in which progressively older rock units are exposed at depth (see figure 2-24, page 2-19). The result is that the map patterns of horizontal strata parallel stream valleys, producing a dendritic pattern on the geologic map. Although all maps do not show topographic contour lines, the contacts of horizontal rock units parallel the contours. Escarpments and gentle slopes generally develop on resistant and nonresistant beds, respectively, producing variations in the width of the map outcrop pattern. The upper and lower contacts are close together on steep cliffs; on gentle slopes of the same formation, the contacts are further apart. The map width of the outcrops of horizontal beds does not indicate the thickness of the strata. Gently dipping beds develop the same basic outcrop pattern as horizontal beds. However, the contacts of gently dipping strata, if traced far enough up a valley, cross topographic contours and form a large V-shaped pattern that points in the direction the beds dip, assuming that the beds do not dip in the direction of the stream gradient, but at a smaller angle.

INCLINED STRATA

2-36. When a sequence of rocks is tilted and cut off by erosion, the outcrop pattern appears as bands that, on a regional basis, are roughly parallel. Where dipping strata cross a valley, they produce a V-shaped outcrop pattern that points in the direction of dip, except in cases where the beds dip in the direction of the stream gradient at smaller angles than the gradient. The size of the V is inversely proportional to the degree of dip.

- Low-angle dip (large V) (see front part of figure 2-25, page 2-20).
- High-angle dip (small V).
- Vertical dip (no V) (see back part of figure 2-25).

2-37. Other relationships that are basic to the interpretation of geologic maps are also shown in figure 2-25. For example, they show that older beds dip toward younger beds unless the sequence has been overturned (as by folding or faulting). Maps also show that outcrop width depends on the thickness of the beds, the dip of the beds (low dip, maximum width), and the slope of the topography (steep slope, minimum width).

DOMES

2-38. Eroded dome-shaped structures form a roughly circular outcrop pattern with beds dipping away from a central area in which the oldest rocks outcrop (see figure 2-26, page 2-20). These structures range from small features only a few meters across to great upwarps covering areas of hundreds or thousands of square kilometers.

2-39. Drainage patterns are helpful in interpreting a domal structure. Radial drainage patterns tend to form on domes. Streams cutting across the resistant beds permit one to apply the "rule of Vs" as explained above to interpret the direction of dip.

BASINS

2-40. Eroded structural basins form an outcrop pattern very similar to that of an eroded dome (see figure 2-27, page 2-21). However, two major features serve to distinguish them: younger rocks outcrop in the center of a basin and, if the structure has been dissected by stream erosion, the outcrop Vs normally point toward the center of a basin, whereas they usually point away from the center of a dome.

PLUNGING FOLDS

2-41. Folding is found in complex mountain ranges and sometimes in lowlands and plateaus. When folds erode, the oldest rocks outcrop in the center of the anticlines (or upfolds) and the youngest rocks outcrop in the center of the synclines (or downfolds). The axes of folded beds are horizontal in some folds, but they are usually inclined. In this case, the fold is said to plunge. Plunging folds form a characteristic zigzag outcrop pattern when eroded (see figure 2-28, page 2-21). A plunging anticline forms a V-shaped pattern with the apex (or nose) of the V pointing in the direction of the plunge. Plunging synclines form a similar pattern, but the limbs of the fold open in the direction of the plunge.

FAULTS

2-42. Fault patterns on geologic maps are distinc- tive in that they abruptly offset structures and terminate contacts (see figure 2-29, page 2-22). They are expressed on the geologic map by heavy lines in order to be readily distinguished. Some common types are—

- Normal.
- Reverse.
- Thrust.

Normal and Reverse

2-43. (See A and B, respectively, in figure 2-29). Both normal and reverse fault planes generally dip at a high angle, so outcrop patterns are relatively straight. Older rocks are usually exposed on the upthrown block. It is thus possible to determine the relative movement on most high-angle faults from map relations alone. Linear streams, offsets, linear scarps, straight valleys, linear-trending springs or ponds, and omitted or repeated strata are common indications of faulting (see paragraph on recognition of faults, page 2-7).

Thrust

2-44. (See C in figure 2-29). Thrust faults are reverse faults that dip at low angles (less than 15 degrees) and have stratigraphic displacements, commonly measured in kilometers (see figure 2-10, page 2-8). The trace of the thrust commonly forms Vs where it intersects the valleys. The Vs point in the direction of the fault plane dip, except in cases where the fault plane dips in the direction of the stream gradient, but a smaller angle. Erosion may form windows (fensters) through the thrust sheet so that underlying rocks are exposed or produce isolation remnants (klippen) above the underlying rocks. Hachure symbols are used to designate the overthrust block that usually contains the oldest rocks.

INTRUSIONS

2-45. Larger igneous intrusions, such as batholiths and stocks, are typically discordant and appear on geologic maps as elliptical or roughly circular areas that cut across the contacts of surrounding formations (see figure 2-30, page 2-22). Smaller discordant intrusions, such as dikes, are usually tabular and appear on geologic maps as straight, usually short, bands. However, some dikes are lenticular and appear as such on the map. Concordant intrusions, such as sills and laccoliths, have contacts that parallel those of the surrounding formations (see figure 1-5, page 1-8).

2-46. The relative age of igneous bodies can be recognized from crosscutting relationships. The younger intrusions cut the older ones. With this in mind, it is clear from the relationships in figure 2-30, that the elliptical stock is the oldest intrusion, the northeast trending dike the next oldest, and the northwest trending dike the youngest. The age of the small discontinuous dikes near the western part of the map is younger than that of the stock, but the age relation with the other dikes is not indicated.

SURFICIAL DEPOSITS

2-47. Surficial deposits are recent accumulations of various types of sediment or volcanic debris on the surface of the landscape (see figure 2-31, page 2-23). The primary types are—

- Windblown sand and loess.
- Stream channel and floodplain deposits.
- Landslide deposits.
- Glacial deposits.
- Present beaches and other shoreline sediments.

SECTION III - ENGINEERING CONSIDERATIONS

ROCK DISTRIBUTION

2-48. Geologic structure controls the distribution of rock bodies and features along and beneath the earth's surface. The presence and orientation of such features as bedding, folding, faulting, and unconformities must be determined before construction begins. Otherwise, foundation, excavation, and groundwater conditions cannot be properly evaluated.

ROCK FRAGMENTATION

2-49. Rocks tend to fracture along existing zones of weakness. The presence and spacing of bedding, foliation, and joint planes can control the size and shape of rock fragments produced in quarries and other excavations. Operational and production costs may be prohibitive if rock fragments are too large, too small, too slabby, or too irregular for aggregate requirements. Advantageous joint or bedding spacings can significantly reduce excavation and aggregate production costs.

2-50. Many weak, thinly bedded, or highly fractured rocks can be excavated without blasting by using ripping devices drawn by heavy crawler tractors. When ripping is used to break up and loosen rock for removal, the work should proceed in the direction of the dip. This prevents the ripping devices from riding up the dip surfaces and out of the rock mass (see figure 2-32).

2-51. Most rock must be drilled and blasted for removal. Where joints or bedding planes incline across the axis of the drill hole, drill bits tend to follow these planes, causing the holes to be misaligned; or, more often, the bits to bind, stick, or break off in the holes (see figure 2-33). Open fractures and layers of weak rock greatly reduce blasting effectiveness by allowing the force of the blast to escape before the surrounding rock has been properly fragmented. Such situations require special drilling and blasting techniques that generally lower the efficiency of quarrying operations.

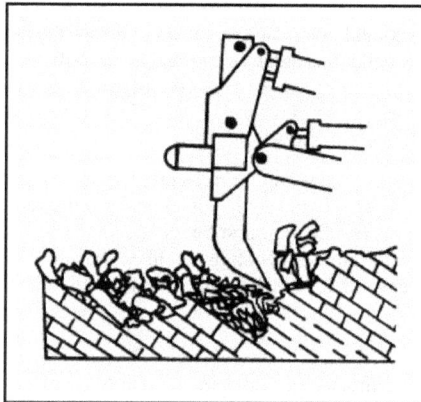

Figure 2-32. Ripping in the direction of dip

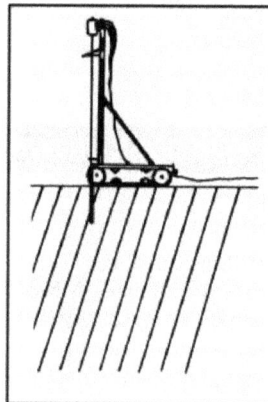

Figure 2-33. Rock drills

ROCK SLIDES AND SLUMPS

2-52. Massive rock slides may occur where unconfined rock masses overlie inclined bedding, foliation, fault, or joint surfaces (see figure 2-34). The risk of such slides is generally greatest over smooth, continuous, water- or clay-lubricated surfaces that dip steeply toward natural or man-made excavations. The following general observations may assist in evaluating hazards (see figure 2-35):

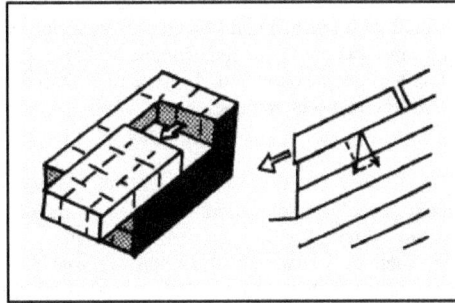

Figure 2-34. Rock slide on inclined bedding plane

- Most rocks are stable above surfaces that dip less than about 18 degrees toward an excavation. Excavation slopes with horizontal to vertical ratios of 1:2 (1 foot horizontal to 2 feet vertical) are usually feasible in such cases unless weak rocks underlie the excavation sidewalls.

- Some rocks may slide on surfaces that dip between about 18 degrees and 35 degrees toward an excavation, particularly if the surfaces are wet, clayey, smooth, and continuous. Side slopes of 1:1 or flatter may be required to stabilize such surfaces. It may be necessary to remove the hazardous rock entirely. Where excessive excavations must be avoided for economic, environmental, or other reasons, artificial supports or drainage works may be employed to stabilize the rock.

- Unless rock surfaces are discontinuous or very rough and uneven, most unconfined rocks will slide over surfaces steeper than about 35 degrees. Excavation side slopes should be cut back to the dipping surface or slightly beyond to assure stability against sliding.

- Rocks along planes of weakness that dip almost vertically toward or away from excavation sidewalls should be cut on horizontal to vertical ratios of 1:4 or 1:2 to prevent toppling failures. This is particularly important if the rocks are weak or the planes of weakness are closely spaced.

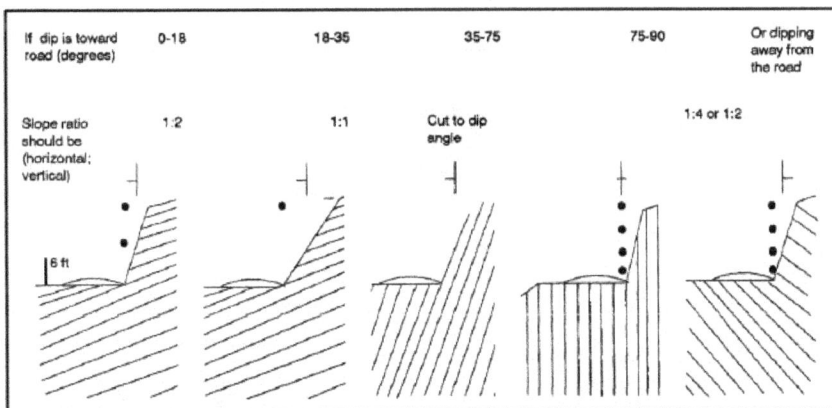

Figure 2-35. Rules of thumb for inclined sedimentary rock cuts

WEAK ROCKS

2-53. Weak rocks, such as shales, may shear or crush under the weight of overlying rock and allow excavation sidewalls to slump or cave in. Such failures can be prevented by installing artificial supports or by using flattened or terraced side slopes to reduce the load on the potential failure zone.

FAULT ZONES

2-54. Fault zones are often filled with crushed and broken rock material. When these materials are water-soaked, they may weaken and cause the fault zone to become unstable. Such zones are extremely hazardous when encountered in tunneling and deep excavations because they frequently slump or cave in. Artificial supports are usually required to stabilize such materials.

GROUNDWATER

2-55. Water entering the ground percolates downward, through open fractures and permeable rocks, until it reaches a subsurface zone below which all void spaces are filled with water. Where such groundwater is intersected in an excavation, such as a road cut or tunnel, drainage problems may occur; rock slides triggered by the weakening and/or lubricating of the rock mass may result. In addition, water trapped under hydrostatic pressure in fault zones, joints, and permeable rock bodies can cause sudden flooding problems when released during excavation. Permeable rock zones may also permit water to escape from canals and reservoirs. However, if properly evaluated, the structural conditions that produce groundwater problems can also provide potential supplies of groundwater or subsurface drainage for engineering projects.

ROAD CUT ALIGNMENT

2-56. The most advantageous alignment for road cuts is generally at right angles (perpendicular) to the strike of the major planes of weakness in the rock (usually the bedding). This allows the rock surfaces to dip along the cut rather than into it (see figure 2-36, a and 2-36, b). Where roads must be aligned parallel to the strike of the major planes of weakness, the most stable alignment is one in which the major planes of weakness dip away from the excavation; however, some overhang should be expected (see figure 2-36, c).

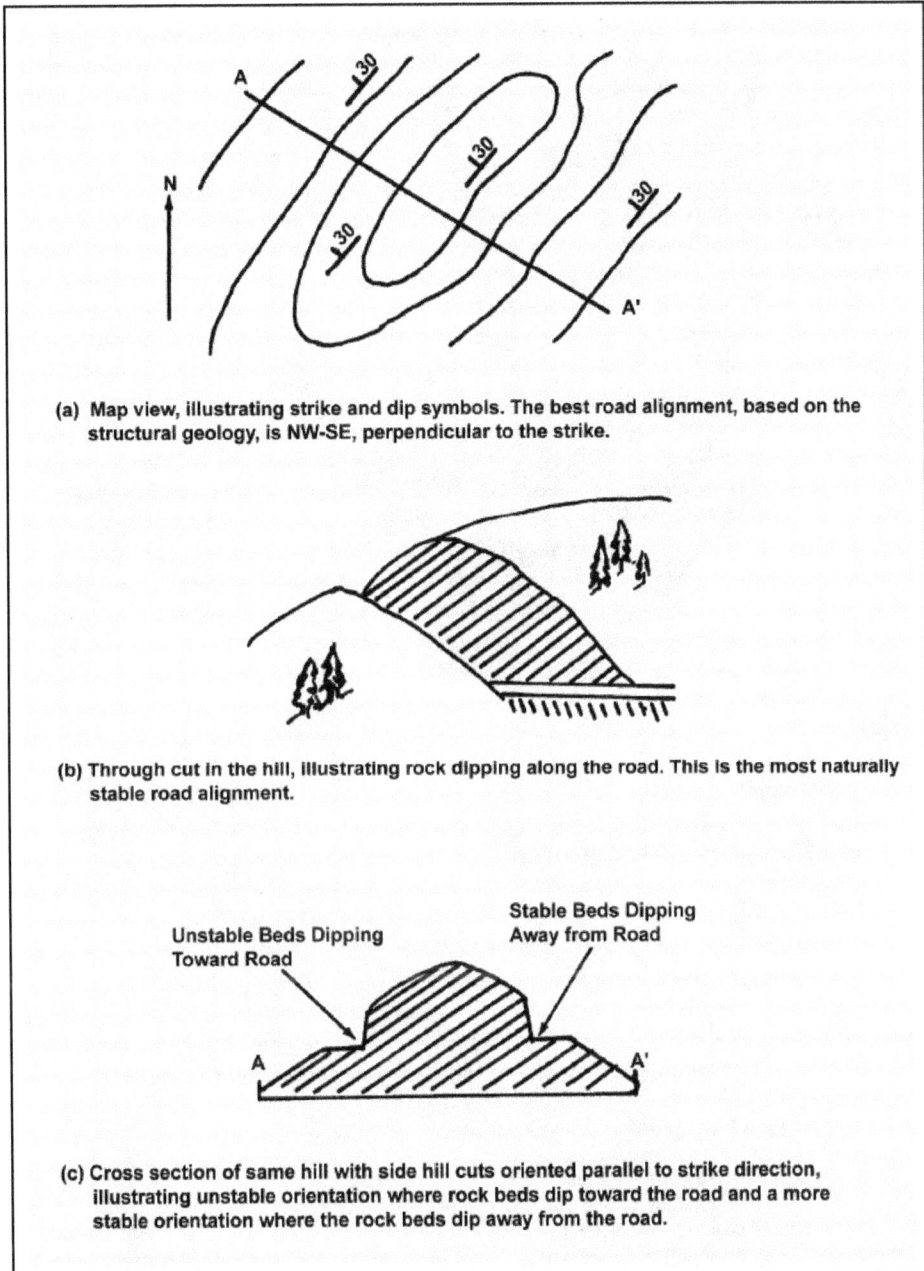

(a) Map view, illustrating strike and dip symbols. The best road alignment, based on the structural geology, is NW-SE, perpendicular to the strike.

(b) Through cut in the hill, illustrating rock dipping along the road. This is the most naturally stable road alignment.

(c) Cross section of same hill with side hill cuts oriented parallel to strike direction, illustrating unstable orientation where rock beds dip toward the road and a more stable orientation where the rock beds dip away from the road.

Figure 2-36. Road cut alignment

QUARRY FACES

2-57. Quarries should normally be developed in the direction of strike so that the quarry face itself is perpendicular to strike (see figure 2-37, a). This particular orientation is especially important where rocks are steeply inclined, because it allows for the optimization of drilling and blasting efforts by creating a vertical or near-vertical rock face after each blast. If necessary, quarries may be worked perpendicular to the strike direction in instances where the rocks are not steeply inclined, but drilling and blasting will prove to be more difficult. In addition, if the rocks dip away from the excavation, overhang and oversized rocks can be expected (See figure 2-37, b). If the rocks dip toward the excavation, problems with slope instability and toeing may result (See figure 2-37, c). In massive igneous rock bodies and horizontal sedimentary rock layers, the direction of quarrying should be chosen based on the most prominent joint set or other discontinuity.

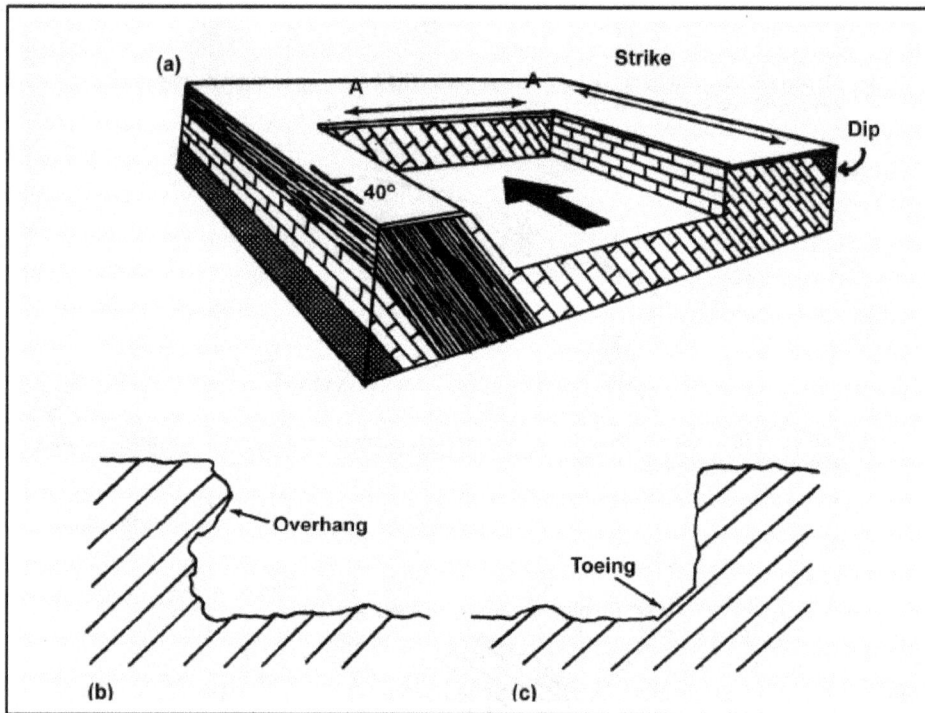

Figure 2-37. Quarry in the direction of the strike

ROCK DEFORMATION

2-58. Rocks may behave as elastic, plastic, or viscous solids under stress. Heavy loads, such as dams, massive fills, tall buildings, or bridge piers, may cause underlying rocks to compress, shear, or squeeze laterally. Particular problems exist where rocks of different strength underlie a site. For example, where weak shale and stronger limestone support different parts of the same structure, the structure may tile or crack due to uneven settlement.

2-59. The removal of confining stresses during excavation may cause rocks to expand or squeeze into the excavated area. Such problems seldom cause more than an increase in excavation or maintenance costs for roads, airfields, and railroads; however, they may cause serious damage to dams, buildings, canals, and tunnels where deformation cannot be tolerated. Weak clays and shales (especially compaction shales) are the most common cause of such problems. Other rocks can also cause trouble if they are weathered or if they have been under great stress. To neutralize the effects of rock flow or rebound, the following may be required:

- Additional excavation.
- Artificial supports or hold-downs.
- Compensating loads.
- Adjustment periods before construction.

EARTHQUAKES (FAULT MOVEMENTS)

2-60. Movement along active faults produces powerful ground vibrations and rock displacements that can seriously disrupt engineering works. Unless proven otherwise by geological or historical evidence, all faults that disrupt recent geologic deposits should be considered active. Many areas suffer earthquakes as a result of deep-seated faults that do not appear at the earth's surface. Consequently, seismic hazards must be thoroughly investigated before any major structure is undertaken. Power lines, dams, canals, tunnels, bridges, and pipelines across active faults must be designed to accommodate earth movements without failure. Buildings and airfields should be located away from known active fault zones. All vertical structures must be designed to accommodate the lateral movements and vibrations associated with earthquakes where seismic hazards exist. Expect increased seismic risk in marginal areas of continental plates (see figure 2-1, page 2-2).

DAMS

2-61. Dam sites are selected on the basis of topography, followed by a thorough geologic investigation. A geologic investigation should include as a minimum—

- Soundness. Determine the soundness of underlying foundation beds and their ability to carry the designated load.
- Water integrity. Determine the degree of watertightness of the foundation beds at the dam location and the necessary measures, if any, to make the underlying geologic strata watertight.
- Duration of water exposure. Study the effect that prolonged exposure to water has on the foundation bedrock.
- Potential for quake activity. Determine the possibility of earth movement occurring at the dam site and what it takes to safeguard against such failures.
- Availability of material. Investigate what natural materials are available near the site (potential quarries, sand, gravel, fill).

2-62. Generally, igneous rocks make the most satisfactory material for a dam foundation. Most igneous rocks are as strong as or stronger than concrete. However, many tuffs and agglomerates are weak. Solution cavities do not occur in igneous rocks because they are relatively insoluble; however, leakage will take place along joints, shear zones, faults, and other fissures. These can usually be sealed with cement grout.

2-63. Most metamorphic rocks have foundation characteristics similar to igneous rocks. Many schists are soft, so they are unsuitable as foundations for large, concrete dams. marble is soluble and sometimes contains large solution cavities. As a rule, metamorphic rocks can be treated with cement grout.

2-64. Sandstones allow seepage through pores, joints, and other fissures. The high porosity and low permeability of many sandstones make them difficult to treat with cement grout. Limestone's solubility creates large underground cavities. Generally, the strength of shale compares favorably with that of concrete; however, its elasticity is greater. Shale is normally watertight.

TUNNELS

2-65. After determining the general location and basic dimensions of a tunnel, consider geological problems before designing and constructing it. Civil engineers dealing with tunnel construction understand the need for geological data in this filed, so failure due to the lack of geological information seldom occurs. However, many failures do occur because engineers improperly interpret the available geological facts.

FOLDED STRATA

2-66. Extensive fracturing often exists along the axis of folded rock. This presents difficulties in tunneling operations. In an anticline, such fractures diverge upward; in a syncline, they diverge downward. If a tunnel is placed along the crest of a fold, the engineer can expect trouble from shattered rock. In such a case, the tunnel may have to be lined its entire length. In a syncline, an engineer could face additional trouble, even with moderate fracturing. The blocks bounded by fracture planes are like inverted keystones and are very likely to drop. When constructing tunnels in areas of folded rocks, the engineer should carefully consider the geologic structure. If a tunnel passes through horizontal beds, the engineer should encounter the same type of rock throughout the entire operation. In folded strata, many series of rock types can be encountered. therefore, carefully mapping geological structures in the construction area is important.

FAULTED STRATA

2-67. As with folded rocks, the importance of having firm, solid rock cannot be stressed enough, not only for safety and convenience in working but also for tunnel maintenance after completion. If rock is shattered by faulting, the tunnel must be lines, at least in the crushed-rock area. Also, if the fault fissure extends to the surface, it may serve as a channel for rainwater and groundwater.

GROUNDWATER PROBLEMS

2-68. Groundwater presence is often the main trouble source in tunnel construction. If tunnel grades cannot facilitate groundwater drainage, it may be necessary to pump throughout the tunnel. Therefore, it is necessary to have accurate information before beginning tunnel construction. Apply grout or cement, when possible, to solve water problems.

BRIDGES

2-69. Geological principles also apply in bridge construction. The weight of the bridge and the loads that it supports must be carried by the underlying foundation bed. In most cases, bridges are constructed for convenience and economy, so they must be located in specified areas. therefore, engineers cannot always choose the best site for piers and abutments. Once construction begins, the bridge location cannot be easily changed and should only be done so under exceptional circumstances.

2-70. As a rule, bridges are constructed to cross over rivers and valleys. An older riverbed or other depression (caused by glaciations or river deposits) could be completely hidden below the existing riverbed. Problems could arise if such a buried valley is not discovered before bridge construction begins. For example, riverbed contain many types of deposits, including large boulders. If preliminary work is not carefully done and correlated with geological principles, existing boulder deposits could be mistaken for solid bedrock.

BUILDINGS

2-71. Ground conditions at a building site may be one of three general types:

- On or near ground surface. Solid rock could exist at ground surface or so close to the surface that the foundation of the building can be placed directly in it.
- Below ground surface. Bedrock could exist below ground surface so that an economic, practical foundation can be used to support the building load.
- Far below ground surface. The nearest rock stratum could be so far below the surface that it cannot be used as a foundation bed. A building's foundation must be built on the material that forms the surface stratum.

2-72. If solid rock is present, its strength and physical properties must be determined. When the foundation consists of loose, unconsolidated sedimentary material, proper steps must be taken to solve the problem of subsidence. Structures that are supported on bedrock, directly or through piles or piers, will settle by extremely small amounts. If a foundation has been supported in unconsolidated strata, appreciable settlement can be expected.

QUARRY OPERATIONS

2-73. Natural sand and gravel are not always available, so it is sometimes necessary to produce aggregate by quarrying and processing rock. Quarrying is normally done only where other materials of adequate quality and size cannot be obtained economically and efficiently.

2-74. Many rock types are suitable for construction, and they exist throughout the world. Therefore, the quality and durability of the rock selected depend on local conditions. The following rock types are usually easy to quarry. They are also durable and resistant to weathering.

- Granite. As a dimension stone, granite is fairly durable and has texture and color that are desirable for polishing. As a construction material for base courses and aggregate, it is not as desirable as some for the denser igneous rocks.
- Felsite and rhyolite. These are durable and make good aggregate for base courses. They are not suitable for concrete aggregate.
- Basalt. The dense, massive variety of basalt if excellent for crushed rock, base course, and bituminous aggregate. It is very strong and durable. Due to basalt's high compressive strength, processing it may be more difficult than processing other rocks.
- Sandstone. Few sedimentary rocks are desirable for construction due to their variable physical properties, but sandstone is generally durable. However, deposits must be evaluated individually, due the variable nature of grains and cement present in the rock.
- Limestone. Limestone is widely used for road surfacing, in concrete, and for lime production. It is good, all-around aggregate.
- Gneiss. Most varieties of gneiss have good strength and durability and make good road aggregate.
- Quartzite. quartzite is hard and durable, so it is an excellent rock for construction. However, it can be difficult to quarry.
- Marble. The texture and color of marble make it desirable for dimension stone. It can also be used for base course and aggregate material.

2-75. Factors that enhance the easy removal of rock often diminish its suitability for construction. Strong, durable, unweathered rock usually serves best for embankment and fill, base and surface course material, concrete aggregate, and riprap. However, these rocks are the most difficult to quarry or excavate.

2-76. Soils continuously change. Weathering, chemical alteration, dissolution, and precipitation of components all occur as soils accumulate and adjust to their environment. Particle coatings and natural cements are added and removed. Soluble components wash downward (leach) or accumulate in surface layers by evaporation. Plants take root and grow as soil profiles develop. These changes tend to become more pronounced with increasing age of the soil deposit. These possible alterations to the sediment may affect its utility in construction. The soils of natural geological deposits are commonly used as construction materials.

SECTION IV - APPLIED MILITARY GEOLOGY

2-77. The science and applied art of geology is an important component of military planning and operations. Today, computers, satellites, geographic information systems, geographic positioning systems, and similar technology have catapulted the art and science of geology as an everyday tool for military commanders and engineers.

MILITARY GEOGRAPHIC INTELLIGENCE

2-78. The purpose of geographic intelligence is to obtain data about terrain and climate. Commanders use the information to make sound decisions and soldiers use it to execute their missions. In planning an operation, commanders and their staff analyze the effects that terrain and climatic conditions will have on the activities of friendly and enemy forces. Knowledge of how geology controls and influences terrain is helpful for classifying and analyzing terrain and terrain effects. When provided with adequate geographic and geologic intelligence, commanders are able to exploit the advantages of the terrain and avoid or minimize its unfavorable aspects. Data on soil movement, the presence of hard rock, and the kind and distribution of vegetation is needed when considering concealment and cover, cross-country travel, and field fortifications. Strategic intelligence studies prepared by Department of Defense agencies provide detailed geographic and terrain information (table 2-2) useful for compiling and analyzing geographic intelligence.

Table 2-2. Reports for geographic/terrain intelligence

Report Title	Report Application
Imagery interpretation	Planning combat and support operations
	Planning recon activities
	Supporting requests for terrain intelligence
	Analyzing areas of operations
	Planning terrain studies
Terrain study on soils	Supporting communications planning
	Executing movement, maneuver operations
	Planning combat operations (construction of landing strips, maintenance of culverts)
	Selecting avenues of approach
Terrain study on rocks	Planning movement, maneuver operations
	Planning combat operations (construction, maintenance, and destruction of roads, bridges, culverts, and defensive installations)
	Selecting avenues of approach
Terrain study on water resources	Selecting locations of and routes to water points
	Planning combat operations (street crossing, bridging)
	Supporting logistics planning
Terrain study on drainage	Supporting communications planning
	Planning combat operations (constructing roads, fortifications, and fjords)
	Supporting river crossings and cross-country movement
Terrain study on surface configuration	Supporting communications planning
	Planning observation posts and recon activity
	Planning tactical operations and executing tactical objectives
	Planning barrier and denial operations
	Planning artillery support

Table 2-2. Reports for geographic/terrain intelligence

Report Title	Report Application
Terrain study on the state of the ground	Planning movement, maneuver operations
	Planning ADM activity
	Planning combat operations (constructing, maintaining, and repairing roads, fjords, landing strips, and fortifications)
	Planning logistics support
Terrain study on construction suitability	Planning combat operations (constructing fortifications, landing strips, camouflage, obstacles, and a CP's supply installations)
	Selecting construction supply-point locations
Terrain study on coasts and landing beaches	Planning amphibious operations (preparing and removing obstacles and fortifications) Planning recon activity
	Planning port construction
Terrain study on cross-country movement	Planning and executing maneuver, movement operations
	Planning logistics support
	Planning barrier and denial operations
	Planning engineer combat operations
Terrain study on airborne landing areas	Planning area-clearing support
	Planning recon activity
	Planning combat operations (constructing and repairing landing strips)
	Selecting helicopter landing zones

MAPS AND TERRAIN MODELS

2-79. Maps are a basic source of terrain information. Geologic maps show the distribution, age, and characteristics of geologic units. They may also contain cross sections that show the subsurface occurrence of rock and soil. Maps may have text explaining physical properties; engineering properties; and other information on natural construction materials, rock formations, and groundwater resources. Soil maps are commonly presented under an agricultural classification but may be used for engineering purposes after they have been converted to engineering nomenclature.

2-80. A terrain model is a three-dimensional graphic representation of an area showing the conformation of the ground to scale. Computer-generated images have recently been developed that display two-dimensional form images in three-dimensional form for viewing from any angle.

REMOTE IMAGERY

2-81. Aerial and ground photography and remote imagery furnish information that is not readily available or immediately apparent by direct ground or aerial observation. Examples are infrared photography and side-looking radar. Imagery permanently preserves information so that it is available for later study. Photographs normally depict more recent terrain features than available maps.

TERRAIN CLASSIFICATION

2-82. Land forms are the physical expression of the land surface. For terrain intelligence purposes, major land forms are arbitrarily described on the basis of local relief (the difference in elevation between land forms in a given area).

GEOLOGY IN RESOLVING MILITARY PROBLEMS

2-83. To apply geology in solving military problems, personnel must first consider geologic techniques and uses and then determine how to acquire the needed information. Commanders can use geology in three ways-geologic and topographic map interpretation, photo interpretation, and ground reconnaissance.

MAPS

2-84. Combination topographic and geologic maps can tell commanders and engineers what the ground looks like. Map information should be made available early in operation planning. The success of many military operations depends on the speed of required advance construction. Speed, in turn, depends on completely understanding the needs and adequately planning to meet those needs in a particular area. Topographic maps are important sources of information on slope and land forms. One problem in using a topographic map is that some relief or roughness may be hidden between the contour lines. Generally, the larger the interval between contour lines and the smaller the scale of the map, the more the relief is hidden. Conversely, the smaller the contour intervals and the larger the scale of the map, the less the relief is hidden.

PHOTOGRAPHIC INTERPRETATION

2-85. Many principles that make a geologic map useful for estimating the terrain situation also apply to the usefulness of aerial photographs. The interpreter's skill is very important. Without a professional photo interpretation and the knowledge of geologic process and forces, a great deal of information may be overlooked. In preparing terrain intelligence, aerial photographs alone will not provide enough information. Geologic maps must also be available. Aerial photographs are completely reliable for preparing tactical terrain intelligence. However, the usefulness of aerial photographs varies with the scale of the report being prepared.

RECONNAISSANCE

2-86. Interpretation of the land for military purposes can be accomplished by studying maps and can be even more accurate by adding photographs. Using maps or photographs is highly effective, but it cannot compare in accuracy to actual ground observation. In a tactical situation, such as when attacking a defended position, the knowledge of the terrain behind the position is vital for planning the next move. Terrain may be different on the side that cannot be seen. Generally, the difference is expressed in the geology of the slope that can be observed. By combining ground observation and reconnaissance by a trained observer (geologist) with aerial and map interpretation, the commander can plan ahead. Reconnaissance of secure territory can be used to an even greater advantage when developing the area of occupation or advancement.

REMOTE IMAGERY

2-87. The major kinds of remotely sensed imagery are photography, radar, and multispectral or digital scanner imagery. Since the 1930s, various worldwide agencies have acquired a large amount of remote imagery. The quality, scale, and nature of the coverage vary considerably because new techniques and equipment are being developed rapidly. Remote imagery can be—

- Generated quickly in a specified time.
- Displayed accurately.
- Produced in a useful and storable format.
- Produced in volume (table 2-3).

2-88. High-altitude aircraft and spacecraft imagery are desirable for regional geologic mapping and delineating major structural features. Stereo coverage of low-, medium-, and high-altitude photography is used for detailed geologic mapping of rock units, structure, soil type, groundwater sources, and geologic hazards such as slope failure, sinkholes, fracturing zones, and flooding.

FIELD DATA COLLECTION

2-89. Field data collection is necessary to—

- Supplement existing information obtained from published and unpublished literature.
- Obtain site-specific, detailed information that describes the local geology.

2-90. Mission constraints, such as time, resources, weather, and political climate, limit the amount of field investigation that can be conducted. The following methods and procedures, which were designed primarily for civilian peacetime projects, serve as a guide for the kinds of geologic information that can be obtained in field investigation.

Table 2-3. Sources of remote imagery

Procurement Platform	Imagery Format	Scale	Coverage	Source Agency
Low-, medium-, and high-altitude aircraft	Black and white or color stereo pairs of aerial photos of high resolution	1:12,000 1:125,000	Limited worldwide low altitude, 1942 (to present; high altitude, 1965 to present)	USGS, EROS data center, Sioux Falls, SD
Low-, medium-, and high-altitude aircraft	Black and white, color, color IR, black and white IR, thermal IR, SLAR, multiband imagery (reproductions of imagery are available to all US military organizations and US government agencies)	1:1,000 1:100,000	Partial to full coverage of most foreign countries from late 1930s to present	DIA, ATTN: DC-6C2, Washington, DC 20301
Low-, medium-, and high-altitude aircraft	Black and white, color IR	Main scales: 1:20,000 1:40,000 Black and white: 20,000 to 1:120,000	Black and white: 80% of US Color IR: Partial US	ASCS, Aerial Photo Field Office, PO Box 30010, Salt Lake City, UT 84130
Unmanned satellites ERTS LAND-SAT 1-5 MSS— Band 4: green (0.5-0.6 m) Band 5: red (0.7-0.8 m) Band 6: near IR (0.7-0.8 m) Band 7: near IR (0.8-1.0 m)	Black and white, color composite (IR), color composition generation, 7 and 9—track computer—compatible tapes (800 and 1,600 bpi)	1:250,000 1:3,369,000	Worldwide coverage, complete earth's surface coverage every 18 days, 1972 to present; every 9 days, 1975 to present	EROS data center

Table 2-3. Sources of remote imagery

Procurement Platform	Imagery Format	Scale	Coverage	Source Agency
Manned spacecraft Skylab with S-190A multispectral cam-era and S-190B system, single lens (earth terrain camera)	S-190A: black and white, black and white IR, color IR, high resolution color S-190B: black and white, color, stereo pairs	S-190A: 1:1,250,000 1:2,850,000 S-190B: 1:125,000 1:950,000	Limited worldwide coverage	EROS data center

2-91. All available geologic information (literature, geologic and topographic maps, remote sensing imagery, boring logs, and seismic data relative to the general area of the project) should be collected, identified, and incorporated into the preliminary study for the project. This preliminary study should be completed before field investigations begin. A preliminary study allows those assigned to the project to become familiar with some of the engineering problems that they may encounter. Geologic information available in published and unpublished sources must be supplemented by data that is gathered in field investigations. Some of the most reliable methods available for field investigation are boring, exploration excavation, and geophysical exploration.

BORING

2-92. Borings are required to characterize the basic geologic materials at a project. They are broadly classified as disturbed, undisturbed, and core. Borings are occasionally made for purposes that do not require the recovery of samples, and they are frequently used for more than one purpose. Therefore, it is important to have a complete log of every boring, even if there is not an immediate use for some of the information. Initial exploration phases should concentrate on providing overall information about the site.

EXPLORATION EXCAVATION

2-93. Test pits and trenches can be constructed quickly and economically using bulldozers, backhoes, draglines, or ditching machines. Depths are normally less than 30 feet. Side excavations may require shoring, if personnel must work in the excavated areas. Exploratory tunnels allow for detailed examination of the composition and the geometry of rock structures such as joints, fractures, faults, and shear zones. Tunnels are helpful in defining the extent of the marginal strength of rock or adverse rock structures that surface mapping and boring information provide.

GEOPHYSICAL EXPLORATION

2-94. Geophysical exploration consists of making indirect measurements from the earth's surface or in boreholes to obtain subsurface information. Boreholes or other subsurface explorations are needed for reference and control when using geophysical methods. Geophysical explorations are appropriate for rapidly locating and correlating geologic features such as stratigraphy, lithology, discontinuities, structure, and groundwater.

EQUIPMENT FOR FIELD DATA COLLECTION

2-95. The type of equipment needed on a field trip depends on the type of survey being conducted. The following items are always required:

- Hammer. This is usually a geologic hammer, and it is used to break or dig rock and soil and to prepare samples for laboratory examination. A hammer is also useful for determining the strength and resistance of the rock and the toughness of the grains.

- Hand lens. A hand lens is a small magnifying glass, usually 10X power. It is used to examine individual mineral grains of rock for identification, shape, and size.
- Dilute hydrochloric (HCl) acid. One part concentrated HCl acid to four parts water is used to determine carbonate rocks. Dilute acid is preferred because the degree of its reaction with different substances is seen more easily.
- Brunton compass. One of the most important pieces of equipment to the field geologist is the Brunton compass. This instrument is an ordinary magnetic compass with folding open sights, a mirror, and a rotating dial graduated in degrees or mils. In some cases, the dial is numbered to 360 degrees, while in others, the dial may be graduated in 90-degree quadrants. A compass helps measure dips, slopes, strikes, and directions of rock formations.
- Base map. A base map is essential in all types of field work except when the plane-table-and-alidade method is used.

GEOLOGICAL SURVEYING

2-96. The instruments and methods of geological surveying are numerous and varied. Instruments used on a particular project depend on the scale, time, detail, and accuracy required.

PACE-AND-COMPASS METHOD

2-97. The pace-and-compass method is probably the least-accurate procedure used. The survey is conducted by pacing the distance to be measured and determining the angle of direction with the compass. The field geologist records elevations on a topographic map, when it is available. When a topographic map or a good equivalent is unavailable, an aneroid barometer or some type of accurate altimeter is used to record elevations.

PLANE-TABLE-AND-ALIDADE METHOD

2-98. When accurate horizontal and vertical measurements are required, the plane-table-and-alidade method is used. The equipment consists of a stadia rod, a tripod, a plane table, and an alidade. Sheets of heavy paper are placed on the plane table to record station readings. After recording the stations, the geologist places formation contacts, faults, and other map symbols on the paper. This information can later be transferred to a finished map. The alidade is a precision instrument consisting of a flat base that can easily be moved on the plane table. The straight edge, along the side of the flat base and parallel to the line of sight, is used to plot directions on the base map. The alidade also consists of a telescopic portion with a lens that contains one vertical hair and three horizontal hairs (stadia hairs). The vertical hair is used to align the stadia rod with the alidade. The horizontal hairs are used to read the distance to the rod. Vertical elevations are determined from the stadia distance and the vertical angle of the points in question.

BRUNTON-COMPASS-AND-AERIAL-PHOTO METHOD

2-99. A Brunton compass and a vertical aerial photo provide a rapid, accurate method for geological surveys. The aerial photo is used in place of a base map. Contact, dips and strikes, faults, and other features may be plotted directly on the aerial photo or on a clear acetate overlay. Detailed notes of the geological features must be kept in a separate field notebook. If topographic maps are available, they will supplement the aerial photo and eliminate the need for an aneroid barometer. A topographic map may be used for the base map and to plot control for horizontal displacements.

PLANE-TABLE-, MOSAIC-, AND ALIDADE METHOD

2-100. A mosaic of aerial photos may be used with a plane table and an alidade, thus eliminating the base map. This procedure is used to eliminate the rod holder and to accelerate the mapping. By using a plane table and an alidade as a base, horizontal distance does not have to be measured; and the surveyor can compute vertical elevation. The surveyor—

- Reads the angle of variation from the horizontal distance.

- Scales the distance from the map.
- Finds the difference by trigonometric calculations.

2-101. This method is faster than the normal plane-table method, but it is not as accurate.

PHOTOGEOLOGY METHOD

2-102. Photogeology is the fastest and least-expensive method of geological surveying. However, a certain amount of accuracy and detail is sacrificed. This type of survey is normally used for reconnaissance of large areas. When a geologic map is constructed from aerial-photo analysis alone, it shows only major structural trends. After a preliminary office investigation, a geologist may go to the field and determine detailed geology of particular areas.

Chapter 3
Surficial Geology

An integral part of the military engineer's mission is the location and processing of materials for construction use. Most construction materials are derived from rocks and soils that occur naturally on or near the surface of the earth. These materials may be obtained by developing a quarry or a borrow pit.

Quarries are sites where open excavations are made into rock masses by drilling, cutting, or blasting for the purpose of producing construction aggregate. These operations require extensive time, manpower, and machinery. Borrow pits are sites where unconsolidated material has been deposited and can be removed easily by common earth-moving machinery, generally without blasting.

This chapter covers the processes that form surficial features which are suitable for potential borrow pit operations and the types of construction materials found in these features.

FLUVIAL PROCESS

3-1. The main process responsible for the erosion and subsequent deposition of weathered material suitable for the development of borrow pits is that of moving water. When water moves very quickly, as over a steep gradient, it picks up weathered material and carries it away. When the stream slows down (for example, when the gradient is reduced), the capacity of the stream to carry the weathered material decreases; then it deposits the material in a variety of possible surficial features.

3-2. Stream deposits are characteristically stratified (layered) and composed of particles within a limited size range. Fluvial deposits are sorted by size based on the velocity of the water. When the velocity of the stream falls below the minimum necessary to carry the load, deposition occurs beginning with the heaviest material. In this way, rivers build gravel and sandbars on the inside of meander loops and dump fine silts and muds outside their levees during floods. This creates deposits of reasonably well-sorted, natural construction materials.

DRAINAGE PATTERNS

3-3. Without the benefit of geologic maps, it is difficult to determine the type and structure of the underlying rocks. However, by studying the drainage patterns as they appear on a topographic map, both the rock structure and composition may be inferred.

3-4. Many drainage patterns exist; however, the more common patterns are—

- Rectangular.
- Parallel.
- Dendritic.
- Trellis.
- Radial.
- Annular.
- Braided.

Rectangular

3-5.   This pattern is characterized by abrupt, nearly 90-degree changes in stream directions. It is caused by faulting or jointing of the underlying bedrock. Rectangular drainage patterns are generally associated with massive igneous and metamorphic rocks, although they may be found in any rock type. Rectangular drainage is a specific type of angular drainage and is usually a minor pattern associated with a major type, such as dendritic (see figure 3-1, a). Angular drainage is characterized by distinct angles of stream juncture.

Parallel

3-6.   This drainage is characterized by major streams trending in the same direction. Parallel streams are indicative of gently dipping beds or uniformly sloping topography. Extensive, uniformly sloping basalt flows and young coastal plains exhibit this type of drainage pattern. On a smaller scale, the slopes of linear ridges may also reflect this pattern (see figure 3-1, b).

Dendritic

3-7.   This is a treelike pattern, composed of branching tributaries to a main stream. It is characteristic of essentially flat- lying and/or relatively homogeneous rocks (see figure 3-1, c).

Trellis

3-8.   This is a modified version of the dendritic pattern. Tributaries generally flow perpendicular to the main streams and join them at right angles. This pattern is found in areas where sedimentary or metamorphic rocks have been folded and the main streams now follow the strike of the rock (see figure 3-1, d).

Radial

3-9.   This pattern, in which streams flow outward from a high central area, is found on domes, volcanic cones, or round hills (see figure 3-1, e).

Annular

3-10.   This pattern is usually associated with radial drainage where sedimentary rocks are upturned by a dome structure. In this case, streams circle around a high central area (see figure 3-1, f).

Braided

3-11.   A braided stream pattern commonly forms in arid areas during flash flooding or from the meltwater of a glacier. The stream attempts to carry more material than it is capable of handling. Much of the gravels and sands are deposited as bars and islands in the stream bed (see figure 3-1, g and figure 3-2, page 3-4). Figure 3-2 shows the vicinity of Valdez, Alaska. Both the Copper and Tonsina Rivers are braided streams.

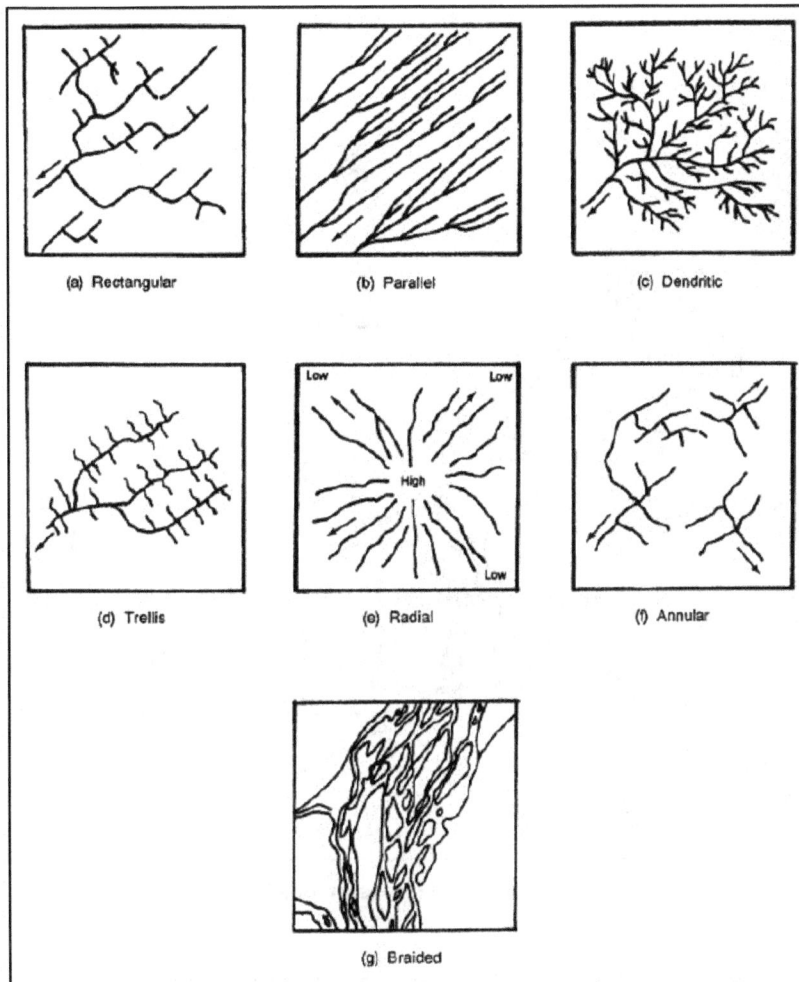

Figure 3-1. Typical drainage patterns

Figure 3-2. Topographic expression of a braided stream

DENSITY

3-12. The nature and density of the drainage pattern in an area provides a strong indicator as to the particle size of the soils that have developed. Sands and gravels are usually both porous and permeable. This means that during periods of precipitation, water percolates down through the sediment. The density of the drainage and the surface runoff are minimal due to this good internal drainage.

3-13. Clays and silts are normally porous but not permeable. Most precipitated water is forced to run off, creating a fine network of stream erosion.

3-14. Sandstone and shale may exhibit the same type of drainage pattern. Sandstone, due to its porosity and permeability, has good internal drainage while shale does not. Therefore, the texture or density of the drainage pattern which develops on the sandstone is coarse while that on shale is fine.

STREAM EVOLUTION

3-15. The likelihood of finding construction materials in a particular stream valley can be characterized by the evolution of that valley. The evolutionary stages are described as—

- Youth.
- Maturity.
- Old age.

Youth

3-16. Youthful stream valleys, which are located in highland areas, are typified by steep gradients, high water velocities with rapids and waterfalls present, downcutting in stream bottoms resulting in the creation of V-shaped valleys, and the filling of the entire valley floor by the stream (see figure 3-3, a). Although there is considerable erosion taking place, there is very little deposition.

Maturity

3-17. A mature system has a developed floodplain and, while the stream no longer fills the entire valley floor, it meanders to both edges of the valley. The stream gradient is medium to low, deposition of materials can be found, and (when compared with the youthful stream) there is less downcutting and more lateral erosion that contributes to widening the valley (see figure 3-3, b).

Old Age

3-18. In an old-age system, the stream gradient is very gentle, and the water velocity is low. The river exhibits little downcutting, and lateral meandering produces an extensive floodplain. Because of the low water velocity, there is a great amount of deposition. The river only occupies a small portion of the floodplain (see figure 3-3, c).

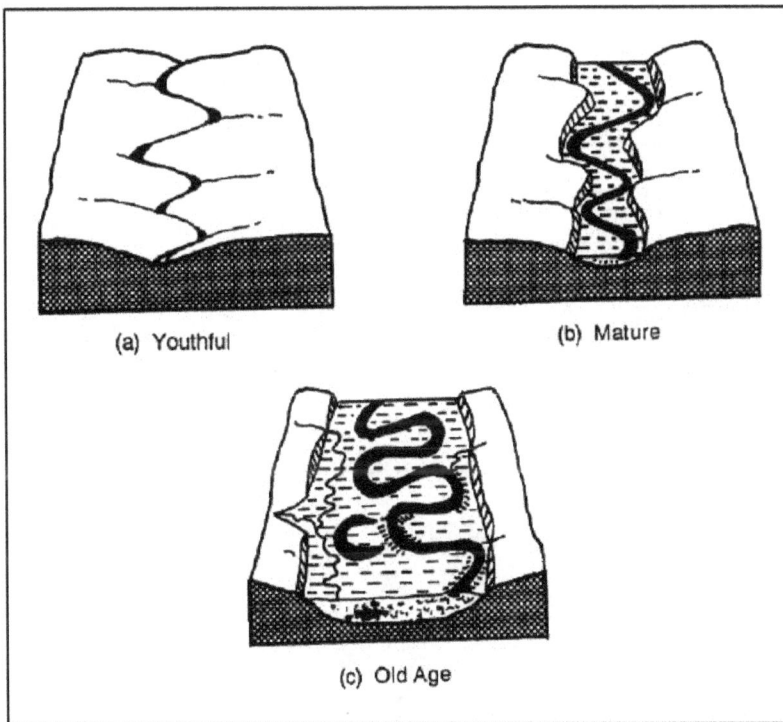

(a) Youthful

(b) Mature

(c) Old Age

Figure 3-3. Stream evolution and valley development

3-19. Recognition of the stream evolution stage of a particular river system is required to develop sources of construction aggregate. Rivers in maturity or old age provide the greatest quantities of aggregate. In youthful rivers, sources of aggregate are often scarce or unobtainable due to the steep gradients and high velocity. Table 3-1 summarizes the characteristics of each stage of stream evolution. Figure 3-4 shows an example of the topographic expression of a youthful stream valley in the vicinity of Portage, Montana.

Figure 3-5, page 3-8, shows a mature stream valley in the vicinity of Fort Leavenworth, Kansas. Figure 3-6, page 3-9, shows an old age stream valley in the vicinity of Philipp, Mississippi.

Table 3-1. Stream evolution process

Characteristic	Youth	Maturity	Old Age
Gradient	Steep, irregular	Moderate, Smooth	Low, smooth
Valley profile	Narrow, V-shaped	Broad, moderately U-shaped	Very broad
Valley depth	Deep	Deep, moderate, shallow	Shallow
Meanders	Absent	Common	Extremely common
Floodplain	Absent or small	Equals width of meander belt	Wider than width of meander belt
Natural levees	Absent	May be present	Abundant
Tributaries	Few, small	Many	Few, large
Velocity	High	Moderate	Sluggish
Waterfalls	Many	Few	None
Erosion	Downward cutting	Downward and lateral cutting equal	Lateral cutting
Deposition	Absent or transitory	Present, but partly transitory	Much and fairly permanent
Culture	Steep-walled valleys are barriers to road and railroads	Flat valley floors are good transportation routes	Large rivers and nearby swamps are barriers
Summary of Regional Erosion Cycle			
Dissection	Partial	Complete	None
Divides	Broad, flat, high	Knife-edged	Low, broad, rounded
Valley development	Youthful to mature	Mostly mature	Old age
Number of streams	Few	Maximum	Few
Relief	Great	Maximum	Minimum

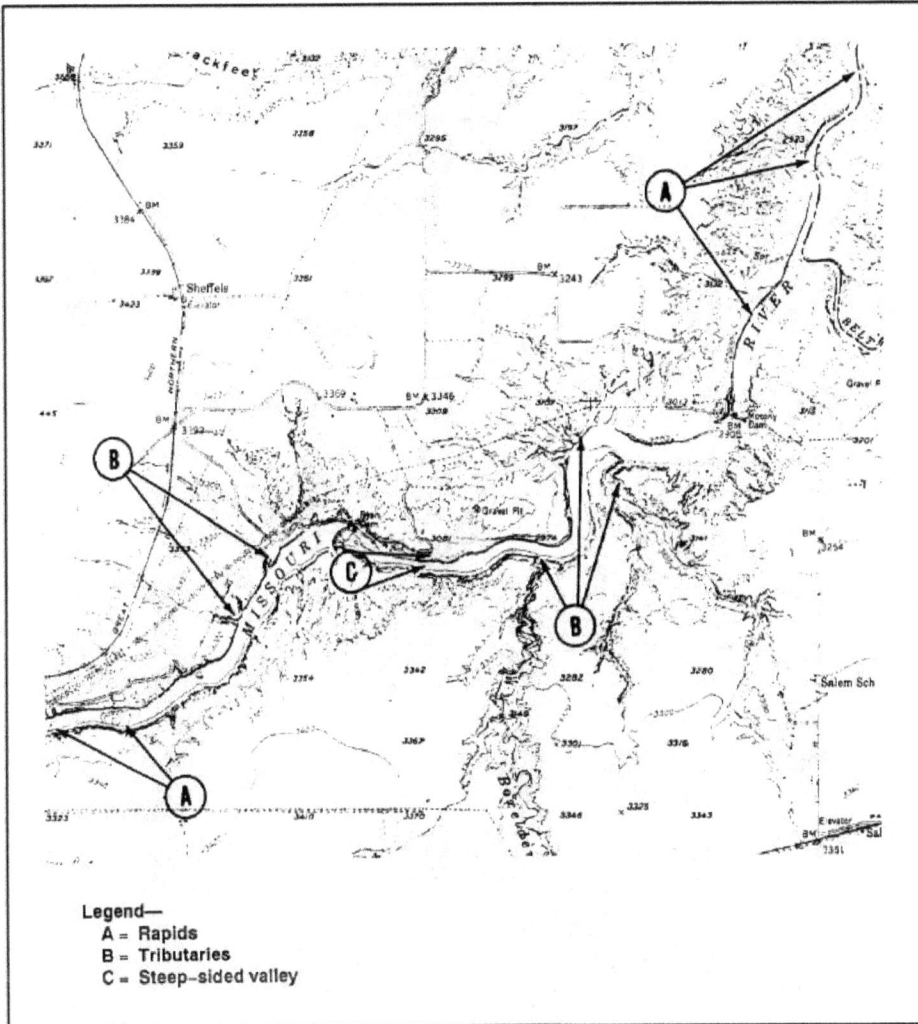

Figure 3-4. A youthful stream valley

Legend—
    A = Rapids
    B = Tributaries
    C = Steep-sided valley

Figure 3-5. A mature stream valley

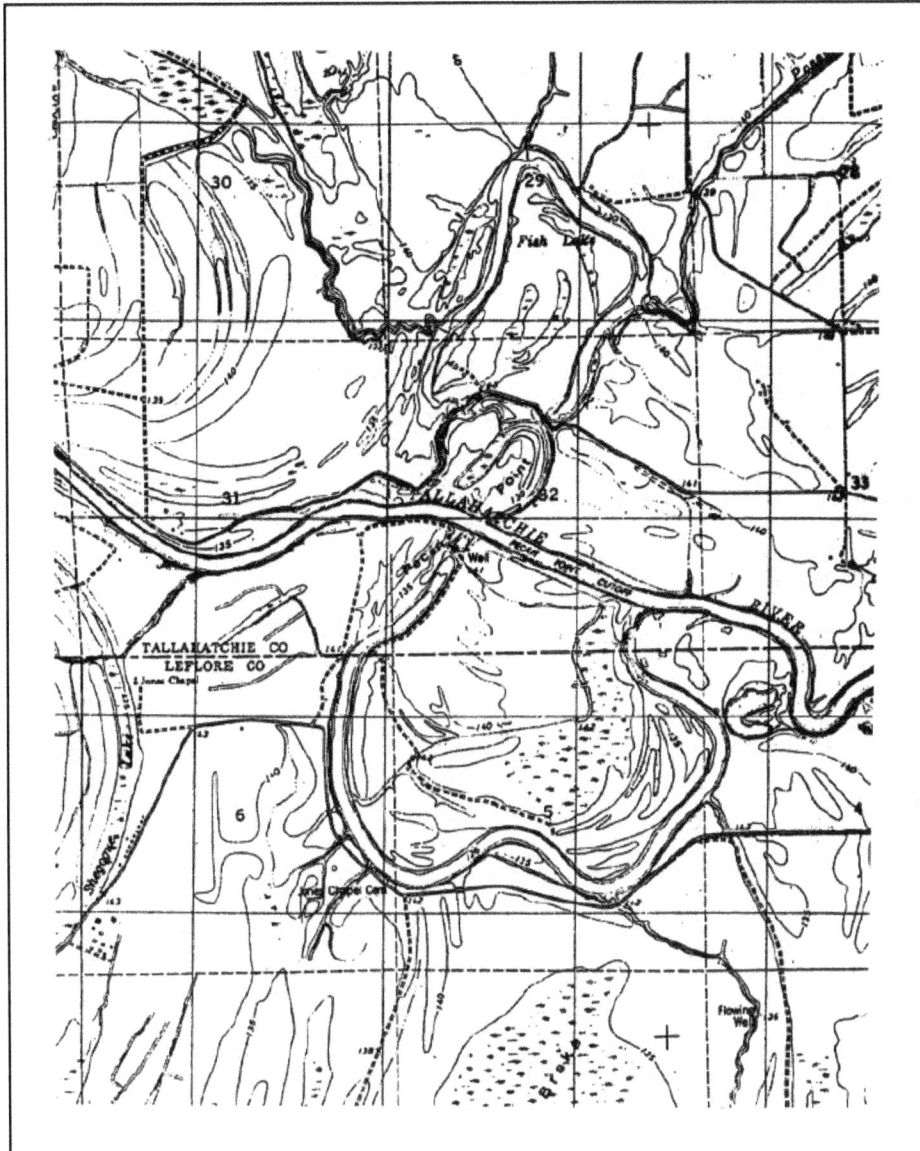

Figure 3-6. An old age stream valley

STREAM DEPOSITS

3-20. Coarse-grained (gravels and sands) and fine-grained (silts and clays) deposits can be found by map reconnaissance. Certain surficial features are comprised of coarse-grained materials, others are made up of medium-sized particles, and still others of fine-grained sediments. However, if the source area for a stream is composed only of fine-grained materials, then the resulting depositional features will also contain fine-grained sediments, regardless of their usual composition.

3-21. The following surficial features can be identified by their topographic expressions on military maps and are likely sources of construction materials.

- Point bars.
- Channel bars.
- Oxbow lakes.
- Natural levees.
- Backswamps/floodplains.
- Alluvial terraces.
- Deltas.
- Alluvial fans.

Point Bars

3-22. Meandering is the process by which a stream is gradually deflected from a straight-line course by slight irregularities. Most streams that flow in wide, flat-floored valleys tend to meander (bend). These streams are alternately cutting and filling their channels, and as the deflection progresses, the force of the flowing water concentrates against the channel wall on the outside of the curve. This causes erosion on that wall (a body in motion tends to remain in motion in the same direction and with the same velocity until acted on by an external force) (see figure 3-7). Consequently, there is a decrease in velocity and carrying power of the water on the inside of the curve, and the gravels and sands are deposited, forming point bar deposits (see figure 3-8). Point bar deposits on many maps will not be apparent but can be inferred to be at the inside of each meander loop.

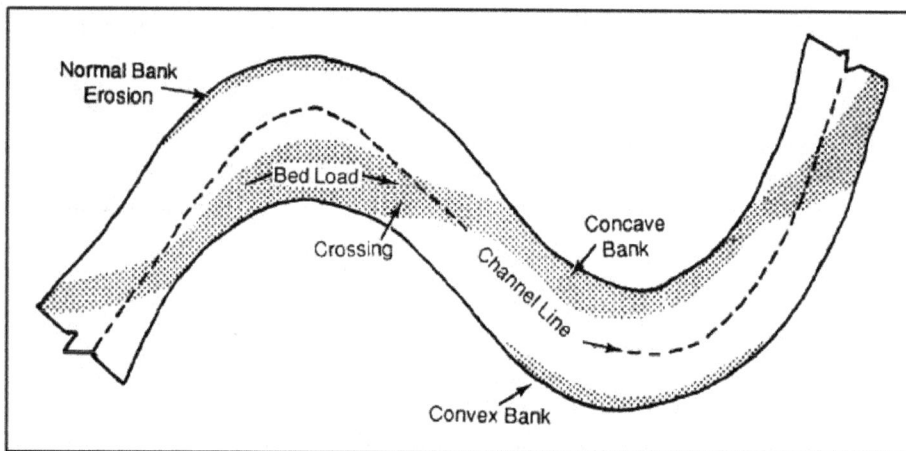

Figure 3-7. Meander erosion and deposition

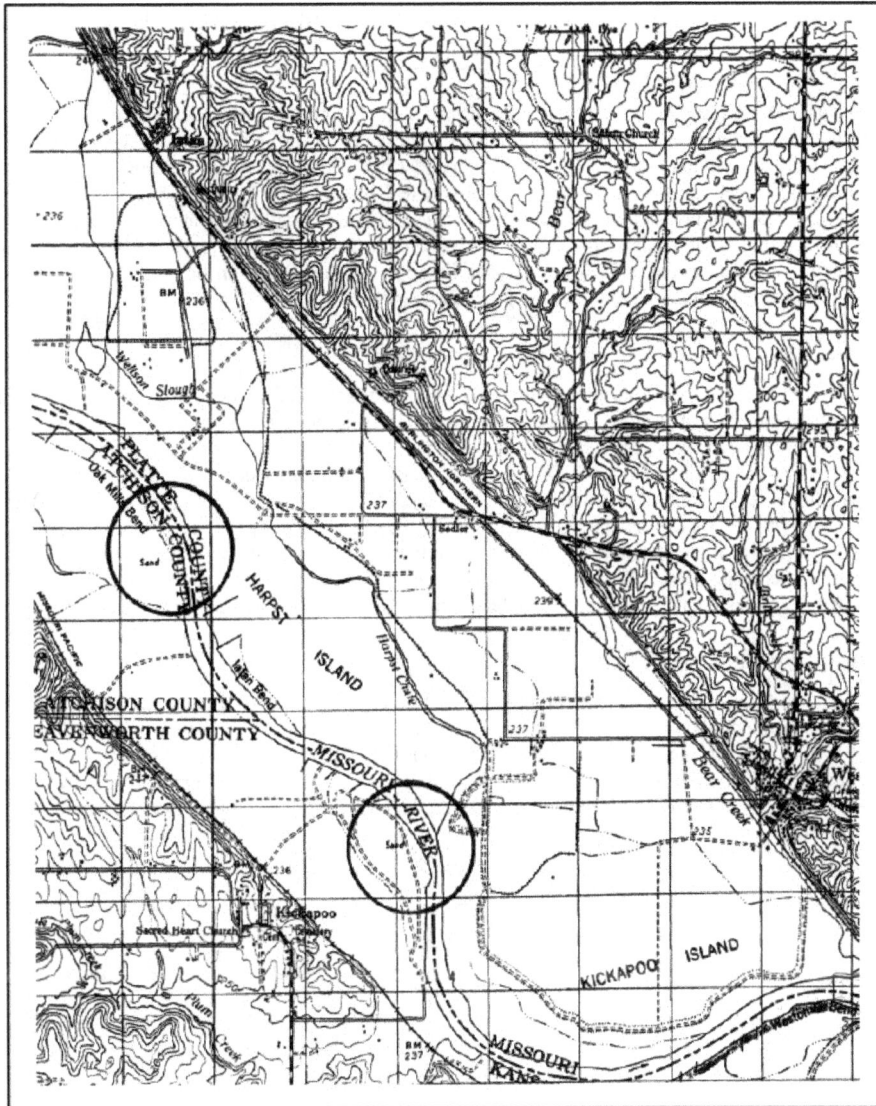

Figure 3-8. Point bar deposits designated by gravel symbols

Channel Bars

3-23. When a stream passes through a meander loop, its speed increases on the outer bank due to the greater volume of water that is forced to flow on the outside of the loop. When the stream leaves the meander and the channel straightens out, the forces that caused the stream to move faster are no longer in control and the stream slows down and deposits materials. These materials are coarse-grained (gravels and sands) and are on the opposite bank and downstream of the point bar. If there is a series of meander loops, these deposits may or may not be present between point bars, depending on the spacing of the meanders. However, a channel bar can be expected after the last meander loop. Figure 3-9, page 3-13, shows channel bar deposits, oxbow lakes, and backswamp/floodplain deposits in the vicinity of Fort Leavenworth, Kansas.

A prominent channel bar is located north of Stigers Island. Mud Lake, Burns Lake, and Horseshoe Lake are oxbow lakes. Backswamps on the floodplain are represented by swampy ground symbols.

Oxbow Lakes

3-24. During high-water stages, a stream that normally flows through a meander loop may cut through the neck of a point, thus separating the loop. When this happens, the stream has taken the path of least resistance and has isolated the bend. The cutoff meander bend is eventually sealed from the main stream by fine deposits. The bend itself then forms an oxbow lake (see figure 3-10, page 3-14). These deserted loops may become stagnant lakes or bogs, or the water may evaporate completely leaving a U-shaped depression in the ground. Fine-grained deposits (silts and clays) are normally located in oxbow lakes. An old point bar deposit can be found on the inside of the U (see figure 3-11, page 3-14). In figure 3-9, Horseshoe Lake is an example of the topographic expression of an oxbow lake.

Natural Levees

3-25. Stream velocity increases during flooding as the stream swells within the confines of its bank to move a greater volume of water. As the stream moves faster, it has the ability to carry more material. If the volume of water becomes so great that the water cannot stay in the channel, the stream spills over its banks onto the surrounding floodplain, which is a flat expanse of land adjacent to a stream or river. Once the stream spills over its banks, the water velocity decreases as the water spreads out to occupy a larger area. As the velocity decreases, sediment carried by the floodwater is deposited. The size of this material depends primarily on the character of the material in the source area upstream and the velocity of the water in the stream channel. Generally, gravels and sands can be found in a natural levee, with the larger material deposited near the stream bank and a gradual gradation to smaller sand particles away from the stream.

Backswamps/Floodplains

3-26. After a flood ends and the stream regresses into its channel, much of the water that spilled over the banks onto the floodplain is trapped on the outside of the natural levees. The fine materials (silts and clays) that are suspended in this water settle onto the floodplain. Consequently, these areas are often used for agricultural production. In the lower-lying areas of the floodplain, a large amount of fines may accumulate, inhibit drainage, and form swamplike conditions called a backswamp (see figure 3-9).

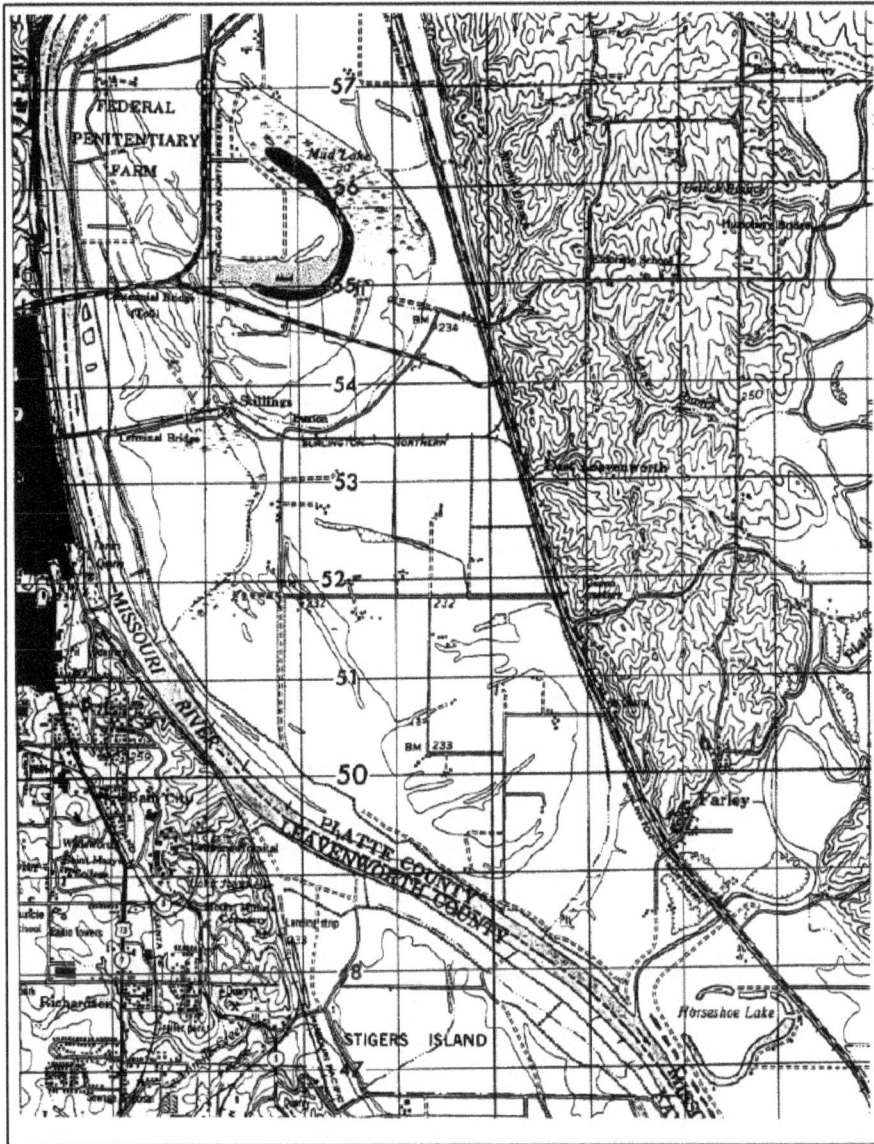

Figure 3-9. Channel bar deposits, oxbow lakes, and backswamp/floodplain deposits

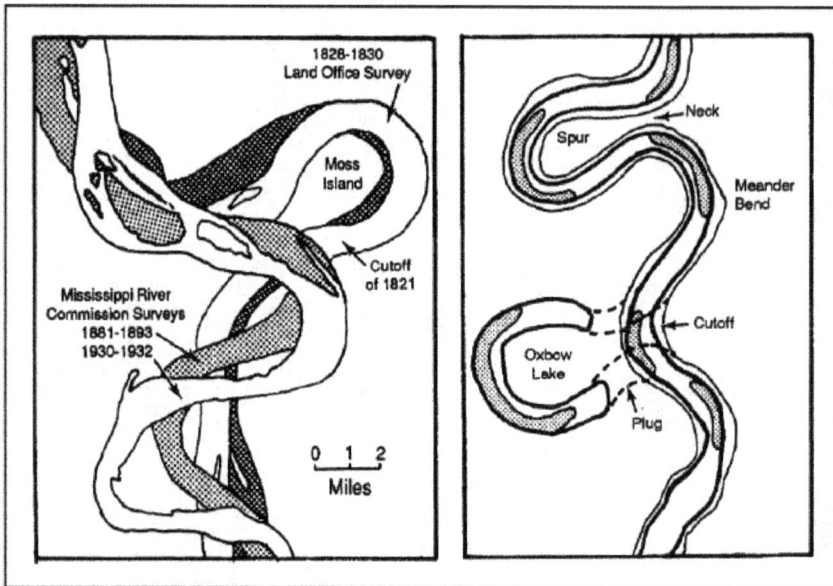

Figure 3-10. Meander development and cutoff

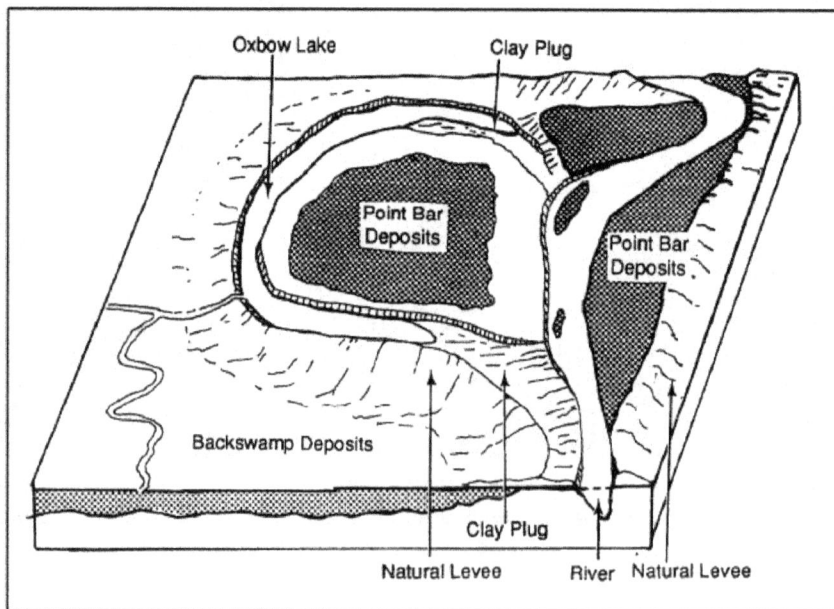

Figure 3-11. Oxbow lake deposits, natural levees, and backswamp deposits

Alluvial Terraces

3-27. A depositing stream tends to fill its valley with a fair amount of granular alluvial material. If a change in the geological situation results in the uplift of a large area or rejuvenation of the stream, an increase in the stream velocity by other means, or a change in the sedimentation and erosion process, the stream may begin to erode away the material it had deposited previously. As the eroding stream meanders about in its new valley, it may leave benchlike remnants of the preexisting valley fill material perched against the valley walls as terraces. This action of renewed downcutting may occur several times, leaving several terrace levels (see figure 3-12). These are easily recognized on a topographic map because they show up as flat areas with no contour lines, alternating with steeply sloping regions with many contour lines. Alluvial terraces usually occur on one side of the stream but can be found on both sides. They are a normal feature of the history of any fluvial valley. They are usually a good source of sands and gravels. Figure 3-13, page 3-16, shows alluvial terraces in the vicinity of Souris River, North Dakota.

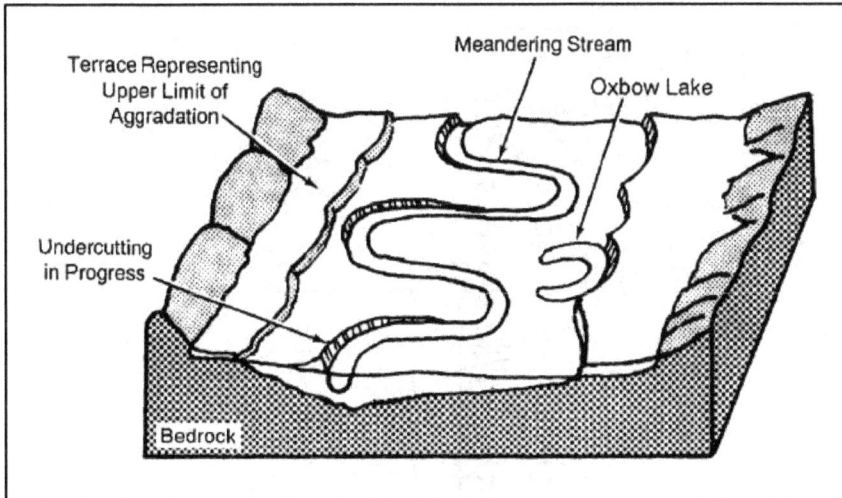

Figure 3-12. Alluvial terraces

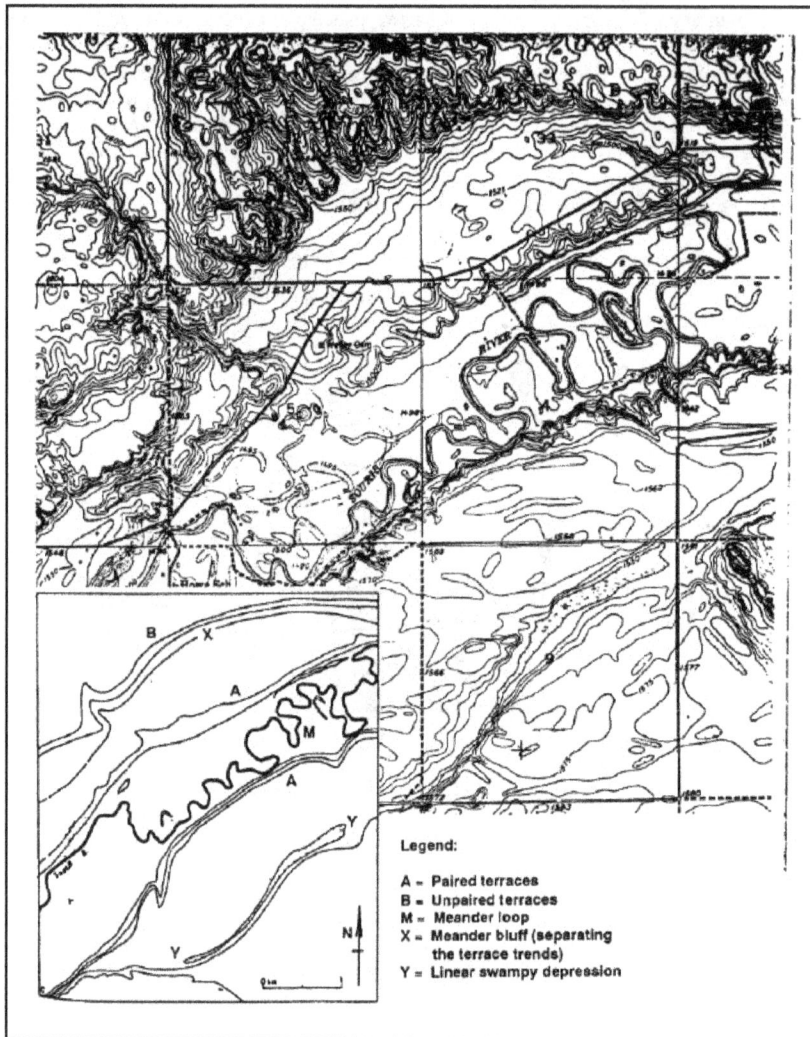

Figure 3-13. Topographic expression of alluvial terraces

Deltas

3-28. When streams carrying sediments in suspension flow into a body of standing water, the velocity of the stream is immediately and drastically reduced. As a result, the sedimentary load begins to settle out of suspension, with the heavier particles settling first. If the conditions in the body of water (sea or lake) are such that these particles are not spread out over a large area by wave action, or if they are not carried away by currents, they continue to accumulate at the mouth of the stream. Large deposits of these sediments gradually build up to just above the water level to form deltas (see figure 3-14). These assume three general forms, depending mainly on the relative influence of waves, fluvial processes, and tides. These forms are—

- Arcuate (see figure 3-15, a and b).
- Bird's-foot (see figure 3-15, c).
- Elongate (see figure 3-15, d).

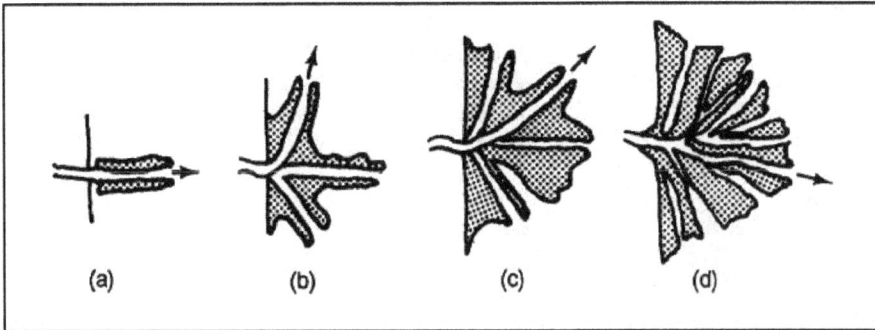

Figure 3-14. Growth of a simple delta

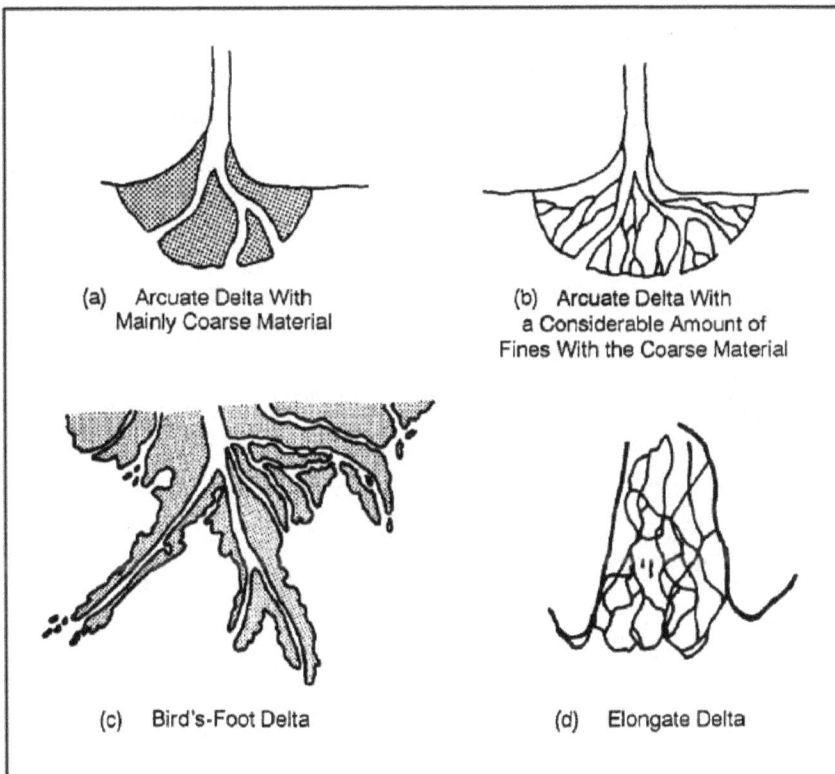

(a) Arcuate Delta With
Mainly Coarse Material

(b) Arcuate Delta With
a Considerable Amount of
Fines With the Coarse Material

(c) Bird's-Foot Delta

(d) Elongate Delta

Figure 3-15. Arcuate, bird's foot, and elongate deltas

3-29. Arcuate deltas are arc- or fan-shaped and are formed when waves are the primary force acting on the deposited material. Arcuate deltas usually result from deposition by streams carrying relatively coarse material (sands and gravels) with some occasional fine material. Arcuate deltas consisting primarily of coarse material have very good internal drainage; therefore, they have few minor channels. On the other hand, an arcuate delta having a considerable amount of fine material (silts and clays) mixed with the coarse material does not have good internal drainage. In this case, a larger number of minor channels develop.

Generally, arcuate deltas are considered good sources of sands and gravels. An example of an arcuate delta is the Nile Delta in Egypt.

3-30. Bird's-foot deltas are formed in situations where fluvial processes have a major influence on deposited sediments. Bird's-foot deltas resemble a bird's foot from the air, hence the name. They are generally composed of fine-grained material and have very poor internal drainage. These deltas are flat with vegetation, have many small outlets, and are a good source of fine materials. The Mississippi Delta is a classic example of this delta type.

3-31. Elongate deltas form where tidal currents have a major impact on sediment deposition. They contain only a few distributaries, but the distributaries that occur are large.

Alluvial Fans

3-32. These are the dry land counterpart of deltas. They are formed by streams flowing from rough terrain, such as mountains or steep faults, onto a flat plain. This type of deposit is found in regions that have an arid to semiarid type of climate, such as the western interior, the Basin and Range Province of the United States, and the desert mountain areas worldwide. The valleys in these areas are normally dry much of the year, with streams resulting only after torrential rainstorms or following the spring snow melt. The mountains themselves are devoid of vegetation, and erosion by the streams is not impeded. These streams rush down a steep gradient, and when they meet the valley floor, there is a sudden reduction in velocity. The sediment load is deposited at the foot of the rough terrain. This deposit is in the form of a broad "semicone" with the apex pointing upstream. Coalescing alluvial fans consist of a series of fans that have joined to form one large feature. This is typical in arid areas. Figure 3-16 depicts alluvial and coalescing alluvial fans. Alluvial fans may be readily identified by their topographic expressions of concentric half-circular contour lines. Figure 3-17 is a topographic map showing the Cedar Creek alluvial fan in the vicinity of Ennis Lake, Montana. This alluvial fan is approximately four miles in radius. Figure 3-18, page 3-20, shows coalescing alluvial fans in the vicinity of Las Vegas, Nevada.

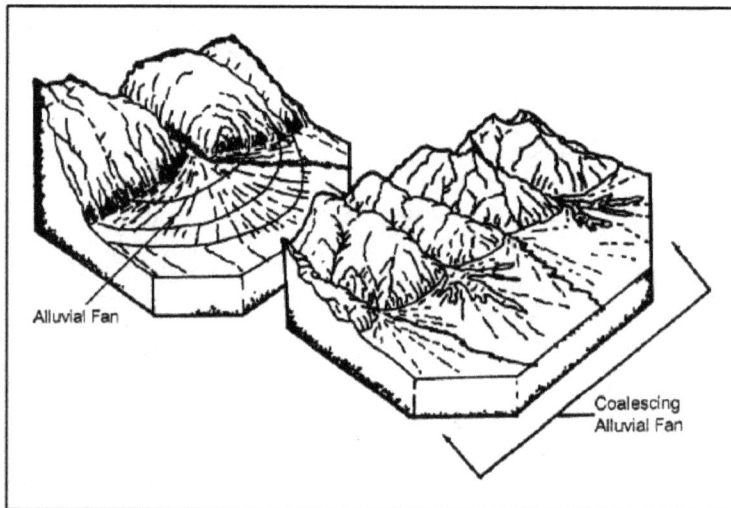

Figure 3-16. Alluvial fan and coalescing alluvial fans

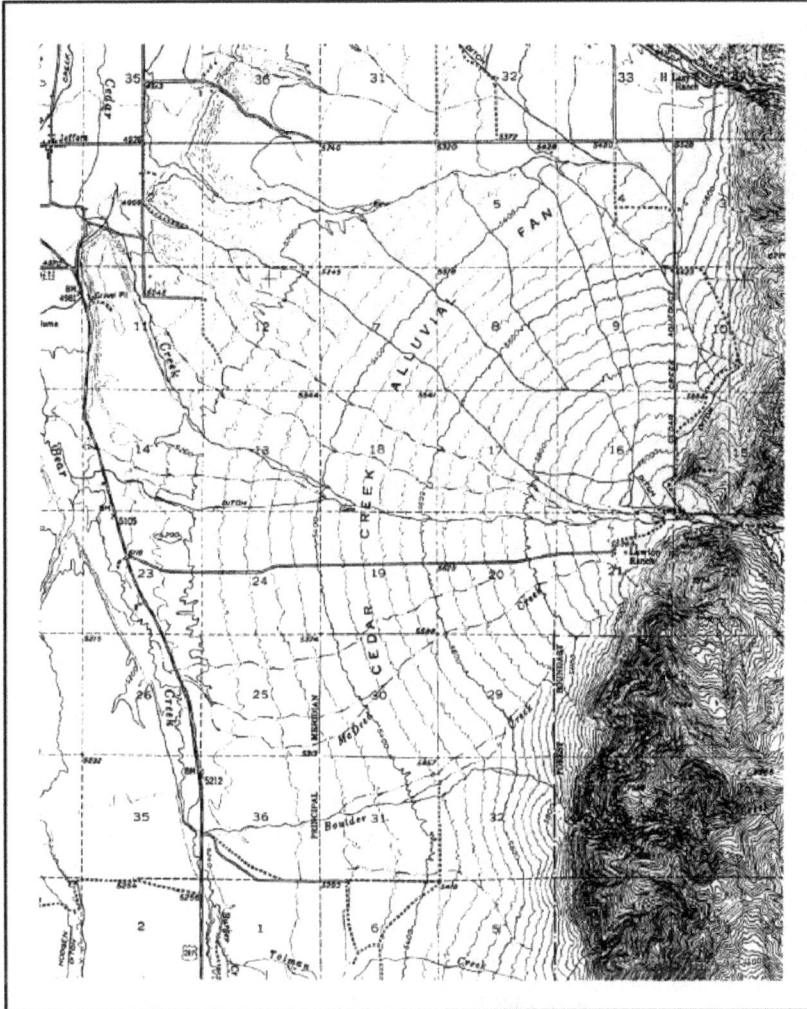

Figure 3-17. Cedar Creek alluvial fan

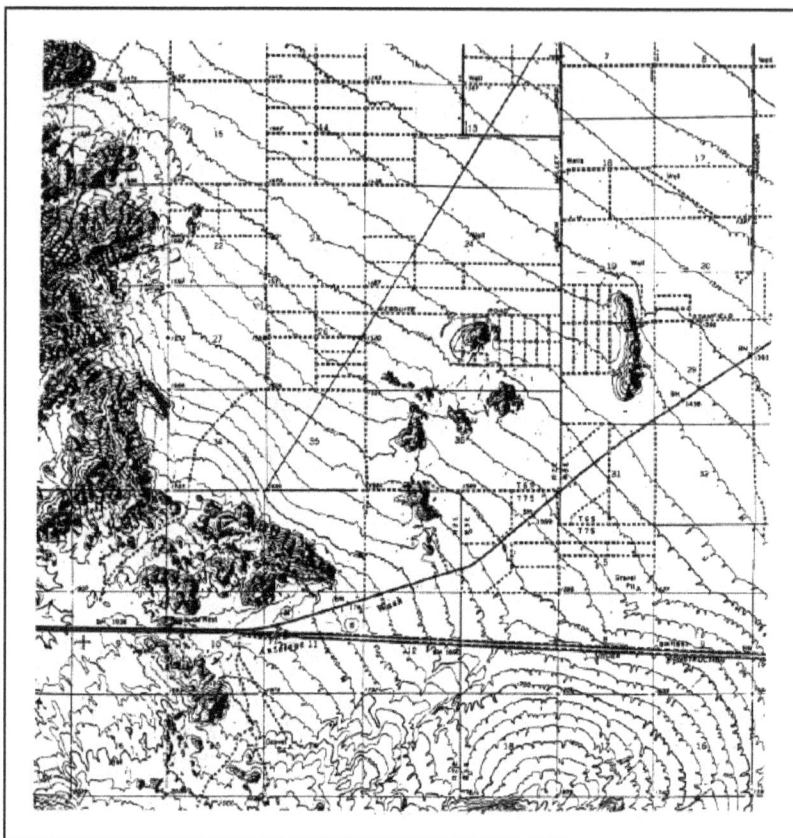

Figure 3-18. Coalescing alluvial fans

3-33. The types of materials found in alluvial fans are gravels, sands, and fines based on a ⅓ rule. The first ⅓, the area adjacent to the highland, is primarily composed of gravels; the middle ⅓ is composed of sands; and the final ⅓, the area farthest from the highland, is composed of fines.

3-34. Fluvial features are found throughout the world and are the primary source of borrow pit materials for military engineers. Table 3-2, and figure 3-19, present a summary of fluvial features. Figure 3-20, page 3-22, shows a generalized distribution of fluvial surficial features throughout the world.

Table 3-2. Fluvial surficial features

Feature	Description
Point bar	A low, crescent-shaped mound located at the inside of many bends in rivers or streams.
Channel bar	A Low, streamlined mound in braided streams or just downstream from point bars and on the opposite bank
Floodplain	A flat valley floor, leveled by back and forth erosion of the river or stream between the valley walls.
Alluvial Terrace	A platform or flat surface higher than the floodplain and generally close to the valley walls. It is all that remains of what was a floodplain many years before.

Table 3-2. Fluvial surficial features

Feature	Description
Oxbow lake	A horseshoe-shaped, abandoned section of a stream or river channel still containing water.
Clay plug	A clay-filled, abandoned section of a stream or river at the ends of horseshoe-shaped oxbow lakes.
Natural levee	Wide, low mounds (5 to 15 feet high), paralleling the river along both banks with sloping sides away from the river.
Backswamp	A swampy, level portion of a floodplain with very poor drainage and a high water table
Delta	Natural land extension into a body of water, visible as a low, almost level, land mass protruding into the body of water.
Alluvial fan	A cone-shaped mound formed against a valley wall, appearing fan-shaped from above.
Lake bed deposit	A layer formed from the settling of sediments to the bottom of lakes. The layers are thick at the center of the lake bed and thin near the lake margins.

Figure 3-19. Major floodplain features

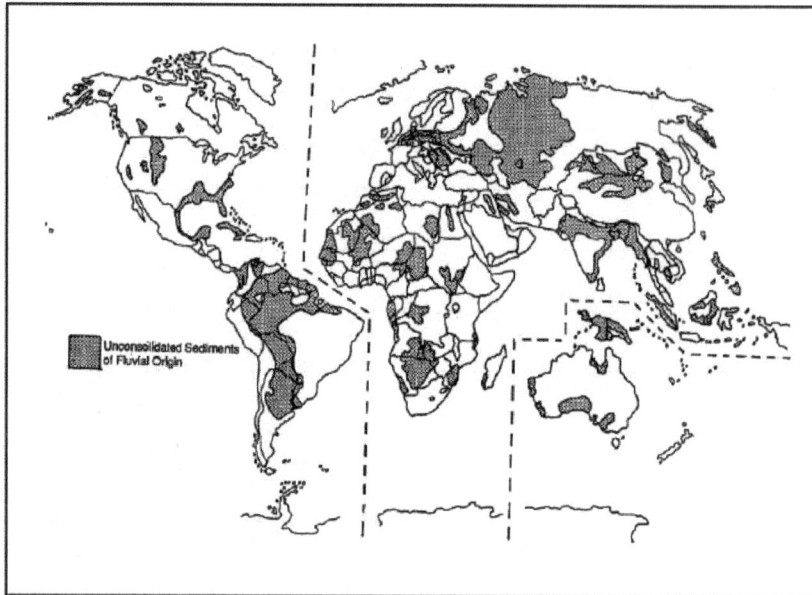

Figure 3-20. World distribution of fluvial landforms

GLACIAL PROCESS

3-35. Between ten and twenty-five thousand years ago, much of North America, Europe, and Northern Asia was covered by glaciers. Significant ice sheets still cover Greenland and Antarctica, and lesser ice sheets can be found at high elevations and latitudes (see figure 3-21).

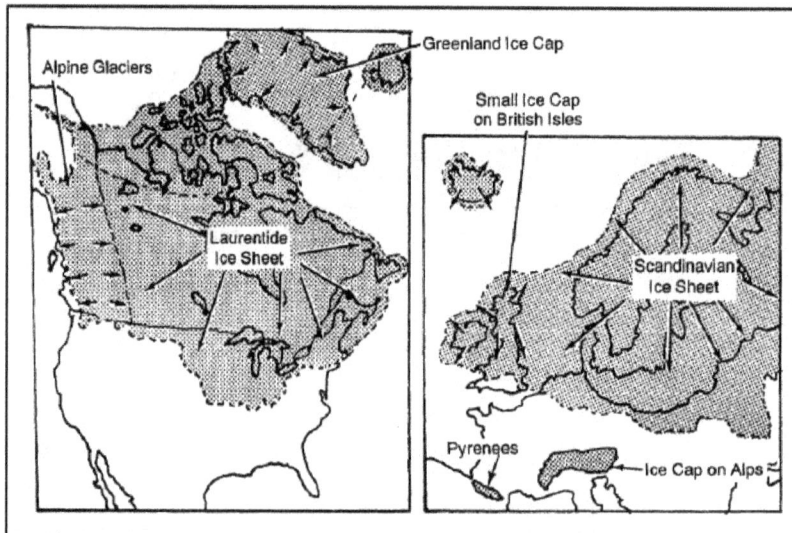

Figure 3-21. Ice sheets of North America and Europe

3-36. Glaciation produces great changes in the existing topography by reshaping the land surface and depositing new surficial features that may serve as a source of construction aggregate for military engineers.

TYPES OF GLACIATION

3-37. The glaciation process may be described as either continental or alpine glaciation.

Continental

3-38. Continental glaciation occurs on a large, regional scale affecting vast areas. It may be characterized by the occurrence of more depositional features than erosional features. Continental glaciers can be of tremendous thickness and extent. They move slowly in a plastic state with the ice churning the soil and rocks beneath it as well as crushing and plucking rocks from the ground and incorporating large amounts of material within the glacier itself. The overall range of particle size of these materials is from clays through cobbles and boulders (see figure 3-22).

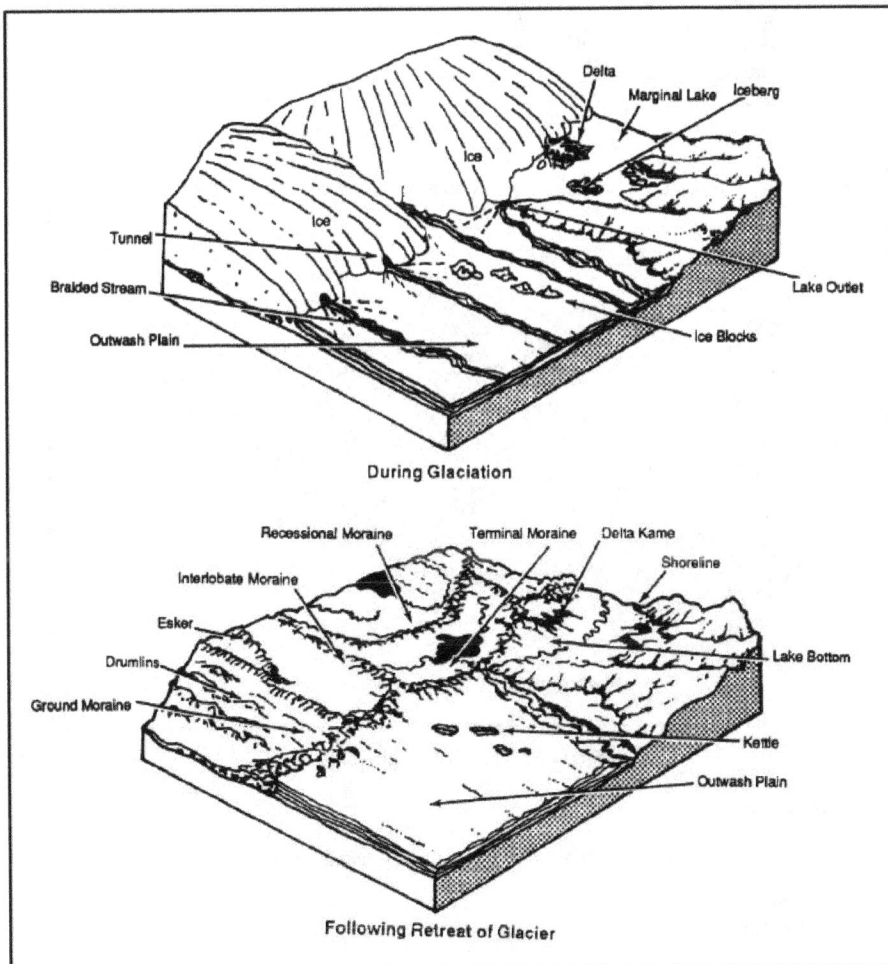

Figure 3-22. Contintental glaciation

Alpine

3-39. Alpine or mountain glaciation takes place in mountainous areas and generally results in the creation of mainly erosional forms. Alpine glacial features are very distinctive and easy to recognize. In the past, glaciers scooped out and widened the valleys through which they moved, producing valleys with a U-shaped profile in contrast to the V-shaped profile produced by fluvial erosion (see figure 3-23).

Area of alpine glaciation showing erosional features and placement of ice.

Same area of alpine glaciation with ice removed.

Figure 3-23. Alpine glaciation

Glacial Deposits

3-40. Materials deposited by glaciers are frequently differentiated into two types. They are—

- Stratified.
- Unstratified.

Stratified

3-41. The features composed of stratified deposits are actually the result of deposition of sediment by glacial streams (glaciofluvial) and not by the movement of the ice itself. These features are—

- Outwash plains.
- Eskers.
- Kames.
- Kame terraces.
- Glacial lake deposits.

3-42. They result when the material in the glacier has been carried and deposited by meltwater from the glacier. The water selectively deposits the coarsest materials, carrying the fines away from the area. The end result is essentially deposits of sands and gravels.

3-43. Outwash plains result when melting ice at the edge of the glacier creates a great volume of water that flows through the end moraine as a number of streams rather than as a continuous sheet of water. Each of the streams builds an alluvial fan and each of the fans joins together and forms a plain that slopes gently away from the end moraine area. The coarsest material is deposited nearest the end moraine, and the fines are deposited at greater distances. Much of the prairie land in the United States consists of outwash plains. Drainage and trafficability in the outwash plains are much better than in a ground moraine; however, kettles can be formed in outwash plains due to large masses of ice left during the recession of the ice front. If the kettles are numerous, the outwash area is called a pitted plain (see figure 3-22, page 3-23).

3-44. Eskers are winding ridges of irregularly stratified sands and gravels that are found within the area of the ground moraine. The ridges are usually several miles long but are rarely more than 45 to 60 feet wide or more than 150 feet high. They are formed by water that flowed in tunnels or ice-walled gorges in or beneath the ice. They branch and wind like stream valleys but are not like ordinary valleys in that they may cross normal drainage patterns at an angle, and they may also pass over hills (see figures 3-22 and figure 3-24, page 3-26). Figure 3-24 shows kettle lakes, swamps, and eskers.

3-45. A similar feature that resembles an esker, but is rarely more than a mile in length, is a ridge known as a crevasse filling. A crevasse is a large, deep crevice or fissure on the surface of a glacier. Unsorted debris washes into the crevasse, and when the surrounding ice melts, a ridge containing a considerable amount of fines is left standing.

3-46. Kames are conical hills of sands and gravels deposited by heavily laden glacial streams that flowed on top of or off of the glacier. They are usually isolated hills that are associated with the end or recessional moraine; kettle lakes are commonly found in the same area. The formation of kames normally occurred when meltwater streams deposited relatively coarse materials in the form of a glacioalluvial fan at the edge of the ice; the fine particles were washed away. This material accumulated along the side of the ice, and when the ice receded, the material slumped back on the side formerly in contact with the glacier.

3-47. Delta kames are another type of kame that may be formed when the meltwater flows into a marginal lake and forms a delta. After the lake and the ice disappear, deltas are left as flat-topped, steep-sided hills of well-sorted sands and gravels (see figure 3-22).

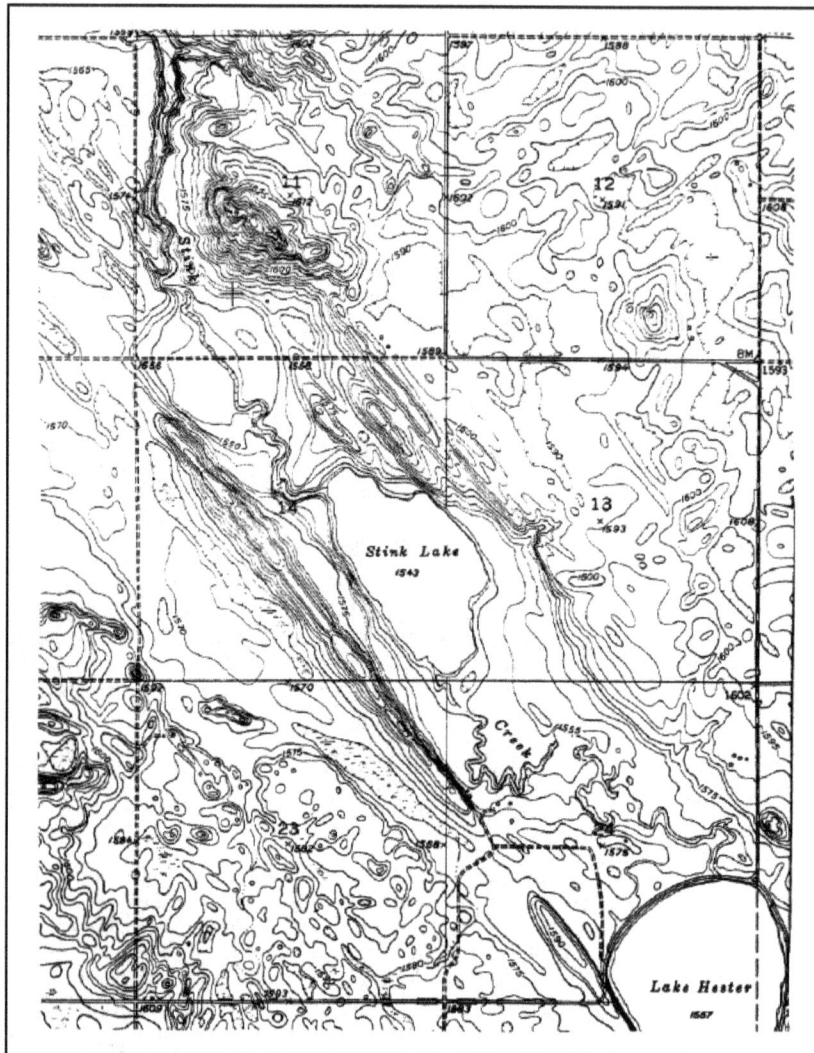

Figure 3-24. Moraine topographic expression with kettle lakes, swamps, and eskers

3-48. Kame terraces are features associated with alpine glaciation. When the ice moves down a valley, it is in contact with the sides of the valley. As the glacier melts away from the valley wall, glacial water flows into the space created between the side of the glacier and the valley wall. The void is filled with gravels and sands, while the fines are carried away by the stream water. A terrace is left where the ice was in contact with the valley; gravels and sands can be found at the base of the terrace (see figure 3-25).

3-49. Glacial lake deposits occur during the melting of the glacier when many lakes and ponds are created by the meltwater in the outwash areas. The streams that fed these waters were laden with glacial material. Most of the gravels and sands that were not deposited before reaching the lake accumulated as a delta (later to be called a delta kame) after melting of the ice. The fines that remained suspended in the water were, on the other hand, deposited throughout the lake. During the summer, a band consisting of light-colored, coarse silt was deposited, whereas a thinner band of darker, finer-grained material was deposited in the winter. The two bands together represent a time span of one year and are referred to as a varve.

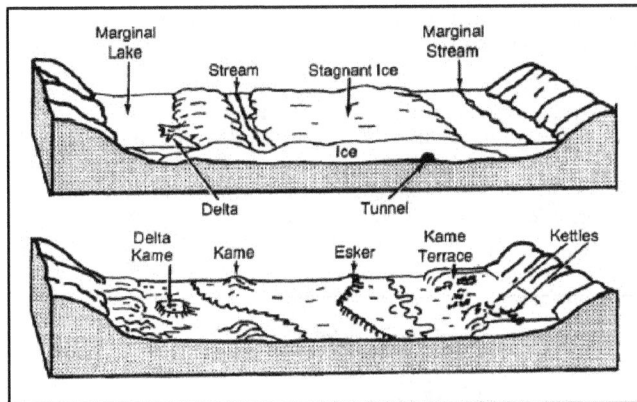

Figure 3-25. Valley deposits from melting ice

Unstratified

3-50. Unstratified glacial deposits (sediments deposited by the ice itself) are the most common of the of glacial deposits. They comprise the following surfical features:

- Ground moraines.
- End moraines.
- Recessional moraines.
- Drumlins.

3-51. Unstratified deposits make up landforms that may be readily identified in the field, on aerial photographs, and from topographic and other maps. Unstratified deposits are composed of a heterogeneous mixture of particle types and sizes ranging from clays to boulders. Till is the name given to this mixture of materials. It is the most widespread of all the forms of glacial debris. In general, features comprised of till are undesirable as sources of military construction aggregate since the material must be washed and screened to provide proper gradation.

3-52. Ground moraines, sometimes called till plains, are deposits that are laid down as the glacier recedes. Melting ice drops material that blankets the area over which the glacier traveled. A deposit of this kind forms gently rolling plains. The deposit itself may be a thin veneer of material lying on the bedrock, or it may be hundreds of feet thick. Moraine soil composition is complex and often indeterminate. This variation in sediment makeup is due to the large variety of rocks and soil picked up by the moving glacier (see figure 3-22, page 3-23).

3-53. Morainic areas have a highly irregular drainage pattern because of the haphazard arrangement of ridges and hills, although older till plains tend to develop dendritic patterns. Frequent features associated with ground moraines are kettle holes and swamps. Kettles are usually formed by the melting of ice that had been surrounded by or embedded in the moraine material. Large amounts of fines in the till prevent water from percolating down through the soil. This may allow for the accumulation of water in the kettle holes forming kettle lakes or, in low-lying areas, swamps. Figure 3-22 and figure 3-24, show ground moraine with an esker.

3-54. End moraines, sometimes called terminal moraines, are ridges of till material that were pushed to their locations at the limit of the glacier's advance by the forceful action of the ice sheet. Generally, there is no one linear element, such as a continuous ridge, evident in either the field or on aerial photos. Normally, this deposit appears as a discontinuous chain of elongated to oval hills. These hills vary in height from tens to hundreds of feet. The till material is, at times, quite clayey. Kettle lakes are sometimes associated with terminal moraine deposits also (see figure 3-22).

3-55. Recessional moraines, which are similar to end moraines, are produced when a receding glacier halts its retreat for a considerable period of time. The stationary action allows for the accumulation of till material along the glacier's edge. A series of these moraines may result during the retreat of a glacier (see figure 3-22, page 3-23).

3-56. Drumlins are asymmetrical, streamlined hills of gravel till deposited at the base of a glacier and oriented in a direction parallel to ice flow. The stoss side (the side from which the ice flowed) of the drumlin is steeper and blunter than the lee side. The overall appearance of a drumlin resembles an inverted spoon if viewed from above. Drumlins commonly occur in groups of two or more. Individual drumlins are seldom more than ½ mile long, and they can rise to heights of 75 to 100 feet (see figure 3-26 and figure 3-27).

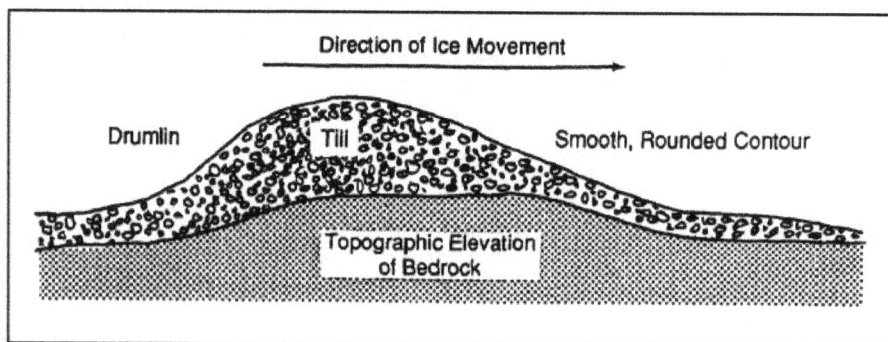

Figure 3-26. Idealized cross section of a drumlin

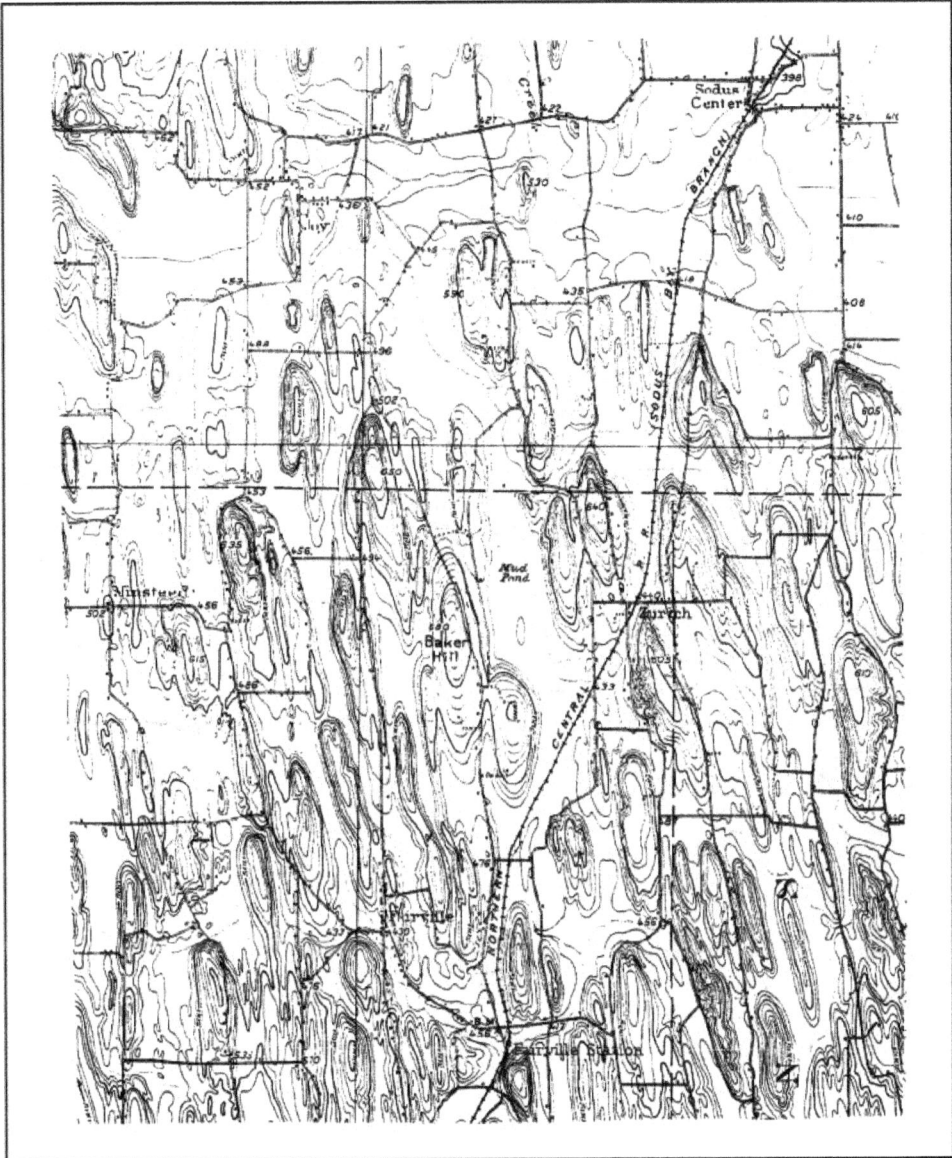

Figure 3-27. Topographic expression of a drumlin field

3-57. It is important to understand that features formed from the glacial process only occur in certain areas of the world. Figures 3-28 and 3-29 illustrate the regions of the United States and the world where glacial landforms occur. Table 3-3, page 3-32, is a summary of glacial surficial features.

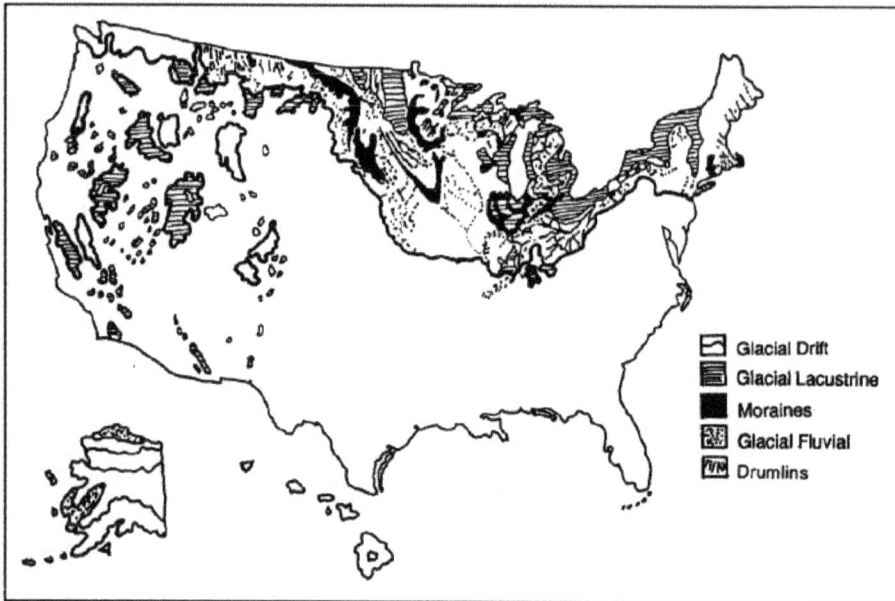

Figure 3-28. Distribution of major groups of glacial landforms across the United States

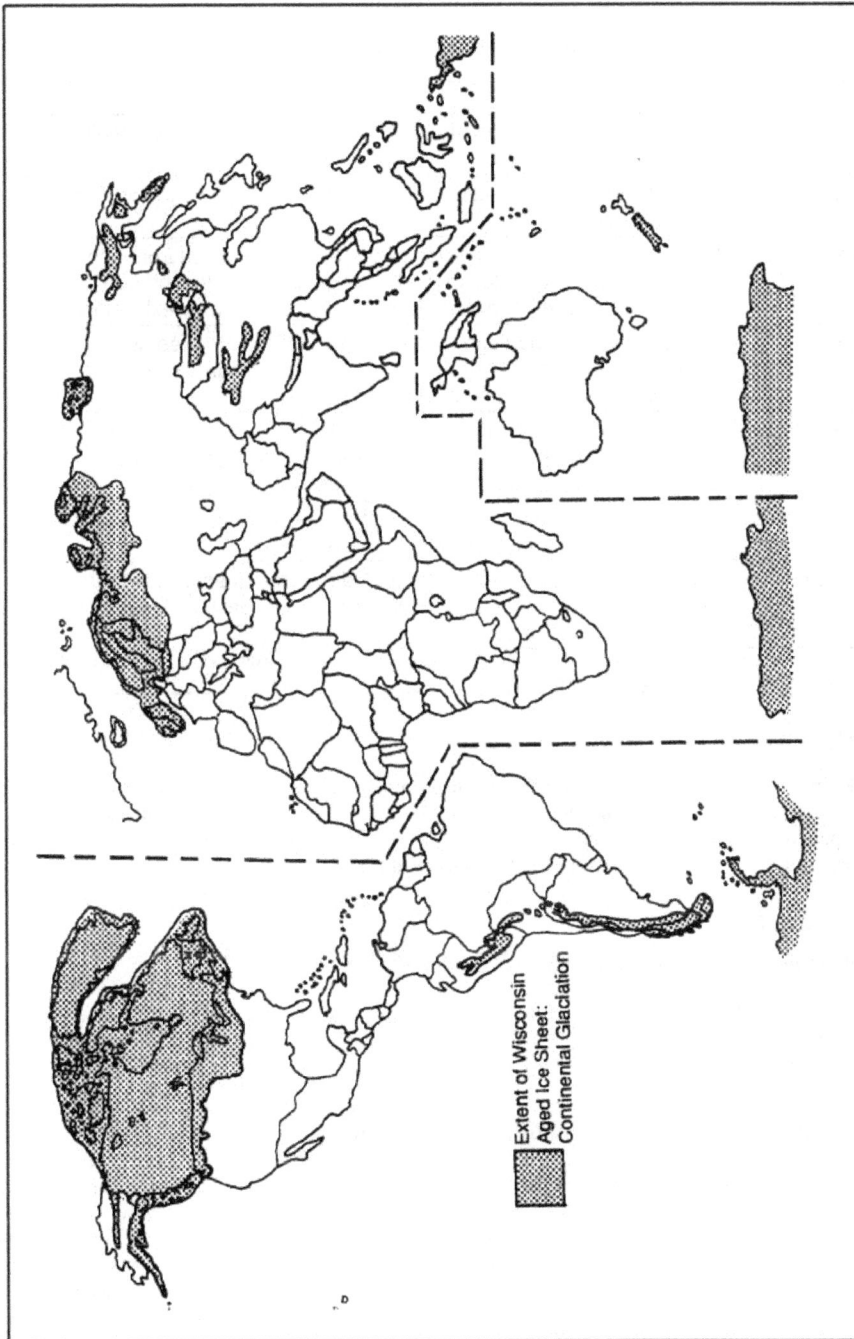

Figure 3-29. World distribution of glacial landforms

Table 3-3. Glacial surficial features

Feature	Description
Grand Moraine	A blanket of till covering the bedrock or an old soil layer extending for miles in all directions
End moraine	A long, irregularly rounded, narrow ridge or parallel ridges of till formed at the edges of a glacier.
Drumlin	A streamlined hill of till (teardrop-shaped when viewed from above), generally from 20 to 200 feet tall, 50 to 400 feet wide, and 500 to 3,000 feet long. Drumlins are commonly found in groups but may exist alone.
Esker	A winding, rounded ridge that is generally 100 to 200 feet wide, with a fairly constant height of 50 to 100 feet, trending for distances of several thousand feet to several miles
Kame	An irregularly shaped mound of sand and gravel found near or in ridge moraines. Distinguished from mounds of till by slightly greater relief, slightly steeper sides, and better drainage (drier).
Kame Terrace	A platform or flat surface (in U-shaped, glaciated valleys) higher than the valley floor and butting against the valley wall.
Outwash Plain	A fairly flat, well-drained plain, formed by large quantities of meltwater running away from the edge of the glacier. Outwash plains are commonly found in front of end moraines
Note. Till is not a feature. It is the material picked up, moved, and deposited by glaciers. It is largely boulders and silt but also contains gravels, sands, and clays.	

EOLIAN PROCESS

3-58. In arid areas where water is scarce, wind takes over as the main erosional agent. When a strong wind passes over a soil, it carries many particles of soil with it. The height and distance the materials are transported is a function of the particle size and the wind velocity. The subsequent decrease in the wind velocity gives rise to a set of wind-borne deposits called eolian features.

TYPES OF EOLIAN EROSION

3-59. There are two types of wind erosion. They are—

- Deflation.
- Abrasion.

Deflation

3-60. Deflation occurs when loose particles are lifted and removed by the wind. This results in a lowering of the land surface as materials are carried away. Unlike stream erosion, in which downcutting is limited by a "base level" (usually sea level), deflation can continue lowering a land surface as long as it has loose material to carry away. Deflation may be terminated if the land surface is cut down to the water table (moist soil is not carried away as easily) or if vegetation is sufficient to hold the soil in place. In addition, deflation may be halted when the supply of fine material has been depleted. This makes a surface of gravel in the area where deflation has taken place. This gravel surface is known as desert pavement (see figure 3-30).

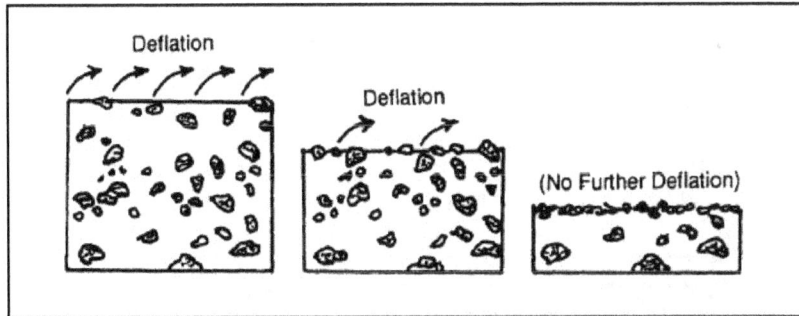

Figure 3-30. Three stages illustrating the development of desert armor

Abrasion

3-61. Abrasion occurs when hard particles are blown against a rock face causing the rocks to break down. As fragments are broken off, they are carried away by the wind. This process can grind down and polish rock surfaces. A rock fragment with facets that have been cut in this way is called a ventifact (see figure 3-31).

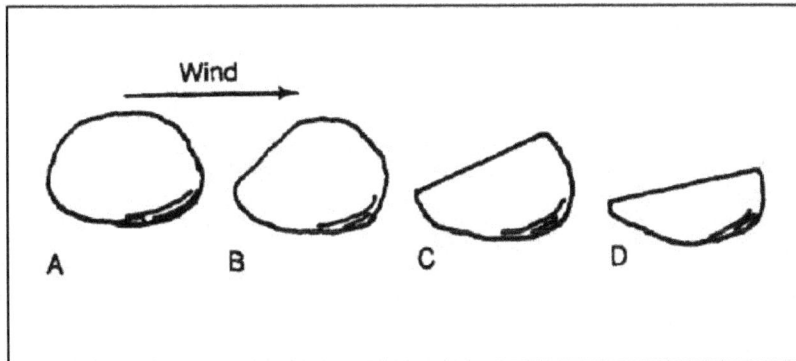

Figure 3-31. Cutting of a ventifact

MODES OF TRANSPORTATION

3-62. Soil particles can be carried by the wind in the following ways:

Bed Load

3-63. Material that is too heavy to be carried by the wind for great distances at a time (mainly sand-sized particles) bounces along the ground, rarely higher than two feet.

Suspended Load

3-64. These are fines (mostly silts) that are easily carried by the wind. Suspended loads extend to high altitudes (sometimes thousands of feet) and can be transported for thousands of miles. During a particularly bad dust storm in the mid-western "dust bowl" on 20 March 1935, the suspended load extended to altitudes of over 12,000 feet. The lowermost mile of the atmosphere was estimated to contain over 166,000 tons of suspended particles per cubic mile. Enough material was transported to bring temporary twilight to New York and New England (over 2,000 miles away) on 21 March.

EOLIAN FEATURES

3-65. Eolian surficial features may consist of gravels, sands, or fines. The three main types of eolian features are—

- Lag deposits or desert pavement.
- Sand dunes.
- Loess deposits.

3-66. Figure 3-32 illustrates the origin of these deposits.

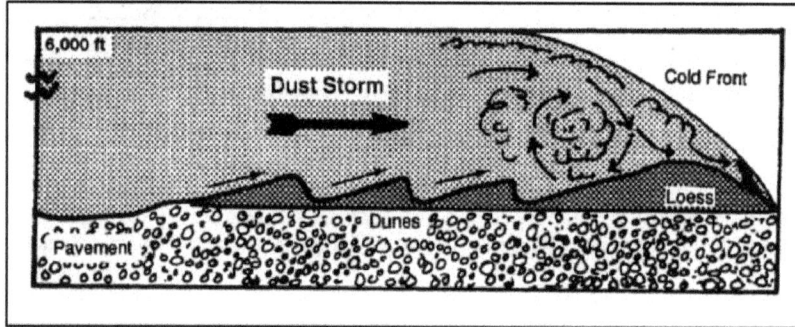

Figure 3-32. Eolian features

Lag Deposits or Desert Pavement

3-67. As the wind billows across the ground, sands and fines are continually removed. Eventually, gravels and pebbles that are too large to be carried by the wind cover the surface. These remnants accumulate into a sheet that ultimately covers the finer-grained material beneath and protects it from further deflation. Desert pavement usually develops rapidly on alluvial fan and alluvial terrace surfaces. The exposed surface of the gravels may become coated with a black, glittery substance termed desert varnish. In some locations, the evaporation of water, brought to the surface by the capillary action of the soil, may leave behind a deposit of calcium carbonate (caliche) or gypsum. It acts as a cement, hardening the pavement into a conglomeratelike slab.

3-68. Although desert pavement contains good gravel material, the layers are normally too thin to supply the quantity required for construction. However, it does provide a rough but very trafficable surface for all types of vehicles and also provides excellent airfields.

Sand Dunes

3-69. Dunes may take several forms, depending on the supply of sand, the lay of the land, vegetation restrictions, and the steady direction of the wind. Their general expressions are as follows:

- Transverse.
- Longitudinal.
- Barchan.
- Parabolic.
- Complex.

3-70. Transverse dunes are wavelike ridges that are separated by troughs; they resemble sea waves during a storm. These dunes, which are oriented perpendicular to the prevailing wind direction, occur in desert locations where a great supply of sand is present over the entire surface. A collection of transverse dunes is known as a sand sea (see figure 3-33, a).

3-71. Longitudinal dunes have been elongated in the direction of the prevailing winds. They usually occur where strong winds blow across areas of meager amounts of sand or where the winds compete with the stabilizing effect of grass or small shrubs (see figure 3-33, b).

3-72. Barchan dunes are the simplest and most common of the dunes. A barchan is usually crescent-shaped, and the windward side has a gentle slope rising to a broad dome that cuts off abruptly to the leeward side. Barchans form in open areas where the direction of the wind is fairly constant and the ground is flat and unrestricted by vegetation and topography (see figure 3-33, c).

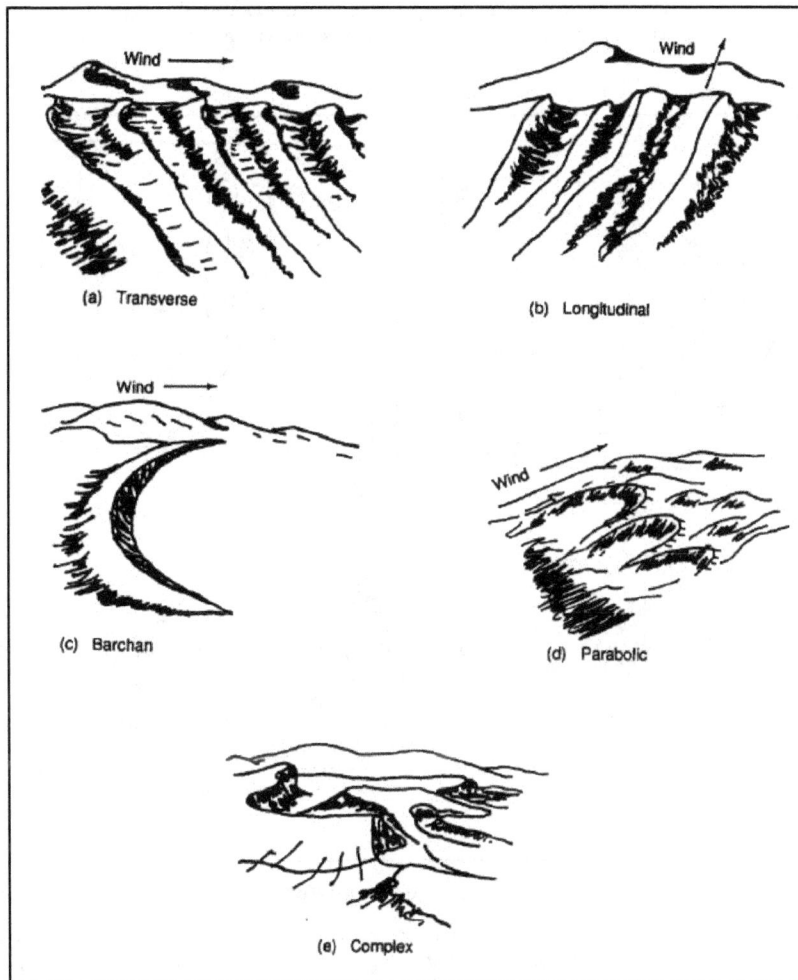

Figure 3-33. Sand dune types

3-73. Parabolic, or U-shaped, dunes have tips that point upwind. They typically form along coastlines where the vegetation partially covers the sand or behind a gap in an obstructing ridge. Later, a parabolic dune may detach itself from the site of formation and migrate independently (see figure 3-33, d).

3-74. Complex dunes lack a distinct form and develop where wind directions vary, sand is abundant, and vegetation may interfere. These can occur locally when other dune types become overcrowded and overlap, thereby losing their characteristic shapes in a disorder of varying slopes (see figure 3-33, e).

Loess Deposits

3-75. In a number of regions of the world, thick accumulations of yellowish-brown material composed primarily of windblown silts make up a substantial amount of surface area. These deposits are known as loess. The material that makes up these deposits originated mainly from dried glacial outwash, floodplains, or desert area fines. Loess is composed of physically ground rock rather than of chemically weathered material. The source and deposition point for the material may be many miles apart, and the deposits may range in thickness from a few feet to hundreds of feet. Thickness tends to decrease with distance from the source. In the United States, most of Kansas, Nebraska, Iowa, and Illinois are covered by loess. After a loess has been laid down, it is rarely picked up again. This is due to a very thin layer of fines that interlock after wetting. While dry loess is trafficable, it loses all strength with a slight amount of water (see figure 3-34).

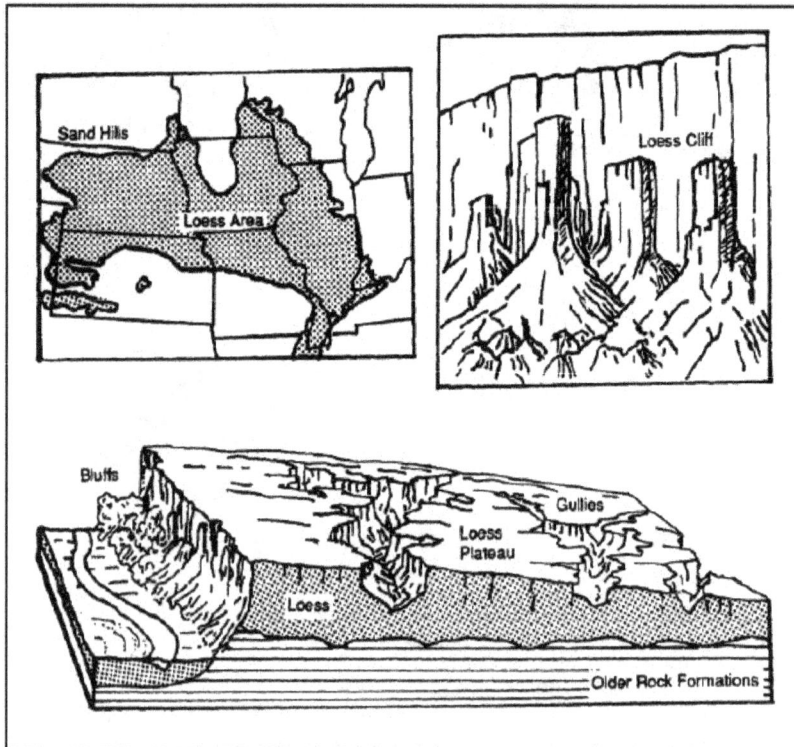

Figure 3-34. Loess landforms

3-76. Eolian features occur worldwide and may consist of areas of sand dunes and desert pavement or loess; however, their topographic expressions vary. In general, dune areas are specified on maps by special topographic symbols since they are continually changing unless stabilized by vegetation. Figure 3-35, is a topographic expression, using special symbols, of sand dunes and desert pavement (Summan). Figure 3-36, page 3-38, shows the generalized distribution of eolian landforms throughout the world.

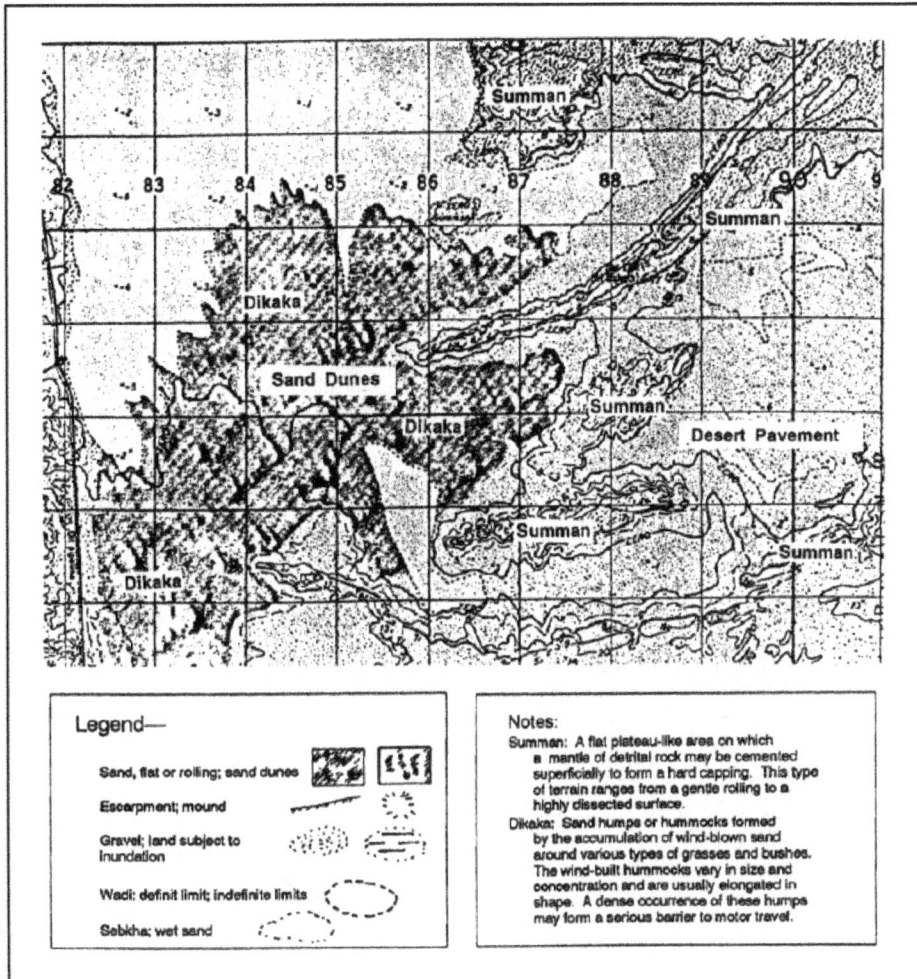

Figure 3-35. Topographic expression of dunes and desert pavement

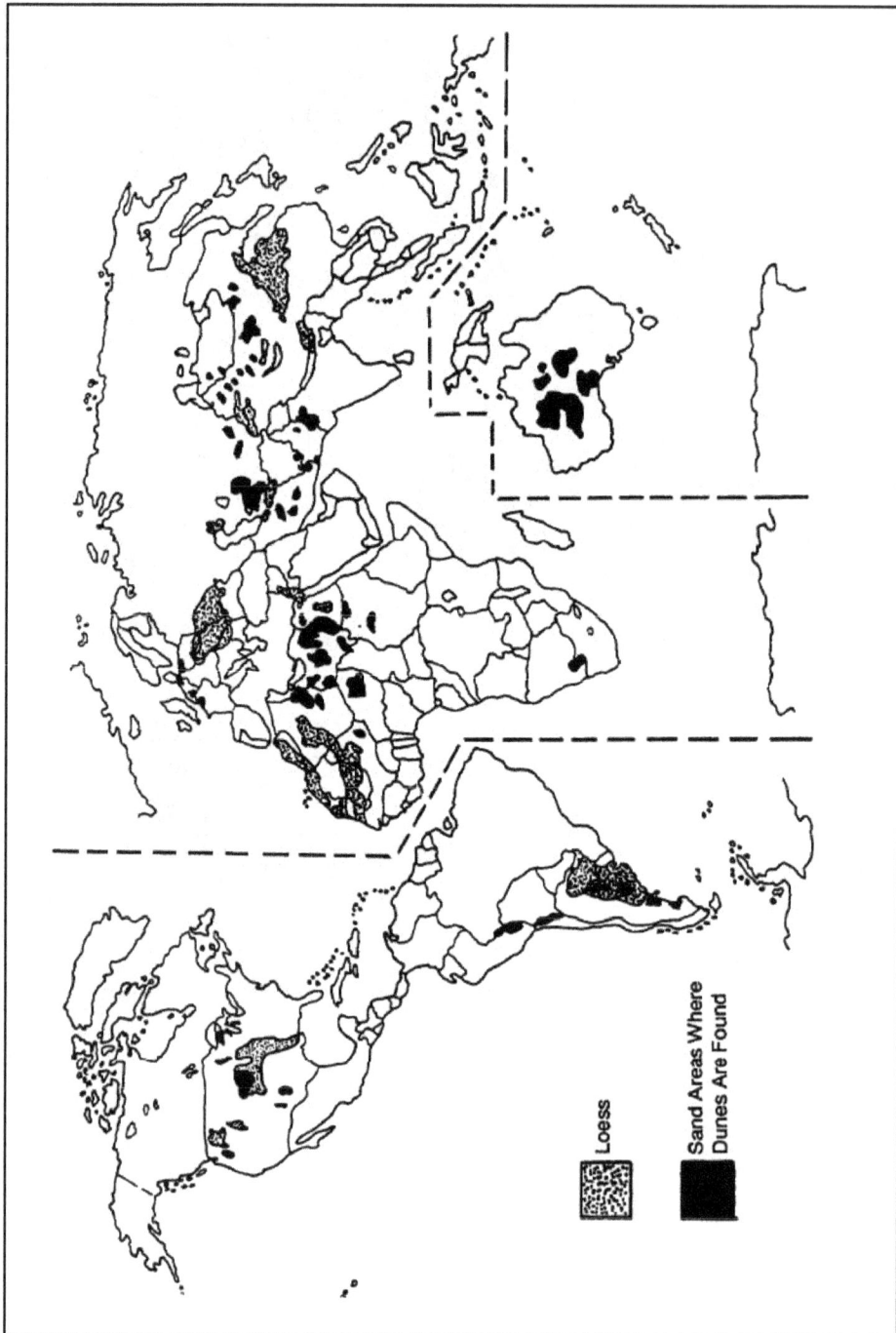

Figure 3-36. Worldwide distribution of eolian landforms

SOURCES OF CONSTRUCTION AGGREGATE

3-77. Military engineers use their knowledge of surficial features to develop borrow pits and provide construction aggregate to meet mission requirements. Generally, engineer units attempt to develop borrow pit operations in fluvial features since they are easy to identify and are normally accessible. In arid and semiarid regions, eolian deposits and alluvial fans provide large amounts of aggregate. In mountainous regions and continentally glaciated regions, fluvial-glacial deposits can provide large quantities of quality aggregate. Therefore, their presence should not necessarily be discounted in preference to fluvial materials. Table 3-4 summarizes the types of aggregates found in common fluvial, glacial, and eolian surficial features.

Table 3-4. Aggregate types by feature

Feature	Cobbles	Gravels	Sands	Silts	Clays
Point bar		X	X		
Channel bar		X	X		
Alluvial Terrace	X	X	X		
Oxbow Lake, Clay plug				X	X
Natural levee		X	X	X	
Backswamp				X	X
Delta, Bird's-foot				X	X
Delta, Arcuate		X	X	X	X
Alluvial fan	X	X	X	X	X
Lake bed deposits				X	X
Esker	X	X	X		
Kame	X	X	X		
Kame Terrace	X	X	X		
Outwash Plains		X	X		
Desert Pavement	X	X			
Sand Dunes			X		
Loess				X	

This page intentionally left blank.

Chapter 4

Soil Formation and Characteristics

The term "soil," as used by the US Army, refers to the entire unconsolidated material that overlies and is distinguishable from bedrock. Soil is composed principally of the disintegrated and decomposed particles of rock. It also contains air and water as well as organic matter derived from the decomposition of plants and animals. Bedrock is considered to be the solid part of the earth's crust, consisting of massive formations broken only by occasional structural failures. Soil is a natural conglomeration of mineral grains ranging from large boulders to single mineral crystals of microscopic size. Highly organic materials, such as river bottom mud and peat, are also considered soil. To help describe soils and predict their behavior, the military engineer should understand the natural processes by which soils are formed from the parent materials of the earth's crust. As soils are created, by the process of rock weathering and often by the additional processes of transportation and disposition, they often acquire distinctive characteristics that are visible both in the field and on maps and photographs.

SECTION I - SOIL FORMATION

WEATHERING

4-1.   Weathering is the physical or chemical breakdown of rock. It is this process by which rock is converted into soil. Weathering is generally thought of as a variety of physical or chemical processes that are dependent on the environmental conditions present.

PHYSICAL PROCESSES

4-2.   Physical weathering is the disintegration of rock. Physical weathering processes break rock masses into smaller and smaller pieces without altering the chemical composition of the pieces. Therefore, the disintegrated fragments of rock exhibit the same physical properties as their sources. Processes that produce physical weathering are—

- Unloading.
- Frost action.
- Organism growth.
- Temperature changes.
- Crystal growth.
- Abrasion.

Unloading

4-3.   When rock layers are buried under the surface, they are under compressive stress from the weight of overlying materials. When these materials are removed, the resulting stress reduction may allow the rock unit to expand, forming tensional cracks (jointing) and causing extensive fracturing.

Frost Action

4-4. Most water systems in rocks are open to the atmosphere, but freezing at the surface can enclose the system. When the enclosed water freezes, it expands nearly one-tenth of its volume, creating pressures up to 4,000 pounds per square inch (psi). The expanding ice fractures the rock.

Organism Growth

4-5. Trees and plants readily grow in the joints of rock masses near the surface. The wedging action caused by their root growth hastens the disintegration process.

Temperature Changes

4-6. Daily or seasonal temperature changes can cause differential expansion and contraction of rocks near the earth's surface. This results in a tensional failure called spalling or exfoliation. As the rock's surface heats up, it expands; as it cools, it contracts. The jointing patterns of igneous rock are often the result of temperature changes.

Crystal Growth

4-7. The growth of minerals precipitating from groundwater can apply pressure similar to that of expanding ice. Soluble minerals, such as halite (salt), readily crystalize out of solution.

Abrasion

4-8. Sediments suspended in wind or fast-moving water can act as abrasives to physically weather rock masses. Rock particles carried by glacial ice can also be very abrasive.

CHEMICAL PROCESSES

4-9. Chemical weathering is the decomposition of rock through chemical processes. Chemical reactions take place between the minerals of the rock and the air, water, or dissolved or suspended chemicals in the atmosphere. Processes that cause chemical weathering are—

- Oxidation.
- Hydration.
- Hydrolysis.
- Carbonation.
- Solution.

Oxidation

4-10. Oxidation is the chemical union of a compound with oxygen. An example is rusting, which is the chemical reaction of oxygen, water, and the iron mineral pyrite (FeS_2) to form ferrous sulfate ($FeSO_4$). Oxidation is responsible for much of the red and yellow coloring of soils and surface rock bodies. This type of reaction is important in the decomposition of rocks, primarily those with metallic minerals.

Hydration

4-11. Hydration is the chemical union of a compound with water. For example, the mineral anhydrite ($CaSO_4$) incorporates water into its structure to form the new mineral gypsum ($CaSO_42H_2O$).

Hydrolysis

4-12. This decomposition reaction is related to hydration in that it involves water. It is a result of the partial dissociation of water during chemical reactions that occur in a moist environment. It is one of the types of weathering in a sequence of chemical reactions that turns feldspars into clays. An example of hydrolysis is the altering of sodium carbonate (Na_2CO_3) to sodium hydroxide ($NaOH$) and carbonic acid (H_2CO_3).

Carbonation

4-13. This is the chemical process in which carbon dioxide from the air unites with various minerals to form carbonates. A copper penny eventually turns green from the union of copper with carbon dioxide in the air to form copper carbonate. Carbonate rocks, in turn, are susceptible to further weathering processes, namely solution.

Solution

4-14. Carbon dioxide dissolved in water forms a weak acid called carbonic acid (H_2CO_3). Carbonic acid acts as a solvent to dissolve carbonates, such as limestone, and carry them away. This creates void spaces, or caves, in the subsurface. Areas that have undergone extensive solutioning are known as karst regions.

DISCONTINUITIES AND WEATHERING

4-15. Jointing and other discontinuities increase the surface area of the rock mass exposed to the elements and thereby enhance chemical weathering. Discontinuities, such as joints, faults, or caverns, act as conduits into the rock mass for the weathering agents (air and water) to enter. Weathering occurs on exposed surfaces, such as excavation walls, road cuts, and the walls of discontinuities. The effect of weathering along discontinuities is a general weakening in a zone surrounding the wall surface.

EFFECTS ON CLIMATE

4-16. The climate determines largely whether a type of rock weathers mostly by chemical or mostly by physical processes. Warm, wet (tropical or subtropical) climates favor chemical weathering. In such climates, there is abundant water to support the various chemical processes. Also, in warm, wet climates, the temperature is high enough to allow the chemical reactions to occur rapidly. Cold, dry climates discourage chemical weathering of rock but not physical weathering. The influence of climate on the weathering of the many rock types varies; however, most rock types weather more rapidly in warm, wet climates than in cold and/or dry climates.

EFFECTS ON RELIEF FEATURES

4-17. Weathering combined with erosion (the transportation of weathered materials) is responsible for most of the relief features on the earth's surface. For example, the subsurface cavities so predominant in karst regions develop along the already existing joints and bedding planes and commonly form an interlacing network of underground channels. If the ceiling of one of these subterranean void spaces should collapse, a sinkhole forms at the earth's surface. The sinkhole may range in size from a few feet to several miles in diameter. It may be more than a hundred feet deep, and it may be dry or contain water. Extensive occurrences of sinkholes result in the formation of karst topography, which is characterized by a pitted or pinnacle ground surface with numerous depressions and a poorly developed drainage pattern. Other features associated with karst topography include:

- Lost or disappearing streams where surface streams disappear or flow underground.
- Rises where underground streams suddenly reappear at the surface to form springs.

4-18. Solution cavities and sinkholes can be detrimental to foundations for horizontal and vertical construction and should be identified and evaluated for military operations.

SOIL FORMATION METHODS

4-19. Soils may be divided into two groups based on the method of formation—residual soils and transported soils.

RESIDUAL SOILS

4-20. Where residual soils are formed, the rock material has been weathered in place. While mechanical weathering may occur, chemical weathering is the dominant factor. As a result of this process, and because the rock material may have an assorted mineral structure, the upper layers of soils are usually fine-grained and relatively impervious to water. Under this fine-grained material is a zone of partially disintegrated parent rock. It may crumble easily and break down rapidly when exposed to loads, abrasions, or further weathering. The boundary line between soil and rock is usually not clearly defined. Lateritic soils (highly weathered tropical soils containing significant amounts of iron or iron and aluminum) are residual. Residual soils generally present both drainage and foundation problems. Residual soil deposits are characteristically erratic and variable in nature. Figure 4-1, shows a typical residual soil.

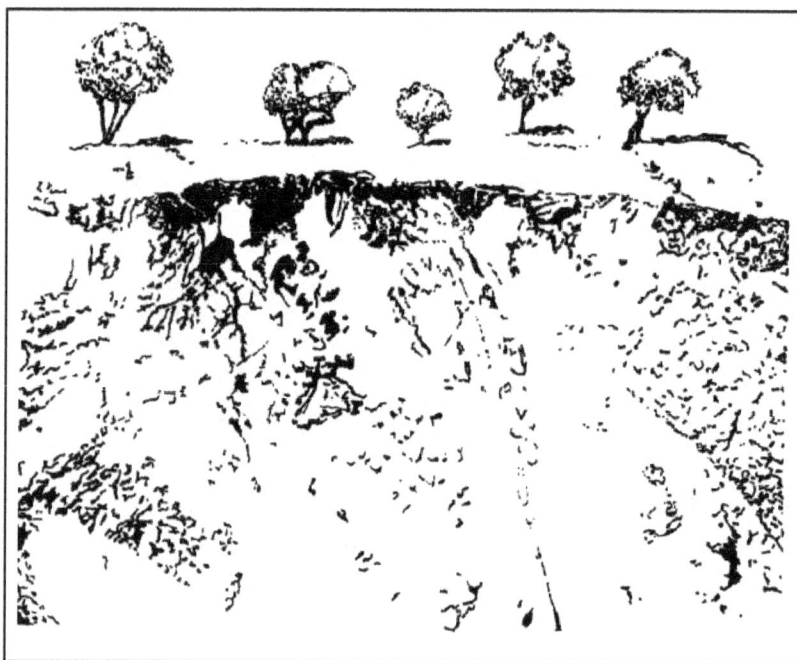

Figure 4-1. Residual soil forming from the in-place weathering of igneous rock

TRANSPORTED SOILS

4-21. By far, most soils the military engineer encounters are materials that have been transported and deposited at a new location. Three major forces—glacial ice, water, and wind—are the transporting agents. These forces have acted in various ways and have produced a wide variety of soil deposits. Resulting foundations and construction problems are equally varied. These soils may be divided into glacial deposits, sedimentary or water-laid deposits, and eolian or wind-laid deposits. Useful construction material can be located by being able to identify these features on the ground or on a map (see chapter 3).

SOIL PROFILES

4-22. As time passes, soil deposits undergo a maturing process. Every soil deposit develops a characteristic profile because of weathering and the leaching action of water as it moves downward from the surface. The profile developed depends not only on the nature of the deposit but also on factors such as temperature, rainfall amounts, and vegetation type. Under certain conditions, complex profiles may be developed, particularly with old soils in humid regions. In dry regions, the profile may be obscured.

4-23. Typical soil profiles have at least three layers, known as horizons (see figure 4-2). They are—

- A horizon.
- B horizon.
- C horizon.

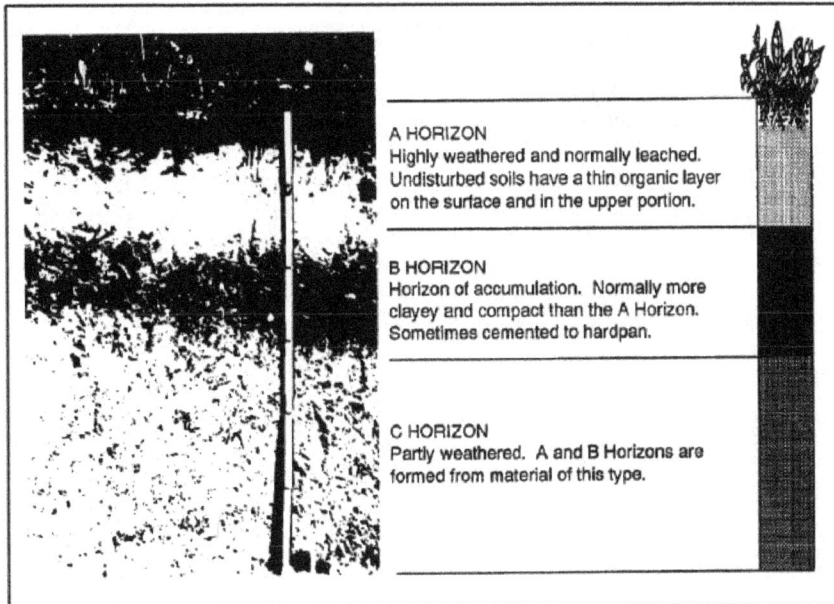

Figure 4-2. Soil profile showing characteristic soil horizons

4-24. The A horizon, or upper layer, contains a zone of accumulation of organic materials in its upper portion and a lower portion of lighter color from which soil colloids and other soluble constituents have been removed. The B horizon represents the layer where soluble materials accumulate that have washed out of the A horizon. This layer frequently contains much clay and may be several feet thick. The C horizon is the weathered parent material. The development of a soil profile depends on the downward movement of water. In arid and semiarid regions, the movement of water may be reversed and water may be brought to the surface because of evaporation. Soluble salts may thus be brought to the surface and deposited.

4-25. The study of the maturing of soils and the relationship of the soil profile to the parent material and its environment is called pedology. As will be explained later, soils may be classified on the basis of their soil profiles. This approach is used by agricultural soil scientists and some engineering agencies. This system is of particular interest to engineers who are concerned with road and airfield problems.

4-26. Soils not only characteristically vary with depth, but several soil types can and often do exist within a relatively small area. These variations may be important from an engineering standpoint. The engineering properties of a soil are a function not only of the kind of soil but also of its conditions.

SECTION II - SOIL CHARACTERISTICS

PHYSICAL PROPERTIES

4-27. The engineering characteristics of soil vary greatly, depending on such physical properties as—

- Grain or particle size.
- Gradation.

- Particle shape.
- Structure.
- Density.
- Consistency.

4-28. These properties are defined, in most cases numerically, as a basis for the systematic classification of soil types. Such a classification system, used in connection with a common descriptive vocabulary, permits the ready identification of soils that may be expected to behave similarly.

4-29. The nature of any given soil can be changed by manipulation. Vibration, for example, can change a loose sand into a dense one. Therefore, the behavior of a soil in the field depends not only on the significant properties of the individual constituents of the soil mass but also on properties due to particle arrangement within the mass.

4-30. Frequently, the available laboratory equipment or other considerations only permit the military engineer or the engineer's soil technician to determine some of the soil's properties and then only approximately. Hasty field identification often permits a sufficiently accurate evaluation for the problem at hand. However, the engineer cannot rely solely on experience and judgment in estimating soil conditions or identifying soils. He must make as detailed a determination of the soil properties as possible and subsequently correlate these identifying properties with the observed behavior of the soil.

GRAIN OR PARTICLE SIZE

4-31. In a natural soil, the soil particles or solids form a discontinuous mass with spaces or voids between the particles. These spaces are normally filled with water and/or air. Organic material may be present in greater or lesser amounts. The following paragraphs are concerned with the soil particles themselves. The terms "particle" and "grain" are used interchangeably.

Determination

4-32. Soils may be grouped on the basis of particle size. Particles are defined according to their sizes by the use of sieves, which are screens attached to metal frames. Figure 4-3 shows sieves used for the Unified Soil Classification System (USCS). If a particle will not pass through the screen with a particular size opening, it is said to be "retained on" that sieve. By passing a soil mixture through several different size sieves, it can be broken into its various particle sizes and defined according to the sieves used. Many different grain-size scales have been proposed and used. Coarse gravel particles are comparable in size to a lemon, an egg, or a walnut, while fine gravel is about the size of a pea. Sand particles range in size from that of rock salt, through table salt or granulated sugar, to powdered sugar. Below a Number 200 sieve, the particles (fines) are designated as silts or clays, depending on their plasticity characteristics.

4-33. Other grain-size scales apply other limits of size to silts and clays. For example, some civil engineers define silt as material less than 0.05 mm in diameter and larger than 0.005 mm. Particles below 0.005 mm are clay sizes. A particle 0.07 mm in diameter is about as small as can be detected by the unaided eye. It must be emphasized that below the Number 200 sieve (0.074-mm openings), particle size is relatively unimportant in most cases compared to other properties. Particles below 0.002 mm (0.001 mm in some grain-size scales) are frequently designated as soil colloids. The organic materials that may be present in a soil mass have no size boundaries.

4-34. Several methods may be used to determine the size of soil particles contained in a soil mass and the distribution of particle sizes. Dry sieve analysis has sieves stacked according to size, the smallest being on the bottom. Numbered sieves designate the number of openings per lineal inch. Dimensioned sieves indicate the actual size of the opening. For example, the Number 4 standard sieve has four openings per lineal inch (or 16 openings per square inch), whereas the ¼-inch sieve has a sieve opening of ¼ inch.

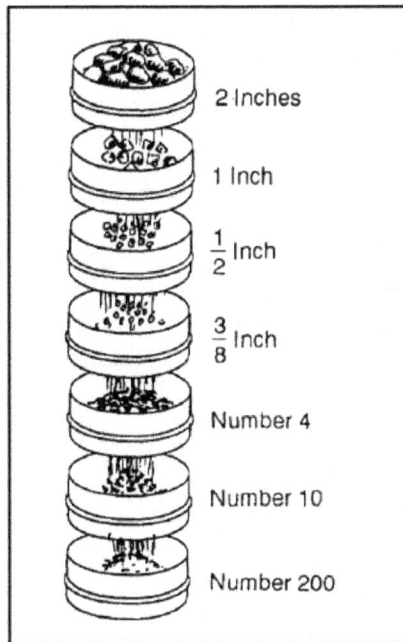

Figure 4-3. Dry sieve analysis

4-35. The practical lower limit for the use of sieves is the Number 200 sieve, with 0.074-millimeter-square openings. In some instances, determining the distribution of particle sizes below the Number 200 sieve is desirable, particularly for frost susceptibility determination. This may be done by a process known as wet mechanical analysis, which employs the principle of sedimentation. Grains of different sizes fall through a liquid at different velocities. The wet mechanical analysis is not a normal field laboratory test. It is not particularly important in military construction, except that the percentage of particles finer than 0.02 mm has a direct bearing on the susceptibility of soil to frost action. A field method for performing a wet mechanical analysis for the determination of the percentage of material finer than 0.02 mm is given in Technical Manual (TM) 5-530. The procedure is called decantation.

4-36. The procedures that have been described above are frequently combined to give a more complete picture of grain-size distribution. The procedure is then designated as a combined mechanical analysis.

4-37. Other methods based on sedimentation are frequently used in soils laboratories, particularly to determine particle distribution below the Number 200 sieve. One such method is the hydrometer analysis. A complete picture of grain-size distribution is frequently obtained by a combined sieve and hydrometer analysis. This method is described in TM 5-530 (section V).

Reports

4-38. Test results may be recorded in one of the following forms:

- Tabular (see figure 4-4, page 4-8).
- Graphic (see figure 4-5, page 4-9).

The form in this publication is obsolete. See http://www.apd.army.mil for current form.

SIEVE ANALYSIS DATA					1. DATE STARTED 22 FEB 91	
2. PROJECT BRAVO AIRFIELD			3. EXCAVATION 1+00		4. DATE COMPLETED 28 FEB 91	
5. SAMPLE DESCRIPTION LIGHT BROWN SANDY SOIL					6. SAMPLE NUMBER 1A	
					7. PREWASHED (x one) XX YES ___ NO	
8 ORIGINAL SAMPLE WEIGHT 2459			9. + #200 SAMPLE WEIGHT 2359		10 −#200 SAMPLE WEIGHT 100	
11. SIEVE SIZE	12. WEIGHT OF SIEVE	13. WEIGHT OF SIEVE + SAMPLE	14 WEIGHT RETAINED	15 CUMULATIVE WEIGHT RETAINED	16 PERCENT RETAINED	17. PERCENT PASSING
1½	202					
1	231					
½	210	210		0	0	100.0
¼	230	624			16.0	84.0
#4	205	332	127		5.2	78.8
#8	225	466		987	19.0	59.8
#20		612		1384	16.2	43.6
#60	235		346	1730	14.1	29.5
#100	250		362	2092	14.7	14.8
#200	260	515	255	2347	10.4	4.4

18. TOTAL WEIGHT RETAINED IN SIEVES (Sum Column 14)	2347	19 ERROR (8 - 23)
20. WEIGHT SIEVED THROUGH #200 (Weight in pan) 270-260	10	2459-2457 = 2
21. WASHING LOSS (8 - (9 + 10)) 2459-(2359+100)	0	
22. TOTAL WEIGHT PASSING #200 (20 + 10) 10+100	110	
23. TOTAL WEIGHT OF FRACTIONS (18 + 22)	2457	

24 REMARKS	25 ERROR (Percent)
USCS ___SP___ PERCENT-G _21.2_ PERCENT-S _74.4_ PERCENT-F _4.4_	$\dfrac{ERROR (19)}{ORIGINAL\ WT (8)} \times 100 =$ $\dfrac{2}{2459} \times 100 = .08$

26 TECHNICIAN Joe Blob PV2	27. COMPUTED BY (Signature) Joe Blob PV2	28 CHECKED BY (Signature) Fred Jones SSG

DD Form 1206, DEC 86   Previous editions are obsolete

Figure 4-4. Data sheet, example of dry sieve analysis

The form in this publication is obsolete. See http://www.apd.army.mil for current form.

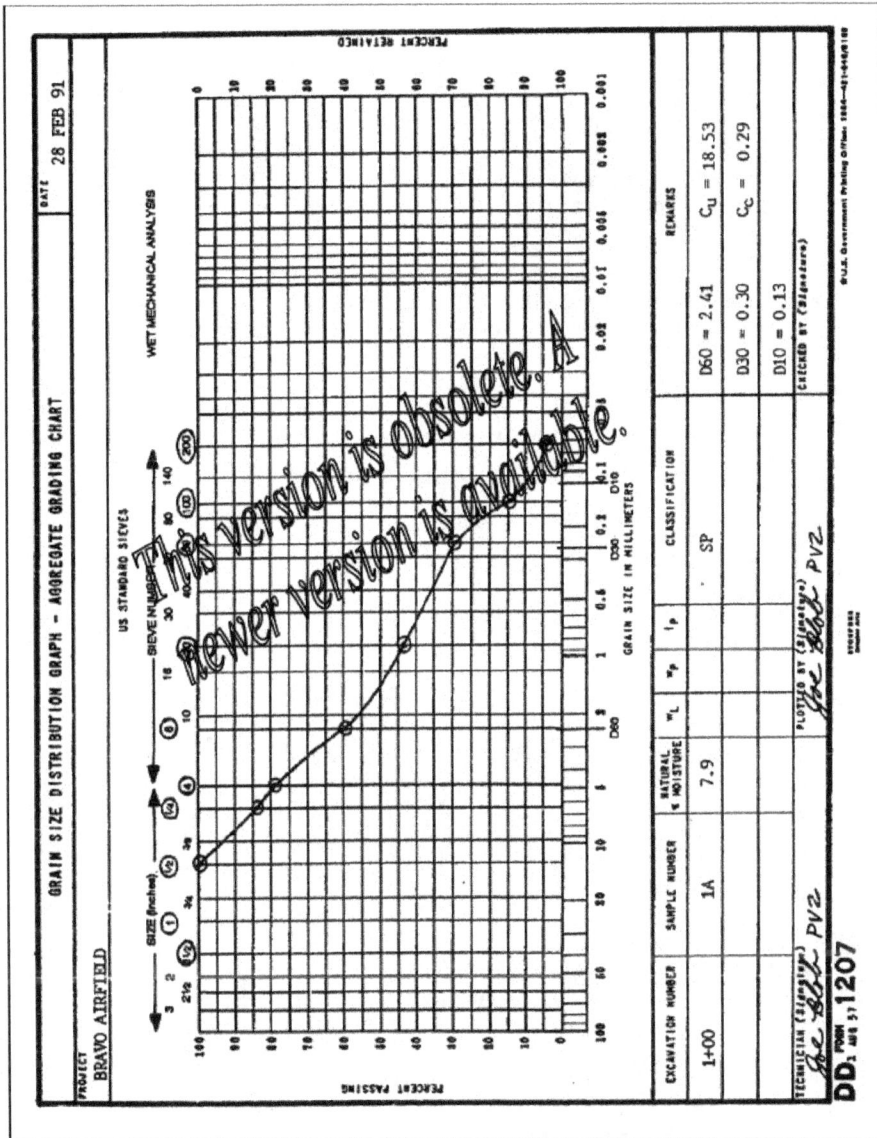

Figure 4-5. Grain-size distribution curve from sieve analysis

4-39. The tabular form is used most often on soil consisting predominantly of coarse particles. This method is frequently used when the soil gradation is being checked for compliance with a standard specification, such as for a gravel base or a wearing course.

4-40. The graphic form permits the plotting of a grain-size distribution curve. This curve affords ready visualization of the distribution and range of particle sizes. It is also particularly helpful in determining the soil classification and the soil's use as a foundation or construction material.

GRADATION

4-41. The distribution of particle sizes in a soil is known as its gradation. Gradation and other associated factors, primarily as applicable to coarse-grained soils, are discussed in the following paragraphs.

Effective Size

4-42. The grain size corresponding to 10 percent passing on a grain-size distribution curve (see figure 4-5, page 4-9) is called Hazen's effective size. It is designated by the symbol D_{10}. For the soil shown, D_{10} is 0.13 mm. The effective sizes of clean sands and gravels can be related to their permeability.

Coefficient of Uniformity

4-43. The coefficient of uniformity (C_u) is defined as the ratio between the grain diameter (in millimeters) corresponding to 60 percent passing on the curve (D_{60}) divided by the diameter of the 10 percent (D_{10}) passing. Hence, $C_u - D_{60}/D_{10}$

4-44. For the soil shown on figure 4-5—

$$D_{60} = 2.4\ mm\ and\ D_{10} = 0.13 mm$$

$$then\ C_u = 2.4/0.13 = 18.5$$

4-45. The uniformity coefficient is used to judge gradation.

Coefficient of Curvature

4-46. Another quantity that may be used to judge the gradation of a soil is the coefficient of curvature, designated by the symbol C_c.

$$C_c = \frac{(D_{30})^2}{(D_{60}\ x\ D_{10})}$$

4-47. D_{10} and D_{60} have been defined, while D_{30} is the grain diameter corresponding to 30 percent passing on the grain-size distribution curve. For the soil shown in figure 4-5:

$$D_{30} = 0.3\ mm$$

$$and\ C_c - \frac{(0.3)^2}{(2.4) x\ 0.13} = .29$$

Well-Graded Soils

4-48. A well-graded soil is defined as having a good representation of all particle sizes from the largest to the smallest (see figure 4-6), and the shape of the grain-size distribution curve is considered "smooth." In the USCS, well-graded gravels must have a C_u value > 4, and well-graded sands must have a C_u value > 6. For well-graded sands and gravels, a C_c value from 1 to 3 is required. Sands and gravels not meeting these conditions are termed poorly graded.

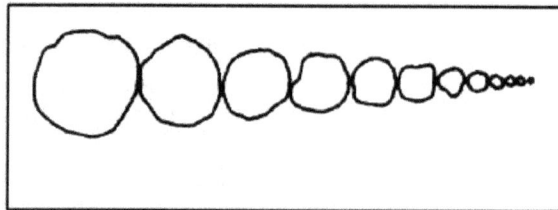

Figure 4-6. Well-graded soil

Poorly Graded Soils

4-49. The two types of poorly graded soils are—

- Uniformly graded.
- Gap-graded.

4-50. A uniformly graded soil consists primarily of particles of nearly the same size (see figure 4-7). A gap-graded soil contains both large and small particles, but the gradation continuity is broken by the absence of some particle sizes (see figure 4-8).

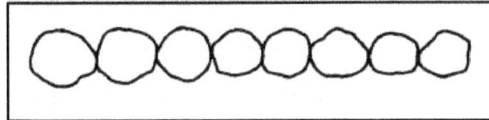

Figure 4-7. Uniformly graded soil

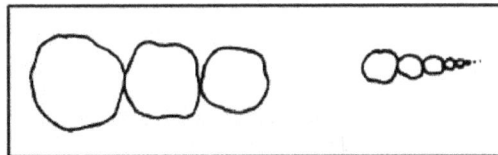

Figure 4-8. Gap-graded soil

4-51. Figure 4-9 shows typical examples of well-graded and poorly graded sands and gravels. Well-graded soils ((GW) and (SW) curves) would be represented by a long curve spanning a wide range of sizes with a constant or gently varying slope. Uniformly graded soils ((SP) curve) would be represented by a steeply sloping curve spanning a narrow range of sizes; the curve for a gap-graded soil ((GP) curve) flattens out in the area of the grain-size deficiency.

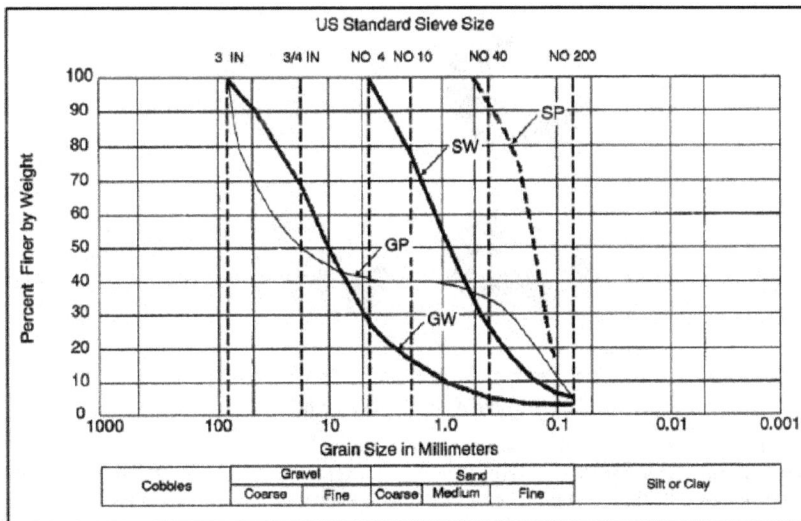

Figure 4-9. Typical grain-size distribution curves for well-graded and poorly graded soils

Bearing Capacity

4-52. Coarse materials that are well-graded are usually preferable for bearing from an engineering standpoint, since good gradation usually results in high density and stability. Specifications for controlling the percentage of the various grain-size groups required for a well-graded soil have been established for engineering performance and testing. By proportioning components to obtain a well-graded soil, it is possible to provide for maximum density. Such proportioning develops an "interlocking" of particles with smaller particles filling the voids between larger particles, making the soil stronger and more capable of supporting heavier loads. Since the particles are "form-fitted", the best load distribution downward will be realized. When each particle is surrounded and "locked" by other particles, the grain-to-grain contact is increased and the tendency for displacement of the individual grains is minimized.

PARTICLE SHAPE

4-53. The shape of individual particles affects the engineering characteristics of soils. Three principal shapes of soil grains have been recognized. They are—

- Bulky.
- Scalelike or platy.
- Needlelike.

Bulky

4-54. Bulky grains are nearly equal in all three dimensions. This shape characterizes sands and gravels and some silts. Bulky grains may be described by such terms as—

- Angular.
- Subangular.
- Subrounded.
- Well-rounded.

4-55. These four subdivisions of the bulky particle shape depend on the amount of weathering that has occurred (see figure 4-10). These subdivisions are discussed in the order of desirability for construction.

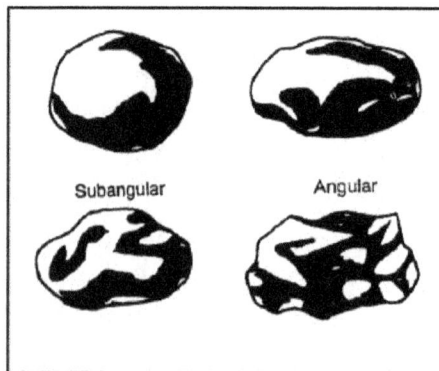

Figure 4-10. Bulky grains

4-56. Angular particles are particles that have recently been broken up. They are characterized by jagged projections, sharp ridges, and flat surfaces. The interlocking characteristics of angular gravels and sands generally make them the best materials for construction. Such particles are seldom found in nature because weathering processes normally wear them down in a relatively short time. Angular material may be produced artificially by crushing, but because of the time and equipment required for such an operation, natural materials with other grain shapes are frequently used.

4-57. Subangular particles have been weathered until the sharper points and ridges of their original angular shape have been worn off. The particles are still very irregular in shape with some flat surfaces and are excellent for construction.

4-58. Subrounded particles are those on which weathering has progressed even further. Still somewhat irregular in shape, they have no sharp corners and few flat areas. Subrounded particles are frequently found in stream beds. They may be composed of hard, durable particles that are adequate for most construction needs.

4-59. Rounded particles are those in which all projections have been removed and few irregularities in shape remain. The particles approach spheres of varying sizes. Rounded particles are usually found in or near stream beds, beaches, or dunes. Perhaps the most extensive deposits exist at beaches where repeated wave action produces almost perfectly rounded particles that may be uniform in size. Rounded particles also exist in arid environments due to wind action and the resulting abrasion between particles. They are not desirable for use in asphalt or concrete construction until the rounded shape is altered by crushing.

Platy

4-60. Platy grains are extremely thin compared to their length and width. They have the general shape of a flake of mica or a sheet of paper. Some coarse particles, particularly those formed by the mechanical breakdown of mica, are flaky or scalelike in shape. However, most particles that fall in the range of clay sizes, including the so-called clay minerals, have this characteristic shape. As will be explained in more detail later, the presence of these extremely small platy grains is generally responsible for the plasticity of clay. This type of soil is also highly compressible under static load.

Needlelike

4-61. These grains rarely occur.

STRUCTURE

4-62. Soils have a three-phase composition, the principal ingredients being the soil particles, water, and air. Organic materials are also found in the surface layer of most soils. Basic concepts regarding volume and weight relationships in a solid mass are shown in figure 4-11. These relationships form the basis of soil testing, since they are used in both quantitative and qualitative reporting of soils. It must be recognized that the diagram merely represents soil mass for studying the relationships of the terms to be discussed. All void and solid volumes cannot be segregated as shown.

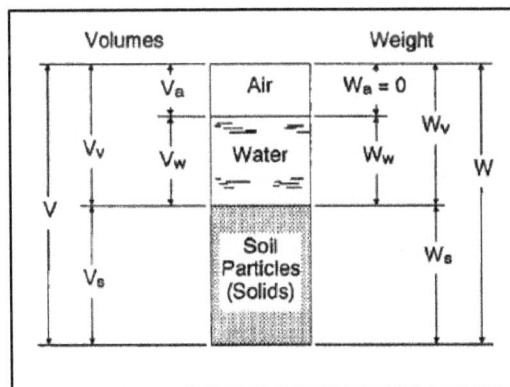

Figure 4-11. Volume-weight relationships of a soil mass

Specific Gravity

4-63. The specific gravity, designated by the symbol G, is defined as the ratio between the weight per unit volume of the material at a stated temperature (usually 20 degrees Celsius (C)) and the weight per unit volume of water.

$$Specific\ gravity = \frac{weight\ of\ sample\ in\ air\ (grams)}{weight\ of\ sample\ in\ air\ (grams)\ -\ weight\ of\ sample\ in\ water\ (grams)}$$

4-64. Test procedures are contained in TM 5-530. The specific gravity of the solid substance of most inorganic soils varies between 2.60 and 2.80. Tropical iron-rich laterite soils generally have a specific gravity of 3.0 or more. Clays can have values as high as 3.50. Most minerals, of which the solid matter of soil particles is composed, have a specific gravity greater than 2.60. Therefore, smaller values of specific gravity indicate the possible presence of organic matter.

Volume Ratios

4-65. The total volume (V) of a soil mass consists of the volume of voids (V_v) and the volume of solids (V_s). The volume of voids in turn consists of the volume of air (V_s), and the volume of water (V_w) (see figure 4-11, page 4-13). The most important volume ratio is the void ratio (e). It is expressed:

$$e = \frac{V_v}{V_v}$$

4-66. The volume of solids is the ratio of the dry weight (W_d) of a soil mass, in pounds, to the product of its specific gravity (G) and the unit weight of water (62.4 pounds per cubic foot (pcf). It is expressed:

$$V_s = \frac{W_d\ (lb)}{62.4\ x\ G}$$

4-67. The volume of the water is the ratio of the weight of the water (Ww), in pounds, to the unit weight of the water. It is expressed:

$$V_w = \frac{W_w}{62.4}$$

4-68. The degree of saturation (S) expresses the relative volume of water in the voids and is always expressed as a percentage. It is expressed:

$$S = \frac{V_w}{V_v}\ x\ 100\ percent$$

4-69. A soil is saturated if S equals 100 percent, which means all void volume is filled with water.

Weight Ratios

4-70. The total weight (W) of a soil mass consists of the weight of the water (W_w) and the weight of the solids (W_s), the weight of the air being negligible. Weight ratios widely used in soil mechanics are moisture content, unit weight, dry unit weight, and wet unit weight.

4-71. Moisture content (w) expressed as a percentage is the ratio of the weight of the water to the weight of the solids. It is expressed:

$$w\ (percent) = \frac{W_w}{W_s}\ x\ 100$$

4-72. The moisture content may exceed 100 percent. By definition, when a soil mass is dried to constant weight in an oven maintained at a temperature of 105 ± 5 degrees C, $W_w = 0$, and the soil is said to be oven dry or dry. If a soil mass is cooled in contact with the atmosphere, it absorbs some water. This water absorbed from the atmosphere is called hygroscopic moisture. TM 5-530 contains testing procedures for determining moisture content.

4-73. Unit weight (γ) is the expression given to the weight per unit volume of a soil mass. It is expressed:

$$\gamma = \frac{W}{V}$$

4-74. In soils terminology, the terms "unit weight" and "density" are used interchangeably.

4-75. Wet unit weight (γ_m), also expressed as wet density, is the term used if the moisture content is anything other than zero. The wet unit weight of natural soils varies widely. Depending on denseness and gradation, a sandy soil may have a wet unit weight or density of 115 to 135 pcf. Some very dense glacial tills may have wet unit weights as high as 145 pcf. Wet unit weights for most clays range from 100 to 125 pcf. Density of soils can be greatly increased by compaction during construction. In foundation problems, the density of a soil is expressed in terms of wet unit weight.

4-76. Dry unit weight (γ_d), also expressed as dry density, is the term used if the moisture content is zero. Since no water is present and the weight of air is negligible, it is written:

$$\gamma_d = \frac{W_s}{V}$$

4-77. Dry unit weight normally is used in construction problems. The general relationship between wet unit weight and dry unit weight is expressed:

$$\gamma_d = \frac{\gamma}{1 + W/100}$$

4-78. A numerical example of volume-weight relationships follows:

GIVEN:

A soil mass with:

wet unit weight = 125 pcf
moisture content = 18 percent
specific gravity = 2.65
volume = 1 cubic foot (cu ft)

FIND:

- γ_d dry unit weight
- (2) e, void ratio

SOLUTION:

- Find dry unit weight

$$\gamma_d = \frac{\gamma_m}{1 + W/100}$$

$$\gamma_d = \frac{125}{1 + 18/100}$$

$$= 125/1.18$$

$$= 105.9 \, pcf$$

- Find the void ratio using the formula:
e = V_v/V_s

$$V_s = W_d/62.4 \, G$$

$$V_s = 105.9/62.4 \, (2.65)$$

$$= 0.64 \, cu \, ft$$

If $V_v = V - V_s$, then
$V_v = 1.00 - 0.64 = 0.36$ cu ft

Thus, if $e = Vv/Vs$, substituting computed values, then

$$e = {0.36}/{0.64} = 0.56$$

4-79. The relationships discussed previously are used in calculations involved in soil construction work. They are used along with the necessary soil tests to classify and to help determine engineering characteristics of soil.

RELATIVE DENSITY

4-80. Use of the void ratio is not very effective in predicting the soil behavior of granular or unrestricted soils. More useful in this respect is the term relative density, expressed as D_r. Relative density is an index of the degree to which a soil has been compacted. Values range from 0 ($e = e_{max}$) to 1.0 ($e = e_{min}$). It is written:

$$D_r = \frac{e_{max} - e}{e_{max} - e_{min}} \; x \; 100$$

where—

e_n = the natural, in-place, void ratio

e_{max} = the void ratio in the loosest possible condition

e_{min} = the void ratio in the most dense condition possible.

4-81. The limiting ranges of e_{max} may be found by pouring the soil loosely into a container and determining its weight and volume. The limiting ranges of e_{min} may be found by tamping and shaking the soil until it reaches a minimum volume and recording its weight and volume at this point. Relative density is important for gravels and sands.

SOIL-MOISTURE CONDITIONS

4-82. Coarse-grained soils are much less affected by moisture than are fine-grained soils. Coarser soils have larger void openings and generally drain more rapidly. Capillarity is no problem in gravels having only very small amounts of fines mixed with them. These soils will not usually retain large amounts of water if they are above the groundwater table. Also, since the particles in sandy and gravelly soils are relatively large (in comparison to silt and clay particles), they are heavy in comparison to the films of moisture that might surround them. Conversely, the small, sometimes microscopic, particles of a fine-grained soil weigh so little that water within the voids has a considerable effect on them. The following phenomena are examples of this effect:

- Clays frequently undergo very large volume changes with variations in moisture content. Evidence of this can be seen in the shrinkage cracks that develop in a lake bed as it dries.
- Unpaved clay roads, although often hard when sunbaked, lose stability and turn into mud in a rainstorm.
- In general, the higher the water content of a clay or silt, the less is its strength and hence its bearing capacity.

4-83. These effects are very important to an engineer and are functions of changing water content. The Army's emphasis on early achievement of proper drainage in horizontal construction stems from these properties of cohesive soils.

Adsorbed Water

4-84. In general terms, adsorbed water is water that may be present as thin films surrounding the separate soil particles. When the soil is in an air-dry condition, the adsorbed water present is called hygroscopic moisture. Adsorbed water is present because the soil particles carry a negative electrical charge. Water is dipolar; it is attracted to the surface of the particle and bound to it (see figure 4-12). The water films are affected by the chemical and physical structure of the soil particle and its relative surface area. The relative surface area of a particle of fine-grained soil, particularly if it has a flaky or needlelike shape, is much greater than for coarse soils composed of bulky grains. The electrical forces that bind adsorbed water to the soil particle also are much greater. Close to the particle, the water contained in the adsorbed layer has properties quite different from ordinary water. In the portion of the layer immediately adjacent to the particle, the water may behave as a solid, while only slightly farther away it behaves as a viscous liquid. The total thickness of the adsorbed layer is very small, perhaps on the order of 0.00005 mm for clay soils. In coarse soils, the adsorbed layer is quite thin compared with the thickness of the soil particle. This, coupled with the fact that the contact area between adjacent grains is quite small, leads to the conclusion that the presence of the adsorbed water has little effect on the physical properties of coarse-grained soils. By contrast, for finer soils and particularly in clays, the adsorbed water film is thick in comparison with particle size. The effect is very pronounced when the particles are of colloidal size.

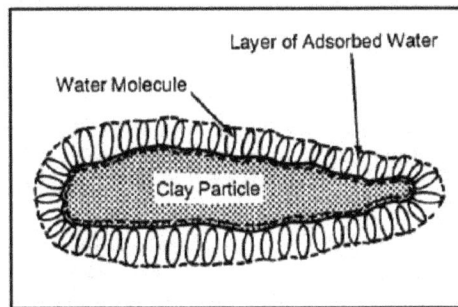

Figure 4-12. Layer of adsorbed water surrounding a soil particle

Plasticity and Cohesion

4-85. Two important aspects of the engineering behavior of fine-grained soils are directly associated with the existence of adsorbed water films. These aspects are plasticity and cohesion.

4-86. Plasticity is the ability of a soil to deform without cracking or breaking. Soils in which the adsorbed films are relatively thick compared to particle size (such as clays) are plastic over a wide range of moisture contents. This is presumably because the particles themselves are not in direct contact with one another. Plastic deformation can take place because of distortion or shearing of the outside layer of viscous liquid in the moisture films. Coarse soils (such as clean sands and gravels) are nonplastic. Silts also are essentially nonplastic materials, since they are usually composed predominantly of bulky grains; if platy grains are present, they may be slightly plastic.

4-87. A plasticity index (PI) is used to determine if soils are cohesive. Not all plastic soils are cohesive; soil is considered cohesive if its PI is 5. That is, they possess some cohesion or resistance to deformation because of the surface tension present in the water films. Thus, wet clays can be molded into various shapes without breaking and will retain these shapes. Gravels, sands, and most silts are not cohesive and are called cohesionless soils. Soils of this general class cannot be molded into permanent shape and have little or no strength when dry and unconfined. Some of these soils may be slightly cohesive when damp. This is attributed to what is sometimes called apparent cohesion, which is also due to the surface tension in the water films between the grains.

Clay Minerals and Base Exchange

4-88. The very fine (colloidal) particles of clay soils consist of clay minerals, which are crystalline in structure. These minerals are complex compounds of hydrous aluminum silicates and are important because their presence greatly influences a soil's physical properties. X rays have been used to identify several different kinds of clay minerals that have somewhat different properties. Two extreme types are kaolinite and montmorillonite. Both have laminated crystalline structures, but they behave differently. Kaolinite has a very rigid crystalline structure, while montmorillonite can swell by taking water directly into its lattice structure. Later, the flakes themselves may decrease in thickness as the water is squeezed out during drying; the flakes are thus subject to detrimental shrinkage and expansion. An example of this type of material is bentonite, largely made up of the montmorillonite type of clay mineral. Because of its swelling characteristics, bentonite is widely used commercially in the construction of slurry walls and temporary dam cores. Most montmorillonites have much thicker films of adsorbed water than do kaolinites. Kaolinites tend to shrink and swell much less than montmorillonites with changes in moisture content. In addition, the adsorbed water film may contain disassociated ions. For example, metallic cations, such as sodium, calcium, or magnesium, may be present. The presence of these cations also affects the physical behavior of the soil. A montmorillonite clay, for example, in which calcium cations predominate in the adsorbed layer may have properties quite different from a similar clay in which sodium cations predominate. The process of replacing cations of one type with cations of another type in the surface of the adsorbed layer is called base (or cation) exchange. It is possible to effect this replacement and thereby alter the physical properties of a clay soil. For example, the soil may swell, the plasticity may be reduced, or the permeability may be increased by this general process.

Capillary Phenomena

4-89. Capillary phenomena in soils are important for two reasons. First, water moves by capillary action into a soil from a free-water surface. This aspect of capillary phenomena is not discussed here but is covered in chapter 8. Second, capillary phenomena are closely associated with the shrinkage and expansion (swelling) of soils.

The capillary rise of water in small tubes is a common phenomenon, which is caused by surface tension (see figure 4-13). The water that rises upward in a small tube is in tension, hanging on the curved boundary between air and water (meniscus) as if from a suspending cable. The tensile force in the meniscus is balanced by a compressive force in the walls of the tube. Capillary phenomena in small tubes can be simply analyzed and equations derived for the radius of the curved meniscus, the capillary stress (force per unit of area), and the height of capillary rise (see h_c in figure 4-13). A soil mass may be regarded as being made up of a bundle of small tubes formed by the interconnected void spaces. These spaces form extremely irregular, tortuous paths for the capillary movement of water. An understanding of capillary action in soils is thus gained by analogy. Theoretical analyses indicate that maximum possible compressive pressure that can be exerted by capillary forces is inversely proportional to the size of the capillary openings.

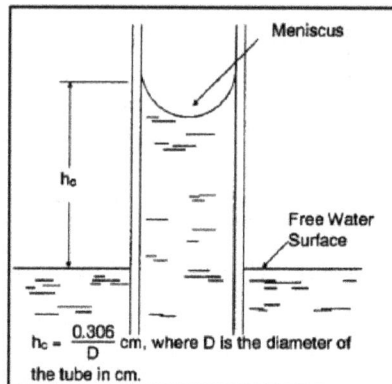

Figure 4-13. Capillary rise of water in small tubes

Shrinkage

4-90. Many soils undergo a very considerable reduction in volume when their moisture content is reduced. The effect is most pronounced when the moisture content is reduced from that corresponding to complete saturation to a very dry condition. This reduction in volume is called shrinkage and is greatest in clays. Some of these soils show a reduction in volume of 50 percent or more while passing from a saturated to an oven-dry condition. Sands and gravels, in general, show very little or no change in volume with change in moisture content. An exception to this is the bulking of sands, which is discussed below. The shrinkage of a clay mass may be attributed to the surface tension existing in the water films created during the drying process. When the soil is saturated, a free-water surface exists on the outside of the soil mass, and the effects of surface tension are not important. As the soil dries out because of evaporation, the surface water disappears and innumerable meniscuses are created in the voids adjacent to the surface of the soil mass. Tensile forces are created in each of these boundaries between water and air. These forces are accompanied by compressive forces that, in a soil mass, act on the soil structure. For the typical, fairly dense structure of a sand or gravel, the compressive forces are of little consequence; very little or no shrinkage results. In fine-grained soils, the soil structure is compressible and the mass shrinks. As drying continues, the mass attains a certain limiting volume. At this point, the soil is still saturated. The moisture content at this stage is called the shrinkage limit. Further drying will not cause a reduction in volume but may cause cracking as the meniscuses retreat into the voids. In clay soils, the internal forces created during drying may become very large. The existence of these forces also principally accounts for the rocklike strength of a dried clay mass. Both silt and clay soils may be subject to detrimental shrinkage with disastrous results in some practical situations. For example, the uneven shrinkage of a clay soil may deprive a concrete pavement of the uniform support for which it is designed; severe cracking or failure may result when wheel loads are applied to the pavement.

Swelling and Slaking

4-91. If water is again made available to a still-saturated clay soil mass that has undergone shrinkage, water enters the soil's voids from the outside and reduces or destroys the internal forces previously described. Thus, a clay mass will absorb water and expand or swell. If expansion is restricted, as by the weight of a concrete pavement, the expansion force may be sufficient to cause severe pavement cracking. If water is made available to the soil after it has dried below the shrinkage limit, the mass generally disintegrates or slakes. Slaking may be observed by putting a piece of dry clay into a glass of water. The mass will fall completely apart, usually in a matter of minutes. Construction problems associated with shrinkage and expansion are generally solved by removing the soils that are subject to these phenomena or by taking steps to prevent excessive changes in moisture content.

Bulking of Sands

4-92. Bulking is a phenomenon that occurs in dry or nearly dry sand when a slight amount of moisture is introduced into the soil and the soil is disturbed. Low moisture contents cause increased surface tension, which pulls the grains together and inhibits compaction. As a result, slightly moist sands can have lower compacted densities than totally dry or saturated sands. Commonly in sands, this problem is made worse because a slight addition of moisture above the totally dry state increases the sliding coefficient of the particles. The U-shaped compaction curve, with characteristic free-draining soils (sands and gravels), illustrates the concept of bulking. Adding sufficient water to saturate the sand eliminates surface tension, and the sand can be compacted to its densest configuration (see figure 4-14, page 4-20).

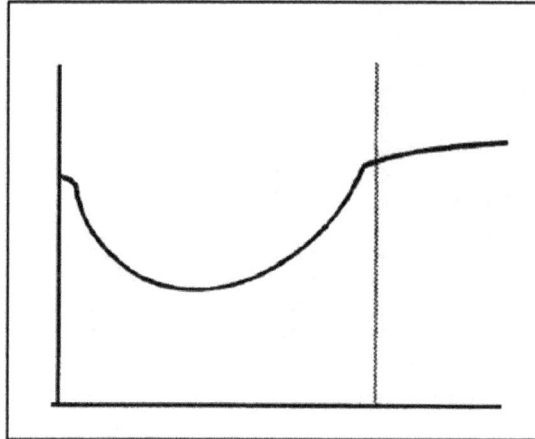

Figure 4-14. U-shaped compaction curve

CONSISTENCY (ATTERBERG) LIMITS

4-93. A fine-grained soil can exist in any one of several different states, depending on the amount of water in the soil. The boundaries between these different soil states are moisture contents called consistency limits. They are also called Atterberg limits after the Swedish soil scientist who first defined them in 1908. The shrinkage limit is the boundary between the semisolid and solid states. The plastic limit (PL) is the boundary between the semisolid and plastic states. The liquid limit (LL) is the boundary between the plastic and liquid states. Above the LL, the soil is presumed to behave as a liquid. The numerical difference between the LL and the PL is called the PI and is the range of moisture content over which the soil is in a plastic condition. The Atterberg limits are important index properties of fine-grained soils. They are particularly important in classification and identification. They are also widely used in specifications to control the properties, compaction, and behavior of soil mixtures.

TEST PROCEDURES

4-94. The limits are defined by more or less arbitrary and standardized test procedures that are performed on the portion of the soil that passes the Number 40 sieve. This portion of soil is sometimes called the soil binder. TM 5-530 contains detailed test procedures to be used in determining the LL and the PL. The tests are performed with the soil in a disturbed condition.

Liquid Limit

4-95. The LL (or wL) is defined as the minimum moisture content at which a soil will flow upon application of a very small shearing force. With only a small amount of energy input, the soil will flow under its own weight. In the laboratory, the LL is usually determined by use of a mechanical device (see figure 4-15). The detailed testing procedure is described in TM 5-530.

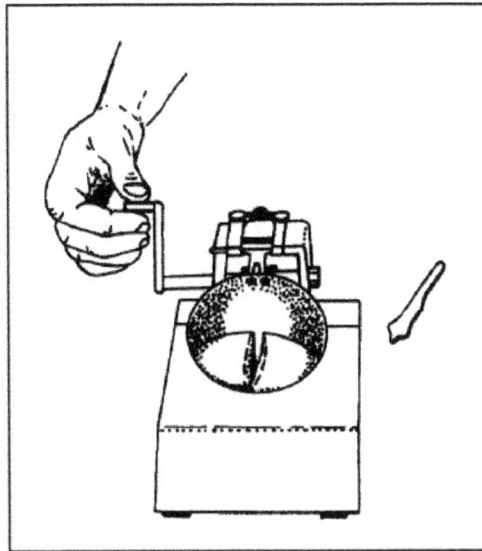

Figure 4-15. Liquid limit test

Plastic Limit

4-96. The PL (or w_p) is arbitrarily defined as the lowest moisture content at which a soil can be rolled into a thread ⅛ inch in diameter without crushing or breaking. If a cohesive soil has a moisture content above the PL, a thread may be rolled to less than ⅛ inch in diameter without breaking. If the moisture content is below the PL, the soil will crumble when attempts are made to roll it into ⅛-inch threads. When the moisture content is equal to the PL, a thread can be rolled out by hand to ⅛ inch in diameter; then it will crumble or break into pieces ⅛ to ⅜ inch long when further rolling is attempted. Some soils (for example, clean sands) are nonplastic and the PL cannot be determined. A clean sand or gravel will progress immediately from the semisolid to the liquid state.

PLASTICITY INDEX

4-97. The PI (or I_p) of a soil is the numerical difference between the LL and the PL. For example, if a soil has a LL of 57 and a PL of 23, then the PI equals 34 (PI = LL - PL). Sandy soils and silts have characteristically low PIs, while most clays have higher values. Soils that have high PI values are highly plastic and are generally highly compressible and highly cohesive. The PI is inversely proportional to the permeability of a soil. Soils that do not have a PL, such as clean sands, are reported as having a PI of zero.

4-98. Relationships between the LLs and PIs of many soils were studied by Arthur Casagrande of Harvard University and led to the development of the plasticity chart. The chart's development and use in classifying and identifying soils and selecting the best of the available soils for a particular construction application are discussed in chapter 5.

This page intentionally left blank.

Chapter 5

Soil Classification

Early attempts to classify soils were based primarily on grain size. These are the textural classification systems. In 1908, a system that recognized other factors was developed by Atterberg in Sweden and primarily used for agricultural purposes. Somewhat later, a similar system was developed and used by the Swedish Geotechnical Commission. In the United States, the Bureau of Public Roads System was developed in the late twenties and was in widespread use by highway agencies by the middle thirties. This system has been revised over time and is widely used today. The Airfield Classification System was developed by Professor Arthur Casagrande of Harvard University during World War II. A modification of this system, the USCS, was adopted by the US Army Corps of Engineers and the Bureau of Reclamation in January 1952. A number of other soil classification systems are in use throughout the world, and the military engineer should be familiar with the most common ones.

The principal objective of any soil classification system is predicting the engineering properties and behavior of a soil based on a few simple laboratory or field tests. Laboratory and/or field test results are then used to identify the soil and put it into a group that has soils with similar engineering characteristics. Probably no existing classification system completely achieves the stated objective of classifying soils by engineering behavior because of the number of variables involved in soil behavior and the variety of soil problems encountered. Considerable progress has been made toward this goal, particularly in relationship to soil problems encountered in highway and airport engineering. Soil classification should not be regarded as an end in itself but as a tool to further your knowledge of soil behavior.

SECTION I - UNIFIED SOIL CLASSIFICATION SYSTEM

SOIL CATEGORIES

5-1.   Soils seldom exist in nature separately as sand, gravel, or any other single component. Usually they occur as mixtures with varying proportions of particles of different sizes. Each component contributes its characteristics to the mixture. The USCS is based on the characteristics of the soil that indicate how it will behave as a construction material.

5-2.   In the USCS, all soils are placed into one of three major categories. They are—

- Coarse-grained.
- Fine-grained.
- Highly organic.

5-3.   The USCS further divides soils that have been classified into the major soil categories by letter symbols, such as—

- S for sand.
- G for gravel

- M for silt.
- C for clay.

5-4. A soil that meets the criteria for a sandy clay would be designated (SC). There are cases of borderline soils that cannot be classified by a single dual symbol, such as GM for silty gravel. These soils may require four letters to fully describe them. For example, (SM-SC) describes a sand that contains appreciable amounts of silt and clay.

COARSE-GRAINED SOILS

5-5. Coarse-grained soils are defined as those in which at least half the material is retained on a Number 200 sieve. They are divided into two major divisions, which are—

- Gravels.
- Sands.

5-6. A coarse-grained soil is classed as gravel if more than half the coarse fraction by weight is retained on a Number 4 sieve. The symbol G is used to denote a gravel and the symbol S to denote a sand. No clearcut boundary exists between gravelly and sandy soils; as far as soil behavior is concerned, the exact point of division is relatively unimportant. Where a mixture occurs, the primary name is the predominant fraction and the minor fraction is used as an adjective. For example, a sandy gravel would be a mixture containing more gravel than sand by weight. Additionally, gravels are further separated into either coarse gravel or fine gravel with the ¾-inch sieve as the dividing line and sands are either coarse, medium, or fine with the Number 10 and Number 40 sieves, respectively. The coarse-grained soils may also be further divided into three groups on the basis of the amount of fines (materials passing a Number 200 sieve) they contain. These amounts are—

- Less than 5 percent.
- More than 12 percent.
- Between 5 and 12 percent.

5-7. Coarse-grained soils with less than 5 percent passing the Number 200 sieve may fall into the following groups:

- (GW) is well-graded gravels and gravel-sand mixtures with little or no fines. The presence of the fines must not notably change the strength characteristics of the coarse-grained fraction and must not interfere with its free- draining characteristics.
- (SW) is well-graded sands and gravelly sands with little or no fines. The grain-size distribution curves for (GW) and (SW) in figure 4-9, page 4-11, are typical of soils included in these groups. Definite laboratory classification criteria have been established to judge if the soil is well-graded (see chapter 4). For the (GW) group, the C_u must be greater than 4; for the (SW) group, greater than 6. For both groups, the C_c must be between 1 and 3.
- (GP) is poorly graded gravels and sandy gravel mixtures with little or no fines.
- (SP) is poorly graded sands and gravelly sands with little or no fines. These soils do not meet the gradation requirements established for the (GW) and (SW) groups. The grain-size distribution curve marked (GP) in figure 4-9, is typical of a poorly graded gravel-sand mixture, while the curve marked (SP) is a poorly graded (uniform) sand.

5-8. Coarse-grained soils containing more than 12 percent passing the Number 200 sieve fall into the following groups:

- (GM) is silty gravel and poorly graded gravel/sand-silt mixtures.
- (SM) is silty sands and poorly graded sand-silt mixtures.

5-9. Gradation of these materials is not considered significant. For both of these groups, the Atterberg limits must plot below the A-line of the plasticity chart shown in figure 5-1. A dual symbol system allows more precise classification of soils based on gradation and Atterberg limits.

- (GC) is clayey gravels and poorly graded gravel-sand-clay mixtures.
- (SC) is clayey sands and poorly graded sand-clay mixtures.

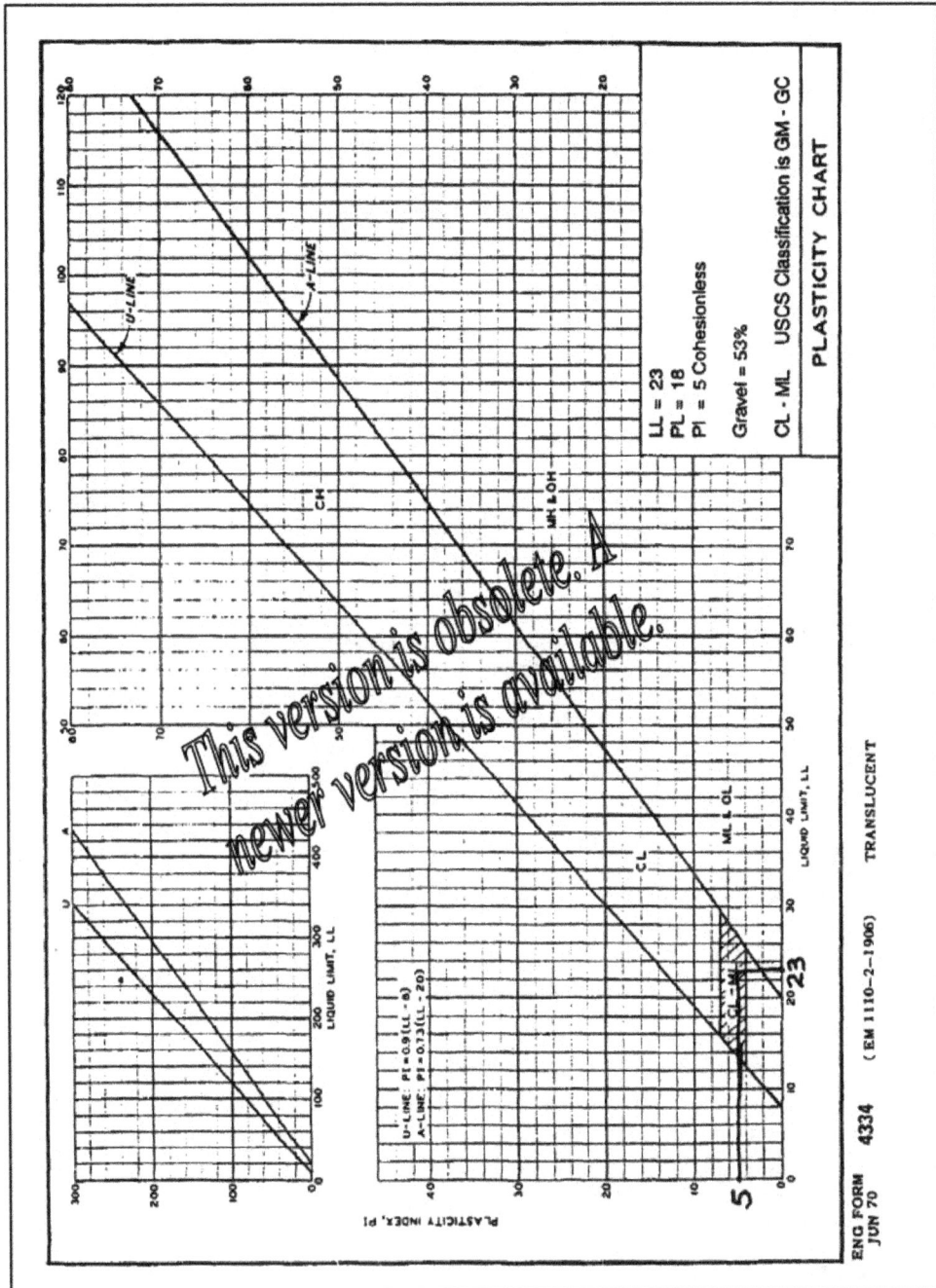

Figure 5-1. Sample plasticity chart

5-10. Gradation of these materials is not considered significant. For both of these groups, the Atterberg limits plot above the A-line.

5-11. The use of the symbols M and C is based on the plasticity characteristics of the material passing the Number 40 sieve. The LL and PI are used in determining the plasticity of the fine materials. If the plasticity chart shown in figure 5-1, page 5-3, is analyzed with the LL and PI, it is possible to determine if the fines are clayey or silty. The symbol M is used to indicate that the material passing the Number 40 sieve is silty in character. M usually designates a fine-grained soil of little or no plasticity. The symbol C is used to indicate that the binder soil is clayey in character. A dual symbol system allows more precise classification of soils based on gradation and Atterberg limits.

5-12. For example, coarse-grained soils with between 5 and 12 percent of material passing the Number 200 sieve, and which meet the criteria for well-graded soil, require a dual symbol, such as—

- (GW-GM).
- (GP-GM).
- (GW-GC).
- (GP-GC).
- (SW-SC).
- (SW-SM).
- (SP-SC).
- (SP-SM).

5-13. Similarly, coarse-grained soils containing more than 12 percent of material passing the Number 200 sieve, and for which the limits plot in the hatched portion of the plasticity chart (see figure 5-1), are borderline between silt and clay and are classified as (SM-SC) or (GM-GC).

5-14. In rare instances, a soil may fall into more than one borderline zone. If appropriate symbols were used for each possible classification, the result would be a multiple designation using three or more symbols. This approach is unnecessarily complicated. It is considered best to use only a double symbol in these cases, selecting the two believed most representative of probable soil behavior. If there is doubt, the symbols representing the poorer of the possible groupings should be used. For example, a well-graded sandy soil with 8 percent passing the Number 200 sieve, with an LL of 28 and a PI of 9, would be designated as (SW-SC). If the Atterberg limits of this soil were such as to plot in the hatched portion of the plasticity chart (for example, an LL of 20 and a PI of 5), the soil could be designated either (SW-SC) or (SW-SM), depending on the judgment of the soils technician.

FINE-GRAINED SOILS

5-15. Fine-grained soils are those in which more than half the material passes a Number 200 sieve. The fine-grained soils are not classified by grain size but according to plasticity and compressibility. Laboratory classification criteria are based on the relationship between the LL and the PI, determined from the plasticity chart shown in figure 5-1. The chart indicates two major groupings of fine-grained soils. These are—

- The L groups, which have LLs < 50.
- The H groups, which have LLs ≥ of 50.

5-16. The symbols L and H represent low and high compressibility, respectively. Fine-grained soils are further divided based on their position above or below the A-line of the plasticity chart.

5-17. Typical soils of the (ML) and (MH) groups are inorganic silts. Those of low plasticity are in the (ML) group; others are in the (MH) group. Atterberg limits of these soils all plot below the A-line. The (ML) group includes—

- Very fine sands.
- Rock flours.
- Silty or clayey fine sands with slight plasticity.

5-18. Micaceous and diatomaceous soils generally fall into the (MH) group but may extend into the (ML) group with LLs < 50. The same statement is true of certain types of kaolin clays, which have low plasticity. Plastic silts fall into the (MH) group.

5-19. In (CL) and (CH) groups, the C stands for clay, with L and H denoting low or high compressibility. These soils plot above the A-line and are principally inorganic clays. The (CL) group includes gravelly clays, sandy clays, silty clays, and lean clays. In the (CH) group are inorganic clays of high plasticity, including fat clays, the gumbo clays of the southern United States, volcanic clays, and bentonite. The glacial clays of the northern United States cover a wide band in the (CL) and (CH) groups.

5-20. Soils in the (OL) and (OH) groups are characterized by the presence of organic matter, hence the symbol O. The Atterberg limits of these soils generally plot below the A-line. Organic silts and organic silt clays of low plasticity fall into the (OL) group, while organic clays plot in the (OH) zone of the plasticity chart. Many organic silts, silt-clays, and clays deposited by rivers along the lower reaches of the Atlantic seaboard have LLs between 40 and 100 and plot below the A-line. Peaty soils may have LLs of several hundred percent and their Atterberg limits generally plot below the A-line.

5-21. Fine-grained soils having limits that plot in the shaded portion of the plasticity chart are given dual symbols (for example, (CL-ML)). Several soil types exhibiting low plasticity plot in this general region on the chart and no definite boundary between silty and clayey soils exists.

HIGHLY ORGANIC SOILS

5-22. A special classification, (Pt), is reserved for the highly organic soils, such as peat, which have many undesirable engineering characteristics. No laboratory criteria are established for these soils, as they generally can be easily identified in the field by their distinctive color and odor, spongy feel, and frequently fibrous texture. Particles of leaves, grass, branches, or other fibrous vegetable matter are common components of these soils.

5-23. Table 5-1, page 5-6, and table 5-2, page 5-8, are major charts which present information applicable to the USCS and procedures to be followed in identifying and classifying soils under this system. Principal categories shown in the chart include—

- Soil groups, soil group symbols, and typical soil names.
- Laboratory classification criteria.
- Field identification procedures.
- Information for describing soils.

5-24. These charts are valuable aids in soil classification problems. They provide a simple systematic means of soil classification.

LABORATORY TESTING

5-25. Usually soil samples are obtained during the soil survey and are tested in the laboratory to determine test properties for classifying the soils. The principal tests are—

- Mechanical analysis.
- Liquid limit.
- Plastic limit.

5-26. These tests are used for all soils except those in the (Pt) group. With the percentages of gravel, sand, and fines and the LL and PI, the group symbol can be obtained from the chart in table 5-2 by reading the diagram from top to bottom. For the gravels and sands containing 5 percent (or less) fines, the shape of the grain-size distribution curve can be used to establish whether the material is well-graded or poorly graded. For the fine-grained soils, it is necessary to plot the LL and PI in the drawing on figure 5-1, to establish the proper symbol. Organic silts or clays (ML) and (MH) are subjected to LL and PL tests before and after oven drying. An organic silt or clay shows a radical drop in these limits as a result of oven drying. An inorganic soil shows a slight drop that is not significant. Where there is an appreciable drop, the predrying values should be used when the classification is determined from table 5-2.

Table 5-1. Unified soil classification (including identification and description)

Major Division (1)	(2)	Group Symbols (3)	Typical Names (4)	Field Identification Procedures (Excluding particles larger than 3 inches and basing fractions on estimated weights) (5)			Information Required for Describing Soils (6)
Coarse grained soils — More than half of the material is larger than No 200 sieve size	**Gravels** — More than half of coarse fraction is larger than No 4 sieve size						For undisturbed soils, add information on stratification, degree of compactness, cementation, moisture conditions, and drainage characteristics.
	Clean gravels (little or no fines)	GW	Well-graded gravels, gravel-sand mixtures, little or no fines	Wide range in grain sizes and substantial amounts of all intermediate particle sizes			Give typical name; indicate approximate percentages of sand and gravel, maximum size; angularity, surface condition, and hardness of the coarse grains; local or geologic name and other pertinent descriptive information; and symbol in parentheses.
		GP	Poorly graded gravels or gravel-sand mixtures, little or no fines	Predominantly one size or a range of sizes with some intermediate particle sizes missing			
	Gravels with fines (appreciable amount of fines)	GM	Silty gravels, gravel-sand-silt mixture	Nonplastic fines or fines with low plasticity (for identification procedures, see ML below)			Example: Silty sand, gravelly; about 20% hard, angular gravel particles 1/2 inch maximum size; rounded and subangular sand grains, coarse to fine; about 15% nonplastic fines with low dry strength; well compacted and moist in place; alluvial sand; (SM)
		GC	Clayey gravels, gravel-sand-clay mixtures	Plastic fines (for identification procedures, see CL below)			
	Sands — More than half of coarse fraction is smaller than No 4 sieve size						For undisturbed soils, add information on structure, stratification, consistency in undisturbed and remolded states, and moisture and drainage conditions.
	Clean sands (little or no fines)	SW	Well-graded sands, gravelly sands, little or no fines	Wide range in grain size and substantial amounts of all intermediate particle sizes			Give typical name; indicate degree and character of plasticity; amount and maximum size of coarse grains; color in wet condition; odor, if any; local or geologic name and other pertinent descriptive information; and symbol in parentheses
		SP	Poorly graded sands or gravelly sands, little or no fines	Predominantly one size or a range of sizes with some intermediate sizes missing			
	Sands with fines (appreciable amount of fines)	SM	Silty sands, sand-silt mixture	Nonplastic fines or fines with low plasticity (for identification procedures, see ML below)			Example: Clayey silt, brown; slightly plastic; small percentage of fine sand; numerous vertical root holes; firm and dry in place; loess; (ML)
		SC	Clayey sands, sand-clay mixtures	Plastic fines (for identification procedures see CL below)			
Fine grained soils — More than half of the material is smaller than No 2 sieve size				**Identification procedures on fraction smaller than No 40 sieve size**			
				Dry strength (crushing characteristics)	Dilatancy (reaction to shaking)	Toughness (consistency near PL)	
Silts and clays LL < 50		ML	Inorganic silts and very fine sands, rock flour, silty or clayey fine sands or clayey silts with slight plasticity	None to slight	Quick to slow	None	
		CL	Inorganic clays of low to medium plasticity, gravelly clays, sandy clays, silty clays, lean clays	Medium to high	None to very slow	Medium	
		OL	Organic silts and organic silty clays of low plasticity	Slight to medium	Slow	Slight	
Silts and clays LL > 50		MH	Inorganic silts, micaceous or diatomaceous fine sandy or silty soils, elastic silts	Slight to medium	Slow to none	Slight to medium	
		CH	Inorganic clays of high plasticity, fat clays	High to very high	None	High	
		OH	Organic clays of medium to high plasticity, organic silts	Medium to high	None to very slow	Slight to medium	
Highly organic soils		Pt	Peat and other highly organic soils	Readily identified by color, odor, spongy feel, and frequently by fibrous texture			

For visual classification, the ½ inch size may be used as equivalent to the No 4 sieve size.

(The No 200 sieve size is about the smallest particle visible to the naked eye)

(1) Boundary classifications: soils possessing characteristics of two groups are designated by combinations of group symbols. For example (GW –OC): well-graded gravel-sand mixture with clay binder

(2) All sieve sizes on this chart are US standard.

Table 5-1. Unified soil classification (including identification and description)

Field Identification Procedures for Fine-Grained Soils or Fractions

These procedures are to be performed on the minus No 40 sieve size particles, approximately 1/64 inch. For field classification purposes, screening is not intended, simply remove by hand the coarse particles that interfere with the tests

Dilatancy (reaction to shaking)
After removing particles larger than Number 40 sieve size, prepare a pat of moist soil with a volume of about 1/2 cubic inch. Add enough water, if necessary, to make the soil soft but not sticky. Place the pat in the open palm of one hand and shake horizontally, striking vigorously against the other hand several times. A positive reaction consists of the appearance of water on the surface of the pat, which changes to a livery consistency and becomes glossy. When the sample is squeezed between the fingers, the water and gloss disappear from the surface, the pat stiffens, and finally it cracks or crumbles. The rapidity of appearance of water during shaking and of its disappearance during squeezing assist in identifying the character of the fines in a soil. Very fine clean sands give the quickest and most distinct reaction, whereas a plastic clay has no reaction. Inorganic silts, such as a typical rock flour, show a moderately quick reaction.

Dry Strength (crushing characteristics)
After removing particles larger than Number 40 sieve size, mold a pat of soil to the consistency of putty, adding water if necessary. Allow the pat to dry completely by oven, sun, or air, and then test its strength by breaking and crumbling between the fingers. This strength is a measure of the character and quantity of the colloidal fraction contained in the soil. The dry strength increases with increasing plasticity.
High dry strength is characteristic for clays of the (CH) group. A typical inorganic silt possesses only very slight dry strength. Silty fine sands and silts have about the same slight dry strength but can be distinguished by the feel when powdering the dried specimen. Fine sand feels gritty, whereas a typical silt has the smooth feel of flour

Toughness (consistency near plastic limit)
After particles larger than the Number 40 sieve size are removed, a specimen of soil about 1/2 cubic inch in size is molded to the consistency of putty. If too dry, water must be added and if sticky, the specimen should be spread out in a thin layer and allowed to lose some moisture by evaporation. Then the specimen is rolled out by hand on a smooth surface or between the palms into a thread about 1/8 inch in diameter. The thread is then folded and rerolled repeatedly. During this manipulation, the moisture content is gradually reduced and the specimen stiffens, finally loses its plasticity, and crumbles when the plastic limit is reached.
After the thread crumbles, the pieces should be lumped together and a slight kneading action continued until the lump crumbles.
The tougher the thread near the plastic limit and the stiffer the lump when it finally crumbles, the more potent is the colloidal clay fraction in the soil. Weakness of the thread at the plastic limit and quick loss of coherence of the lump below the plastic limit indicate either inorganic clay of low plasticity or materials such as kaolin-type clays and organic clays that occur below the A-line.
Highly organic clays have a very weak and spongy feel at the plastic limit.

Laboratory Classification Criteria

7

$$C_u = \frac{D_{60}}{D_{10}} \quad \text{Greater than 4}$$

$$C_c = \frac{(D_{30})^2}{D_{10} \times D_{60}} \quad \text{Between 1 and 3}$$

Not meeting all gradation requirements for (GW)

Atterberg limits below A-line with PI between 4 and 7 are borderline cases requiring use of dual symbols

Atterberg limits below A-line PI < 4

Atterberg limits above A-line or PI >7

$$C_u = \frac{D_{60}}{D_{10}} \quad \text{Greater than 6}$$

$$C_c = \frac{(D_{30})^2}{D_{10} \times D_{60}} \quad \text{Between 1 and 3}$$

Not meeting all gradation requirements for (SW)

Atterberg limits below A-line or are borderline cases requiring use of dual symbols

Atterberg limits below A-line PI < 4

Atterberg limits above A-line or PI >7

Depending on percentage of fines (fraction smaller than No 200 sieve size) coarse-grained soils are classified as follows:

Less than 5% = GW, GP, SW, SP
More than 12% = GM, GC, SM, SC
5% TO 12% = borderline cases requiring use of dual symbols

Determine percentages of gravel and sand from grain–size curve.

Use grains-size curve in identifying the fractions as given under field identification.

Plasticity Chart
For laboratory classification of fine-grained soil
(CH, CL, ML, MH, CL-ML, OH & MH, A-line)
Plasticity Index vs Liquid Limit

Comparing soils at equal liquid limit toughness and dry strength increase with increasing plasticity index

Table 5-2. Auxiliary laboratory identification procedure

DESIRABLE SOIL PROPERTIES FOR ROAD AND AIRFIELDS

5-27. The properties desired in soils for foundations under roads and airfields are—

- Adequate strength.
- Resistance to frost action (in areas where frost is a factor).
- Acceptable compression and expansion.
- Adequate drainage.
- Good compaction.

5-28. Some of these properties may be supplied by proper construction methods. For instance, materials having good drainage characteristics are desirable, but if such materials are not available locally, adequate drainage may be obtained by installing a properly designed water-collecting system. Strength requirements for base course materials are high, and only good quality materials are acceptable. However, low strengths in subgrade materials may be compensated for in many cases by increasing the thickness of overlying base materials or using a geotextile (see chapter 11). Proper design of road and airfield pavements requires the evaluation of soil properties in more detail than possible by use of the general soils classification system. However, the grouping of soils in the classification system gives an initial indication of their behavior in road and airfield construction, which is useful in site or route selection and borrow source reconnaissance.

5-29. General characteristics of the soil groups pertinent to roads and airfields are in the soil classification sheet in table 5-3, page 5-13, as follows:

- Columns 1 through 5 show major soil divisions, group symbols, hatching, and color symbols.
- Column 6 gives names of soil types.
- Column 7 evaluates the performance (strength) of the soil groups when used as subgrade materials that are not subject to frost action.
- Columns 8 and 9 make a similar evaluation for the soils when used as subbase and base materials.
- Column 10 shows potential frost action.
- Column 11 shows compressibility and expansion characteristics.
- Column 12 presents drainage characteristics.
- Column 13 shows types of compaction equipment that perform satisfactorily on the various soil groups.
- Column 14 shows ranges of unit dry weight for compacted soils.
- Column 15 shows ranges of typical California Bearing Ratio (CBR) values to be anticipated for use in airfield design.
- Column 16 gives ranges of modulus of subgrade reaction, k.

5-30. The various features are discussed in the following paragraphs.

STRENGTH

5-31. In column 3 of table 5-3 the basic soil groups (GM) and (SM) have each been subdivided into two groups designated by the following suffixes:

- d (represents desirable base and subbase materials).
- u (represents undesirable base and subbase materials).

5-32. This subdivision applies to roads and airfields only and is based on field observation and laboratory tests on soil behavior in these groups. The basis for the subdivision is the LL and PI of the fraction of the soil passing the Number 40 sieve. The suffix d is used when the LL is 25 and the PI is 5; the suffix u is used otherwise.

5-33. The descriptions in columns 7, 8, and 9 generally indicate the suitability of the soil groups for use as subgrade, subbase, or base materials not subjected to frost action. In areas where frost heaving is a problem, the value of materials as subgrades is reduced, depending on the potential frost action of the material (see column 10). Proper design procedures should be used in situations where frost action is a problem.

Coarse-Grained Soils

5-34. Generally, the coarse-grained soils make the best subgrade, subbase, and base materials. The (GW) group has excellent qualities as a base material. The adjective "excellent" is not used for any of these soils for base courses, because "excellent" should only be used to describe a high quality processed crushed stone. Poorly graded gravels and some silty gravels (groups (GP) and (GMd)) are usually only slightly less desirable as subgrade or subbase materials. Under favorable conditions, these gravels may be used as base materials; however, poor gradation and other factors sometimes reduce the value of these soils so they offer only moderate strength. For example—

- The (GMu), (GC), and (SW) groups are reasonably good subgrade or select materials but are generally poor to not suitable as base materials.
- The (SP) and (SMd) soils usually are considered fair to good subgrade and subbase materials but are generally poor to not suitable as base materials.

Fine-Grained Soils

5-35. The fine-grained soils range from fair to very poor subgrade materials as follows—

- Silts and lean clays (ML) and (CL) are fair to poor.
- Organic silts, lean organic clays, and micaceous or diatomaceous soils (OL) and (MH) are poor.
- Fat clays and fat organic clays (CH) and (OH) are poor to very poor.

5-36. These qualities are compensated for in flexible pavement design by increasing the thickness of overlying base material. In rigid pavement design, these qualifications are compensated for by increasing the pavement thickness or by adding a base course layer. None of the fine-grained soils are suitable as a subbase under bituminous pavements, but soils in the (ML) and (CL) groups may be used as select material. The fibrous organic soils (group (Pt)) are very poor subgrade materials and should be removed wherever possible; otherwise, special construction measures should be adopted. They are not suitable as subbase and base materials. The CBR values shown in column 15 give a relative indication of the strength of the various soil groups when used in flexible pavement design. Similarly, values of subgrade modulus (k) in column 16 are relative indications of strengths from plate-bearing tests when used in rigid pavement design. Actual test values should be used for this purpose instead of the approximate values shown in the tabulation.

5-37. For wearing surfaces on unsurfaced roads, slightly plastic sand-clay-gravel mixtures (GC) are generally considered the most satisfactory. However, they should not contain too large a percentage of fines, and the PI should be in the range of 5 to about 15.

FROST ACTION

5-38. The relative effects of frost action on the various soil groups are shown in column 10. Regardless of the frost susceptibility of the various soil groups, two conditions must be present simultaneously before frost action is a major consideration. These are—

- A source of water during the freezing period.
- A sufficient period for the freezing temperature to penetrate the ground.

5-39. Water necessary for the formation of ice lenses may become available from a high groundwater table, a capillary supply, water held within the soil voids, or through infiltration. The degree of ice formation that will occur is markedly influenced by physical factors, such as—

- Topographic position.
- Stratification of the parent soil.
- Transitions into a cut section.

- Lateral flow of water from side cuts.
- Localized pockets of perched groundwater.
- Drainage conditions.

5-40. In general, the silts and fine silty sands are most susceptible to frost. Coarse-grained materials with little or no fines are affected only slightly or not at all. Clays ((CL) and (CH)) are subject to frost action, but the loss of strength of such materials may not be as great as for silty soils. Inorganic soils containing less than 3 percent (by weight) of grains finer than 0.02 mm in diameter are considered nonfrost-susceptible. Where frost-susceptible soils occur in subgrades and frost is a problem, two acceptable methods of pavement design are available:

- Place a sufficient depth of acceptable granular material over the soils to limit the depth of freezing in the subgrade and thereby prevent the detrimental effects of frost action.
- Use a design load capacity during the period of the year when freezing conditions are expected.

5-41. In the second case, design is based on the reduced strength of the subgrade during the frost-melting period. Often an appropriate drainage measure to prevent the accumulation of water in the soil pores helps limit ice development in the subgrade and subbase.

COMPRESSION

5-42. The compression or consolidation of soils becomes a design factor primarily when heavy fills are made on compressible soils. The two types of compression are—

- Relatively long-term compression or consolidation under the dead weight of the structure.
- Short-term compression and rebound under moving wheel loads.

5-43. If adequate provision is made for this type of settlement during construction, it will have little influence on the load-carrying capacity of the pavement. However, when elastic soils subject to compression and rebound under wheel loads are encountered, adequate protection must be provided. Even small movements of this type soil may be detrimental to the base and wearing course of pavements. Fortunately, the free-draining, coarse-grained soils ((GW), (GP), (SW), and (SP)), which generally make the best subgrade and subbase materials, exhibit almost no tendency toward high compressibility or expansion. In general, the compressibility of soil increases with an increasing LL. However, compressibility is also influenced by soil structure, grain shape, previous loading history, and other factors not evaluated in the classification system. Undesirable compression or expansion characteristics may be reduced by distributing the load through a greater thickness of overlying material. These factors are adequately handled by the CBR method of design for flexible pavements. However, rigid pavements may require the addition of an acceptable base course under the pavement.

DRAINAGE

5-44. The drainage characteristics of soils are a direct reflection of their permeability. The evaluation of drainage characteristics for use in roads and runways is shown in column 12 of table 5-3, page 5-13. The presence of water in base, subbase, and subgrade materials, except for free-draining, coarse-grained soils, may cause pore water pressures to develop resulting in a loss of strength. The water may come from infiltration of groundwater or rainwater or by capillary rise from an underlying water table. While free-draining materials permit rapid draining of water, they also permit rapid ingress of water. If free-draining materials are adjacent to less pervious materials and become inundated with water, they may serve as reservoirs. Adjacent, poorly drained soils may become saturated. The gravelly and sandy soils with little or no fines (groups (GW), (GP), (SW), (SP)) have excellent drainage characteristics. The (GMd) and (SMd) groups have fair to poor drain-age characteristics, whereas the (GMu), (GC), (SMu), and (SC) groups have very poor drainage characteristics or are practically impervious. Soils of the (ML), (MH), and (Pt) groups have fair to poor drainage characteristics. All other groups have poor drainage characteristics or are practically impervious.

COMPACTION

5-45. Compacting soils for roads and airfields requires attaining a high degree of density during construction to prevent detrimental consolidation from occurring under an embankment's weight or under traffic. In addition, compaction reduces the detrimental effects of water. Processed materials, such as crushed rock, are often used as a base course and require special treatment during compaction. Types of compaction equipment that may be used to achieve the desired soil densities are shown in table 5-3, column 13. For some of the soil groups, several types of equipment are listed because variations in soil type within a group may require the use of a specific type of compaction equipment. On some construction projects, more than one type of compaction equipment may be necessary to produce the desired densities. For example, recommendations include—

- Steel-wheeled rollers for angular materials with limited amounts of fines.
- Crawler-type tractor or rubber-tired rollers for gravels and sand.
- Sheepsfoot rollers for coarse-grained or fine-grained soils having some cohesive qualities.
- Rubber-tired rollers for final compaction operations for most soil except those with a high LL (group H).

5-46. Suggested minimum weights of the various types of equipment are shown in note 2 of table 5-3. Column 14 shows ranges of unit dry weight for soil compacted according to the moisture-density testing procedures outlined in Military Standard 621A, method 100. These values are included primarily for guidance; base design or control of construction should be based on laboratory test results.

DESIRABLE SOIL PROPERTIES FOR EMBANKMENTS AND FOUNDATIONS

5-47. Table 5-4, page 5-14, lists the soil characteristics pertinent to embankment and foundation construction. After the soil has been classified, look at column 3 and follow it downward to the soil class. Table 5-4 contains the same type of information as table 5-3 except that column 8 lists the soil permeability and column 12 lists possible measures to control seepage. Material not pertinent to embankments and foundations, such as probable CBR values, are not contained in table 5-4. Both tables are used in the same manner. Read the notes at the bottom of both tables carefully.

Table 5-3. Characteristics pertinent to roads and airfields

Major Division (1)	Letter (3)	Color (5)	Name (6)	Value as Subgrade When not Subject to Frost Action (7)	Value as Subbase When not Subject to Frost Action (8)	Value as base When not Subject to Frost Action (9)	Potential Frost Action (10)	Compressibility and Expression (11)	Drainage Characteristics (12)	Compaction Equipment (13)	Unit Dry Weight Pounds Per Cubic Foot (14)	CBR (15)	Subgrade Modulus k, Pounds Per Cubic Inch (16)
Coarse-grained soils — Gravel and gravelly soils	GW	Red	Well-graded gravels or gravel sand mixtures, little or no fines	Excellent	Excellent	Good	None to very slight	Almost none	Excellent	Crawler-type tractor, rubber-tired roller, steel-wheeled roller	125-140	40-80	300-500
	GP	Red	Poorly graded gravels or gravel sand mixtures, little or no fines	Good to excellent	Good	Fair to good	None to very slight	Almost none	Excellent	Crawler-type tractor, rubber-tired roller, steel-wheeled roller	110-140	30-60	300-500
	GM d	Yellow	Silty gravels, gravel-sand-silt mixtures	Good to excellent	Good	Good to fair	Slight to medium	Very Slight	Fair to poor	Rubber-tired roller, sheepsfoot roller, close control of moisture	125-145	40-60	300-500
	GM u	Yellow		Good	Fair	Poor to not suitable	Slight to medium	Slight	Poor to practically impervious	Rubber tired roller, sheepsfoot roller	115-135	20-30	200-500
	GC	Yellow	Clayey gravels, gravel-sand-clay mixtures	Good	Fair	Poor to not suitable	Slight to medium	Slight	Poor to practically impervious	Rubber tired roller, sheepsfoot roller	130-145	20-30	200-500
Coarse-grained soils — Sand and sandy soils	SW	Red	Well-graded sands or gravelly sands, little or no fines	Good	Fair to good	Poor	None to very slight	Almost none	Excellent	Crawler-type tractor, rubber-tired roller	110-130	20-40	200-400
	SP	Red	Poorly graded sands or gravelly sands, little or no fines	Fair to good	Fair	Poor to not suitable	None to very slight	Almost none	Excellent	Crawler-type tractor, rubber-tired roller	105-135	10-40	150-400
	SM d	Yellow	Silty sands, sand-silt mixture	Fair to good	Fair to good	Poor	Slight to high	Very slight	Fair to poor	Rubber-tired roller, sheepsfoot roller, close control of moisture	120-135	15-40	150-400
	SM u	Yellow		Fair	Poor to fair	Not suitable	Slight to high	Slight to medium	Poor to practically impervious	Rubber tired roller, sheepsfoot roller	100-130	10-20	100-300
	SC	Yellow	Clayey sands, sand-silt mixtures	Poor to fair	Poor	Not suitable	Slight to high	Slight to medium	Poor to practically impervious	Rubber tired roller, sheepsfoot roller	100-135	5-20	100-300
Fine-grained soils — Silts and clays LL<50	ML	Green	Inorganic silts and very fine sands, rock flour, silty or clayey fine sands, or clayey silts with slight plasticity	Poor to fair	Poor	Not suitable	Medium to very high	Slight to medium	Fair to poor	Rubber-tired roller, sheepsfoot roller, close control of moisture	90-130	15 or less	100-200
	CL	Green	Inorganic clays of low to medium plasticity, gravelly clays, sandy clays, silty clays, lean clays	Poor to fair	Not suitable	Not suitable	Medium to high	Medium	Practically impervious	Rubber tired roller, sheepsfoot roller	90-130	15 or less	100-200
	OL	Green	Organic silts and organic silt-clays of low plasticity	Poor	Not suitable	Not suitable	Medium to high	Medium to high	Poor	Rubber tired roller, sheepsfoot roller	90-105	5 or less	50-150
Fine-grained soils — Silts and clays LL>50	MH	Blue	Inorganic silts, micaceous or diatomaceous fine sandy or silty soils, elastic silts	Poor to fair	Not suitable	Not suitable	Medium to high	High	Fair to poor	Sheepsfoot roller, rubber tired roller	80-105	10 or less	50-100
	CH	Blue	Inorganic clays of high plasticity, fat clays	Poor to fair	Not suitable	Not suitable	Medium to high	High	Practically impervious	Sheepsfoot roller, rubber tired roller	90-115	15 or less	50-100
	OH	Blue	Organic clays of medium to high plasticity, organic silts	Poor to very poor	Not suitable	Not suitable	Medium to high	High	Practically impervious	Sheepsfoot roller, rubber tired roller	80-110	5 or less	25-100
High organic soils	Pt	Orange	Peat and other highly organic soils	Not suitable	Not suitable	Not suitable	Medium to very high	Very high	Fair to poor	Compaction not practical	—	—	—

Notes.

1. In column 3, the division (GM) and (SM) groups into subdivisions of d and u are for roads and airfields only. Subdivision is on the basis of Atterberg limits; suffix d (for example GM_d) will be used when the liquid limit is 25 or less and the plasticity index is 5 or less; the suffix u will be used otherwise.

2. In column 13, the equipment listed will usually produce the required densities with a reasonable number of passes when moisture conditions and thickness of lift are properly controlled. In some instances, several types of equipment are listed because variable soil characteristics within a given soil group may require different equipment. In some instances, a combination of two types may be necessary.

   a. Processed base materials and other angular materials. Steel-wheeled and rubber tired rollers are recommended for hard, angular materials with limited fines and screenings. Rubber-tired equipment is recommended for softer materials subject to degradation.

   b. Finishing. Rubber-tired equipment is recommended for rolling during final shaping operations for most soils and processed materials.

   c. Equipment size. The following sizes of equipment are necessary to assure the high densities required for airfield construction:
   Crawler-type tractor – total weight in excess of 30,000 pounds.
   Rubber-tired equipment – wheel load in excess of 15,000 pounds, wheel loads as high as 40,000 pounds may be necessary to obtain the required densities for some materials (based on contact pressure of approximately 65 to 150 psi).
   Sheepsfoot roller – unit pressure (on 6- to 12-square-inch foot) to be in excess of 250 psi and unit pressures as high as 650 psi may be necessary to obtain the required densities for some materials.
   The area of the feet should be at least 5 percent of the total peripheral area of the drum, using the diameter measured to the faces of the feet.

3. In column 14, unit dry weights are for compacted soil at optimum moisture content for modified American Association of the State Highway and Transportation Official (AASHTO) (standard Proctor) compactive effort CE 55.

4. In column 15, the maximum value that can be used in the design of airfields is, in some cases, limited by gradation and plasticity requirements.

Table 5-4. Classifications pertinent to embankment and foundation construction

Major Division (1)	(2)	Letter (3)	Symbol Hatching (4)	Symbol Color (5)	Name (6)	Value for Embankment (7)	Permeability Centimeters Per Second (8)	Compaction Characteristics (9)	Standard AASHTO Maximum Unit Dry Weight Pounds Per Cubic Foot (10)	Value for Foundations (11)	Requirements for Seepage Control (12)
Coarse grained soils	Gravels and gravelly soils	GW		Red	Well-graded gravels, gravel-sand mixtures, little or no fines	Very stable, pervious shells of dikes and dams	$k > 10^{-2}$	Good, tractor, rubber-tired roller, steel-wheeled roller	125-135	Good bearing value	Positive cutoff
		GP		Red	Poorly graded gravels or gravel-sand mixtures, little or no fines	Reasonably stable, pervious shells of dikes and dams	$k > 10^{-2}$	Good, tractor, rubber-tired roller, steel-wheeled roller	115-125	Good bearing value	Positive cutoff
		GM		Yellow	Silty gravels, gravel-sand-silt mixture	Reasonably stable, not particularly suited to shells but may be used for impervious cores and blankets	$k - 10^{-3}$ to 10^{-6}	Good, with close control, rubber-tired roller, sheepsfoot roller	120-135	Good bearing value	Toe trench to none
		GC		Yellow	Clayey gravels, gravel-sand-clay mixtures	Fairly stable, may be used for impervious cores	$k - 10^{-6} - 10^{-8}$	Fair, rubber-tired roller, sheepsfoot roller	115-130	Good bearing value	None
	Sands and sandy soils	SW		Red	Well-graded sands, gravelly sands, little or no fines	Very stable, pervious sections, slope protection required	$k > 10^{-3}$	Good, tractor	110-130	Good bearing value	Upstream blanket and toe drainage or wells
		SP		Red	Poorly graded sands or gravelly sands, little or no fines	Reasonably stable, may be used in dike section with flat slopes	$k > 10^{-3}$	Good, tractor	100-120	Good to poor bearing value depending on density	Upstream blanket and toe drainage or wells
		SM		Yellow	Silty sands, sand-silt mixture	Fairly stable, not particularly suited to shells but may be used for impervious cores and dikes	$k - 10^{-3}$ to 10^{-6}	Good, with close control, rubber-tired roller, sheepsfoot roller	110-125	Good to poor bearing value depending on density	Upstream blanket and toe drainage or wells
		SC		Yellow	Clayey sands, sand-silt mixtures	Fairly stable, may be used for impervious core for flood-control structures	$k - 10^{-6} - 10^{-8}$	Fair, sheepsfoot, roller, rubber-tired roller	105-125	Good to poor bearing value	None
Fine grained soils	Silts and clays LL < 50	ML		Green	Inorganic silts and very fine sands, rock flour, silty or clayey fine sands, or clayey silts with slight plasticity	Poor stability, may be used for embankments with proper control	$k - 10^{-3} - 10^{-6}$	Good to poor, close control essential, rubber-tired roller, sheepsfoot roller	95-120	Very poor, susceptible to liquefaction	Toe trench to none
		CL		Green	Inorganic clays of low to medium plasticity, gravelly clays, sandy clays, silty clays, lean clays	Stable, impervious cores, and blankets	$k - 10^{-6} - 10^{-8}$	Fair to good, sheepsfoot roller, rubber-tired roller	95-120	Good to poor bearing value	None
		OL		Green	Organic silts and organic silty clays of low plasticity	Not suitable for embankments	$k - 10^{-4} - 10^{-6}$	Fair to poor sheepsfoot roller	80-100	Fair to poor bearing value, may have excessive settlements	None
	Silts and clays LL > 50	MH		Blue	Inorganic silts, micaceous or diatomaceous fine sandy or silty soils, elastic silts	Poor stability, core of hydraulic fill dam, not desirable in rolled fill construction	$k - 10^{-4} - 10^{-6}$	Poor to very poor, sheepsfoot roller	70-95	Poor bearing value	None
		CH		Blue	Inorganic clays of high plasticity, fat clays	Fair stability with flat slopes, thin cores, blankets and dike sections	$k - 10^{-6} - 10^{-8}$	Fair to poor, sheepsfoot roller	75-105	Fair to poor bearing value	None
		OH		Blue	Organic clays of medium to high plasticity, organic silts	Not suitable for embankments	$k - 10^{-6} - 10^{-8}$	Poor to very poor, sheepsfoot roller	65-100	Very poor bearing value	None
Highly organic soils		Pt		Orange	Peat and other highly organic soils	Not used for construction		Compaction not practical		Remove from foundations	

Notes.
1. Values in columns 7 and 11 are for guidance only. Design should be based on test results.
2. In column 9, the equipment listed will usually produce the desired densities with a reasonable number of passes when moisture conditions and the thickness of lift are properly controlled.
3. In column 10, unit dry weights are for compacted soil optimum moisture content for standard AASHTO (standard Proctor) compactive effort CE 55.

SOIL GRAPHICS

5-48. It is customary to present the results of soils explorations on drawings as schematic representations of the borings or test pits or on soil profiles with the various soils encountered shown by appropriate symbols. One approach is to write the group letter symbol in the appropriate section of the log. As an alternative, hatching symbols shown in column 4 of table 5-3, page 5-13, may be used. In addition, show the natural water content of fine-grained soils along the side of the log. Use other descriptive remarks as appropriate. Colors may be used to delineate soil types on maps and drawings. A suggested color scheme to show the major soil groups is described in column 5. Boring logs are discussed in more detail in chapter 3. Soil graphics generated in terrain studies usually use numeric symbols, each of which represents a USCS soil type.

FIELD IDENTIFICATION

5-49. The soil types of an area are an important factor in selecting the exact location of airfields and roads. The military engineer, construction foreman, and members of engineer reconnaissance parties must be able to identify soils in the field so that the engineering characteristics of the various soil types encountered can be compared. Because of the need to be economical in time, personnel, equipment, materiel, and money, selection of the project site must be made with these factors in mind. Lack of time and facilities often make laboratory soil testing impossible in military construction. Even where laboratory tests are to follow, field identification tests must be made during the soil exploration to distinguish between the different soil types encountered so that duplication of samples for laboratory testing is minimized. Several simple field identification tests are described in this manual. Each test may be performed with a minimum of time and equipment, although seldom will all of them be required to identify a given soil. The number of tests required depends on the type of soil and the experience of the individual performing them. By using these tests, soil properties can be estimated and materials can be classified. Such classifications are approximations and should not be used for designing permanent or semipermanent construction.

PROCEDURES

5-50. The best way to learn field identification is under the guidance of an experienced soils technician. To learn without such assistance, systematically compare laboratory test results for typical soils in each group with the "feel" of these soils at various moisture contents.

5-51. An approximate identification of a coarse-grained soil can be made by spreading a dry sample on a flat surface and examining it, noting particularly grain size, gradation, grain shape, and particle hardness. All lumps in the sample must be thoroughly pulverized to expose individual grains and to obtain a uniform mixture when water is added to the fine-grained portion. A rubber-faced or wooden pestle and a mixing bowl is recommended for pulverizing. Lumps may also be pulverized by placing a portion of the sample on a firm, smooth surface and using the foot to mash it. If an iron pestle is used for pulverizing, it will break up the mineral grains and change the character of the soil; therefore, using an iron pestle is discouraged.

5-52. Tests for identification of the fine-grained portion of any soil are performed on the portion of the material that passes a Number 40 sieve. This is the same soil fraction used in the laboratory for Atterberg limits tests, such as plasticity. If this sieve is not available, a rough separation may be made by spreading the material on a flat surface and removing the gravel and larger sand particles. Fine-grained soils are examined primarily for characteristics related to plasticity.

EQUIPMENT

5-53. Practically all the tests to be described may be performed with no equipment or accessories other than a small amount of water. However, the accuracy and uniformity of results is greatly increased by the proper use of certain equipment. The following equipment is available in nearly all engineer units (or may be improvised) and is easily transported:

- A Number 40 US standard sieve. Any screen with about 40 openings per lineal inch could be used, or an approximate separation may be used by sorting the materials by hand. Number 4 and Number 200 sieves are useful for separating gravels, sands, and fines.
- A pick and shovel or a set of entrenching tools for obtaining samples. A hand earth auger is useful if samples are desired from depths more than a few feet below the surface.
- A spoon issued as part of a mess equipment for obtaining samples and for mixing materials with water to desired consistency.
- A bayonet or pocket knife for obtaining samples and trimming them to the desired size.
- A small mixing bowl with a rubber-faced or wooden pestle for pulverizing the fine-grained portion of the soil. Both may be improvised by using a canteen cup and a wooden dowel.
- Several sheets of heavy nonabsorbent paper for rolling samples.
- A pan and a heating element for drying samples.
- A balance or scales for weighing samples.

FACTORS

5-54. The soil properties that form the basis for the Unified Soil Classification System are the—

- Percentage of gravels, sands, and fines.
- Shape of the grain-size distribution curve.
- Plasticity.

5-55. These same properties are to be considered in field identification. Other characteristics observed should also be included in describing the soil, whether the identification is made by field or laboratory methods.

5-56. Properties normally included in a description of a soil are—

- Color.
- Grain size, including estimated maximum grain size and estimated percent by weight of fines (material passing the Number 200 sieve).
- Gradation.
- Grain shape.
- Plasticity.
- Predominant type.
- Secondary components.
- Classification symbol.
- Other remarks, such as organic, chemical, or metallic content; compactness; consistency; cohesiveness near PL; dry strength; and source— residual or transported (such as eolian, water-borne, or glacial deposit).

5-57. An example of a soil description using the sequence and considering the properties referred to above might be—

- Dark brown to white.
- Coarse-grained soil, maximum particle size 2 3/4 inches, estimating 60 percent gravel, 36 percent sand, and 4 percent passing the Number 200 sieve.
- Poorly graded (insufficient fine gravel, gap-graded).
- Gravel particles subrounded to rounded.
- Nonplastic.
- Predominantly gravel.
- Considerable sand and a small amount of nonplastic fines (silt).
- (GP).
- Slightly calcareous, no dry strength, dense in the undisturbed state.

5-58. A complete description with the proper classification symbol conveys much more to the reader than the symbol or any other isolated portion of the description used alone.

TESTS

5-59. The following tests can be performed to aid in field identification of soils:

Visual Examination Test

5-60. Determine the color, grain size, and grain shape of the coarse-grained portion of a soil by visual examination. The grain-size distribution may be estimated. To observe these properties, dry a sample of the material and spread it on a flat surface.

5-61. In soil surveys in the field, color is often helpful in distinguishing among various soil strata, and from experience with local soils, color may aid in identifying soil types. Since the color of a soil often varies with its moisture content, the condition of the soil when color is determined must always be recorded. Generally, more contrast occurs in these colors when the soil is moist, with all the colors becoming lighter as the moisture contents are reduced. In fine-grained soils, certain dark or drab shades of gray or brown (including almost-black colors) are indicative of organic colloidal matter ((OL) and (OH)). In contrast, clean and bright-looking colors (including medium and light gray, olive green, brown, red, yellow, and white) are usually associated with inorganic soils. Soil color may also indicate the presence of certain chemicals. Red, yellow, and yellowish-brown soil may be a result of the presence of iron oxides. White to pinkish colors may indicate the presence of considerable silica, calcium carbonate, or (in some cases) aluminum compounds . Grayish-blue, gray, and yellow mottled colors frequently indicate poor drainage.

5-62. Estimate the maximum particle size for each sample, thereby establishing the upper limit of the grain-size distribution curve for that sample. The naked eye can normally distinguish the individual grains of soil down to about 0.07 mm. All particles in the gravel and sand ranges are visible to the naked eye. Most of the silt particles are smaller than this size and are invisible to the naked eye. Material smaller than 0.75 mm will pass the Number 200 sieve.

5-63. Perform the laboratory mechanical analysis whenever the grain-size distribution of a soil sample must be determined accurately; however, the grain-size distribution can be approximated by visual inspection. The best way to evaluate a material without using laboratory equipment is to spread a portion of the dry sample on a flat surface. Then, using your hands or a piece of paper, separate the material into its various grain-size components. By this method, the gravel particles and some of the sand particles can be separated from the remainder. This will at least give you an opportunity to estimate whether the total sample is to be considered coarse-grained or fine-grained, depending on whether or not more than 50 percent of the material would pass the Number 200 sieve. Percentage of values refers to the dry weight of the soil fractions indicated as compared to the dry weight of the original sample. A graphical summary of the procedure is shown in figure 5-2, page 5-18.

5-64. If you believe the material is coarse-grained, then consider the following criteria:

- Does less than 5 percent pass the Number 200 sieve?
- Are the fines nonplastic?

5-65. If both criteria can be satisfied and there appears to be a good representation of all grain sizes from largest to smallest, without an excessive deficiency of any one size, the material may be said to be well-graded ((GW) or (SW)). If any intermediate sizes appear to be missing or if there is too much of any one size, then the material is poorly graded ((GP) or (SP)). In some cases, it may only be possible to take a few of the standard sieves into the field. When this is the case, take the Number 4, Number 40, and Number 200 sieves. The sample may be separated into gravels, sands, and fines by use of the Number 4 and Number 200 sieves. However, if there is a considerable quantity of fines, particularly clay particles, separation of the fines can only be readily accomplished by washing them through the Number 200 sieve. In such cases, a determination of the percentage of fines is made by comparing the dry weight of the original sample with that retained on the Number 200 sieve after washing. The difference between these two is the weight of the fines lost in the washing process. To determine the plasticity, use only that portion of the soil passing through a Number 40 sieve.

The form in this publication is obsolete. See http://www.apd.army.mil for current form.

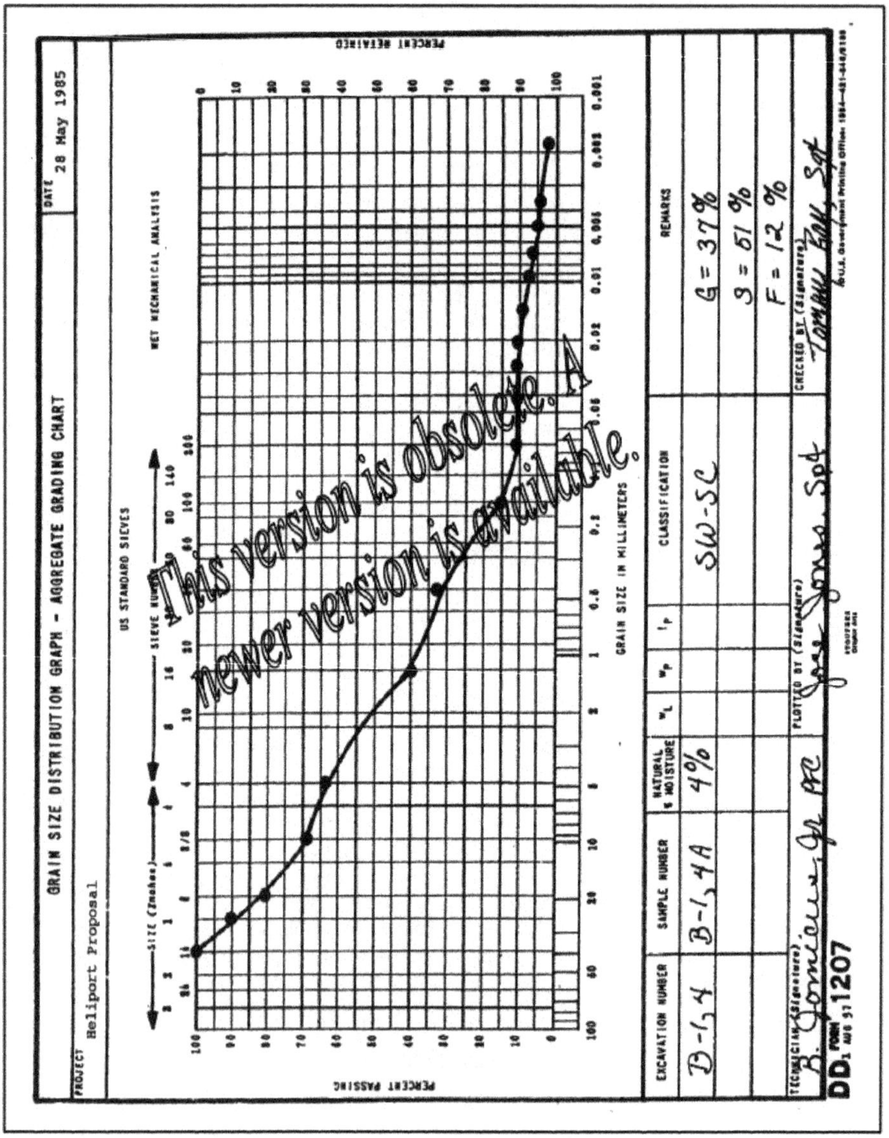

Figure 5-2. Graphical summary of grain-size distribution

5-66. Estimating the grain-size distribution of a sample using no equipment is probably the most difficult part of field identification and places great importance on the experience of the individual making the estimate. A better approximation of the relative proportions of the components of the finer soil fraction may sometimes be obtained by shaking a portion of this sample into a jar of water and allowing the material to settle. It will settle in layers, with the gravel and coarse sand particles settling out almost immediately. The fine sand particles settle within a minute; the silt particles require as much as an hour; and the clay particles remain in suspension indefinitely or until the water is clear. In using this method, remember that the gravels and sands settle into a much more dense formation than either the silts or clays.

5-67. The grain shape of the sand and gravel particles can be determined by close examination of the individual grains. The grain shape affects soil stability because of the increased resistance to displacement found in the more irregular particles. A material with rounded grains has only the friction between the surfaces of the particles to help hold them in place. An angular material has this same friction force, which is increased by the roughness of the surface. In addition, an interlocking action is developed between the particles which gives the soil much greater stability.

5-68. A complete description of a soil should include prominent characteristics of the undisturbed material. The aggregate properties of sand and gravel are described qualitatively by the terms "loose," "medium," and "dense." Clays are described as "hard," "stiff," "medium," and "soft."

5-69. These characteristics are usually evaluated on the basis of several factors, including the relative ease or difficulty of advancing the drilling and sampling tools and the consistency of the samples. In soils that are described as "soft," there should be an indication of whether the material is loose and compressible, as in an area under cultivation, or spongy (elastic), as in highly organic soils. The moisture condition at the time of evaluation influences these characteristics and should be included in the report.

Breaking or Dry Strength Test

5-70. The breaking test is performed only on material passing the Number 40 sieve. This test, as well as the roll test and the ribbon test, is used to measure the cohesive and plastic characteristics of the soil. The test is normally made on a small pat of soil about ½ inch thick and about 2 inches in diameter. The pat is prepared by molding a portion of the soil in the wet plastic state into the size and shape desired and then allowing the pat to dry completely. Samples may be tested for dry strength in their natural conditions. Such a test may be used as an approximation; however, it should be verified later by testing a carefully prepared sample.

5-71. After the prepared sample is thoroughly dry, attempt to break it using the thumb and forefingers of both hands (see figure 5-3). If it can be broken, try to powder it by rubbing it with the thumb and fingers of one hand.

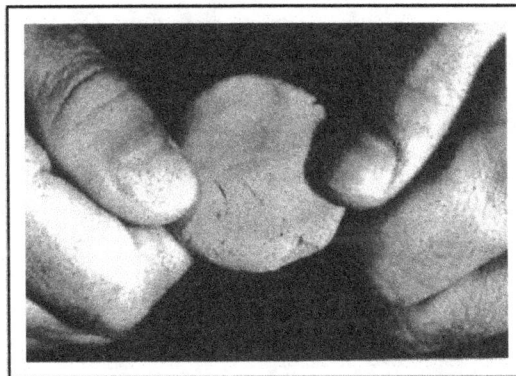

Figure 5-3. Breaking or dry strength test

5-72. Typical reactions obtained in this test for various types of soils are described below.

- Very highly plastic soils, (CH); very high dry strength. Samples cannot be broken or powdered using finger pressure.
- Highly plastic soils, (CH); high dry strength. Samples can be broken with great effort but cannot be powdered.
- Medium plastic soils, (CL); medium dry strength. Samples can be broken and powdered with some effort.
- Slightly plastic soils, (ML), (MH), or (CL); low dry strength. Samples can be broken quite easily and powdered readily.

● Nonplastic soils, (ML) or (MH); very little or no dry strength. Samples crumble and powder on being picked up in the hands.

5-73. The breaking or dry strength test is one of the best tests for distinguishing between plastic clays and nonplastic silts or fine sands. However, a word of caution is appropriate. Dry pats of highly plastic clays quite often display shrinkage cracks. Breaking the sample along one of these cracks gives an indication of only a very small part of the true dry strength of the clay. It is important to distinguish between a break along such a crack and a clean, fresh break that indicates the true dry strength of the soil.

Roll or Thread Test

5-74. The roll or thread test is performed only on material passing the Number 40 sieve. Prepare the soil sample by adding water to the soil until the moisture content allows easy remolding of the soil without sticking to the fingers. This is sometimes referred to as being just below the "sticky limit." Using a nonabsorbent surface, such as glass or a sheet of heavy coated paper, rapidly roll the sample into a thread approximately ⅛ inch in diameter figure 5-4.

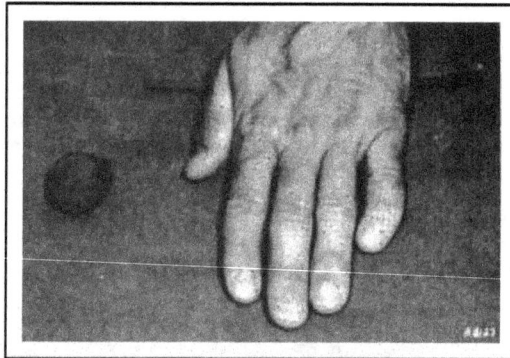

Figure 5-4. Roll or thread test

5-75. A soil that can be rolled into a ⅛-inch- diameter thread at some moisture content has some plasticity. Materials that cannot be rolled in this manner are nonplastic or have very low plasticity. The number of times that the thread may be lumped together and the rolling process repeated without crumbling and breaking is a measure of the degree of plasticity of the soil. After the PL is reached, the degree of plasticity may be described as follows:

● Highly plastic soils, (CH). The soil may be remolded into a ball and the ball deformed under extreme pressure by the fingers without cracking or crumbling.
● Medium plastic soils, (CL). The soil may be remolded into a ball, but the ball will crack and easily crumble under pressure of the fingers.
● Low plastic soils, (CL), (ML), or (MH). The soil cannot be lumped together into a ball without completely breaking up.
● Organic materials, (OL) or (OH). Soils containing organic materials or mica particles will form soft spongy threads or balls when remolded.
● Nonplastic soils, (ML) or (MH). Non- plastic soils cannot be rolled into a thread at any moisture content.

5-76. From this test, the cohesiveness of the material near the PL may also be described as weak, firm, or tough. The higher the position of a soil on the plasticity chart, the stiffer are the threads as they dry out and the tougher are the lumps if the soil is remolded after rolling.

Ribbon Test

5-77. The ribbon test is performed only on the material passing the Number 40 sieve. The sample prepared for use in this test should have a moisture content slightly below the sticky limit. Using this material, form a roll of soil about ½ or ¾ inch in diameter and about 3 to 5 inches long. Place the material in the palm of the hand and, starting with one end, flatten the roll, forming a ribbon ⅛ to ¼ inch thick by squeezing it between the thumb and forefinger (see figure 5-5). The sample should be handled carefully to form the maximum length of ribbon that can be supported by the cohesive properties of the material. If the soil sample holds together for a length of 8 to 10 inches without breaking, the material is considered to be both plastic and highly compressive (CH).

5-78. If soil cannot be ribboned, it is nonplastic (ML) or (MH). If it can be ribboned only with difficulty into short lengths, the soil is considered to have low plasticity (CL). The roll test and the ribbon test complement each other in giving a clearer picture of the degree of plasticity of a soil.

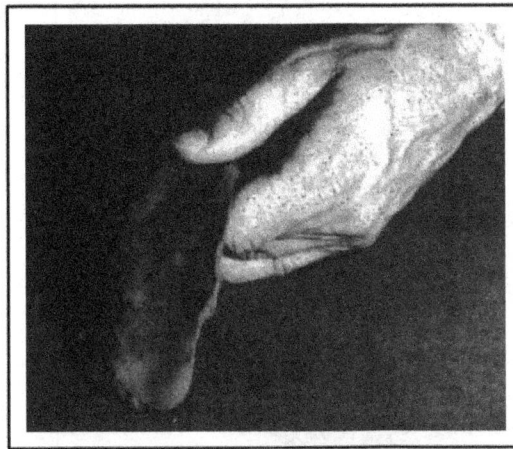

Figure 5-5. Ribbon test (highly plastic clay)

Wet Shaking Test

5-79. The wet shaking test is performed only on material passing the Number 40 sieve. In preparing a portion of the sample for use in this test, moisten enough material with water to form a ball of material about ¾ inch in diameter. This sample should be just wet enough so that the soil will not stick to the fingers when remolding (just below the sticky limit) (see figure 5-6, page 5-22, a).

5-80. Place the sample in the palm of your hand and shake vigorously (see figure 5-6, b). Do this by jarring the hand on the table or some other firm object or by jarring it against the other hand. The soil has reacted to this test when, on shaking, water comes to the surface of the sample producing a smooth, shiny appearance (see figure 5-6, c). This appearance is frequently described as "livery." Then, squeeze the sample between the thumb and forefinger of the other hand. The surface water will quickly disappear, and the surface will become dull. The material will become firm and resist deformation. Cracks will occur as pressure is continued, with the sample finally crumbling like a brittle material. The vibration caused by the shaking of the soil sample tends to reorient the soil grains, decrease voids, and force water that had been within these voids to the surface. Pressing the sample between the fingers tends to disarrange the soil grains, increase the void space, and draw the water into the soil. If the water content is still adequate, shaking the broken pieces will cause them to liquefy again and flow together. This process only occurs when the soil grains are bulky and cohesionless.

5-81. Very fine sands and silts are readily identified by the wet shaking test. Since it is rare that fine sands and silts occur without some amount of clay mixed with them, there are varying degrees of reaction to this test. Even a small amount of clay tends to greatly retard this reaction. Descriptive terms applied to the different rates of reaction to this test are—

- Sudden or rapid.
- Sluggish or slow.
- No reaction.

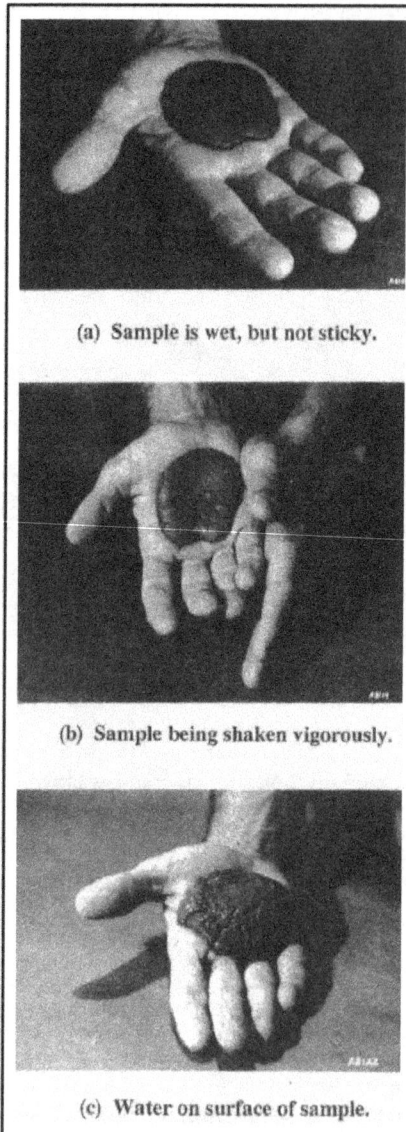

(a) Sample is wet, but not sticky.

(b) Sample being shaken vigorously.

(c) Water on surface of sample.

Figure 5-6. Wet shaking test

5-82. A sudden or rapid reaction to the shaking test is typical of nonplastic fine sands and silts. A material known as rock flour, which has the same size range as silt, also gives this type of reaction.

5-83. A sluggish or slow reaction indicates slight plasticity, such as that which might be found from a test of some organic or inorganic silts or from silts containing a small amount of clay.

5-84. Obtaining no reaction at all to this test does not indicate a complete absence of silt or fine sand. Even a slight content of colloidal clay imparts some plasticity and slows the reaction to the shaking test. Extremely slow or no reaction is typical of all inorganic clays and highly plastic clays.

Odor Test

5-85. Organic soils of the (OL) and (OH) groups have a distinctive musty, slightly offensive odor, which can be used as an aid in identifying such material. This odor is especially apparent from fresh samples. Exposure to air gradually reduces the odor, but heating a wet sample rejuvenates the odor. Organic soils are undesirable as foundation or base course material and are usually removed from the construction site.

Bite or Grit Test

5-86. The bite or grit test is a quick and useful method of distinguishing among sands, silts, or clays. In this test, a small pinch of the soil material is ground lightly between the teeth and identified.

5-87. A sandy soil may be identified because the sharp, hard particles of sand grate very harshly between the teeth and will be highly objectionable. This is true even of the fine sand.

5-88. The silt grains of a silty soil are so much smaller than sand grains that they do not feel nearly so harsh between the teeth. Silt grains are not particularly objectionable, although their presence is still easily detected.

5-89. The clay grains of a clayey soil are not gritty but feel smooth and powdery, like flour, between the teeth. Dry lumps of clayey soils stick when lightly touched with the tongue.

Slaking Test

5-90. The slaking test is useful in determining the quality of certain shales and other soft rocklike materials. The test is performed by placing the soil in the sun or in an oven to dry and then allowing it to soak in water for a period of at least 24 hours. The strength of the soil is then examined. Certain types of shale completely disintegrate, losing all strength.

5-91. Other materials that appear to be durable rocks may be crumbled and readily broken by hand after such soaking. Materials that have a considerable reduction in strength are undesirable for use as base course materials.

Acid Test

5-92. The acid test is used to determine the presence of calcium carbonate. It is performed by placing a few drops of HCl on a piece of soil. A fizzing reaction (effervescence) to this test indicates the presence of calcium carbonate.

CAUTION

HC1 may cause burns. Use appropriate measures to protect the skin and eyes. If it is splashed on the skin or in the eyes, immediately flush with water and seek medical attention.

5-93. Calcium carbonate is normally desirable in a soil because of the cementing action it provides to add to the soil's stability. In some very dry noncalcareous soils, the absorption of the acid creates the illusion of effervescence. This effect can be eliminated in dry soils by moistening the soil before applying the acid.

5-94. Since cementation is normally developed only after a considerable curing period, it cannot be counted on for strength in most military construction. This test permits better understanding of what appears to be abnormally high strength values on fine-grained soils that are tested in-place where this property may exert considerable influence.

Shine Test

5-95. The shine test is another means of measuring the plasticity characteristics of clays. A slightly moist or dry piece of highly plastic clay will give a definite shine when rubbed with a fingernail, a pocket knife blade, or any smooth metal surface. On the other hand, a piece of lean clay will not display any shine but will remain dull.

Feel Test

5-96. The feel test is a general-purpose test and one that requires considerable experience and practice before reliable results can be obtained. This test will be used more as familiarity with soils increases. Moisture content and texture can be readily estimated by using the feel test.

5-97. The natural moisture content of a soil indicates drainage characteristics, nearness to a water table, or other factors that may affect this property. A piece of undisturbed soil is tested by squeezing it between the thumb and forefinger to determine its consistency. The consistency is described by such terms as "hard," stiff," "brittle," "friable," "sticky," "plastic," or "soft." Remold the soil by working it in the hands and observing any changes. By this test, the natural water content is estimated relative to the LL or PL of the soil. Clays that turn almost liquid on remolding are probably near or above the LL. If the clay re-mains stiff and crumbles on being remolded, the natural water content is below the PL.

5-98. The term "texture," as applied to the fine-grained portion of a soil, refers to the degree of fineness and uniformity. Texture is described by such expressions as "floury," "smooth," "gritty," or "sharp," depending on the feel when the soil is rubbed between the fingers. Sensitivity to this sensation may be increased by rubbing some of the material on a more tender skin area, such as the inside of the wrist. Fine sand will feel gritty. Typical dry silts will dust readily and feel relatively soft and silky to the touch. Clay soils are powdered only with difficulty but become smooth and gritless like flour.

HASTY FIELD IDENTIFICATION

5-99. With the standard methods of field identification supplemented with a few simplified field tests, an approximate and hasty classification of almost any soil can be obtained. The simple or hasty tests outlined in figure 5-7, pages 5-25 through 5-27, will, for the most part, eliminate the need for specialized equipment such as sieves. The results will give at least a tentative classification to almost any soil. The schematic diagram in figure 5-7, may be used as a guide to the testing sequence in the process of assigning a symbol to a sample of soil.

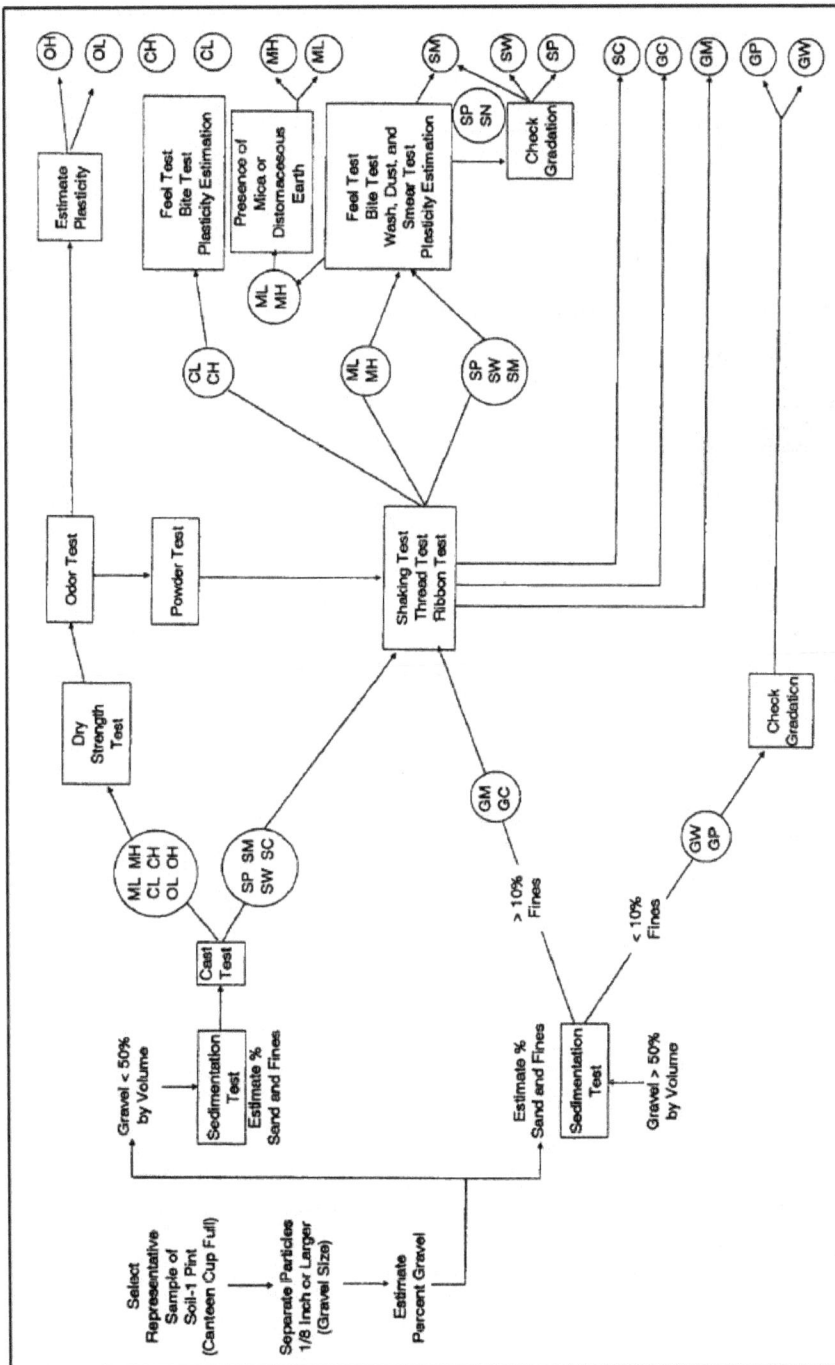

Figure 5-7. Suggested procedure for hasty field identification

1. Select a random but typical sample of soil.

2. Separate the gravel.
   (a) Remove from the sample all particles larger than 1/4 inch in diameter.
   (b) Estimate the percent of gravel.

3. Use the sedimentation test to determine the percent of sand.
   (a) Place the sample (less gravel) in a canteen cup and fill it with water.
   (b) Shake the mixture vigorously.
   (c) Allow the mixture to stand for 30 seconds to settle out.
   (d) Pour the water containing the suspended fines into another container.
   (e) Repeat steps (b) through (d) until the water poured off is clear.
   (f) Dry the soil in the cup (sand).
   (g) Estimate the percent of sand.

4. Compare the gravels, sands, and fines.
   (a) The gravels have been estimated in test (2), step (b).
   (b) The sands have been estimated in test (3), step (g).
   (c) Dry the soil remaining in the second container (fines).
   (d) Estimate the percent of fines.

5. Cast test.
   (a) Compress a handful of moist (but not sticky) soil into a ball or cigar-shaped cast.
   (b) Observe the ability of the cast to withstand handling without crumbling.
   (c) If the cast crumbles when touched, the sample is a sand with little or not fines (SW) or (SP).
   (d) If the cast withstands careful handling, the sample is a sand with an appreciable amount of fines (SM) or (SC).
   (e) If the cast can be handled freely or withstands rough handling, the sample is either silt, clay, or organic.

6. Dry strength test.*
   (a) Form a moist pat 2 inches in diameter by 1/2 inch thick.
   (b) Allow it to dry with low heat.
   (c) Place the dry pat between the thumb and index finger only and attempt to break it.
   (d) If breakage is easy, it is a slightly plastic silt (ML).
   (e) If breakage is difficult, it is a medium plastic and medium compressible clay (CL) or a highly compressible silt (MH).
   (f) If breakage is impossible, it is a highly plastic and highly compressible clay (CH).

7. Odor test.
   (a) Heat the sample with a match or open flame.
   (b) If the odor becomes musty or foul smelling, there is a strong indication that organic material is present.

8. Powder test.*
   (a) Rub a portion of the broken pat with the thumb and attempt to flake particles off.
   (b) If the pat powders, it is silt.
   (c) If the pat does not powder, it is clay.

9. Wet shaking test.*
   (a) Place the pat of moist (not sticky) coil in the palm of the hand (the volume is about 1/2 cubic inch).
   (b) Shake the hand vigorously and strike it against the other hand.
   (c) Observe how rapidly water rises to the surface.
   (d) If it is fast, the sample is silty. If there is not reaction, the sample is clayey (C).

* Tests conducted on material smaller than 1/32 inch in diameter (passes Number 40 sieve).

Figure 5-7. Suggested procedure for hasty field identification (continued)

10. Thread test.*
    (a) Form a ball of moist soil (marble size).
    (b) Attempt to roll the ball into a thread 1/8 inch in diameter.
    (c) If a thread is easily obtained, it is clay
    (d) If a thread cannot be obtained, it is silt.

11. Ribbon test.*
    (a) Form a cylinder of soil that is approximately the shape and size of a cigar.
    (b) Flatten the cylinder over the index finger with the thumb, attempting to form a ribbon 8 to 9 inches long, 1/8 to 1/4 inch thick, and 1 inch wide.
    (c) If 8 to 9 inches is obtained, it is (CH); if less than 8 inches is obtained, it is (CL); if there is no ribbon, it is silt (ML) or sand.

12. Shine test.*
    (a) Draw a smooth surface, such as a knife blade or thumbnail, over a pat of slightly moist soil.
    (b) If the surface becomes shiny and lighter in texture, the sample is a highly plastic compressible clay (CL).
    (c) If the surface remains dull, the sample is a low plasticity compressible clay (CL).
    (d) If the surface remains very dull or granular, the sample is silt or sand.

13. Feel test.*
    (a) Rub a portion of dry soil over a sensitive portion of the skin, such as the inside of the wrist.
    (b) If the feel is harsh and irritating, the sample is silt (ML) or sand.
    (c) If the feel is smooth and floury, the sample is clay.

14. Grit or bite test.*
    (a) Place a pinch of the sample between the teeth and bite.
    (b) If the sample feels gritty, the sample is silt (ML) or sand.
    (c) If the sample feels floury, the sample is clay.

15. Wash, dust, and smear test.*
    (a) Drop a completely dry sample of soil from a height of 1 to 2 feet onto a solid surface.
    (b) If a fairly large amount of dust is produced, the sample is a silty sand (SM).
    (c) If very little dust is produced, the sample is a clean sand (SW or SP). Check the gradation.
    (d) Smear a moistened (just below "sticky limit") sample of soil between the thumb and forefinger.
    (e) If a gritty, harsh feel is produced, the soil contains a small amount of silt (SW or SP). Check the gradation.
    (f) If a rough, less harsh feel is produced, the soil contains about 10 percent silt (SM).

16. Plasticity estimation.
    (a) Remove all particles coarser than a grain of sugar or table salt from the sample.
    (b) From the remaining soil particles, mold a small cube tot he consistency of stiff putty.
    (c) Dry the cube in the air or sunlight.
    (d) Crush the cube between the fingers.
    (e) If the cube falls apart easily, the soil is nonplastic; PI range 0 to 3.
    (f) If the cube is easily crushed, the soil is slightly plastic; PI range, 4 to 8.
    (g) If the cube is difficult to crush he soil is medium plastic; PI range, 9 to 30.
    (h) If the cube is impossible to crush the soil is highly plastic; PI range, 31 or more.

* Tests conducted on material smaller than 1/32 inch in diameter (passes Number 40 sieve).

Figure 5-7. Suggested procedure for hasty field identification (continued)

OPTIMUM MOISTURE CONTENT (OMC)

5-100.   To determine whether a soil is at or near OMC, mold a golf-ball-size sample of the soil with your hands. Then squeeze the ball between your thumb and forefinger. If the ball shatters into several fragments of rather uniform size, the soil is near or at OMC. If the ball flattens out without breaking, the soil is wetter than OMC. If, on the other hand, the soil is difficult to roll into a ball or crumbles under very little pressure, the soil is drier than OMC.

SECTION II - OTHER SOIL CLASSIFICATION SYSTEMS

COMMONLY USED SYSTEMS

5-101.   Information about soils is available from many sources, including publications, maps, and reports. These sources may be of value to the military engineer in studying soils in a given area. For this reason, it is important that the military engineer have some knowledge of other commonly used systems. Table 5-5 gives approximate equivalent groups for the USCS, Revised Public Roads System, and the Federal Aviation Administration (FAA) System.

Table 5-5. Comparison of the USCS, revised public roads system, and FAA system

USCS	Revised Public Roads System	Federal Aviation Administration System	
GW	A-1-a	Gravelly soils not included directly	
GP	A-1-a		
GM	A-1-a, A-2-4 or 5		
GC	A-2-6 or 7		
SW	A-1-b	E-1, 2 or 3	E-4 or 5 (usually SM or SC)
SP	A-3		
SM	A-1-b, A-2-4 or 5		
SC	A-2-6 or 7		
MI	A-4	E-6	
CL	A-6, A-7-5		
OL	A-4, A-7-5	E-6	E-8 (usually L group) E-9 (usually not CH)
MH	A-5		
CH	A-7	E-10, 11, or 12	
OH	A-7		
Pt		E-13	

Note. Groups are only approximately equivalent, since different limiting values are used in each system.

REVISED PUBLIC ROADS SYSTEM

5-102.   Most civil agencies concerned with highways in the United States classify soil by the Revised Public Roads System. This includes the Bureau of Public Roads and most of the state highway departments. The Public Roads System was originated in 1931. Part of the original system, which applied to uniform subgrade soils, used a number of tables and charts based on several routine soil tests to permit placing of a given soil into one of eight principal groups, designated A-1 through A-8. The system was put into use by many agencies. As time passed, it became apparent that some of the groups were too broad in coverage because somewhat different soils were classed in the same group. A number of the agencies using the system modified it to suit their purposes. Principal modifications included breaking down some of the broad groups into subgroups of more limited scope. The revisions culminated in a comprehensive committee report that appeared in the Proceedings of the 25th Annual Meeting of the Highway Research Board (1945). This same report contains detailed information relative to the Airfield Classification System and the Federal Aviation Administration System. The Revised Public Roads System is primarily designed for the evaluation of subgrade soils, although it is useful for other purposes also.

Basis

5-103.   Table 5-6, page 5-30, shows the basis of the Revised Public Roads System. Soils are classed into one of two very broad groups. They are—

- Granular materials, which contain < 35 percent of material passing a Number 200 sieve.
- Silt-clay materials, which contain > 35 percent of material passing a Number 200 sieve.

5-104.   There are seven major groups, numbered A-1 through A-7, together with a number of suggested subgroups. The A-8 group of the original system, which contained the highly organic soils such as peat, is not included in the revised system. The committee felt that no group was needed for these soils because of their ready identification by appearance and odor. Whether a soil is silty or clayey depends on its PI. "Silty" is applied to material that has a PI < 10 and "clayey" is applied to a material that has a PI ≥ 10.

5-105.   Figure 5-8, page 5-31, shows the formula for group index and charts to facilitate its computation. The group index was devised to provide a basis for approximating within-group evaluations. Group indexes range from 0 for the best subgrade soils to 20 for the poorest. Increasing values of the group index within each basic soil group reflect the combined effects of increasing LLs and PIs and decreasing percentages of coarse material in decreasing the load-carrying capacity of subgrades. Figure 5-9, page 5-32, graphically shows the ranges of LL and PI for the silt-clay groups. It is particularly useful for subdividing the soils of the A-7 group.

Table 5-6. Revised public roads system of soil classification

General Classifications	Granular Materials (35 percent or less of total sample passing No 200 sieve)							Silt-Clay Materials (more than 35 percent of total sample passing No 200 sieve)			
	A-1		A-3	A-2				A-4	A-5	A-6	A-7
Group classification	A-1-a	A-1-b	A-3	A-2-4	A-2-5	A-2-6	A-2-7	A-4	A-5	A-6	A-7-5 A-7-6
Sieve analysis, percent passing:											
No 10 sieve	50 max	—	—	—	—	—	—	—	—	—	—
No 40 sieve	30 max	50 max	51 min	—	—	—	—	—	—	—	—
No 200 sieve	15 max	25 max	10 max	35 max	35 max	35 max	35 max	36 min	36 min	36 min	36 min
Characteristics of portion passing:											
Liquid limit		—	—	40 max	41 min	40 max	41 min	40 max	41 min	40 max	41 min*
Plasticity index		6 max	NP**	10 max	10 max	11 min	11 min	10 max	10 max	10 max	10 max
Group index[6]		0	0	0	0	4 max	4 max	8 max	12 max	16 max	20 max

* PI of A-7-5 subgroup is ≤ LL minus 30. PI of A-7-6 subgroup is > LL minus 30.
** Nonplastic

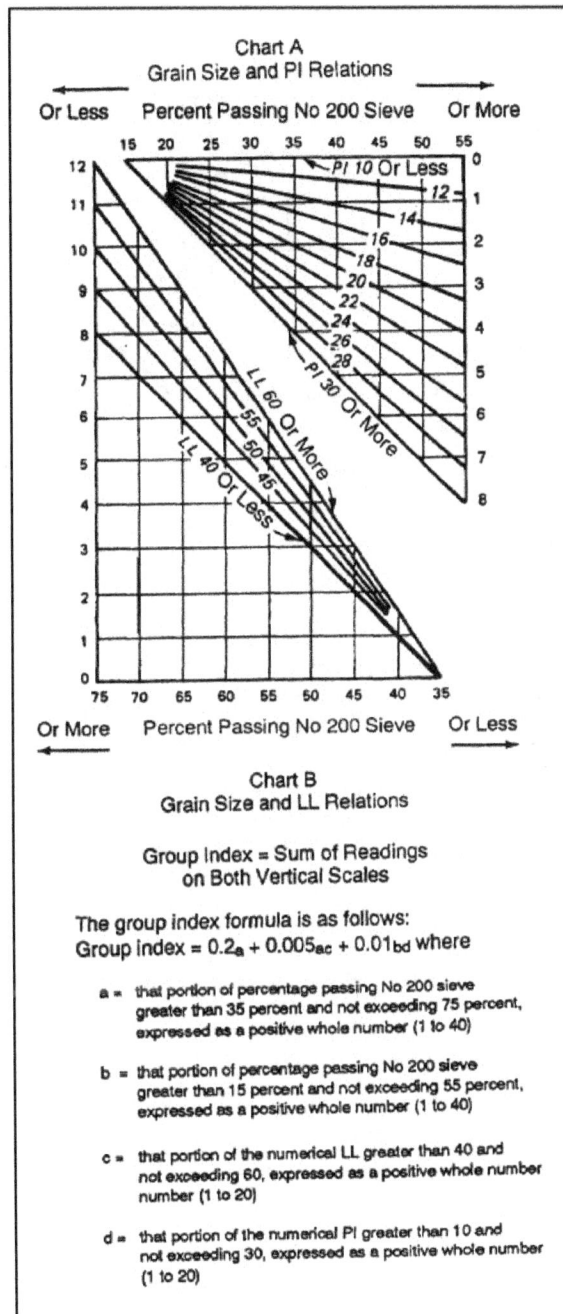

Chart A
Grain Size and PI Relations

Or Less   Percent Passing No 200 Sieve   Or More

Chart B
Grain Size and LL Relations

Group Index = Sum of Readings
on Both Vertical Scales

The group index formula is as follows:
Group index = $0.2_a + 0.005_{ac} + 0.01_{bd}$ where

a = that portion of percentage passing No 200 sieve
greater than 35 percent and not exceeding 75 percent,
expressed as a positive whole number (1 to 40)

b = that portion of percentage passing No 200 sieve
greater than 15 percent and not exceeding 55 percent,
expressed as a positive whole number (1 to 40)

c = that portion of the numerical LL greater than 40 and
not exceeding 60, expressed as a positive whole number
number (1 to 20)

d = that portion of the numerical PI greater than 10 and
not exceeding 30, expressed as a positive whole number
(1 to 20)

Figure 5-8. Group index formula and charts,
revised public roads system

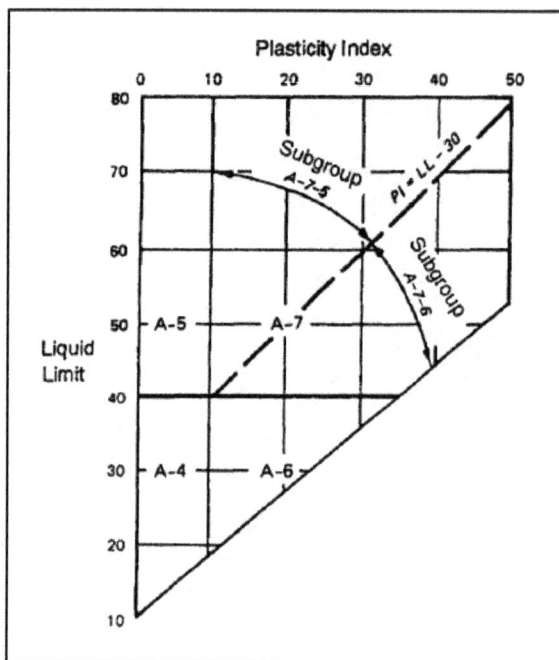

Figure 5-9. Relationship between LL and PI for silt-clay groups, revised public roads system

Procedure

5-106.   Table 5-6, page 5-30, is used in a left-to-right elimination process, and the given soil is placed into the first group or subgroup in which it fits. In order to distinguish the revised from the old system, the group symbol is given, followed by the group index in parentheses (for example, A-4(5)). The fact that the A-3 group is placed ahead of the A-2 group does not imply that it is a better subgrade material. This arrangement is used to facilitate the elimination process. The classification of some borderline soils requires judgment and experience. The assignment of the group designation is often accompanied by the writing of a careful description, as in the USCS. Detailed examples of the classification procedure are given later in this chapter.

AGRICULTURAL SOIL CLASSIFICATION SYSTEM

5-107.   Many reports published for agricultural purposes can be useful to the military engineer. Two phases of the soil classification system used by agricultural soil scientists are discussed here. These are—

- Grain-size classification (textural classification).
- Pedological classification.

Textural Classification

5-108.   Information about the two textural classification systems of the US Department of Agriculture is contained in figure 5-10 and table 5-7, page 5-36. The chart and table are largely self-explanatory. The grain-size limits, which are applicable to the categories shown, are as follows:

- Coarse gravel, retained on Number 4 sieve.
- Fine gravel, passing Number 4 and retained on Number 10 sieve.

- Coarse sand, passing Number 10 and retained on Number 60 sieve.
- Fine sand, passing Number 60 and retained on Number 270 sieve (0.05 mm).
- Silt, 0.05 to 0.005 mm.
- Clay, below 0.005 mm.

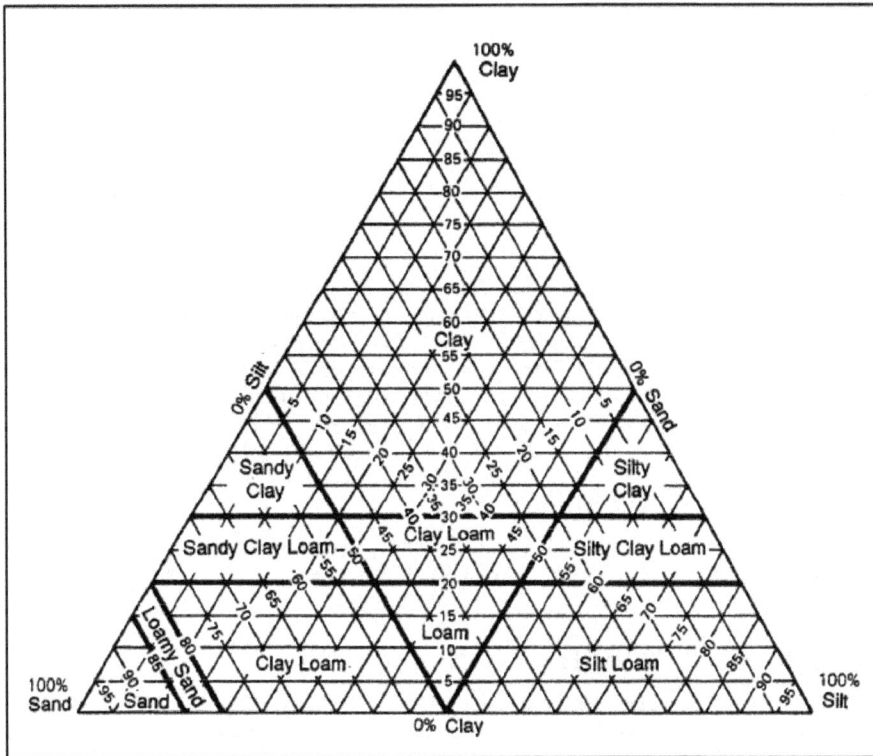

Figure 5-10. US Department of Agriculture textural classification chart

Pedological Classification

5-109. Soil profile and pedology were discussed in chapter 1. Agricultural soil scientists have devised a complete and complex system for describing and classifying surface soils. No attempt will be made to discuss this system in detail here. The portion of the system in which engineers are principally interested refers to the terms used in mapping limited areas. Mapping is based on—

- Series.
- Type.
- Phase.

5-110. The designation known as soil series is applied to soils that have the same genetic horizons, possess similar characteristics and profiles, and are derived from the same parent material. Series names follow no particular pattern but are generally taken from the geographical place near where they were first found. The soil type refers to the texture of the upper portion of the soil profile. Several types may, and usually do, exist within a soil series. Phase is variation, usually of minor importance, in the soil type.

5-111. A mapping unit may be "Emmet loamy sand, gravelly phase" or any one of the large number of similar designations. Soils given the same designation generally have the same agricultural properties wherever they are encountered. Many of their engineering properties may also be the same. The mapping unit in which a given soil is placed is determined by a careful examination of the soil sample obtained by using an auger to bore or by observing highway cuts, natural slopes, and other places where the soil profile is exposed. Particularly important factors are—

- Color.
- Texture.
- Organic material.
- Consistency.

5-112. Other important factors are—

- Slope.
- Drainage.
- Vegetation.
- Land use.

5-113. Agricultural soil maps prepared from field surveys show the extent of each important soil type and its geographical location. Reports that accompany the maps contain word descriptions of the various types, some laboratory test results, typical profiles, and soil properties important to agricultural use. Frequently prepared on a county basis, soil surveys (maps and reports) are available for many areas in the United States and in many foreign countries. For installation projects, check with the Natural Resources Branch or Land Management Branch of the Directorate of Engineering and Housing for a copy of the soil survey. For off-post projects, request a copy of the soil survey from the county Soil Conservation Service (SCS). The information contained in the soil survey is directly useful to engineers.

GEOLOGICAL SOIL CLASSIFICATION

5-114. Geologists classify soils according to their origin (process of formation) following a pattern similar to that used in chapter 1. TM 5-545 gives a geological classification of soil deposits and related information.

TYPICAL SOIL CLASSIFICATION

5-115. The following paragraphs concern the classification of four inorganic soil types on the basis of laboratory test data. Each soil is classified under the Unified Soil Classification System, the Revised Public Roads System, and the Agricultural Soil Classification System. Table 5-8, page 5-37, shows the information known about each soil.

UNIFIED SOIL CLASSIFICATION SYSTEM

5-116. The results of the classification of the four soil types under this system are as follows:

Soil Number 1

5-117. The soil is fine-grained since more than half passes the Number 200 sieve. The LL is less than 50; therefore, it must be (ML) or (CL) since it is inorganic. On the plasticity chart, it falls below the A-line; therefore, it is a sandy silt (ML).

Soil Number 2

5-118. The soil is fine-grained since more than half passes the Number 200 sieve. The LL is more than 50; therefore, it must be (MH) or (CH). On the plasticity chart, it falls above the A-line; therefore, it is a sandy clay (CH).

Soil Number 3

5-119. The soil is coarse-grained since very little passes the Number 200 sieve. It must be a sand since it all passes the Number 10 sieve. The soil contains less than 5 percent passing the Number 200 sieve;

therefore, it must be either an (SW) or an (SP) (see table 5-2, page 5-8). The value of $C_u = 2$ will not meet requirements for (SW); therefore, it is a poorly graded sand (SP).

Soil Number 4

5-120. The soil is coarse-grained since again very little passes the Number 200 sieve. It must be a gravel since more than half of the coarse fraction is larger than a Number 4 sieve. Since the soil contains less than 5 percent passing the Number 200 sieve, it is either (GW) or (GP) (see table 5-2). It meets the gradation requirements relative to C_u and C_c; therefore, it is a well-graded gravel (GW).

REVISED PUBLIC ROADS CLASSIFICATION SYSTEM

5-121. The results of the classification of the four soil types under this system are as follows:

Soil Number 1

5-122. To calculate the group index, refer to figure 5-8, page 5-31. From chart A, read 0; from chart B, read 3; therefore, the group index = 0 + 3 = 3. Table 5-6, page 5-30, shows by a left-to-right elimination process, that the soil cannot be in one of the granular materials groups, since more than 35 percent passes a Number 200 sieve. It meets the requirements of the A-4 group; therefore, it is A-4(3).

Soil Number 2

5-123. The group index = 8 + 10 = 18. Table 5-6 shows that the soil falls into the A-7 group, since this is the only group that will permit a group index value as high as 18. Figure 5-9, page 5-32, shows that it falls in the A-7-6 subgroup; therefore, A-7-6(18).

Soil Number 3

5-124. The group index = 0 + 0 = 0. This is one of the soils described as granular material. It will not meet the requirements of an A-1 soil, since it contains practically no fines. It does not meet the requirements of the A-3 group; therefore, it is A-3(0) (see table 5-6).

Soil Number 4

5-125. The group index = 0 + 0 = 0. This is obviously a granular material and meets the requirements of the A-1-a (0) (see table 5-6).

AGRICULTURAL SOIL CLASSIFICATION SYSTEM

5-126. Although the values are not given in the previous tabulation, assume that 12 percent of Soil Number 1 and 35 percent of Soil Number 2 are in the range of clay sizes that is below 0.005 mm.

Soil Number 1

5-127. This soil contains 100 - 48.2 = 51.8 percent sand, since the opening of a Number 270 sieve is 0.05 mm. The soil is then composed of 52 percent sand, 36 percent silt, and 12 percent clay. Figure 5-10, page 5-33, classifies this soil as a sandy loam.

Soil Number 2

5-128. This soil contains approximately 35 percent sand, 30 percent silt, and 35 percent clay. Figure 5-10 classifies this soil as clay.

Soil Number 3

5-129. This soil is 99 percent sand; therefore, it can only be classified as sand.

Soil Number 4

5-130. This soil contains approximately 70 percent coarse gravel, 14 percent fine gravel, 13 percent sand, and 3 percent silt and clay combined. It cannot be classified by using figure 5-10, page 5-33, because the chart does not cover gravels and gravelly sands. Table 5-7, classifies the material as gravel and sand.

Table 5-7. Agricultural soil classification system

Gradation Limits of Textural Soil Groups*							
First place the soil in Textural Group I, II, or III according to clay content (column 7) or silt and clay content combined (column 5); and finally the sand content and gravel content (columns 1 to 4 inclusive). When clay content approaches the upper limit for that group, it is called "heavy," such as "heavy clay loam," and a "light" soil when approaching the lower clay limit.							
Soil texture	(1) Fine and coarse gravel, percent	(2) Coarse sand and gravel, percent	(3) Sand and gravel, percent	(4) Fine sand, percent	(5) Silt and clay, percent	(6) Silt, percent	(7) Clay, percent
Group IA: Sands and Gravels							
Gravel	85 to 100	—	—	—	0 to 15	—	0 to 20
Gravel and sand	50 to 85	—	—	—	0 to 15	—	0 to 20
Sand and gravel	25 to 50	—	—	—	0 to 15	—	0 to 20
Coarse sand	0 to 25	50 to 100	—	—	0 to 15	—	0 to 20
Sand	0 to 25	0 to 50	—	0 to 50	0 to 15	—	0 to 20
Fine sand	0 to 25	—	—	0 to 100	0 to 15	—	0 to 20
Group IB: Loamy Sands							
Gravelly, loamy, coarse sand	25 to 85	50 to 85	—	—	15 to 20	—	0 to 20
Gravelly, loamy sand	25 to 50	25 to 50	—	0 to 50	15 to 20	—	0 to 20
Gravelly, loamy, fine sand	25 to 35	—	—	50 to 60	15 to 20	—	0 to 20
Loamy, coarse sand	0 to 25	50 to 85	—	—	15 to 20	—	0 to 20
Loamy sand	0 to 25	0 to 50	—	0 to 50	15 to 20	—	0 to 20
Loamy, fine sand	0 to 25	—	—	50 to 85	15 to 20	—	0 to 20
Group IC Sandy Loams							
Gravelly, coarse, sandy loam	25 to 85	30 to 80	—	—	20 to 50	—	0 to 20
Gravelly, sandy loam	25 to 50	25 to 50	—	0 to 50	20 to 50	—	0 to 20
Gravelly, fine, sandy loam	25 to 30	—	—	50 to 55	20 to 50	—	0 to 20
Coarse, sandy loam	0 to 25	50 to 80	—	—	20 to 50	—	0 to 20
Sandy loam	0 to 25	0 to 50	—	0 to 50	20 to 50	—	0 to 20
Fine, sandy loam	0 to 25	—	—	50 to 80	20 to 50	—	0 to 20
Groups ID: Loams and Silt Loams							
Gravelly loam	25 to 50	—	—	—	50 to 70	30 to 50	0 to 20
Gravelly, silt loam	25 to 50	—	—	—	50 to 100	50 to 80	0 to 20
Loam	0 to 25	—	—	—	50 to 70	30 to 50	0 to 20
Silt loam	0 to 25	—	—	—	50 to 100	50 to 80	0 to 20
Silt	0 to 25	—	—	—	50 to 100	80 to 100	0 to 20
Groups II: Clay Loams							
Gravelly, sandy, clay loam	25 to 80	—	50 to 80	—	—	—	20 to 30
Gravelly, clay loam	50 to 50	—	25 to 50	—	—	20 to 50	20 to 30
Gravelly, silty, clay loam	25 to 30	—	0 to 30	—	—	50 to 55	20 to 30
Sandy, clay loam	0 to 25	—	50 to 80	—	—	0 to 30	20 to 30
Clay loam	0 to 25	—	20 to 50	—	—	20 to 50	20 to 30
Silty, clay loam	0 to 25	—	0 to 30	—	—	50 to 80	20 to 30
Group III: Clays							
Gravelly, sandy clay	25 to 70	—	50 to 70	—	—	—	30 to 100
Gravelly clay	25 to 50	—	25 to 50	—	—	0 to 45	30 to 100
Sandy clay	0 to 25	—	50 to 70	—	—	0 to 20	30 to 100
Silty clay	0 to 25	—	0 to 20	—	—	50 to 70	30 to 100
Clay	0 to 25	—	0 to 50	—	—	0 to 20	30 to 100
• The basic concept for this text from the US Bureau of Chemistry and Soils.							

Table 5-8. Classification of four inorganic soil types

Soil Number				
	1	*2*	*3*	*4*
Mechanical Analysis (Percent Passing by Weight)				
3-inch sieve	—	—	—	100.0
¾-inch sieve	—	—	—	56.0
Number 4 sieve	—	—	—	30.0
Number 10 sieve	100.0	100.0	100.0	16.4
Number 40 sieve	85.2	97.6	85.0	7.2
Number 60 sieve	—	—	20.0	5.0
Number 200 sieve	52.1	69.8	1.2	3.5
Number 270 sieve	48.2	65.0	—	—
Numerical Values				
C_u	—	—	2.0	12.5
C_c	—	—	—	2.2
Plasticity Characteristics				
Liquid limit	29	67	21	—
Plasticity index	5	39	NP*	—
* Nonplastic				

COMPARISON OF CLASSIFICATION SYSTEMS

5-131. Table 5-9 is a summary of the classification of the soils in question under the three different classification systems considered.

Table 5-9. Comparison of soils under three classification system

Soil Number	Unified Soil Classification System	Revised Public Roads System	Agricultural Soil Classification System
1	ML	A-4(3)	Sandy loam
2	CH	A-7-6 (18)	Clay
3	SP	A-3(0)	Sand
4	GW	A-1-a (0)	Gravel and sand

This page intentionally left blank.

Chapter 6

Concepts of Soil Engineering

The military engineer encounters a wide variety of soils in varying physical states. Because of the inherent variability of the physical properties of soil, several tests and measurements within soils engineering were developed to quantify these differences and to enable the engineer to apply the knowledge to economical design and construction. This chapter deals with soil engineering concepts, to include settlement, shearing resistance, and bearing capacity, and their application in the military construction arena.

SECTION I - SETTLEMENT

FACTORS

6-1. The magnitude of a soil's settlement depends on several factors, including—

- Density.
- Void ratio.
- Grain size and shape.
- Structure.
- Past loading history of the soil deposit.
- Magnitude and method of application of the load.
- Degree of confinement of the soil mass.

6-2. Unless otherwise stated, it is assumed that the soil mass undergoing settlement is completely confined, generally by the soil that surrounds it.

COMPRESSIBILITY

6-3. Compressibility is the property of a soil that permits it to deform under the action of an external compressive load. Loads discussed in this chapter are primarily static loads that act, or may be assumed to act, vertically downward. Brief mention will be made of the effects of vibration in causing compression. The principal concern here is with the property of a soil that permits a reduction in thickness (volume) under a load like that applied by the weight of a highway or airfield. The compressibility of the underlying soil may lead to the settlement of such a structure.

COMPRESSIVE LOAD BEHAVIOR

6-4. In a general sense, all soils are compressible. That is, they undergo a greater or lesser reduction in volume under compressive static loads. This reduction in volume is attributed to a reduction in volume of the void spaces in the soil rather than to any reduction in size of the individual soil particles or water contained in the voids.

6-5. If the soil is saturated before the load is applied, some water must be forced from the voids before settlement can take place. This process is called consolidation. The rate of consolidation depends on how quickly the water can escape, which is a function of the soil's permeability.

COHESIONLESS SOILS

6-6.   The compressibility of confined coarse-grained cohesionless soils, such as sand and gravel, is rarely a practical concern. This is because the amount of compression is likely to be small in a typical case, and any settlement will occur rapidly after the load is applied. Where these soils are located below the water table, water must be able to escape from the stratum. In the case of coarse materials existing above the water table and under less than saturated conditions, the application of a static load results in the rearrangement of soil particles. This produces deformation without regard to moisture escape. So, generally speaking, settlement occurs during the period of load application (construction). Deformations that are thus produced in sands and gravels are essentially permanent in character. There is little tendency for the soil to return to its original dimensions or rebound when the load is removed. A sand mass in a compact condition may eventually attain some degree of elasticity with repeated applications of load.

6-7.   The compressibility of a loose sand deposit is much greater than that of the same sand in a relatively dense condition. Generally, structures should not be located on loose sand deposits. Avoid loose sand deposits if possible, or compact to a greater density before the load is applied. Some cohesionless soils, including certain very fine sands and silts, have loose structures with medium settlement characteristics. Both gradation and grain shape influence the compressibility of a cohesionless soil. Gradation is of indirect importance in that a well-graded soil generally has a greater natural density than one of uniform gradation. Soils that contain platy particles are more compressible than those composed entirely of bulky grains. A fine sand or silt that contains mica flakes may be quite compressible.

6-8.   Although soils under static loads are emphasized here, the effects of vibration should also be mentioned. Vibration during construction may greatly increase the density of cohesionless soils. A loose sand deposit subjected to vibration after construction may also change to a dense condition. The latter change in density may have disastrous effects on the structures involved. Cohesionless soils are usually compacted or "densified" as a planned part of construction operations. Cohesive soils are usually insensitive to the effects of vibrations.

CONSOLIDATION

6-9.   Consolidation is the time-dependent change in volume of a soil mass under compressive load that occurs when water slowly escapes from the pores or voids of the soil. The soil skeleton is unable to support the load and changes structure, reducing its volume and producing vertical settlement.

COHESIVE SOILS

6-10.   The consolidation of cohesive, fine-grained soils (particularly clays) is quite different from the compression of cohesionless soils. Under comparable static loads, the consolidation of a clay may be much greater than coarse-grained soils and settlement may take a very long time to occur. Structures often settle due to consolidation of a saturated clay stratum. The consolidation of thick, compressible clay layers is serious and may cause structural damage. In uniform settlement, the various parts of a structure settle approximately equal amounts. Such uniform settlement may not be critical. Nonuniform, or differential, settlement of parts of a structure due to consolidation causes serious structural damage. A highway or airfield pavement may be badly damaged by the nonuniform settlement of an embankment founded on a compressible soil.

CONSOLIDATION TESTS

6-11.   The consolidation characteristics of a compressible soil should be determined for rational design of many large structures founded on or above soils of this type. Consolidation characteristics generally are determined by laboratory consolidation tests performed on undisturbed samples. The natural structure, void ratio, and moisture content are preserved as carefully as possible for undisturbed samples. However, military soils analysts are not equipped or trained to perform consolidation tests.

6-12.   Information on consolidation tests and settlement calculations for the design of structures to be built on compressible soils may be found in Naval Facilities Engineering Command Design Manual (NAVFAC DM)-7.1.

SECTION II - SHEARING RESISTANCE

IMPORTANCE

6-13. From an engineering viewpoint, one of the most important properties a soil possesses is shearing resistance or shear strength. A soil's shearing resistance under given conditions is related to its ability to withstand loads. The shearing resistance is especially important in its relation to the supporting strength, or bearing capacity, of a soil used as a base or subgrade beneath a road, runway, or other structure. The shearing resistance is also important in determining the stability of the slopes used in a highway or airfield cut or embankment and in estimating the pressures exerted against an earth-retaining structure, such as a retaining wall.

LABORATORY TESTS

6-14. Three test procedures are commonly used in soil mechanics laboratories to determine the shear strength of a soil. These are—

- Direct shear test.
- Triaxial compression test.
- Unconfined compression test.

6-15. The basic principles involved in each of these tests are illustrated in the simplified drawings of figure 6-1, page 6-4. Military soils analysts are not equipped or trained to perform the direct shear or triaxial compression tests. For most military applications, the CBR value of a soil is used as a measure of shear strength (see TM 5-530).

6-16. A variation of the unconfined compression test can be performed by military soils analysts, but the results are ordinarily used only in evaluation of soil stabilization. Shear strength for a soil is expressed as a combination of an apparent internal angle of friction (normally associated with cohesive soils).

CALIFORNIA BEARING RATIO

6-17. The CBR is a measure of the shearing resistance of a soil under carefully controlled conditions of density and moisture. The CBR is determined by a penetration shear test and is used with empirical curves for designing flexible pavements. Recommended design procedures for flexible pavements are presented in TM 5-330. The CBR test procedure for use in design consists of the following steps:

- Prepare soil test specimens.
- Perform penetration test on the prepared soil samples.
- Perform swell test on soil test specimens.

6-18. Although a standardized procedure has been established for the penetration portion of the test, one standard procedure for the preparation of test specimens cannot be established because soil conditions and construction methods vary widely. The soil test specimen is compacted so it duplicates as nearly as possible the soil conditions in the field.

6-19. In a desert environment, soil may be compacted and tested almost completely dry. In a wet area, soil should probably be tested at 100 percent saturation. Although penetration tests are most frequently performed on laboratory-compacted test specimens, they may also be performed on undisturbed soil samples or on in-place soil in the field. Detailed procedures for conducting CBR tests and analyzing the data are in TM 5-330. Appendix A describes the procedure for applying CBR test data in designing roads and airfields.

6-20. Column 15, table 5-3, page 5-13, shows typical ranges in value of the field CBR for soils in the USCS. Values of the field CBR may range from as low as 3 for highly plastic, inorganic clays (CH) and some organic clays and silts (OH) to as high as 80 for well-graded gravel and gravel-sand mixtures.

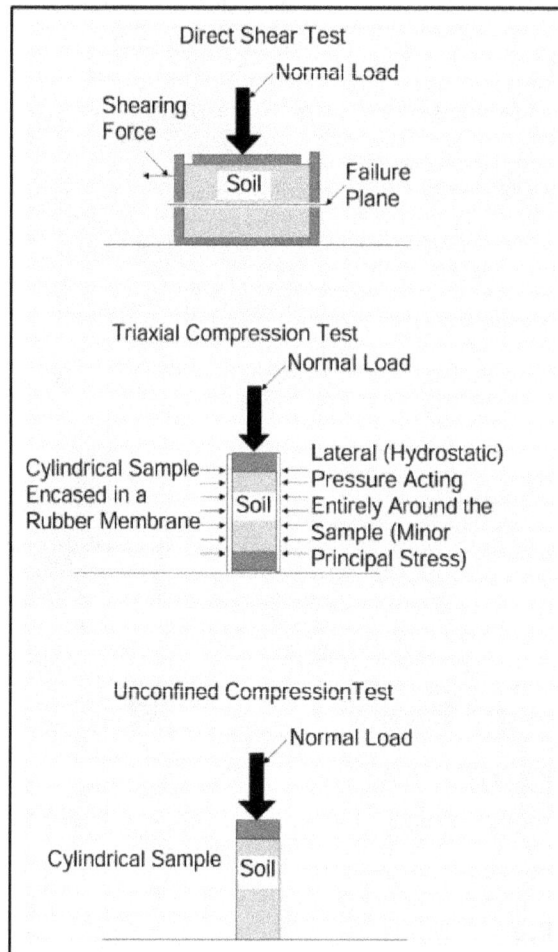

Figure 6-1. Laboratory shear tests

AIRFIELD INDEX (AI)

6-21. Engineering personnel use the airfield cone penetrometer to determine an index of soil strengths (called Airfield Index) for various military applications.

AIRFIELD CONE PENETROMETER

6-22. The airfield cone penetrometer is compact, sturdy, and simple enough to be used by military personnel inexperienced in soil strength determination. If used correctly, it can serve as an aid in maintaining field control during construction operations; however, this use is not recommended, because more accurate methods are available for use during construction.

Description

6-23. The airfield cone penetrometer is a probe-type instrument consisting of a right circular cone with a base diameter of ½ inch mounted on a graduated staff. On the opposite end of the staff are a spring, a load indicator, and a handle. The overall length of the assembled penetrometer is 36 ⅛ inches. For ease in carrying, the penetrometer can be disassembled into three main pieces. They are—

- Two extension staffs, each 12 ⅝ inches long.
- One piece 14 ¾ inches long containing the cone, handle, spring, and load indicator.

6-24. The airfield cone penetrometer has a range of zero to 15.

6-25. The airfield cone penetrometer must not be confused with the trafficability penetrometer, a standard military item included in the Soil Test Set. The cone penetrometer used for trafficability has a dial-type load indicator (zero to 300 range) and is equipped with a cone ½ inch in diameter and a cross-sectional area of 0.2 square inch and another cone 0.8 inch in diameter and a cross-sectional area of 0.5 square inch. If the trafficability penetrometer is used to measure the AI, the readings obtained with the 0.2-square-inch cone must be divided by 20; the reading obtained with the 0.5-square-inch cone must be divided by 50.

Operation

6-26. Before the penetrometer is used, inspect the instrument to see that all joints are tight and that the load indicator reads zero. To operate the penetrometer, place your palms down symmetrically on the handle. Steady your arms against your thighs and apply force to the handle until a slow, steady, downward movement of the instrument occurs. Read the load indicator at the moment the base of the cone enters the ground (surface reading) and at desired depths at the moment the corresponding depth mark on the shaft reaches the soil surface. The reading is made by shifting the line of vision from the soil surface to the indicator just a moment before the desired depth is reached. Maximum efficiency is obtained with a two-person team in which one person operates and reads the instrument while the other acts as a recorder. One person can operate the instrument and record the measurements by stopping the penetration at any intermediate depth, recording previous readings, and then resuming penetration. Observe the following rules to obtain accurate data:

- Make sure the instrument reads zero when suspended by the handle and 15 when a 150-pound load is applied.
- Keep the instrument in a vertical position while it is in use.
- Control the rate of penetration at about ½ to 1 inch per second. (Slightly faster or slower rates will not materially affect the readings, however.)
- If you suspect the cone is encountering a stone or other foreign body at the depth where a reading is desired, make another penetration nearby.
- Take readings at the proper depths. (Carelessness in determining depth is one significant source of error in the use of the penetrometer.)

Maintenance

6-27. The airfield cone penetrometer is simply constructed of durable metals and needs little care other than cleaning and oiling. The calibration should be checked occasionally. If an error in excess of about 5 percent is noted, recalibrate the penetrometer.

SOIL-STRENGTH EVALUATION

6-28. The number of measurements to be made, the location of the measurements, and other such details vary with each area to be examined and with the time available. For this reason, hard and fast rules for evaluating an airfield are not practical, but the following instructions are useful:

Fine-Grained Soils

6-29. A reading near zero can occur in a very wet soil; it cannot support traffic. A reading approaching 15 occurs in dry, compact clays or silts and tightly packed sands or gravels. Most aircraft that might be required to use an unpaved area could easily be supported for a substantial number of landings and takeoffs on a soil having an AI of 15.

6-30. Soil conditions are extremely variable. As many penetrometer measurements should be taken as time and circumstances permit. The strength range and uniformity of an area controls the number of measurements necessary. Areas obviously too soft for emergency landing strips will be indicated after a few measurements, as will areas with strengths that are more than adequate. In all areas, the spots that appear to be softest should be tested first, since the softest condition of an area controls suitability. Soft spots are not always readily apparent. If the first test results indicate barely adequate strength, the entire area should be examined. Penetrations in areas that appear to be firm and uniform may be few and widely spaced, perhaps every 50 feet along the proposed centerline. In areas of doubtful strength, the penetrations should be more closely spaced, and areas on both sides of the centerline should be investigated. No fewer than three penetrations should be made at each location and usually five are desirable. If time permits, or if inconsistencies are apparent, as many as 10 penetrations should be made at each test location.

6-31. Soil strength usually increases with depth; but in some cases, a soil has a thin, hard crust over a deep, soft layer or has thin layers of hard and soft material. For this reason, each penetration should be made to a 24-inch depth unless prevented by a very firm condition at a lesser depth. When penetration cannot be made to the full 24-inch depth, a hole should be dug or augured through the firm materials, and penetrometer readings should be taken in the bottom of the hole to ensure that no soft layer underlies the firm layer. If possible, readings should be taken every 2 inches from the surface to a depth of 24 inches. Generally, the surface reading should be disregarded when figures are averaged to obtain a representative AI.

6-32. In the normal soil condition, where strength increases with depth, the readings at the 2- to 8-inch depths (4 to 10 inches for dry sands and for larger aircraft) should be used to designate the soil strength for airfield evaluation. If readings in this critical layer at any one test location do not differ more than 3 or 4 units, the arithmetic average of these readings can be taken as the AI for the areas represented by the readings. When the range between the highest and lowest readings is more than about 4, the engineer must use judgment in arriving at a rating figure. For conservatism, the engineer should lean toward the low readings.

6-33. In an area in which hard crust less than about 4 inches thick overlies a much softer soil, the readings in the crust should not be used in evaluating the airfield. For example, if a 3-inch-thick crust shows average readings of 10 at the 2-inch depth and average readings of 5 below 3 inches, the area should be evaluated at 5. If the crust is more than about 4 inches thick, it will probably play an important part in aircraft support. If the crust in the above instance is 5 inches thick, the rating of the field would then be about halfway between the 10 of the crust and the 5 of the underlying soil or, conservatively, 7. Innumerable combinations of crust thickness and strength and underlying soil strength can occur. Sound reasoning and engineering judgment should be used in evaluating such areas.

6-34. In an area in which a very soft, thin layer is underlain by a firmer layer, the evaluation also is a matter of judgment. If, for example, there are 1 to 2 inches of soil with an index averaging about 5 overlying a soil with an index of 10, rate the field at 10; but if this soft layer is more than about 4 inches thick, rate the field at 5. Areas of fine-grained soils with very low readings in the top 1 inch or more are likely to be slippery or sticky, especially if the soil is a clay.

Coarse-Grained Soils

6-35. When relatively dry, many sands show increasing AIs with depth, but the 2-inch depth index will often be low, perhaps about 3 or 4. Such sands usually are capable of supporting aircraft that require a much higher AI than 3 to 4, because the strength of the sand actually increases under the confining action of the aircraft tires. Generally, any dry sand or gravel is adequate for aircraft in the C-130 class, regardless of the penetrometer readings. All sands and gravel in a "quick" condition (water moving upward through them) must be avoided. Evaluation of wet sands should be based on the penetrometer readings obtained as described earlier.

6-36. Once the strength of the soil, in terms of AI, has been established by use of the airfield cone penetrometer, the load-carrying capability of this soil can be determined for each kind of forward, support, or rear-area airfield through use of the subgrade strength requirements curves. These curves are based on correlations of aircraft performance and AIs. Unfortunately, these are not exact correlations uniquely relating aircraft performance to AI. As soils vary in type and condition from site to site, so varies the relation of AI to aircraft performance. For this reason, the curves may not accurately reflect performance in all cases. These relations were selected so that in nearly all cases aircraft performance will be equal to or better than that indicated.

CORRELATION BETWEEN CBR AND AI

6-37. Expedient soil strength measurements in this manual are treated in terms of AI. Measurement procedures using the airfield penetrometer are explained; however, in the references listed at the end of this manual, which cover less expedient construction methods, soil strength is treated in terms of CBR. To permit translation between the CBR and the AI, a correlation is presented in figure 6-2. This figure can be used for estimating CBR values from AI determinations. This correlation has been established to yield values of CBR that generally are conservative. The tendency toward conservatism is necessary because there is no unique relationship between these measurements over a wide range of soil types. It follows that the curve should not be used to estimate AI values from CBR determinations since these would not be conservative.

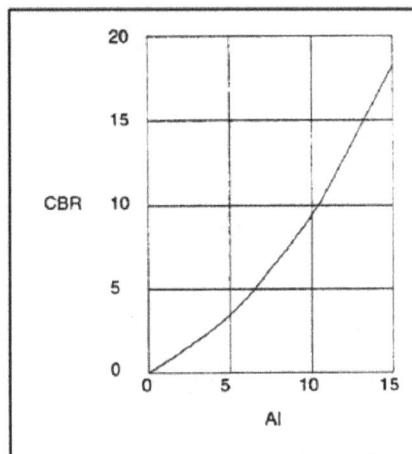

Figure 6-2. Correlation of CBR and AI

SECTION III - BEARING CAPACITY

IMPORTANCE

6-38. The bearing capacity of a soil is its ability to support loads that may be applied to it by an engineering structure, such as—

- A building, a bridge, a highway pavement, or an airport runway and the moving loads that may be carried thereon.
- An embankment.
- Other types of load.

6-39. A soil with insufficient bearing capacity to support the loads applied to it may simply fail by shear, allowing the structure to move or sink into the ground. Such a soil may fail because it undergoes excessive deformation, with consequent damage to the structure. Sometimes the ability of a soil to support loads is simply called its stability. Bearing capacity is directly related to the allowable load that may be safely placed on a soil. This allowable load is sometimes called the allowable soil pressure.

6-40. Types of failure that may take place when the ultimate bearing capacity is exceeded are illustrated in figure 6-3. Such a failure may involve tipping of the structure, with a bulge at the ground surface on one side of the structure. Failure may also take place on a number of surfaces within the soil, usually accompanied by bulging of the ground around the foundation. The ultimate bearing capacity not only is a function of the nature and condition of the soil involved but also depends on the method of application of the load.

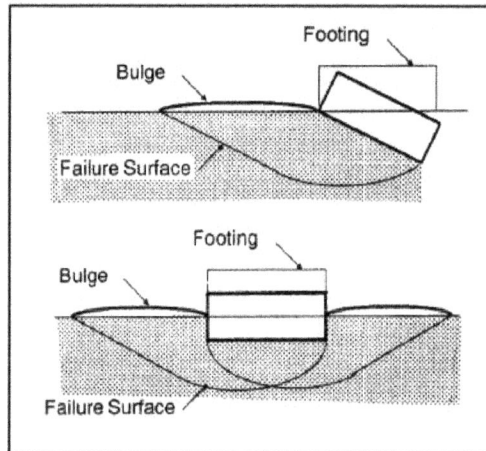

Figure 6-3. Typical failure surfaces beneath shallow foundations

FOUNDATIONS

6-41. The principle function of a foundation is to transmit the weight of a structure and the loads that it carries to the underlying soil or rock. A foundation must be designed to be safe against a shear failure in the underlying soil. This means that the load placed on the soil must not exceed its ultimate bearing capacity.

SHALLOW FOUNDATIONS

6-42. A shallow foundation is one that is located at, or slightly below, the surface of the ground. A typical foundation of this type is seen in the shallow footings, either of plain or reinforced concrete, which may support a building. Footings are generally square or rectangular. Long continuous or strip footings are also used, particularly beneath basement or retaining walls. Another type of shallow foundation is the raft or mat; it may cover a large area, perhaps the entire area occupied by a structure.

DEEP FOUNDATIONS

6-43. When the surface soils at the site of a proposed structure are too weak and compressible to provide adequate support, deep foundations are frequently used to transfer the load to underlying suitable soils. Two common types of deep foundations are—

- Pile.
- Pier.

Piles

6-44. Piles and pile foundations are very commonly used in both military and civil construction. By common usage, a pile is a load-bearing member made of timber, concrete, or steel, which is generally forced into the ground. Piles are used in a variety of forms and for a variety of purposes. A pile foundation is one or more piles used to support a pier, or column, or a row of piles under a wall. Piles of this type are normally used to support vertical loads, although they may also be used to support inclined or lateral forces.

6-45. Piles driven vertically and used for the direct support of vertical loads are commonly called bearing piles. They may be used to transfer the load through a soft soil to an underlying firm stratum. These are called end-bearing piles. Bearing piles may also be used to distribute the load through relatively soft soils that are not capable of supporting concentrated surface loads. These are called friction piles. A bearing pile may sometimes receive its support from a combination of end bearing and friction. Bearing piles also may be used where a shallow foundation would likely be undetermined by scour, as in the case of bridge piers. Bearing piles are illustrated in figure 6-4, page 6-10.

6-46. A typical illustration of an end-bearing pile is when a pile driven through a very soft soil, such as a loose silt or the mud of a river bottom, comes to rest on firm stratum beneath. The firm stratum may, for example, be rock, sand, or gravel. In such cases, the pile derives practically all its support from the underlying firm stratum.

6-47. A friction pile develops its load-carrying capacity entirely, or principally, from skin friction along the sides of the pile. The load is transferred to the adjoining soil by friction between the pile and the surrounding soil. The load is thus transferred downward and laterally to the soil. The soil surrounding the pile or group of piles, as well as that beneath the points of the piles is stressed by the load.

6-48. Some piles carry a load by a combination of friction and end bearing. A pile of this sort may pass through a fairly soft soil that provides some frictional resistance; then it may pass into a firm layer that develops load-carrying capacity through a combination of friction over a relatively short length of embedment and end bearing.

6-49. Piles are used for many purposes other than support for vertical loads. Piles that are driven at an angle with the vertical are commonly called batter piles. They may be used to support inclined loads or to provide lateral loads. Piles are sometimes used to support lateral loads directly, as in the pile fenders that may be provided along waterfront structures to take the wear and shock of docking ships. Sometimes piles are used to resist upward, tensile forces. These are frequently called anchor piles. Anchor piles may be used, for example, as anchors for bulkheads, retaining walls, or guy wires. Vertical piles are sometimes driven for the purpose of compacting loose cohesionless deposits. Closely spaced piles, or sheet piles, may be driven to form a wall or a bulkhead that restrains a soil mass.

Piers

6-50. Piers are much less common than piles and are normally used only for the support of very heavy loads.

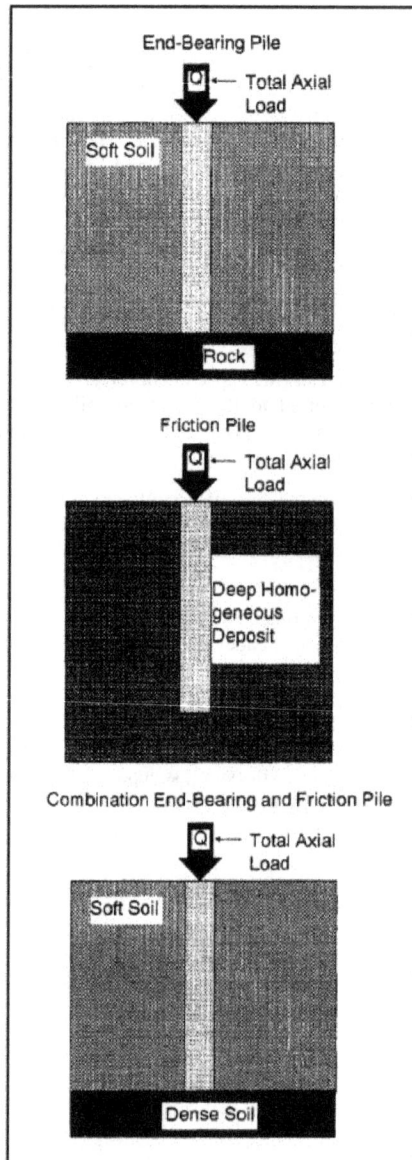

Figure 6-4. Bearing piles

SECTION IV – EARTH-RETAINING STRUCTURES

PURPOSE

6-51. Earth-retaining structures must be used to restrain a mass of earth that will not stand unsupported. Such structures are commonly required when a cut is made or when an embankment is formed with slopes too steep to stand alone.

6-52. Earth-retaining structures are subjected to lateral thrust from the earth masses that they support. The pressure of the earth on such a structure is commonly called lateral earth pressure. The lateral earth pressure that may be exerted by a given soil on a given structure is a function of many variables. It must be estimated with a reasonable degree of accuracy before an earth-retaining structure may be properly designed. In many cases, the lateral earth pressure may be assumed to be acting in a horizontal direction, or nearly so.

TYPES

6-53. Earth-retaining structures discussed in this section are retaining walls and bracing systems used in temporary excavations.

RETAINING WALLS

6-54. A retaining wall is a wall constructed to support a vertical or nearly vertical earth bank that, in turn, may support vertical loads. Generally, retaining walls are classified into the following five types (see figure 6-5, page 6-12):

- Gravity.
- Cantilever.
- Counterfort.
- Buttressed.
- Crib.

6-55. When a retaining wall is used to support the end of a bridge span, as well as retain the earth backfill, it is called an abutment. There are several types of gravity retaining walls, such as—

- Timber.
- Plain concrete.
- Sheet piling.
- Rubble.
- Stone or brick masonry.
- Crib.
- Gabions.

6-56. Retaining walls are used in many applications. For example, a structure of this sort may be used in a highway or railroad cut to permit the use of a steep slope and avoid excessive amounts of excavation. Retaining walls are similarly used on the embankment side of sidehill sections to avoid excessive volumes of fill. Bridge abutments and the headwalls of culverts frequently function as retaining walls. In the construction of buildings and various industrial structures, retaining walls are often used to provide support for the side of deep, permanent excavations.

6-57. Permanent retaining walls are generally constructed from plain or reinforced concrete; stone masonry walls are also used occasionally. In military construction, timber crib retaining walls are important. Their design is discussed later.

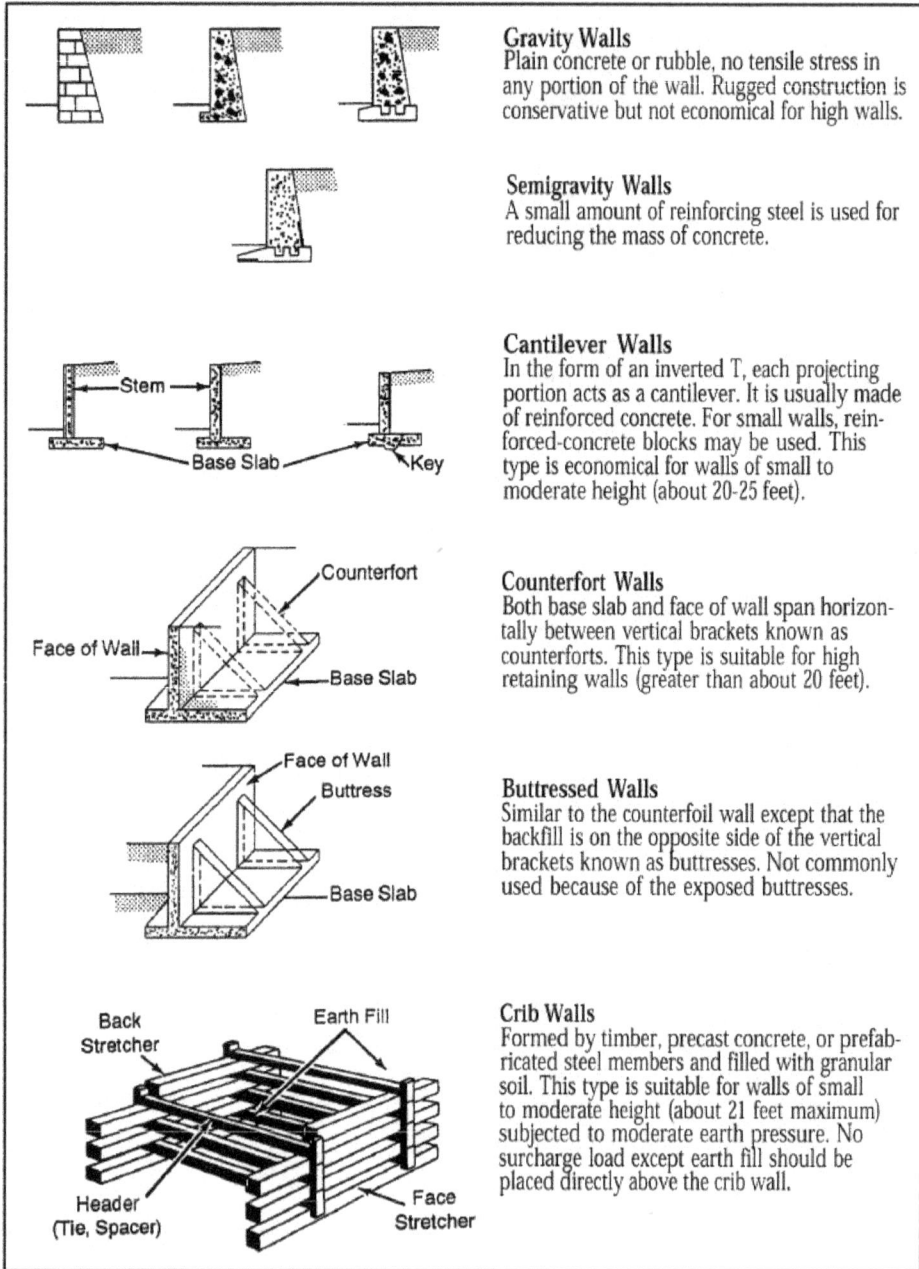

Gravity Walls
Plain concrete or rubble, no tensile stress in any portion of the wall. Rugged construction is conservative but not economical for high walls.

Semigravity Walls
A small amount of reinforcing steel is used for reducing the mass of concrete.

Cantilever Walls
In the form of an inverted T, each projecting portion acts as a cantilever. It is usually made of reinforced concrete. For small walls, reinforced-concrete blocks may be used. This type is economical for walls of small to moderate height (about 20-25 feet).

Counterfort Walls
Both base slab and face of wall span horizontally between vertical brackets known as counterforts. This type is suitable for high retaining walls (greater than about 20 feet).

Buttressed Walls
Similar to the counterfoil wall except that the backfill is on the opposite side of the vertical brackets known as buttresses. Not commonly used because of the exposed buttresses.

Crib Walls
Formed by timber, precast concrete, or prefabricated steel members and filled with granular soil. This type is suitable for walls of small to moderate height (about 21 feet maximum) subjected to moderate earth pressure. No surcharge load except earth fill should be placed directly above the crib wall.

Figure 6-5. Principal types of retaining walls

Backfills

6-58. The design of the backfill for a retaining wall is as important as the design of the wall itself. The backfill must be materials that are—

- Reasonably clean.
- Granular.
- Essentially cohesionless.
- Easily drained.
- Not susceptible to frost action.

6-59. The best materials for backfills behind retaining walls are clean sands, gravels, and crushed rock. In the USCS, the (GW) and (SW) soils are preferred, if they are available. The (GP) and (SP) soils are also satisfactory. These granular materials require compaction to make them stable against the effects of vibration. Compaction also generally increases the angle of internal friction, which is desirable in that it decreases the lateral pressure exerted on the wall. Materials of the (GM), (GC), (SM), and (SC) groups may be used for backfills behind retaining walls, but they must be protected against frost action and may require elaborate drainage provisions. Fine-grained soils are not desirable as backfills because they are difficult to drain. If clay soil must be used, the wall should be designed to resist earth pressures at rest. Ideal backfill materials are purely granular soils containing < 5 percent of fines.

6-60. Backfills behind retaining walls are commonly put in place after the structure has been built. The method of compaction depends on the—

- Soil.
- Equipment available.
- Working space.

6-61. Since most backfills are essentially cohesionless, they are best compacted by vibration. Equipment suitable for use with these soils is discussed in chapter 8. Common practice calls for the backfill to be placed in layers of loose material that, when compacted, results in a compacted layer thickness of from 6 to 8 inches. Each layer is compacted to a satisfactory density. In areas inaccessible to rollers or similar compacting equipment, compaction may be done by the use of mechanical air tampers or hand tools.

Drainage

6-62. Drainage of the backfill is essential to keep the wall from being subjected to water pressure and to prevent frost action. Common drainage provisions used on concrete walls are shown in figure 6-6, page 6-14.

6-63. When the backfill is composed of clean, easily drained materials, it is customary to provide for drainage by making weep holes through the wall. Weep holes are commonly made by embedding pipes 4 to 6 inches in diameter into the wall. These holes are spaced from 5 to 10 feet center to center both horizontally and vertically. A filter of granular material should be provided around the entrance to each weep hole to prevent the soil from washing out or the drain from becoming clogged. If possible, this material should conform to the requirements previously given for filter materials.

6-64. Weep holes have the disadvantage of discharging the water that seeps through the backfill at the toe of the wall where the soil pressures are greatest. The water may weaken the soil at this point and cause the wall to fail. A more effective solution, which is also more expensive, is to provide a longitudinal back drain along the base of the wall (see figure 6-6). A regular pipe drain should be used, surrounded with a suitable filter material. The drainage may be discharged away from the ends of the wall.

6-65. If a granular soil, which contains considerable fine material and is poorly drained (such as an (SC) soil) is used in the backfill, then more elaborate provisions may be installed to ensure drainage. One such approach is to use a drainage blanket (see figure 6-6). If necessary, a blanket of impervious soil or bituminous material may be used on top of the backfill to prevent water from entering the fill from the top. Such treatments are relatively expensive.

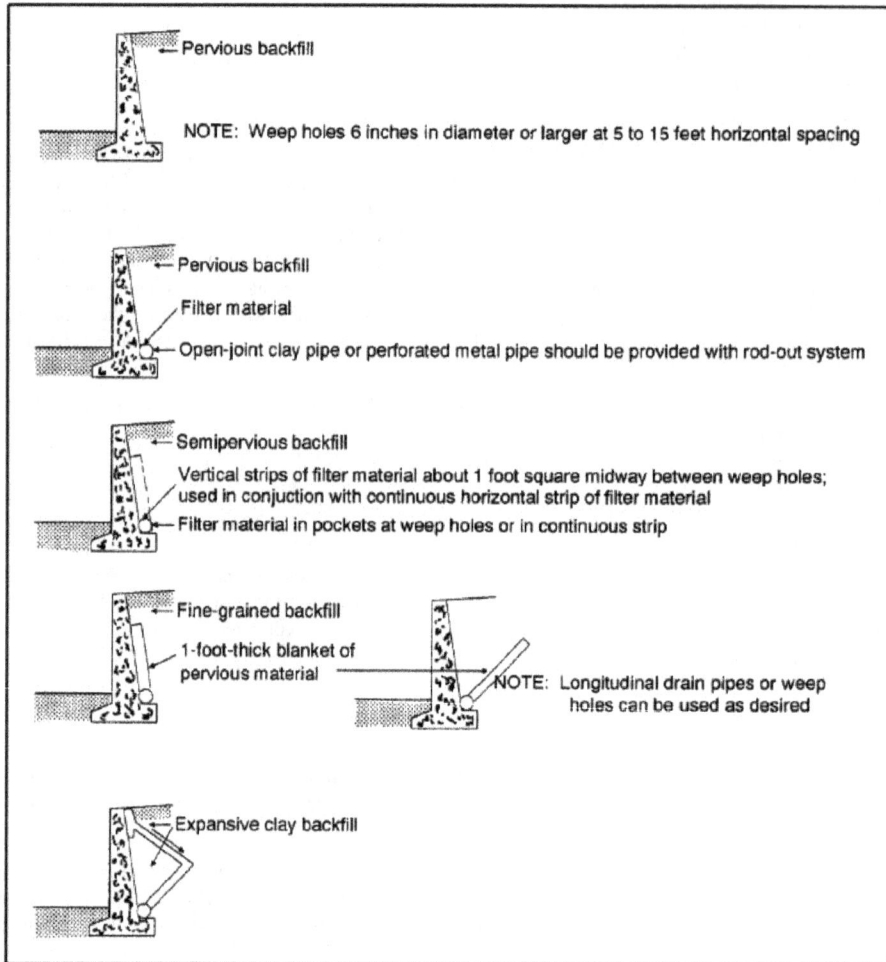

Figure 6-6. Common types of retaining wall drainage

Frost Action

6-66. Conditions for detrimental frost action on retaining walls include the following:

- A frost-susceptible soil.
- Availability of water.
- Freezing temperatures.

6-67. If these conditions are present in the backfill, steps must be taken to prevent the formation of ice lenses and the resultant severe lateral pressures that may be exerted against the wall. The usual way to prevent frost action is to substitute a thick layer of clean, granular, nonfrost-susceptible soil for the backfill material immediately adjacent to the wall. The width of the layer should be as great as the maximum depth of frost penetration in the area (see figure 6-7). As with other structures, the bottom of a retaining wall should be located beneath the line of frost penetration.

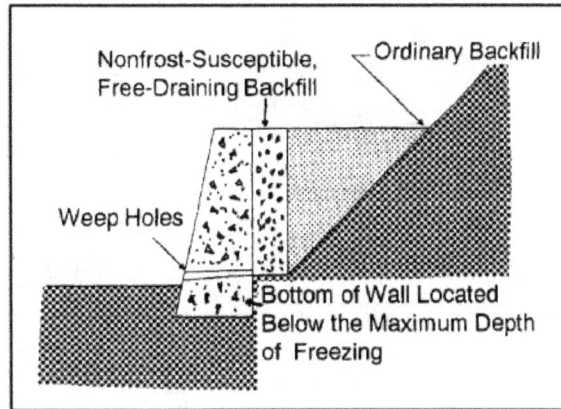

Figure 6-7. Eliminating frost action behind retaining walls

Timber Crib

6-68. A very useful type of retaining wall for military purposes in a theater of operations is timber cribbing. The crib or cells are filled with earth, preferably clean, coarse, granular material. A wall of this sort gains its stability through the weight of the material used to fill the cells, along with the weight of the crib units themselves. The longitudinal member in a timber crib is called a stretcher, while a transverse member is a header.

6-69. A principal advantage of a timber crib retaining wall is that it may be constructed with unskilled labor and a minimum of equipment. Suitable timber is available in many military situations. Little foundation excavation is usually required and may be limited to shallow trenching for the lower part of the crib walls. The crib may be built in short sections, one or two cribs at a time. Where the amount of excavation is sufficient and suitable, the excavated soil may be used for filling the cells.

6-70. A crib of this sort may be used on foundation soils that are weak and might not be able to support a heavy wall, since the crib is fairly flexible and able to undergo some settlement and shifting without distress. However, this should not be misunderstood, as the foundation soil must not be so soft as to permit excessive differential settlement that would destroy the alignment of the crib.

6-71. Experience indicates that a satisfactory design will generally be achieved if the base width is a minimum of 4 feet at the top and bottom or 50 percent of the height of the wall, provided that the wall does not carry a surcharge and is on a reasonably firm foundation. If the wall carries a heavy surcharge, the base width should be increased to a minimum of 65 percent of the height. In any case, the width of the crib at the top and bottom should not be less than 4 feet.

6-72. Timber crib walls may be built with any desired batter (receding upward slope) or even vertical. The batter most often used and recommended is one horizontal to four vertical (see figure 6-8, page 6-16). If less batter is used, the base width must be increased to ensure that the resultant pressure falls within the middle third of the base. The desired batter is normally achieved by placing the base on a slope equal to the batter. The toe may be placed on sills; this is frequently done with high walls. Sometimes double-cell construction is used to obtain the necessary base width of high walls. The wall is then decreased in width, or "stepped-back," in the upper portions of the wall, above one third height. Additional rows of bottom stretchers may be used to decrease the pressure on the soil or to avoid detrimental settlement.

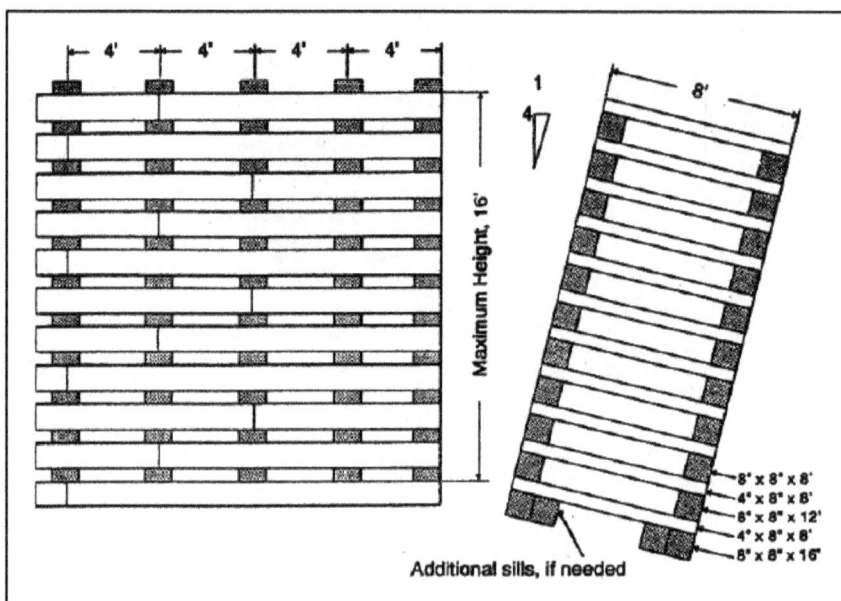

Figure 6-8. Typical timber crib retaining wall

6-73. The front and rear wall of the crib should be connected at each panel point. The crib must be kept an essentially flexible structure and must be free to move somewhat in any direction, so as to adjust itself to thrusts and settlements.

6-74. The material used in filling the cells should be placed in thin layers and should be well compacted. Backfill behind the wall should also be compacted and kept close to, but not above, the level of the material in the cribs. Drainage behind timber crib walls may or may not be required, depending on local conditions and wall construction.

6-75. Figure 6-8 shows the elevation and cross section of a timber crib retaining wall, which may be used to a maximum height of about 16 feet. A similar arrangement may be used for heights up to about 8 feet, with a minimum width of 4 feet. For heights above 16 feet, the headers are usually 6-inch by 12-inch timbers and the stretchers 12-inch by 12-inch timbers. Timbers are normally connected by means of heavy (¾-inch diameter) driftpins.

Other Timber Walls

6-76. Other types of timber retaining walls are used for low heights, particularly in connection with culverts and bridges. A wall of this sort may be built by driving timber posts into the ground and attaching planks or logs. Details on retaining walls, used in conjunction with bridge abutments, are given in TM 5-312. Figure 6-9 illustrates two other types of timber retaining walls.

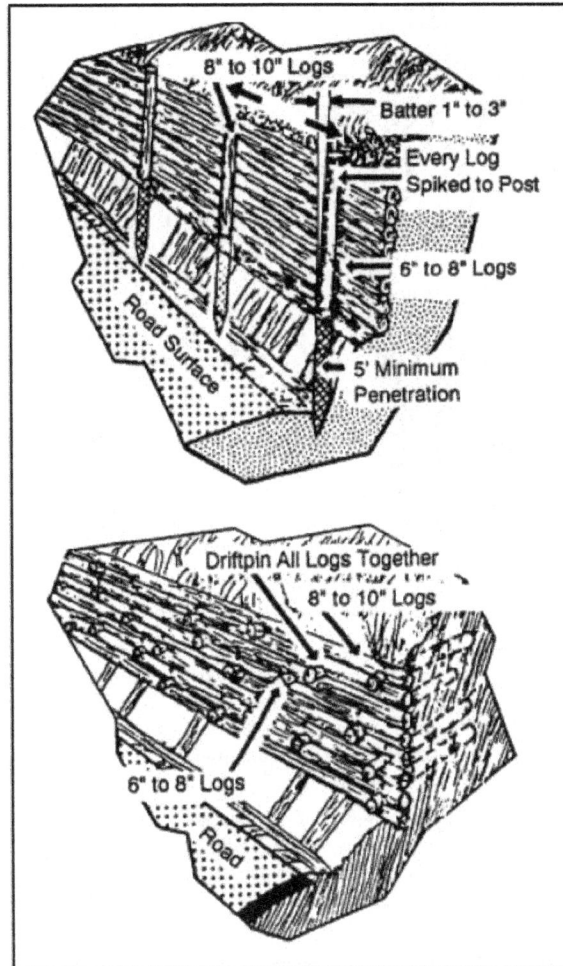

Figure 6-9. Other timber retaining walls

Gabions

6-77. Gabions are large, steel-wire-mesh baskets usually rectangular in shape and variable in size (see figure 6-10, page 6-18). They are designed to solve the problem of erosion at a low cost. Gabions were used in sixteenth century fortifications, and experience in construction with factory-produced prefabricated gabions dates back to 1894 in Italy. Gabions have been widely used in Europe and are now becoming accepted in the United States as a valuable and practical construction tool. They can be used in place of sheet piling, masonry construction, or cribbing. Gabions may be used as—

- Protective and antierosion structures on rivers (as revetments, groynes, or spurs).
- Channel linings.
- Seashore protection.
- Retaining walls for roads or railroads.
- Antierosion structures (such as weirs, drop structures, and check dams).
- Low-water bridges or fords.

- Culvert headwall and outlet structures.
- Bridge abutments and wing walls.

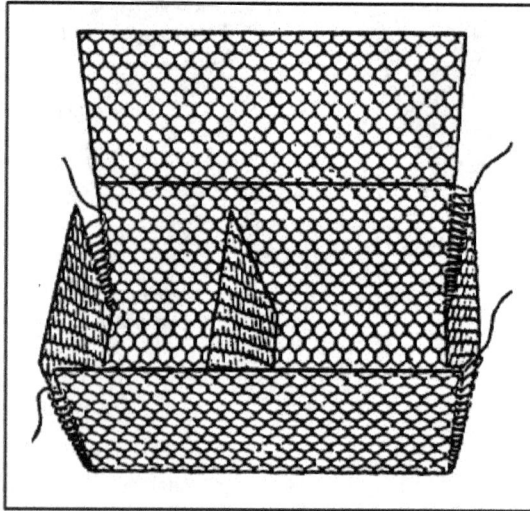

Figure 6-10. Typical gabion

6-78. The best use of gabions as retaining walls is where flexibility and permeability are important considerations, especially where unstable ground and drainage conditions impose problems difficult to solve with rigid and impervious material. Use of gabions does require ready access to large-size stones, such as those found in mountainous areas. Areas that are prone to landslides have used gabions successfully. Gabion walls have been erected in mountainous country to trap falling rocks and debris and in some areas to act as longitudinal drainage collectors.

6-79. The best filling material for a gabion is one that allows flexibility in the structure but also fills the gabion compartments with the minimum of voids and with the maximum weight. Ideally, the stone should be small, just slightly larger than the size of the mesh. The stone must be clean, hard, and durable to withstand abrasion and resistance to weathering and frost action. The gabions are filled in three lifts, one foot at a time. Rounded stone, if available, reduces the possibility of damage to the wire during mechanical filling as compared with sharp quarry stone. If stone is not available, gabions can be filled with a good quality soil. To hold soil, hardware cloth inserts must be placed inside the gabions. For use in gabions, backfill material should meet the following Federal Highway Administration criteria:

- For a 6-inch sieve, 100 percent of the material should pass through.
- For a 3-inch sieve, 75 to 100 percent of the material should pass through.
- For a Number 200 sieve, zero to 25 percent of the material should pass through.
- The PI should be 6 or less.

EXCAVATION BRACING SYSTEMS

6-80. Bracing systems may be required to protect the sides of temporary excavations during construction operations. Such temporary excavations may be required for several purposes but are most often needed in connection with the construction of foundations for structures and the placing of utility lines, such as sewer and water pipes.

Shallow Excavations

6-81. The term "shallow excavation" refers to excavations made to depths of 12 to 20 feet below the surface, depending principally on the soil involved. The lower limit applies to fairly soft clay soils, while the upper limit generally applies to sands and sandy soils.

6-82. Shallow excavations may be made as open cuts with unsupported slopes, particularly when the excavation is being done above the water table. Chapter 10 gives recommendations previously given relative to safe slopes in cuts that are applicable here if the excavation is to remain open for any length of time. If the excavation is purely temporary in nature, most sandy soils above the water table will stand at somewhat steeper slopes (as much as ½ to 1 for brief periods), although some small slides may take place. Clays may be excavated to shallow depths with vertical slopes and will remain stable briefly. Generally, bracing cuts in clay that extend to depths of 5 feet or more below the surface are safer unless flat slopes are used.

6-83. Even for relatively shallow excavations, using unsupported cuts may be unsatisfactory for several reasons. Cohesive soils may stand on steep slopes temporarily, but bracing is frequently needed to protect against a sudden cave-in. Required side slopes, particularly in loose, granular soils, may be so flat as to require an excessive amount of excavation. If the excavation is being done close to other structures, space may be limited, or the consequences of the failure of a side slope may be very serious. Considerable subsidence of the adjacent ground may take place, even though the slope does not actually fail. Finally, if the work is being done below the water table, the excavation may have to be surrounded with a temporary structure that permits the excavation to be unwatered.

Narrow Shallow Excavations

6-84. Several different schemes may be used to brace the sides of a narrow shallow excavation. Two of these schemes are shown in figure 6-11, page 6-20).

6-85. In the first scheme, timber planks are driven around the boundary of the excavation to form what is called vertical sheeting. The bottom of the sheeting is kept at or near the bottom of the pit or trench as excavation proceeds. The sheeting is held in place by means of horizontal beams called wales. These wales are usually supported against each other by means of horizontal members called struts, which extend from one side of the excavation to the other. The struts may be cut slightly long, driven into place, and held by nails or cleats. They may also be held in position by wedges or shims. Hydraulic or screw-type jacks can be used as struts.

6-86. The second scheme uses horizontal timber planks to form what is called horizontal lagging. The lagging, in turn, is supported by vertical solid beams and struts. If the excavation is quite wide, struts may have to be braced horizontally or vertically or both.

6-87. Bracing systems for shallow excavations are commonly designed on the basis of experience. Systems of this sort represent cases of incomplete deformation, since the bracing system prevents deformation at some points while permitting some deformation at others.

6-88. Members used in bracing systems should be strong and stiff. In ordinary work, struts vary from 4-inch to 6-inch timbers for narrow cuts up to 8-inch by 8-inch timbers for excavations 10 or 12 feet wide. Heavier timbers are used if additional safety is desired. Struts are commonly spaced about 8 feet horizontally and from 5 to 6 feet vertically. Lagging or sheeting is customarily made from planks from 6 to 12 inches wide, with the minimum thickness usually being about 2 inches.

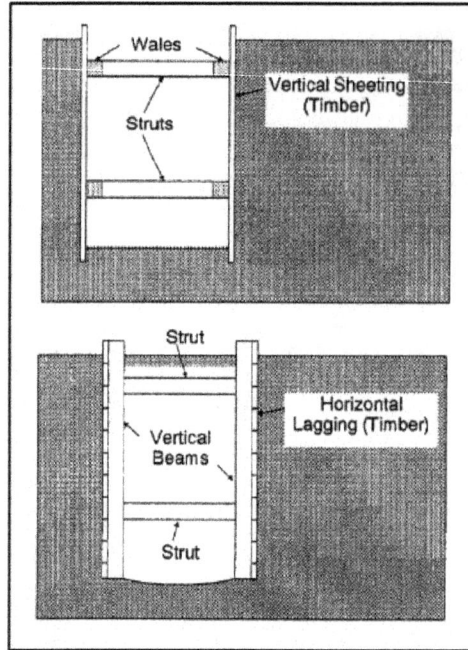

Figure 6-11. Bracing a narrow shallow excavation

Wide Shallow Excavations

6-89. If the excavation is too wide to be cross braced by the use of struts, vertical sheeting may be used (see figure 6-12). The wales are supported by inclined braces, which are sometimes called rakes. The rakes, in turn, react against kicker blocks that are embedded in the soil. As the excavation is deepened, additional wales and braces may be added as necessary to hold the sheeting firmly in position. The success of this system depends on the soil in the bottom of the excavation being firm enough to provide adequate support for the blocks.

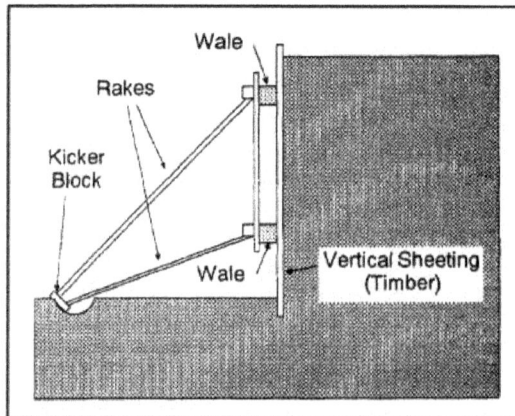

Figure 6-12. Bracing a wide shallow excavation

Chapter 7

Movement of Water Through Soils

The movement of water into or through a soil mass is a phenomenon of great practical importance in engineering design and construction. It is probably the largest single factor causing soil failures. For example, water may be drawn by capillarity from a free water surface or infiltrate through surface cracks into the subgrade beneath a road or runway. Water then accumulated may greatly reduce the bearing capacity of subgrade soil, allowing the pavement to fail under wheel loads if precautions have not been taken in design. Seepage flow may be responsible for the erosion or failure of an open cut slope or the failure of an earth embankment. This chapter concerns the movement of water into and through soils (and, to some extent, about the practical measures undertaken to control this movement) and the problems associated with frost action.

SECTION I - WATER

7-1. Knowledge of the earth's topography and characteristics of geologic formations help the engineer find and evaluate water sources. Water in the soil may be from a surface or a subsurface source. Surface water sources (streams, lakes, springs) are easy to find. Finding subsurface water sources could require extensive searching. Applying geologic principles can help eliminate areas where no large groundwater supplies are present and can indicate where to concentrate a search.

HYDROLOGIC CYCLE

7-2. Water covers 75 percent of the earth's surface. This water represents vast storage reservoirs that hold most of the earth's water. Direct radiation from the sun causes water at the surface of rivers, lakes, oceans, and other bodies to change from a liquid to a vapor. This process is called *evaporation*. Water vapor rises in the atmosphere and accumulates in clouds. When enough moisture accumulates in the clouds and the conditions are right, water is released as *precipitation* (rain, sleet, hail, snow). Some precipitation occurs over land surfaces and represents the early stage of the land hydrologic cycle. Precipitation that falls on land surfaces is stored on the surface, flows along the surface, or flows into the ground as *infiltration*. Infiltration is a major source of groundwater and is often referred to as *recharge*, because it replenishes or recharges groundwater resources.

7-3. The hydrologic cycle (figure 7a, page 7-2) consists of several processes. It does not usually progress through a regular sequence and can be interrupted or bypassed at any point. For example, rain might fall in an area of thick vegetation, and a certain amount of this moisture will remain on the plants and not reach the ground. The moisture could return to the atmosphere by direct evaporation, thereby causing a break in the hydrologic cycle. For the military engineer, the two most important states of this cycle are those pertaining to surface runoff and water infiltration.

SURFACE WATER

7-4. Streams and lakes are the most available and most commonly used sources for military water supplies. However, other subsurface water sources should be considered because long-term droughts could occur. Water naturally enters through soil surfaces unless they are sealed; and sealed surfaces may have cracks, joints, or fissures that allow water penetration. Surface water may also enter from the sides of construction projects, such as roads or airfields.

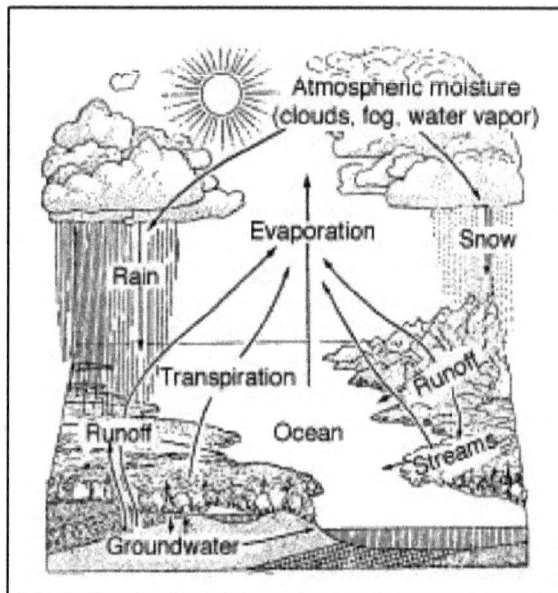

Figure 7a. Hydrologic cycle

7-5. Surface water is not chemically pure. It may contain sediment, bacteria, or dissolved salts that make the water unfit for consumption. Natural contamination and the pollution that man causes are also dangerous and occur in surface water. Test all surface water for purity and take proper precautions before using it.

STREAMS

7-6. Streams normally supply an abundant quantity of water for the initial phase of a field operation. The water supply needs to be adequate for the time of year the operation is planned. For a long-term supply, you must learn the permanent status of the stream flow.

LAKES

7-7. Most lakes are excellent sources of water. They serve as natural reservoirs for storing large amounts of water. Lakes are usually more constant in quality than the streams that feed them. Large lakes are preferable because the water is usually purer. Shallow lakes and small ponds are more likely to be polluted or contaminated. Lakes located in humid regions are generally fresh and permanent. Lakes in desert regions are rare but can occur in basins between mountains. These lakes could have a high percentage of dissolved salts and should not be considered as a permanent source of water.

SWAMPS

7-8. Swamps are likely to occur where wide, flat, poorly drained land and an abundant supply of water exist. A large quantity of water is usually available in swamps; but it may be poor in quality, brackish, or salty.

SPRINGS AND SEEPS

7-9. Water that naturally emerges at the surface is called a *spring* if there is a distinctive current, and a *seep* if there is no current. Most springs and seeps consist of water that has slowly gravitated from nearby higher ground. The water's underground course depends on the permeability and structure of the material through which it moves. Any spring that has a temperature higher than the yearly average temperature of a given region is termed a *thermal spring* and indicates a source of heat other than the surface climate.

GRAVITY SPRINGS AND SEEPS

7-10. In gravity springs and seeps, subsurface water flows by gravity, not by hydrostatic pressure, from a high point of intake to a lower point of issue. Water-table springs and seeps are normally found around the margin of depressions, along the slope of valleys, and at the foot of alluvial fans. Contact springs appear along slopes, and may be found at almost any elevation depending on the position of the rock formations.

ARTESIAN SPRINGS

7-11. When water is confined in a rock layer under hydrostatic pressure, an artesian condition is said to exist. A well drilled into an aquifer where this condition is present, is called an *artesian well* (figure 7b). If the water rises to the surface, it is called an *artesian spring*. Certain situations are necessary for an artesian condition to exist—

- There must be a permeable aquifer that has impervious layers above and below it to confine the water.
- There must be an intake area where water can enter the aquifer.
- A structural dip must exist so that hydrostatic pressure is produced in the water at the lower areas of the aquifer.

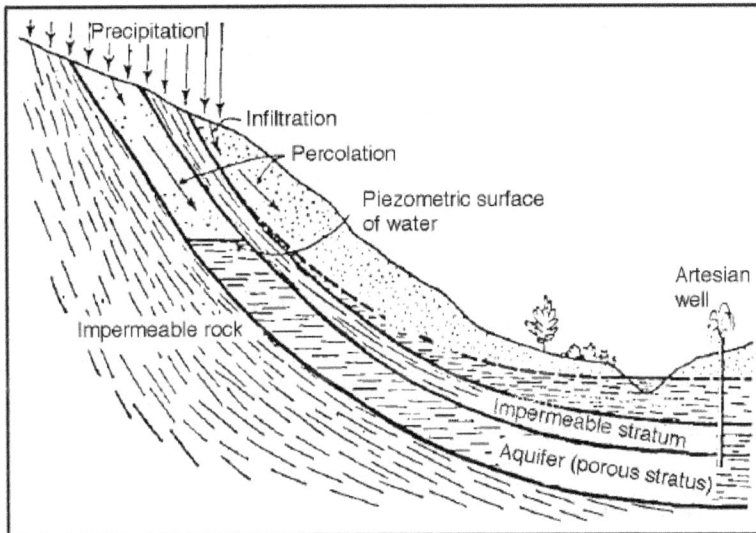

Figure 7b. Artesian groundwater

GROUNDWATER

7-12. Groundwater or *subsurface* water is any water that exists below the earth's surface. Groundwater is located in two principle zones in the earth's surface—aeration and saturation (figure 7c, page 7-4).

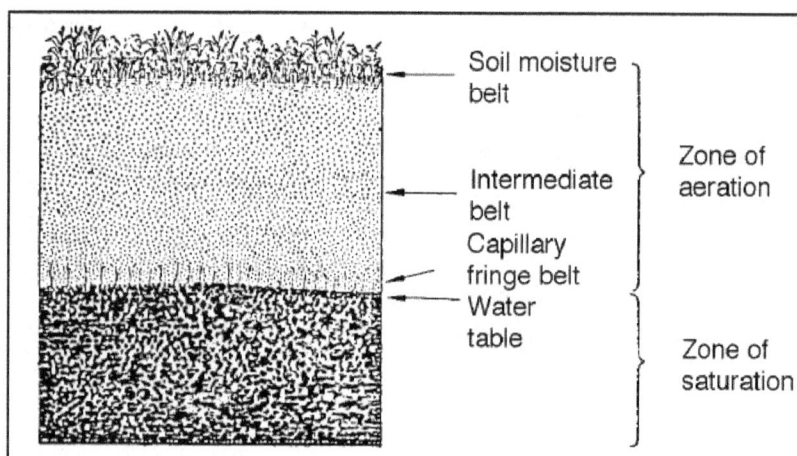

Figure 7c. Groundwater zones

7-13. The zone of aeration consists of three major belts—soil moisture, intermediate, and capillary fringe. As water starts infiltrating the ground surface, it encounters a layer of organic matter. The root systems of plants, decaying organic material, and the small pores found within the upper soil zone hold some of the water in suspension. This shallow layer is called the *soil moisture* belt. The water passes through this belt and continues downward through the *intermediate* belt. The pore spaces in this belt are generally larger than those in the soil moisture belt, and the amount of organic material is considerably reduced. The intermediate belt contains voids so it does not hold water, and the water gradually drains downward. The next belt is called the *capillary fringe*. Most deep-rooted plants sink roots into this area.

7-14. The water table is the contact between the zone of aeration and the zone of saturation. It fluctuates up and down, depending on the recharge rate and the rate of flow away from the area. The pores are filled with water in the zone of saturation.

FREE, OR GRAVITATIONAL, WATER

7-15. Water that percolates down from the surface eventually reaches a depth where there is some medium that restricts (to varying degrees) the further percolation of the moisture. This medium may be bedrock or a layer of soil, not wholly solid but with such small void spaces that the water which leaves this zone is not as great as the volume or supply of water added. In time, the accumulating water completely saturates the soil above the restricting medium and fills all voids with water. When this zone of saturation is under no pressure except from the atmosphere, the water it contains is called free, or gravitational, water. It will flow through the soil and be resisted only by the friction between the soil grains and the free water. This movement of free water through a soil mass frequently is termed see page. The upper limit of the saturated zone of free water is called the groundwater table, which varies with climatic conditions. During a wet winter, the groundwater table rises. However, a dry summer might remove the source of further accumulation of water. This results in a decreased height of the saturated zone, for the free water then flows downward, through, or along its restricting layer. The presence of impervious soil layers may result in an area of saturated soil above the normal groundwater table. This is called a "perched" water table.

HYGROSCOPIC MOISTURE

7-16. When wet soil is air-dried, moisture is removed by evaporation until the hygroscopic moisture in the soil is in equilibrium with the moisture vapor in the air. The amount of moisture in air-dried soil, expressed as a percentage of the weight of the dry soil, is called the hygroscopic moisture content. Hygroscopic moisture films may be driven off from air-dried soil by heating the material in an oven at 100 to 110 degrees Centigrade (C) (210 to 230 degrees Fahrenheit (F)) for 24 hours or until constant weight is attained.

CAPILLARY MOISTURE

7-17. Another source of moisture in soils results from what might be termed the capillary potential of a soil. Dry soil grains attract moisture in a manner similar to the way clean glass does. Outward evidence of this attraction of water and glass is seen by observing the meniscus (curved upper surface of a water column). Where the meniscus is more confined (for example, as in a small glass tube), it will support a column of water to a considerable height. The diagram in figure 7-1 shows that the more the meniscus is confined, the greater the height of the capillary rise.

7-18. Capillary action in a soil results in the "capillary fringe" immediately above the groundwater table. The height of the capillary rise depends on numerous factors. One factor worth mentioning is the type of soil. Since the pore openings in a soil vary with the grain size, a fine-grained soil develops a higher capillary fringe area than a coarse-grained soil. This is because the fine-grained soil can act as many very small glass tubes, each having a greatly confined meniscus. In clays, capillary water rises sometimes as high as 30 feet, and in silts the rise is often as high as 10 feet. Capillary rise may vary from practically zero to a few inches in coarse sands and gravels.

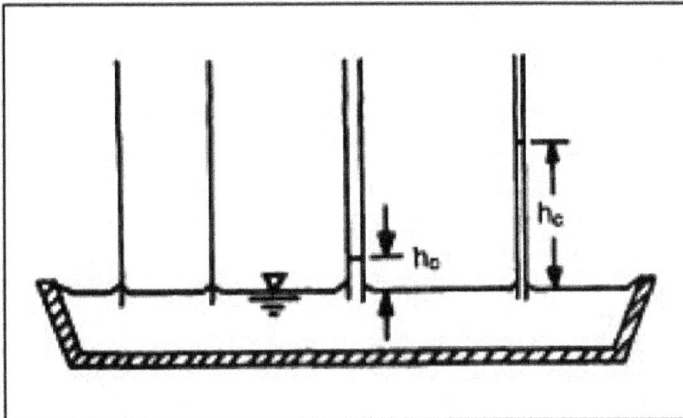

Figure 7-1. Capillary rise of moisture

7-19. When the capillary fringe extends to the natural ground surface, winds and high temperatures help carry this moisture away and reduce its effects on the soil. Once a pavement of watertight surface is applied, however, the effect of the wind and temperature is reduced. This explains the accumulation of moisture often found directly beneath an impervious pavement.

7-20. Capillary moisture in soils located above the water table may be visualized as occurring in the following three zones:

- Capillary saturation.
- Partial capillary saturation.
- Contact moisture.

7-21. In the zone of capillary saturation, the soil is essentially saturated. The height of this zone depends not only on the soil but also on the history of the water table, since the height will be greater if the soil mass has been saturated previously.

7-22. The height of the zone of partial capillary saturation is likely to be considered greater than that of the zone of capillary saturation; it also depends on the water-table history. Its existence is the result of a few large voids serving effectively to stop capillary rise in some parts of the soil mass. Capillary water in this zone is still interconnected or "continuous," while the air voids may not be.

7-23. Above the zones of capillary and partial capillary saturation, water that percolates downward from the surface may be held in the soil by surface tension. It may fill the smaller voids or be present in the form of water films between the points of contact of the soil grains. Water may also be brought into this zone from the water table by evaporation and condensation. This moisture is termed "contact moisture."

7-24. One effect of contact moisture is apparent cohesion. An example of this is the behavior of sand on certain beaches. On these beaches, the dry sand located back from the edge of the water and above the height of capillary rise is generally dry and very loose and has little supporting power when unconfined. Closer to the water's edge, and particularly during periods of low tide, the sand is very firm and capable of supporting stationary or moving automobiles and other vehicles. This apparent strength is due primarily to the existence of contact moisture left in the voids of the soil when the tide went out. The surface soil may be within the zone of partial or complete capillary saturation very close to the edge of the water. Somewhat similarly, capillary forces may be used to consolidate loose cohesionless deposits of very fine sands or silts in which the water table is at or near the ground surface. This consolidation is accomplished by lowering the water table by means of drains or well points. If the operation is properly carried out within the limits of the height of capillary rise, the soil above the lowered water table remains saturated by capillary moisture. The effect is to place the soil structure under capillary forces (such as tension in the water) that compress it. The soil may be compressed as effectively as though an equivalent external load had been placed on the surface of the soil mass.

7-25. Methods commonly used to control the detrimental effects of capillarity, particularly concerning roads and airport pavements, are mentioned briefly here. Additional attention is given to this subject in section II, which is devoted to the closely allied subject of frost action.

7-26. As has been noted, if the water table is closer to the surface than the height of capillary rise, water will be brought up to the surface to replace water removed by evaporation. If evaporation is wholly or partially prevented, as by the construction of impervious pavement, water accumulates and may cause a reduction in shearing strength or cause swelling of the soil. This is true particularly when a fine-grained soil or a coarse soil that contains a detrimental amount of plastic fines is involved.

7-27. One obvious solution is to excavate the material that is subject to capillary action and replace it with a granular material. This is frequently quite expensive and usually may be justified only in areas where frost action is a factor.

7-28. Another approach is to include in the pavement structure a layer that is unaffected by capillary action. This is one of the functions of the base that is invariably used in flexible pavements. The base serves to interrupt the flow of capillary moisture, in addition to its other functions. Under certain circumstances, the base itself may have to be drained to ensure the removal of capillary water (see figure 7-2). This also is usually not justified unless other circumstances, such as frost action, are of importance.

7-29. Still another approach is to lower the water table, which may sometimes be accomplished by the use of side ditches. Subdrains may be installed for the same purpose (see figure 7-3, page 7-8). This approach is particularly effective in relatively pervious or free-draining soils. Some difficulty may be experienced in lowering the water table by this method in flat country because finding outlets for the drains is difficult. An alternative, used in many areas where the permanent water table is at or near the ground surface, is simply to build the highway or runway on a fill. Material that is not subject to detrimental capillarity is used to form a shallow fill. The bottom of the base is normally kept a minimum of 3 or 4 feet above the natural ground surface, depending on the soil used in the fill and other factors. A layer of sand, known as a sand blanket, or a geotextile fabric may be used to intercept capillary moisture, preventing its intrusion into the base course.

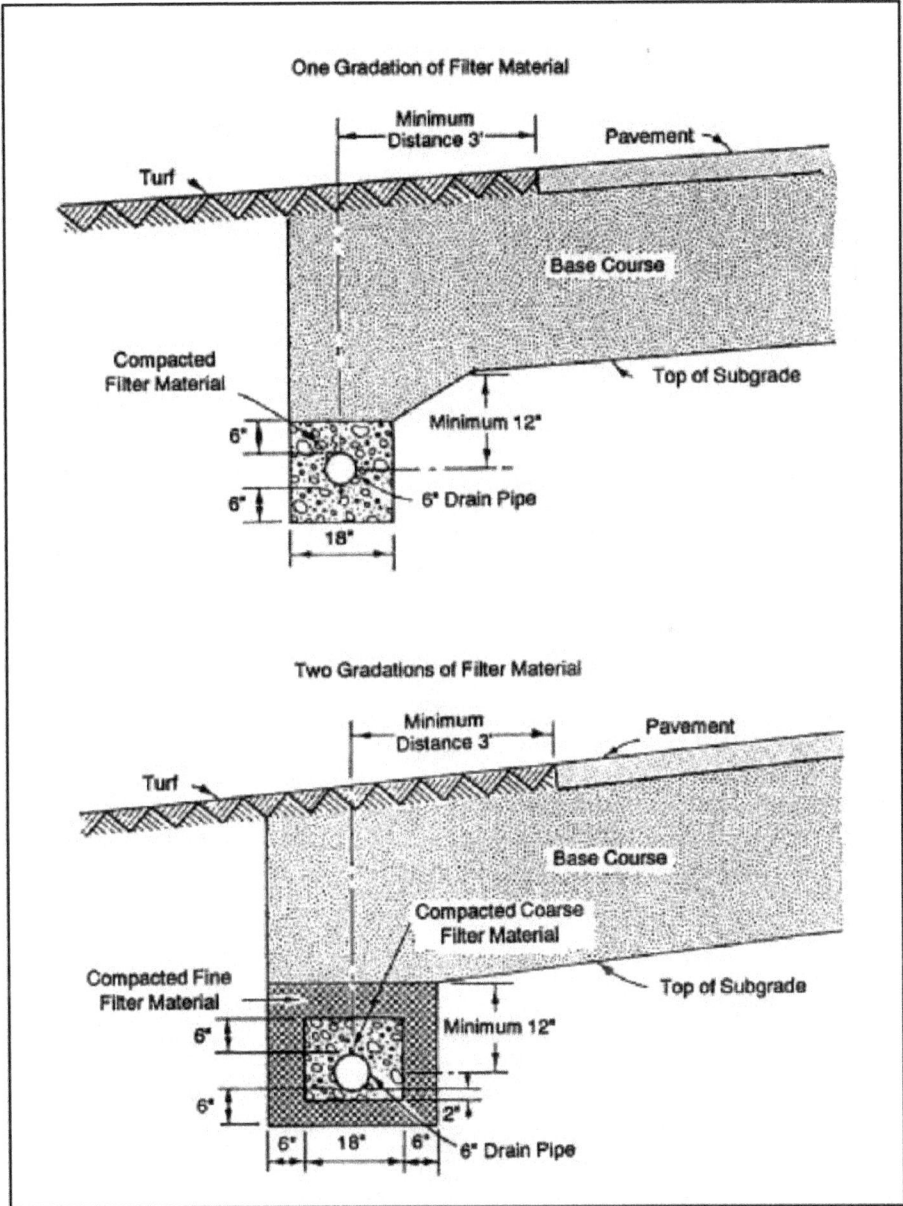

Figure 7-2. Base drains in an airfield pavement

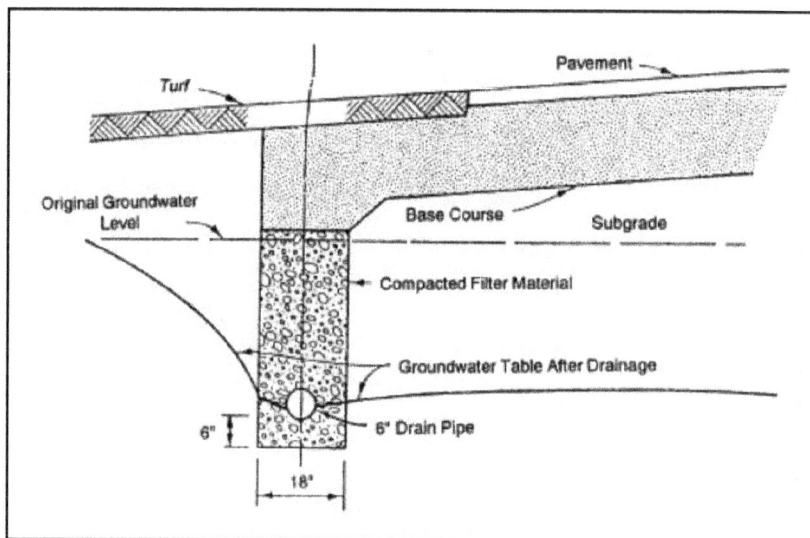

Figure 7-3. Typical subgrade drainage installation

LOCATING GROUNDWATER SOURCES

7-30. Consider exploring rock aquifers only when soil aquifers are not present or when the soil aquifer cannot provide the required water supply. Identifying suitable well sites in rock aquifers is much more difficult than in soil aquifers. Also, water development is usually more time-consuming and costly and has a higher risk of failure. However, in some areas, rock aquifers are the only potential source of groundwater.

Hydrogeologic Indicators

7-31. Indicators that help identify groundwater sources are referred to as hydrogeologic indicators (table 7a). They are divided into three major groups—reservoir, groundwater, and boundary. Groundwater indicators are those conditions or characteristics that directly or indirectly indicate groundwater occurrence. No indicator is 100 percent reliable, but the presence or absence of certain indicators or associations of indicators is fairly reliable.

Table 7a. Hydrogeologic indicators for groundwater exploration

Reservoir Indicators	Groundwater Indicators	Boundary Indicators
Rock type/geometry	Springs and seeps	Location of recharge areas
Stratigraphic sequence	Soil moisture	Location of discharge areas
Degree of lithification	Vegetation type	Impermeable barriers
Grain size	Vegetation density	Semipermeable barriers
Fracture density	Wetlands	Surface-water divides
Dissolution potential	Playas	
Cumulative structure density	Wells	
Drainage basin size	Reservoirs	
Drainage basin elevation and relief	Crop irrigation	
Drainage pattern	Salt encrustations	

Table 7a. Hydrogeologic indicators for groundwater exploration

Reservoir Indicators	Groundwater Indicators	Boundary Indicators
Drainage density	Population distribution	
Landforms	Streams/rivers	
	Snow-melt patterns	
	Karst topography	

Geologic Indicators

7-32. The type of rock or soil present is an important indicator because it usually defines the types of aquifers present and their water-producing characteristics. For field reconnaissance, the engineer need only recognize igneous, metamorphic, and sedimentary rocks and alluvium soil.

- Igneous rock. Usually a poor aquifer except where rock has been disturbed by faulting or fracturing. Igneous rock is normally incapable of storing or transmitting groundwater and acts as a barrier to groundwater flow.
- Metamorphic rock. Rarely capable of producing sufficient groundwater and has poor potential for groundwater development. Metamorphic rock is considered to be an effective barrier to groundwater flow.
- Sedimentary rock. Has the greatest capacity for holding groundwater. Unfractured sedimentary rock is capable of supplying low well yields, and fractured sedimentary rock is capable of supplying moderate to high well yields.
- Alluvium soil. Groundwater is most readily available in areas that are underlaid with alluvium. This is largely because uncemented or slightly cemented, compacted materials have maximum pore space and are relatively shallow and easily penetrated. The size of particles, the percentage of fines, and the degree of gradation have an important bearing on the yield of groundwater in soils. Clay yields almost no water; silt slowly yields some water, and well-sorted, clean, coarse sand and gravel freely yield water. Alluvial valleys are among the most productive terrains for recovering groundwater.

PERMEABILITY

7-33. Permeability is the property of soil that permits water to flow through it. Water may move through the continuous voids of a soil in much the same way as it moves through pipes and other conduits. As has been indicated, this movement of water through soils is frequently termed seepage and may also be called percolation. Soils vary greatly in their resistance to the flow of water through them. Relatively coarse soils, such as clean sands and gravels, offer comparatively little resistance to the flow of water; these are said to be permeable or pervious soils. Fine-grained soils, particularly clays, offer great resistance to the movement of water through them and are said to be relatively impermeable or impervious. Some water does move through these soils, however. The permeability of a soil reflects the ease with which it can be drained; therefore, soils are sometimes classed as well-drained, poorly drained, or impervious. Permeability is closely related to frost action and to the settlement of soils under load.

7-34. The term k is called the coefficient of permeability. It has units of velocity and may be regarded as the discharge velocity under a unit hydraulic gradient. The coefficient of permeability depends on the properties of the fluid involved and on the soil. Since water is the fluid normally involved in soil problems, and since its properties do not vary enough to affect most practical problems, the coefficient of permeability is regarded as a property of the soil. Principal factors that determine the coefficient of permeability for a given soil include—

- Grain size.
- Void ratio.
- Structure.

7-35. The relationships among these different variables for typical soils are quite complex and preclude the development of formulas for the coefficient of permeability, except for the simplest cases. For the usual soil, k is determined experimentally, either in the laboratory or in the field. These methods are discussed briefly in the next paragraph. Typical values of the coefficient of permeability for the soil groups of the USCS are given in column 8 of table 5-4, page 5-14.

DRAINAGE CHARACTERISTICS

7-36. The general drainage characteristics of soils classified under the USCS are given in column 12 of table 5-3, page 5-12. Soils may be divided into three general groups on the basis of their drainage characteristics. They are—

- Well-drained.
- Poorly drained.
- Impervious.

WELL-DRAINED SOILS

7-37. Clean sands and gravels, such as those included in the (GW), (GP), (SW), or (SP) groups, fall into the classification of well-drained soils. These soils may be drained readily by gravity systems. In road and airfield construction, for example, open ditches may be used in these soils to intercept and carry away water that comes in from surrounding areas. This approach is very effective when used in combination with the sealing of the surface to reduce infiltration into the base or subgrade. In general, if the groundwater table around the site of a construction project is controlled in these soils, then it will be controlled under the site also.

POORLY DRAINED SOILS

7-38. Poorly drained soils include inorganic and organic fine sands and silts, organic clays of low compressibility, and coarse-grained soils that contain an excess of nonplastic fines. Soils in the (ML), (OL), (MH), (GM), (GC), (SC), and (SM) groups, and many from the (Pt) group, generally fall into this category. Drainage by gravity alone is likely to be quite difficult for these soils.

IMPERVIOUS SOILS

7-39. Fine-grained, homogeneous, plastic soils and coarse-grained soils that contain plastic fines are considered impervious soils. This normally includes (CL) and (CH) soils and some in the (OH) groups. Subsurface drainage is so slow on these items that it is of little value in improving their condition. Any drainage process is apt to be difficult and expensive.

FILTER DESIGN

7-40. The selection of the proper filter material is of great importance since it largely determines the success or failure of the drainage system. A layer of filter material approximately 6 inches deep should be placed around all subsurface piping systems. The improper selection of a filter material can cause the drainage system to become inoperative in one of three ways:

- The pipe may become clogged by the infiltration of small soil particles.
- Particles in the protected soil may move into or through the filters, causing instability of the surface.
- Free groundwater may not be able to reach the pipe.

7-41. To prevent these failures from occurring, criteria have been developed based on the soil's gradation curve (see chapter 5).

7-42. To prevent the clogging of a pipe by filter material moving through the perforations or openings, the following limiting requirements must be satisfied (see Engineer Manual (EM) 1110-2-1901):

- For slotted openings:

$$\frac{50 \ percent \ size \ of \ filter \ material}{slot \ width} > 1.2$$

- For circular holes:

$$\frac{50 \ percent \ size \ of \ filter \ material}{hole \ diameter} > 1$$

- For porous concrete pipes:

$$\frac{D_{85} \ filter \ (mm)}{D_{15} \ aggregate \ (mm)} > 5$$

- For woven filter cloths:

$$\frac{D_{85} \ surrounding \ soil}{Equivalent \ opeing \ size \ (EOS) of \ cloth} > 1$$

7-43. To prevent the movement of particles from the protected soil into or through the filters, the following conditions must be satisfied:

$$\frac{15 \ percent \ of \ filter \ material}{85 \ percent \ size \ of \ protected \ soil} < 5$$

and

$$\frac{50 \ percent \ of \ filter \ material}{50 \ percent \ size \ of \ protected \ soil} < 25$$

7-44. To permit free water to reach the pipe, the filter material must be many times more pervious than the protected soil. This condition is fulfilled when the following requirement is met:

$$\frac{15 \ percent \ size \ of \ filter \ material}{15 \ percent \ size \ of \ protected \ soil} > 5$$

7-45. If it is not possible to obtain a mechanical analysis of available filter materials and protected soils, concrete sand with mechanical analysis limits as shown in figure 7-4, page 7-12, may be used. Experience indicates that a well-graded concrete sand is satisfactory as a filter material in most sandy, silty soils.

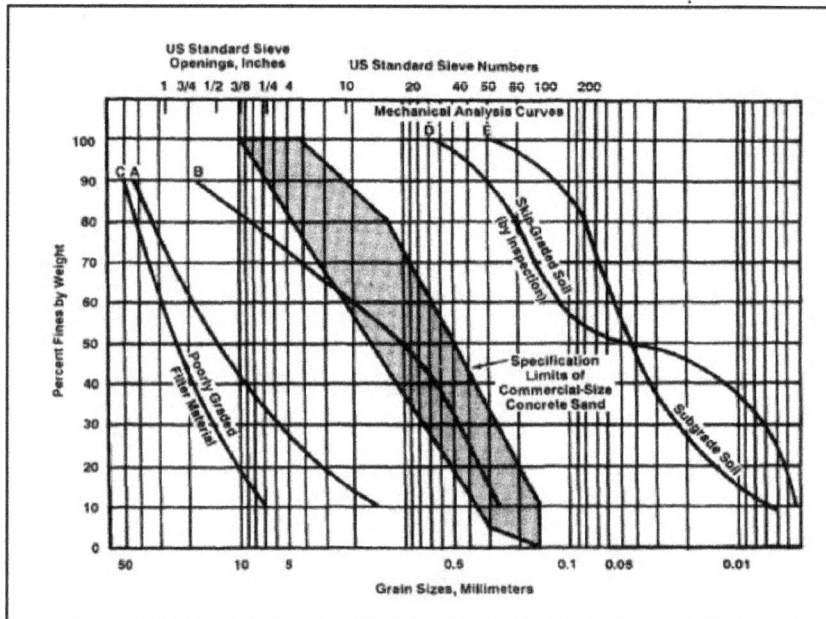

Figure 7-4. Mechanical analysis curves for filter material

POROSITY AND PERMEABILITY OF ROCKS

7-46. Porosity and permeability determine the water-bearing capability of a natural material.

POROSITY

7-47. The amount of water that rocks can contain depends on the open spaces in the rock. Porosity is the percentage of the total volume of the rock that is occupied by voids. Rock types vary greatly in size, number, and arrangement of their pore spaces and, consequently, in their ability to contain and yield water. The following list explains the porosity values of the types of rock displayed in figure 7-4a.

- A and B—A decrease in porosity due to compaction.
- C—A natural sand with high porosity due to good sorting.
- D—A natural sand with low porosity due to poor sorting and a matrix of silt and clay.
- E—Low porosity due to segmentation.
- G—Porous zone between lava flows.
- H—Limestone made porous by solution along joints.
- I—Massive rock made porous by fracturing.

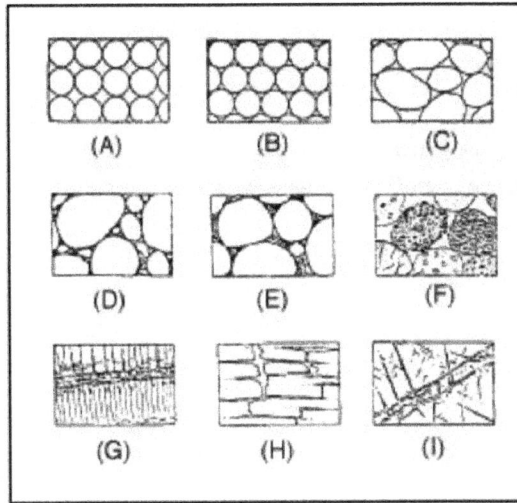

Figure 7-4a. Porosity in rocks

PERMEABILITY

7-48. The permeability of rock is its capacity for transmitting a fluid. The amount of permeability depends on the degree of porosity, the size and the shape of interconnections between pores, and the extent of the pore system.

WATER TABLE

7-49. In most regions, the depth that rocks are saturated with water depends largely on the permeability of the rocks, the amount of precipitation, and the topography. In permeable rocks, the surface below where the rocks are saturated is called the water table (figure 7b, page 7-3). The water table is—

- Not a level surface.
- Irregular and reflects the surface topography.
- Relatively high beneath hills.
- Closer to the surface or approaching the surface in valleys.

PERCHED WATER TABLE

7-50. If impermeable layers are present, descending water stops at their upper surfaces. If a water table lies well below the surface, a mass of impermeable rock may intercept the descending water and hold it suspended above the normal saturated zone. This isolated, saturated zone then has its own water table. Wells drilled into this zone are poor quality, because the well could quickly be drained of its water supply.

AQUIFER

7-51. An aquifer is a layer of rock below the water table. It is also called a water-bearing formation or a water-bearing stratum. Aquifers can be found in almost any area; however, they are difficult to locate in areas that do not have sedimentary rocks. Sands and sandstones usually constitute the best aquifers, but any rock with porosity and permeability can serve as a good water aquifer.

SALTWATER INTRUSION

7-52. There is always a danger of saltwater intrusion into groundwater sources along coastal areas and on islands. Because saltwater is unfit for most human needs, contamination can cause serious problems. When saltwater intrusion is discovered in the groundwater supply, determine the cause and mitigate it as soon as possible.

SECTION II - FROST ACTION

PROBLEMS

7-53. Frost action refers to any process that affects the ability of the soil to support a structure as a result of—

- Freezing.
- Thawing.

7-54. A difficult problem resulting from frost action is that pavements are frequently broken up or severely damaged as subgrades freeze during winter and thaw in the spring. In addition to the physical damage to pavements during freezing and thawing and the high cost of time, equipment, and personnel required in maintenance, the damage to communications routes or airfields may be great and, in some instances, intolerable strategically. In the spring or at other warm periods, thawing subgrades may become extremely unstable. In some severely affected areas, facilities have been closed to traffic until the subgrade recovered its stability.

7-55. The freezing index is a measure of the combined duration and magnitude of below-freezing temperature occurring during any given freezing season. Figure 7-5 shows the freezing index for a specific winter.

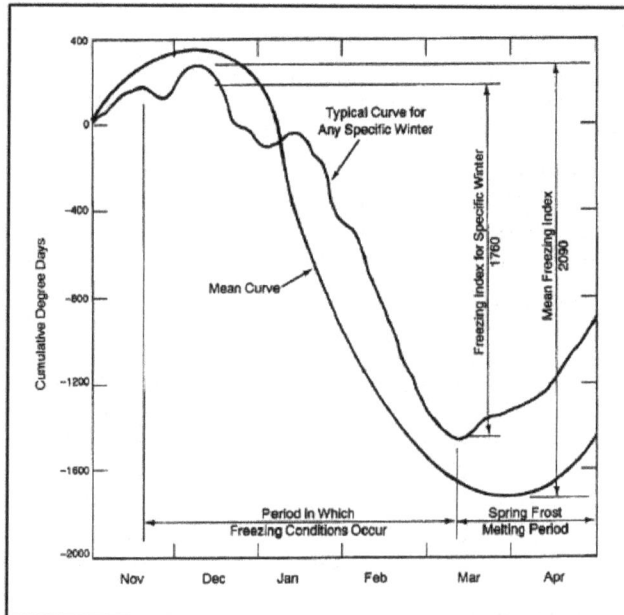

Figure 7-5. Determination of freezing index

FREEZING

7-56. Early theories attributed frost heaves to the expansion of water contained in soil voids upon freezing. However, this expansion would only be about 9 percent of the thickness of a frozen layer if caused by the water in the soil changing from the liquid to the solid state. It is not uncommon to note heaves as great as 60 percent; under laboratory conditions, heaves of as much as 300 percent have been recorded. These facts clearly indicate that heaving is due to the freezing of additional water that is attracted from the nonfrozen soil layers. Later studies have shown that frost heaves are primarily due to the growth of ice lenses in the soil at the plane of freezing temperatures.

7-57. The process of ice segregation may be pictured as follows: the thin layers of water adhering to soil grains become supercooled, meaning that this water remains liquid below 32 degrees Fahrenheit. A strong attraction exists between this water and the ice crystals that form in larger void spaces. This supercooled water flows by capillary action toward the already-formed crystals and freezes on contact. Continued crystal growth leads to the formation of an ice lens, which continues to grow in thickness and width until the source of water is cut off or the temperature rises above the normal freezing point (see figure 7-6).

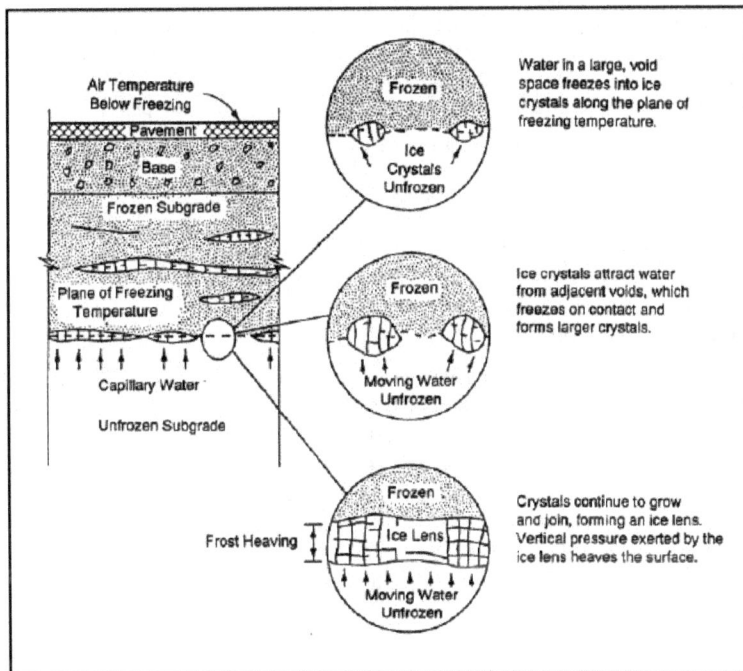

Figure 7-6. Formation of ice crystals on frost line

THAWING

7-58. The second phase of frost damage occurs toward the end of winter or in early spring when thawing begins. The frozen subgrade thaws both from the top and the bottom. For the latter case, if the air temperature remains barely below the freezing point for a sufficient length of time, deeply frozen soils gradually thaw from the bottom upward because of the outward conduction of heat from the earth's interior. An insulating blanket of snow tends to encourage this type of thawing. From an engineering standpoint, this thawing condition is desirable, because it permits melted water from thawed ice lenses to seep back through the lower soil layers to the water table from which it was drawn during the freezing process. Such dissipation of the melted water places no load on the surface drainage system, and no tendency exists to reduce subgrade stability by reason of saturation. Therefore, there is little difficulty in maintaining unpaved roads in a passable condition.

7-59. Thawing occurs from the top downward if the surface temperature rises from below the freezing point to well above that point and remains there for an appreciable time. This leaves a frozen layer beneath the thawed subgrade. The thawed soil between the pavement and this frozen layer contains an excessive amount of moisture resulting from the melting of the ice it contained. Since the frozen soil layer is impervious to the water, adequate drainage is almost impossible. The poor stability of the resulting supersaturated road or airfield subgrade accounts for many pavement failures. Unsurfaced earthen roads may become impassable when supersaturated.

7-60. Thawing from both the top and bottom occurs when the air temperature remains barely above the freezing point for a sufficient time. Such thawing results in reduced soil stability, the duration of which would be less than for soil where the thaw is only from the top downward.

CONDITIONS

7-61. Temperatures below 32 degrees Fahrenheit must penetrate the soil to cause freezing. In general, the thickness of ice layers (and the amount of consequent heaving) is inversely proportional to the rate of penetration of freezing temperature into the soil. Thus, winters with fluctuating air temperatures at the beginning of the freezing season produce more damaging heaves than extremely cold, harsh winters where the water is more likely to be frozen in place before ice segregation can take place.

7-62. A source of water must be available to promote the accumulation of ice lenses. Water may come from—

- A high groundwater table.
- A capillary supply from an adjoining water table.
- Infiltration at the surface.
- A water-bearing system (aquifer).
- Voids of fine-grained soils.

7-63. Ice segregation usually occurs in soils when a favorable source of water and freezing temperatures are present. The potential intensity of ice segregation in a soil depends largely on the size of the void space and may be expressed as an empirical function of grain size.

7-64. Inorganic soils containing 3 percent or more by weight of grains finer than 0.02 mm in diameter are generally considered frost susceptible. Although soils may have as high as 10 percent by weight of grains finer than 0.02 mm without being frost susceptible, the tendency of these soils to occur interbedded with other soils makes it impractical to consider them separately.

7-65. Frost-susceptible soils are classified in the following groups:

- F-1.
- F-2.
- F-3.
- F-4.

7-66. They are listed approximately in the order of increasing susceptibility to frost heaving or weakening as a result of frost melting (see table 7-1). The order of listing of subgroups under groups F-3 and F-4 does not necessarily indicate the order of susceptibility to frost heaving or weakening of these subgroups. There is some overlapping of frost susceptibility between groups. The soils in group F-4 are of especially high frost susceptibility. Soil names are defined in the USCS.

7-67. Varved clays consist of alternating layers of medium-gray inorganic silt and darker silty clay. The thickness of the layers rarely exceeds ½ inch, but occasionally much thicker varves are encountered. They are likely to combine the undesirable properties of both silts and soft clays. Varved clays are likely to soften more readily than homogeneous clays with equal water content. However, local experience and conditions should be taken into account since, under favorable conditions (as when insufficient moisture is available for significant ice segregation), little or no detrimental frost action may occur. Some evidence exists that pavements in the seasonal frost zone, constructed on varved clay subgrades in which the deposit and depth to groundwater are relatively uniform, have performed satisfactorily. Where subgrade conditions are uniform and local evidence indicates that the degree of heave is not exceptional, the varved clay subgrade soil should be assigned a group F-4 frost-susceptibility classification.

Table 7-1. Frost-susceptible soil groups

Frost Group	Kind of Soil	Percentage Finer Than 0.02 mm by Weight	Typical Soil Types Under Unified Soil Classification System
F-1	Gravelly Soils	6 to 10	GM, GW-GM, GP-GM
F-2	Gravelly Soils	10 to 20	GM, GW-GM, GP-GM
	Sands	6 to 15	SM, SW-SM, SP-SM
F-3	Gravelly Soils	Over 20	GM, GC
	Sands, except very fine silty sands	Over 15	SM, SC
	Clays, PI >12	—	CL, CH
F-4	All silts	—	ML, MH
	Very fine silty sands	Over 15	SM
	Clays, PI > 12	—	CL, CL-ML
	Varved clays and other fine-grained, banded sediments	—	CL and ML; CL, ML; and SM; CL, CH, and ML; CL, CH, ML, and SM

EFFECTS

7-68. Frost action can cause severe damage to roads and airfields. The problems include heaving and the resultant loss of pavement strength.

HEAVING

7-69. Frost heave, indicated by the raising of the pavement, is directly associated with ice segregation and is visible evidence on the surface that ice lenses have formed in the subgrade material. Heave may be uniform or nonuniform, depending on variations in the character of the soils and the groundwater conditions underlying the pavement.

7-70. The tendency of the ice layers to develop and grow increases rapidly with decreasing grain size. On the other hand, the rate at which the water flows in an open system toward the zone of freezing decreases with decreasing grain size. Therefore, it is reasonable to expect that the worst frost heave conditions would be encountered in soils having an intermediate grain size. Silt soils, silty sands, and silty gravels tend to exhibit the greatest frost heave.

7-71. Uniform heave is the raising of adjacent areas of pavement surface by approximately equal amounts. In this type of heave, the initial shape and smoothness of the surface remains substantially unchanged. When nonuniform heave occurs, the heave of adjacent areas is appreciably different, resulting in objectionable unevenness or abrupt changes in the grade at the pavement surface.

7-72. Conditions conducive to uniform heave may exist, for example, in a section of pavement constructed with a fairly uniform stripping or fill depth, uniform depth to groundwater table, and uniform soil characteristics. Conditions conducive to irregular heave occur typically at locations where subgrades vary between clean sand and silty soils or at abrupt transitions from cut to fill sections with groundwater close to the surface.

7-73. Lateral drains, culverts, or utility lines placed under pavements on frost-susceptible subgrades frequently cause abrupt differential heaving. Wherever possible, such facilities should not be placed beneath these pavements, or transitions should be provided so as to moderate the roughening of the pavement during the period of heave.

LOSS OF PAVEMENT STRENGTH

7-74. When ice segregation occurs in a frost-susceptible soil, the soil's strength is reduced as is the load-supporting capacity of the pavement during prolonged frost-melting periods. This often occurs during winter and spring thawing periods, because near-surface ice melts and water from melting snow or rain may infiltrate through the surface causing an excess of water. This water cannot drain through the still-frozen soil below, or through the shoulders, or redistribute itself readily. The soil is thus softened.

7-75. Supporting capacity may be reduced in clay subgrades even through significant heave has not occurred. This may occur because water for ice segregation is extracted from the clay lattice below, and the resulting shrinkage of the lattice largely balances the volume of the ice lenses formed.

7-76. Further, traffic may cause remolding or develop hydrostatic pressure within the pores of the soil during the period of weakening, thus resulting in further-reduced subgrade strength. The degree to which a soil loses strength during a frost-melting period and the length of the period during which the strength of the soil is reduced depend on—

- The type of soil.
- Temperature conditions during freezing and thawing periods.
- The amount and type of traffic during the frost-melting periods.
- The availability of water during the freezing and thawing periods.
- Drainage conditions.

Rigid Pavements (Concrete)

7-77. Concrete alone has only a little tensile strength, and a slab is designed to resist loads from above while receiving uniform support from the subgrade and base course. Therefore, slabs have a tendency to break up as a result of the upthrust from nonuniform heaving soils causing a point bearing. As a rule, if rigid pavements survive the ill effects of upheaval, they will generally not fail during thawing. Reinforced concrete will carry a load by beam action over a subgrade having either frozen or supersaturated areas. Rigid pavements will carry a load over subgrades that are both frozen and supersaturated. The capacity to bear the design load is reduced, however, when the rigid slab is supported entirely by supersaturated, semiliquid subgrades.

Flexible Bituminous Pavements

7-78. The ductility of flexible pavements helps them to deflect with heaving and later resume their original positions. While heaving may produce severe bumps and cracks, usually it is not too serious for flexible pavements. By contrast, a load applied to poorly supported pavements during the thawing period normally results in rapid failure.

Slopes

7-79. Exposed back slopes and side slopes of cuts and fills in fine-grained soil have a tendency to slough off during the thawing process. The additional weight of water plus the soil exceeds the shearing strength of the soil, and the hydrostatic head of water exerts the greatest pressure at the foot of the slope. This causes sloughing at the toe of the slope, which multiplies the failure by consecutive shear failures due to inadequate stability of the altered slopes. Flatter slopes reduce this problem. Sustained traffic over severely weakened areas afflicted with frost boils initiates a pumping action that results in complete pavement failure in the immediate vicinity.

INVESTIGATIONAL PROCEDURES

7-80. The field and laboratory investigations conducted according to Chapter 5 of this manual usually provide sufficient information to determine whether a given combination of soil and water conditions beneath the pavement are conducive to frost action. This procedure for determining whether the conditions necessary for ice segregation are present at a proposed site are discussed in the following paragraphs. As stated earlier in this chapter, inorganic soils containing 3 percent or more by weight of grains finer than 0.02 mm are generally considered susceptible to ice segregation. Thus, examination of the fine portion of the gradation curve obtained from hydrometer analysis or the decantation process for these materials indicates whether they should be assumed frost susceptible. In borderline cases, or where unusual materials are involved, slow laboratory freezing tests may be performed to measure the relative frost susceptibility.

7-81. The freezing index value should be computed from actual daily air temperatures, if possible. Obtain the air temperatures from a weather station located as close as possible to the construction site. Differences in elevations, topographical positions, and nearness of cities, bodies of water, or other sources of heat may cause considerable variations in freezing indexes over short distances. These variations are of greater importance to the design in areas of a mean design freezing index of less than 100 (that is, a design freezing index of less than 500) than they are farther north.

7-82. The depth to which freezing temperatures penetrate below the surface of a pavement depends principally on the magnitude and duration of below-freezing air temperatures and on the amount of water present in the base, subbase, and subgrade.

7-83. A potentially troublesome water supply for ice segregation is present if the highest groundwater at any time of the year is within 5 feet of the proposed subgrade surface or the top of any frost-susceptible base materials. When the depth to the uppermost water table is in excess of 10 feet throughout the year, a source of water for substantial ice segregation is usually not present unless the soil contains a significant percentage of silt. In homogeneous clay soils, the water content that the clay subgrade will attain under a pavement is usually sufficient to provide water for some ice segregation even with a remote water table. Water may also enter a frost-susceptible subgrade by surface infiltration through pavement areas. Figure 7-7, page 7-20, illustrates sources of water that feed growing ice lenses, causing frost action.

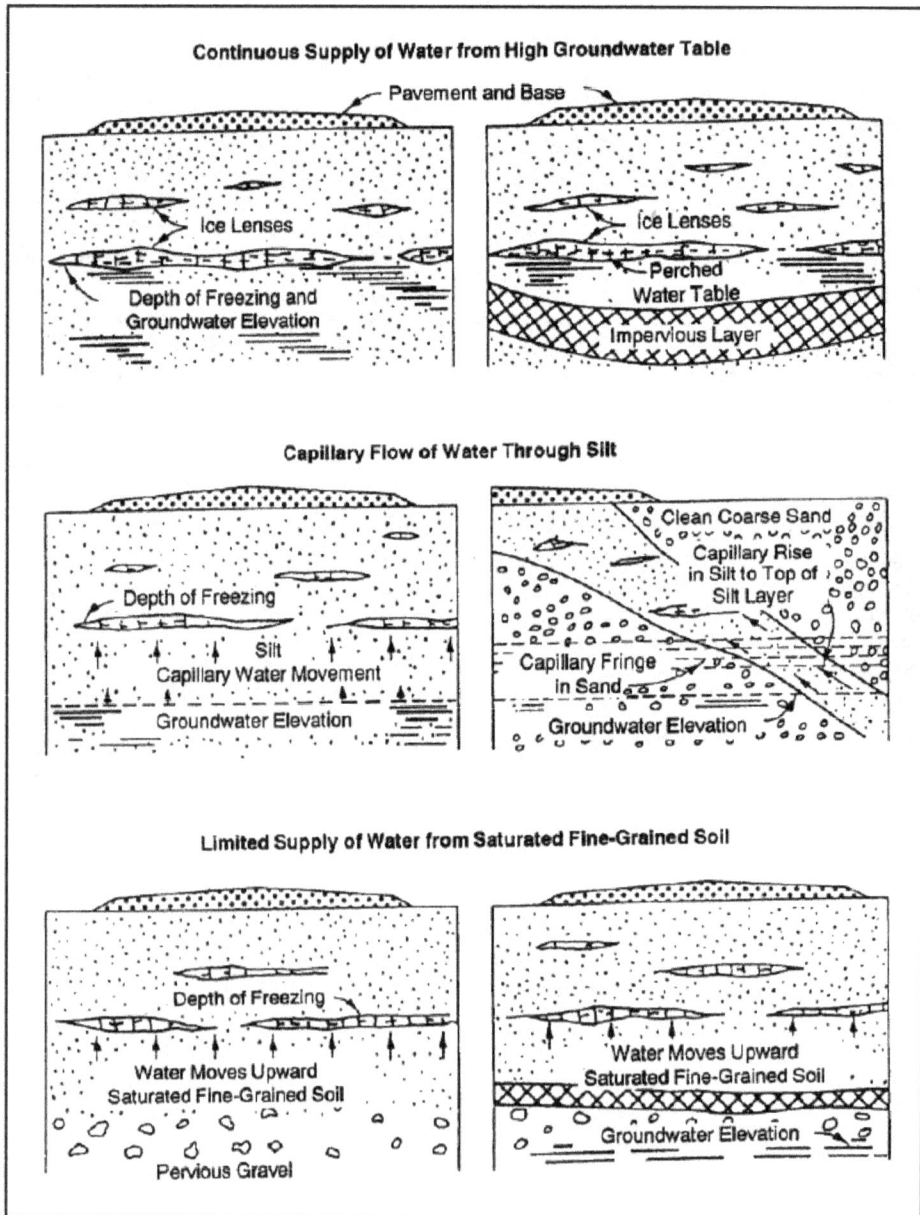

Figure 7-7. Sources of water that feed growing ice lenses

CONTROL

7-84. An engineer cannot prevent the temperatures that cause frost action. If a road or runway is constructed in a climate where freezing temperatures occur in winter, in all probability the soil beneath the pavement will freeze unless the period of lowered temperatures is very short. However, several construction techniques may be applied to counteract the presence of water and frost-susceptible soils.

7-85. Every effort should be made to lower the groundwater table in relation to the grade of the road or runway. This may be accomplished by installing subsurface drains or open side ditches, provided suitable outlets are available and that the subgrade soil is drainable. The same result may be achieved by raising the grade line in relation to the water table. Whatever means are employed for producing the condition, the distance from the top of the proposed subgrade surface (or any frost-susceptible base material used) to the highest probably elevation of the water table should not be less than 5 feet. Distances greater than this are very desirable if they can be obtained at a reasonable cost.

7-86. Where it is possible, upward water movement should be prevented. In many cases, lowering the water table may not be practical. An example is in swampy areas where an outlets for subsurface drains might not be present. One method of preventing the rise of water would be to place a 6-inch layer of pervious, coarse-grained soil 2 or 3 feet beneath the surface. This layer would be designed as a filter to prevent clogging the pores with finer material, which would defeat the original purpose. If the depth of frost penetration is not too great, it may be less expensive to backfill completely with granular material. Another successful method, though expensive, is to excavate to the frost line and backfill with granular material. In some cases, soil cement and asphalt-stabilized mixtures 6 inches thick have been used effectively to cut off the upward movement of water.

7-87. Even though the site selected may be on ideal soil, invariably on long stretches of roads or on wide expanses of runways, localized areas will be subject to frost action. These areas should be removed and replaced with select granular material. Unless this is meticulously carried out, differential heaving during freezing and severe strength loss upon thawing, may result.

7-88. The most generally accepted method of preventing subgrade failure due to frost action is to provide a suitable insulating cover to keep freezing temperatures from penetrating the subgrade to a significant depth. This insulating cover consists of a suitable thick pavement and a thick nonfrost-susceptible base course.

7-89. If the wearing surface is cleared of snow during freezing weather, the shoulders should also be kept free of snow. Where this is not the case, freezing will set in first beneath the wearing surface. This permits water to be drawn into and accumulate in the subgrade from the unfrozen shoulder area, which is protected by the insulating snow. If both areas are free of snow, then freezing will begin in the shoulder area because it is not protected by a pavement. Under this condition, water is drawn from the subgrade to the shoulder area. As freezing progresses to include the subgrade, there will be little frost action unless more water is available from groundwater or seepage.

BASE COMPOSITION REQUIREMENTS

7-90. All base and subbase course materials lying within design depth of frost penetration should be nonfrost-susceptible. Where the combined thickness of pavement and base or subbase over a frost-susceptible subgrade is less than the design depth of frost penetration, the following additional design requirements apply.

7-91. For both flexible and rigid pavements, the bottom 4 inches of base or subbase in contact with the subgrade, as a minimum, will consist of any nonfrost-susceptible gravel, sand, screening, or similar material. This bottom of the base or subbase will be designed as a filter between the subgrade soil and the overlying material to prevent mixing of the frost-susceptible subgrade with the nonfrost-susceptible base during and immediately after the frost-melting period. The gradation of this filter material shall be determined using these guidelines:

- To prevent the movement of particles from the frost-susceptible subgrade soil into or through the filter blanket, all of these must be satisfied:

$$\frac{15\ percent\ size\ of\ filter\ blanket}{85\ percent\ of\ subgrade\ soil} \leq 5$$

$$\frac{50\ percent\ size\ of\ filter\ blanket}{50\ percent\ of\ subgrade\ soil} \leq 25$$

- The filter blanket in the above case prevents the frost-susceptible soil from penetrating; however, the filter material itself must also not penetrate the nonfrost-susceptible base course material. Therefore, the filter material must also meet the following requirements:

$$\frac{15\ percent\ size\ of\ base\ course}{85\ percent\ of\ filter\ blanket} \leq 5$$

$$\frac{50\ percent\ size\ of\ base\ course}{50\ percent\ of\ filter\ blanket} \leq 25$$

- In addition to the above requirements, the filter material will, in no case, have 3 percent or more by weight of grains finer than 0.02 mm.

7-92. A major difficulty in the construction of the filter material is the tendency of the grain-size particles to segregate during placing; therefore, a $C_u > 20$ is usually not desirable. For the same reason, filter materials should not be skip- or gap-graded. Segregation of coarse particles results in the formation of voids through which fine particles may wash away from the subgrade soil. Segregation can best be prevented during placement by placing the material in the moist state. Using water while installing the filter blanket also aids in compaction and helps form satisfactory transition zones between the various materials. Experience indicates that nonfrost-susceptible sand is particularly suitable for use as filter course material. Also fine-grained subgraded soil may work up into an improperly graded overlying gravel or crushed stone base course. This will occur under the kneading action of traffic during the frost-melting period if a filter course is not provided between the subgrade and base course.

7-93. For rigid pavements, the 85-percent size of filter or regular base course material placed directly beneath pavement should be ≥ 2.00 mm in diameter (Number 10 US standard sieve size) for a minimum thickness of 4 inches. The purpose of this requirement is to prevent loss of support by pumping soil through the joints.

PAVEMENT DESIGN

7-94. Pavement may be designed according to either of two basic concepts. The design may be based primarily on—

- Control of surface deformation caused by frost action.
- Provision of adequate bearing capacity during the most critical climatic period.

Control of Surface Deformation

7-95. In this method of pavement design, a sufficient combined thickness of pavement and nonfrost-susceptible base is provided to reduce subgrade frost penetration. Consequently, this reduces pavement heave and subgrade weakening to a low, acceptable level.

Provision of Adequate Bearing Capacity

7-96. In this method, the amount of heave that will result is neglected and the pavement is designed solely on the anticipated reduced subgrade strength during the frost-melting period.

7-97. Detailed design methods used in determining the required thickness of pavement, base, and subbase for given traffic and soil conditions where frost action is a factor are described in FM 5-430.

Chapter 8

Soil Compaction

Soil compaction is one of the most critical components in the construction of roads, airfields, embankments, and foundations. The durability and stability of a structure are related to the achievement of proper soil compaction. Structural failure of roads and airfields and the damage caused by foundation settlement can often be traced back to the failure to achieve proper soil compaction.

Compaction is the process of mechanically densifying a soil. Densification is accomplished by pressing the soil particles together into a close state of contact with air being expelled from the soil mass in the process. Compaction, as used here, implies dynamic compaction or densification by the application of moving loads to the soil mass. This is in contrast to the consolidation process for fine-grained soil in which the soil is gradually made more dense as a result of the application of a static load. With relation to compaction, the density of a soil is normally expressed in terms of dry density or dry unit weight. The common unit of measurement is pcf. Occasionally, the wet density or wet unit weight is used.

SECTION I - SOIL PROPERTIES AFFECTED BY COMPACTION

ADVANTAGES OF SOIL COMPACTION

8-1. Certain advantages resulting from soil compaction have made it a standard procedure in the construction of earth structures, such as embankments, subgrades, and bases for road and airfield pavements. No other construction process that is applied to natural soils produces so marked a change in their physical properties at so low a cost as compaction (when it is properly controlled to produce the desired results). Principal soil properties affected by compaction include—

- Settlement.
- Shearing resistance.
- Movement of water.
- Volume change.

8-2. Compaction does not improve the desirable properties of all soils to the same degree. In certain cases, the engineer must carefully consider the effect of compaction on these properties. For example, with certain soils the desire to hold volume change to a minimum may be more important than just an increase in shearing resistance.

SETTLEMENT

8-3. A principal advantage resulting from the compaction of soils used in embankments is that it reduces settlement that might be caused by consolidation of the soil within the body of the embankment. This is true because compaction and consolidation both bring about a closer arrangement of soil particles.

8-4. Densification by compaction prevents later consolidation and settlement of an embankment. This does not necessarily mean that the embankment will be free of settlement; its weight may cause consolidation of compressible soil layers that form the embankment foundation.

SHEARING RESISTANCE

8-5.   Increasing density by compaction usually increases shearing resistance. This effect is highly desirable in that it may allow the use of a thinner pavement structure over a compacted subgrade or the use of steeper side slopes for an embankment than would otherwise be possible. For the same density, the highest strengths are frequently obtained by using greater compactive efforts with water contents somewhat below OMC. Large-scale experiments have indicated that the unconfined compressive strength of a clayey sand could be doubled by compaction, within the range of practical field compaction procedures.

MOVEMENT OF WATER

8-6.   When soil particles are forced together by compaction, both the number of voids contained in the soil mass and the size of the individual void spaces are reduced. This change in voids has an obvious effect on the movement of water through the soil. One effect is to reduce the permeability, thus reducing the seepage of water. Similarly, if the compaction is accomplished with proper moisture control, the movement of capillary water is minimized. This reduces the tendency for the soil to take up water and suffer later reductions in shearing resistance.

VOLUME CHANGE

8-7.   Change in volume (shrinkage and swelling) is an important soil property, which is critical when soils are used as subgrades for roads and airfield pavements. Volume change is generally not a great concern in relation to compaction except for clay soils where compaction does have a marked influence. For these soils, the greater the density, the greater the potential volume change due to swelling, unless the soil is restrained. An expansive clay soil should be compacted at a moisture content at which swelling will not exceed 3 percent. Although the conditions corresponding to a minimum swell and minimum shrinkage may not be exactly the same, soils in which volume change is a factor generally may be compacted so that these effects are minimized. The effect of swelling on bearing capacity is important and is evaluated by the standard method used by the US Army Corps of Engineers in preparing samples for the CBR test.

SECTION II - DESIGN CONSIDERATIONS

MOISTURE-DENSITY RELATIONSHIPS

8-8.   Nearly all soils exhibit a similar relationship between moisture content and dry density when subjected to a given compactive effort (see figure 8-1). For each soil, a maximum dry density develops at an OMC for the compactive effort used. The OMC at which maximum density is obtained is the moisture content at which the soil becomes sufficiently workable under a given compactive effort to cause the soil particles to become so closely packed that most of the air is expelled. For most soils (except cohesionless sands), when the moisture content is less than optimum, the soil is more difficult to compact. Beyond optimum, most soils are not as dense under a given effort because the water interferes with the close packing of the soil particles. Beyond optimum and for the stated conditions, the air content of most soils remains essentially the same, even though the moisture content is increased.

8-9.   The moisture-density relationship shown in figure 8-1 is indicative of the workability of the soil over a range of water contents for the compactive effort used. The relationship is valid for laboratory and field compaction. The maximum dry density is frequently visualized as corresponding to 100 percent compaction for the given soil under the given compactive effort.

8-10. The curve on figure 8-1 is valid only for one compactive effort, as established in the laboratory. The standardized laboratory compactive effort is the compactive effort (CE) 55 compaction procedure, which has been adopted by the US Army Corp of Engineers. Detailed procedures for performing the CE 55 compaction test are given in TM 5-530. The maximum dry density ($\gamma dmax$) at the 100 percent compaction mark is usually termed the CE 55 maximum dry density, and the corresponding moisture content is the optimum moisture content. Table 8-1 shows the relationship between the US Army Corps of Engineers compaction tests and their civilian counterparts. Many times the names of these tests are used interchangeably in publications.

Table 8-1. Compaction test comparisons

Test Designation	Blow Per Layer	No of Layers	Hammer Weight lb	Hammer Drop in	Mold	
					Volume cu ft	Diameter in
US Army Corps of Engineers (MIL-STD-631A)						
CE 55	55	5	10	18	0.07636	6
CE 12	12	5	10	18	0.07636	6
ASTM						
D-1557 Modified Proctor	25	5	10	18	0.0333	4
	56	5	10	18	0.0750	6
Standard Proctor	25	3	5.5	12	0.0333	4
American Association of State Highway and Transportation Officials (AASHTO)						
T-180 Modified AASHTO	25	5	10	18	0.0333	4
	56	5	10	18	0.0750	6
T-99 Standard AASHTO	25	3	5.5	12	0.0333	4

8-11. Figure 8-1 shows the zero air-voids curve for the soil involved. This curve is obtained by plotting the dry densities corresponding to complete saturation at different moisture contents. The zero air-voids curve represents theoretical maximum densities for given water contents. These densities are practically unattainable because removing all the air contained in the voids of the soil by compaction alone is not possible. Typically, at moisture contents beyond optimum for any compactive effort, the actual compaction curve closely parallels the zero air-voids curve. Any values of the dry density curve that plot to the right of the zero air-voids curve are in error. The specific calculation necessary to plot the zero air-voids curve are in TM 5-530.

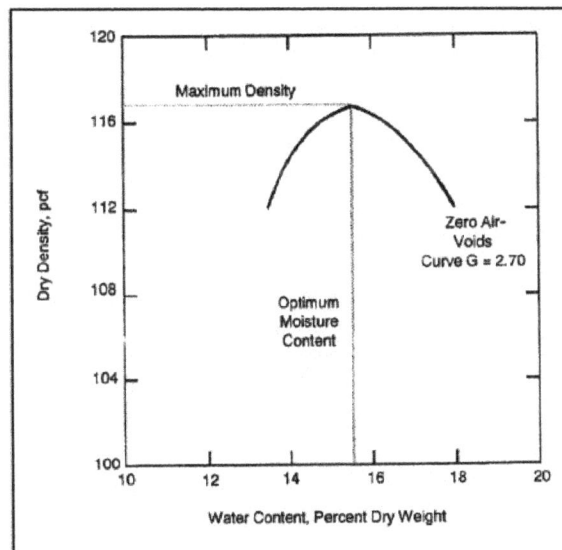

Figure 8-1. Typical moisture-density relationship

COMPACTION CHARACTERISTICS OF VARIOUS SOILS

8-12. The nature of a soil itself has a great effect on its response to a given compactive effort. Soils that are extremely light in weight, such as diatomaceous earths and some volcanic soils, may have maximum densities under a given compactive effort as low as 60 pcf. Under the same compactive effort, the maximum density of a clay may be in the range of 90 to 100 pcf, while that of a well-graded, coarse granular soil may be as high as 135 pcf. Moisture-density relationships for seven different soils are shown in figure 8-2. Compacted dry-unit weights of the soil groups of the Unified Soil Classification System are given in table 5-2, page 5-8. Dry-unit weights given in column 14 are based on compaction at OMC for the CE 55 compactive effort.

8-13. The curves of figure 8-2 indicate that soils with moisture contents somewhat less than optimum react differently to compaction. Moisture content is less critical for heavy clays (CH) than for the slightly plastic, clayey sands (SM) and silty sands (SC). Heavy clays may be compacted through a relatively wide range of moisture contents below optimum with comparatively small change in dry density. However, if heavy clays are compacted wetter than the OMC (plus 2 percent), the soil becomes similar in texture to peanut butter and nearly unworkable. The relatively clean, poorly graded sands also are relatively unaffected by changes in moisture. On the other hand, granular soils that have better grading and higher densities under the same compactive effort react sharply to slight changes in moisture, producing sizable changes in dry density.

8-14. There is no generally accepted and universally applicable relationship between the OMC under a given compactive effort and the Atterberg limit tests described in chapter 4. OMC varies from about 12 to 25 percent for fine-grained soils and from 7 to 12 percent for well-graded granular soils. For some clay soils, the OMC and the PL will be approximately the same.

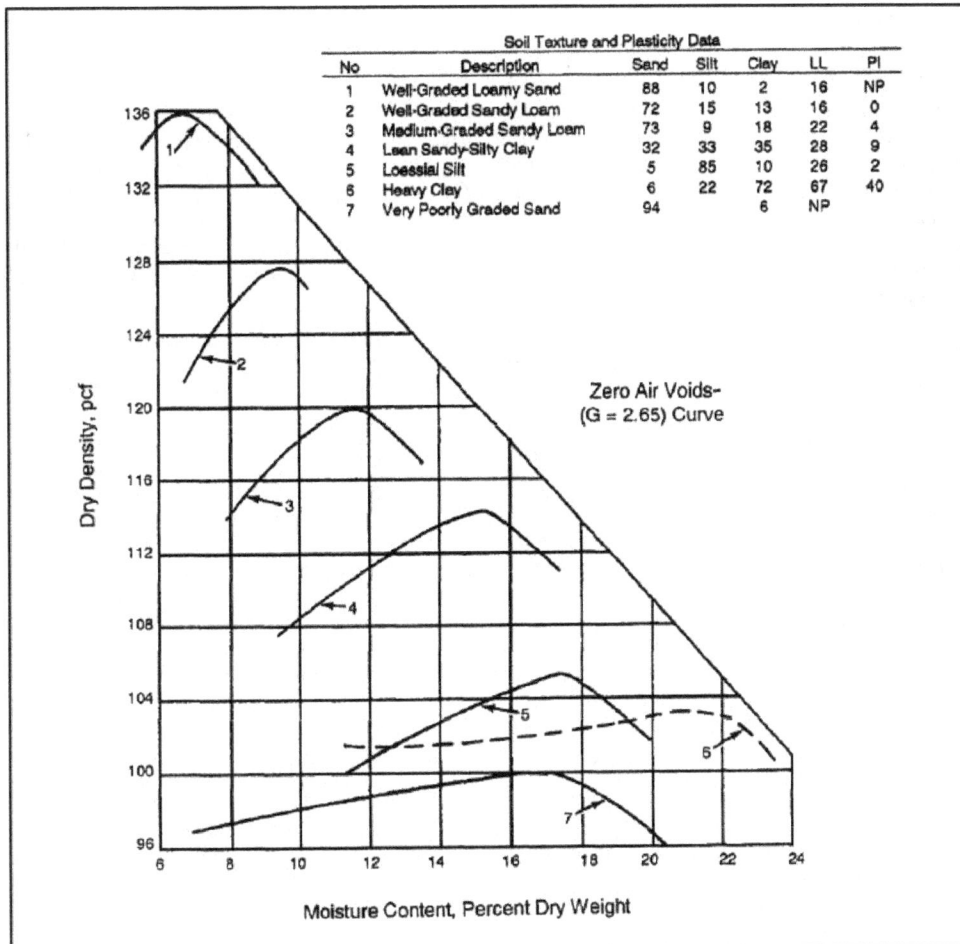

Figure 8-2. Moisture-density relationships of seven soils

OTHER FACTORS THAT INFLUENCE DENSITY

8-15. In addition to those factors previously discussed, several others influence soil density, to a smaller degree. For example, temperature is a factor in the compaction of soils that have a high clay content; both density and OMC may be altered by a great change in temperature. Some clay soils are sensitive to manipulation; that is, the more they are worked, the lower the density for a given compactive effort. Manipulation has little effect on the degree of compaction of silty or clean sands. Curing, or drying, of a soil following compaction may increase the strength of subgrade and base materials, particularly if cohesive soils are involved.

ADDITION OF WATER TO SOIL

8-16. Often water must be added to soils being incorporated in embankments, subgrades, and bases to obtain the desired degree of compaction and to achieve uniformity. The soil can be watered in the borrow pit or in place. After the water is added, it must be thoroughly and uniformly mixed with the soil. Even if additional water is not needed, mixing may still be desirable to ensure uniformity. In processing granular materials, the best results are generally obtained by sprinkling and mixing in place. Any good mixing equipment should be satisfactory. The more friable sandy and silty soils are easily mixed with water. They may be handled by sprinkling and mixing, either on the grade or in the pit. Mixing can be done with motor graders, rotary mixers, and commercial harrows to a depth of 8 inches or more without difficulty.

8-17. If time is available, water may also be added to these soils by diking or ponding the pit and flooding until the desired depth of penetration has taken place. This method usually requires several days to accomplish uniform moisture distribution. Medium clayey soils can be worked in the pit or in place as conditions dictate. The best results are obtained by sprinkling and mixing with cultivators and rotary mixers. These soils can be worked in lifts up to 8 inches or more without great difficulty. Heavy clay soils present many difficulties and should never be used as fill in an embankment foundation. They should be left alone without disturbance since usually no compactive effort or equipment is capable of increasing the in-place condition with reference to consolidation and shear strength.

8-18. The length of the section being rolled may have a great effect on densities in hot weather when water evaporates quickly. When this condition occurs, quick handling of the soil may mean the difference between obtaining adequate density with a few passes and requiring extra effort to add and mix water.

HANDLING OF WET SOILS

8-19. When the moisture content of the soil to be compacted greatly exceeds that necessary for the desired density, some water must be removed. In some cases, the use of excessively wet soils is possible without detrimental effects. These soils (coarse aggregates) are called free-draining soils, and their maximum dry density is unaffected by moisture content over a broad range of moisture. Most often, these soils must be dried; this can be a slow and costly process. The soil is usually dried by manipulating and exposing it to aeration and to the rays of the sun. Manipulation is most often done with cultivators, plows, graders, and rotary mixers. Rotary mixers, with the tail-hood section raised, permit good aeration and are very effective in drying excessively wet soils. An excellent method that may be useful when both wet and dry soils are available is simply to mix them together.

VARIATION OF COMPACTIVE EFFORT

8-20. For each compactive effort used in compacting a given soil, there is a corresponding OMC and maximum density. If the compactive effort is increased, the maximum density is increased and the OMC is decreased. This fact is illustrated in figure 8-3. It shows moisture-density relationships for two different soils, each of which was compacted using two different compactive efforts in the laboratory. When the same soil is compacted under several different compactive efforts, a relationship between density and compactive effort may be developed for that soil.

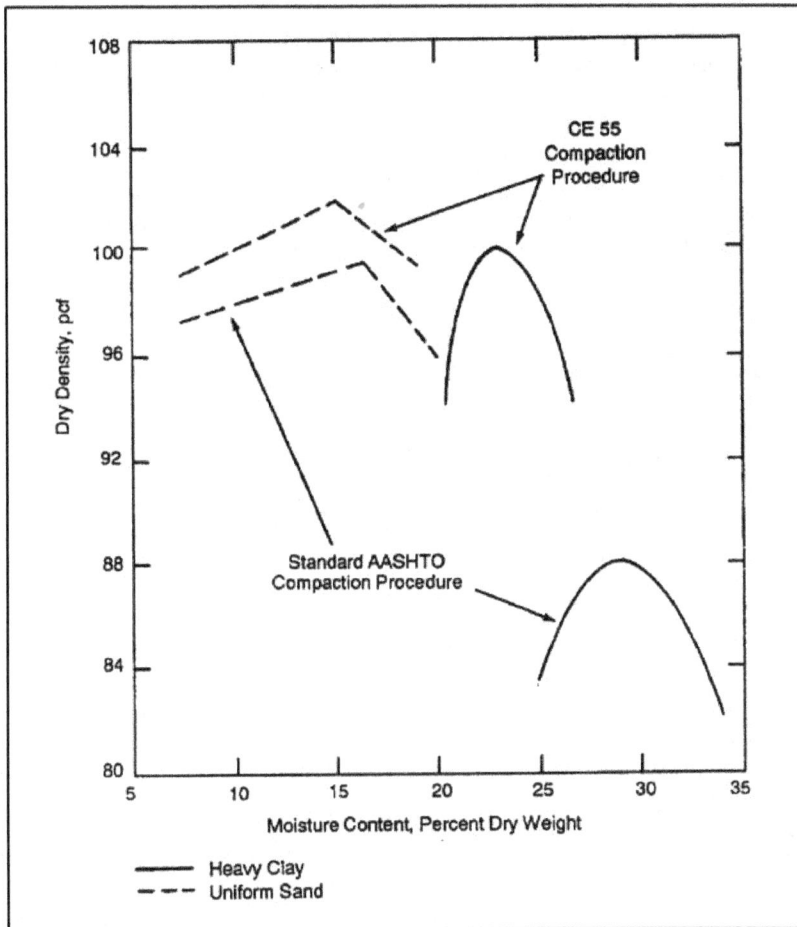

Figure 8-3. Moisture-density relationships of two soils

8-21. This information is of particular interest to the engineer who is preparing specifications for compaction and to the inspector who must interpret the density test results made in the field during compaction. The relationship between compactive effort and density is not linear. A considerably greater increase in compactive effort will be required to increase the density of a clay soil from 90 to 95 percent of CE 55 maximum density than is required to effect the same changes in the density of a sand. The effect of variation in the compactive effort is as significant in the field rolling process as it is in the laboratory compaction procedure. In the field, the compactive effort is a function of the weight of the roller and the number of passes for the width and depth of the area of soil that is being rolled. Increasing the weight of the roller or the number of passes generally increases the compactive effort. Other factors that may be of consequence include—

- Lift thickness.
- Contact pressure.
- Size and length of the tamping feet (in the case of sheepsfoot rollers).
- Frequency and amplitude (in the case of vibratory compactors).

8-22. To achieve the best results, laboratory and field compaction must be carefully correlated.

COMPACTION SPECIFICATIONS

8-23. To prevent detrimental settlement under traffic, a definite degree of compaction of the underlying soil is needed. The degree depends on the wheel load and the depth below the surface. For other airfield construction and most road construction in the theater of operations, greater settlement can be accepted, although the amount of maintenance will generally increase. In these cases, the minimum compaction requirements of table 8-2, should be met. However, strength can possibly decrease with increased compaction, particularly with cohesive materials. As a result, normally a 5 percent compaction range is established for density and a 4 percent range for moisture. Commonly, this "window" of density and moisture ranges is plotted directly on the CE 55 compaction curve and is referred to as the specifications block. Figure 8-4, shows a density range of 90 to 95 percent compaction and a moisture range of 12 to 16 percent.

Table 8-2. Minimum compaction requirements

Soil Placement	Soil Definition	Remarks
Base	Cohesionless, CBR > 80	100 percent compaction
Subbase	Cohesionless, CBR 20-50	100 percent compaction
Select Subgrade	CBR < 20, any in-place soil	Cohesionless: 95 percent compaction Cohesive: 90 percent compaction **Note**. If subgrade CBR > 20, 100 percent compaction

Embankment fill 50 ft > H		Traffic areas
		    Cohesionless: 95 percent compaction
		    Cohesive: 90 percent compaction
		Nontraffic areas
		    Cohesionless: 90 percent compaction
		    Cohesive: 85 percent compaction

Backfill for trenches		Under pavement: Same requirement as base through subgrade Nontraffic area: Subtract 5 percentage points for each
Top 6 inches of sidewalk		Cohesionless: 90 percent compaction Cohesive: 85 percent compaction
Small water-retaining structures		Cohesionless: 95 percent compaction Cohesive: 90 percent compaction
Note. Cohesionless: PI ≤, LL < 25 Cohesive: PI > 5		

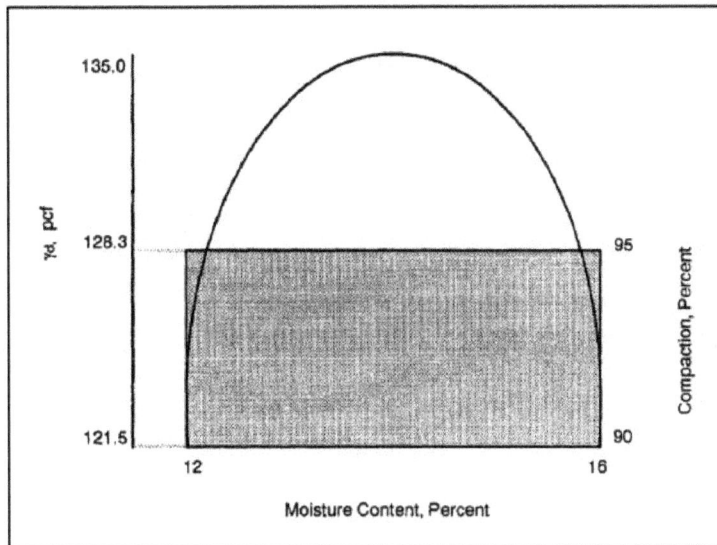

Figure 8-4. Density, compaction, and moisture content

CBR DESIGN PROCEDURE

8-24. The concept of the CBR analysis was introduced in chapter 6. In the following procedures, the CBR analytical process will be applied to develop soil compaction specifications. Figure 8-5, page 8-10, outlines the CBR design process. The first step is to look at the CE 55 compaction curve on a DD Form 2463, page 1. If it is U-shaped, the soil is classified as "free draining" for CBR analysis and the left-hand column of the flowchart should be used through the design process. If it is bell-shaped, use the swell data graphically displayed on a DD Form 1211. Soils that, when saturated, increase in volume more than 3 percent at any initial moisture content are classified as swelling soils. If the percentage of swelling is 3 percent, the soil is considered nonswelling.

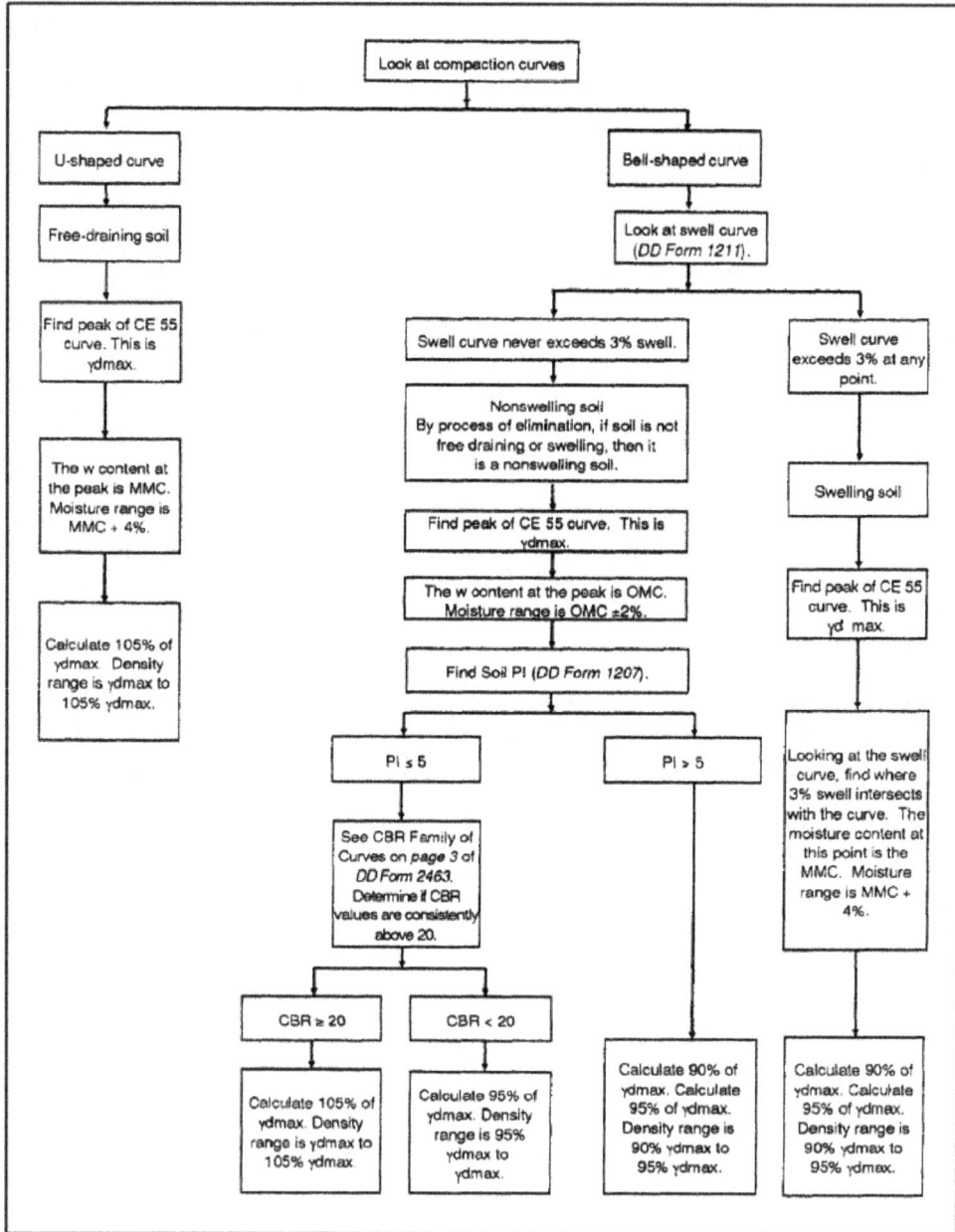

Figure 8-5. Density and moisture determination by CBR design method

8-25. Regardless of the CBR classification of the soil, the density value from the peak of the CE 55 moisture density curve is γdmax. The next step is to determine the design moisture content range. For nonswelling soils, the OMC is used. When the OMC is used, the design moisture content range is ±2 percent. For swelling and free-draining soils, the minimum moisture content (MMC) is used. The MMC is determined differently for swelling soils than it is for free-draining soils. The MMC for swelling soils is determined by finding the point at which the 3 percent swell occurs. The soil moisture content that corresponds to the 3 percent swell is the MMC. Free-draining soils exhibit an increase in density in response to increased soil moisture up to a certain moisture content, at which point no further increase in density is achieved by increasing moisture. The moisture content that corresponds to γdmax is the MMC. For both swelling and free-draining CBR soil classes, the design moisture-content range is MMC + 4 percent.

8-26. For swelling and free-draining soils, the final step in determining design compaction requirements is to determine the density range. Free-draining soils are compacted to 100-105 percent γdmax. Swelling soils are compacted to 90-95 percent γdmax.

8-27. Compaction requirement determinations for nonswelling soils require several additional steps. Once the OMC and design moisture content range have been determined, look at a DD Form 1207 for the PI of the soil. If PI > 5, the soil is cohesive and is compacted to 90-95 percent γdmax. If the PI 5, refer to the CBR Family of Curves on page 3 of DD Form 2463. If the CBR values are consistently above 20, compact the soil to 100-105 percent γdmax. If the CBR values are not above 20, compact the soil to 95-100 percent γdmax.

8-28. Once you have determined the design density range and the moisture content range, you have the tools necessary to specify the requirements for and manage the compaction operations. However, placing a particular soil in a construction project is determined by its gradation, Atterberg limits, and design CBR value. Appendix A contains a discussion of the CBR design process.

8-29. A detailed discussion of placing soils and aggregates in an aggregate surface or a flexible pavement design is in FM 5-430 (for theater-of-operations construction), TM 5-822-2 (for permanent airfield design), and TM 5-822-5 (for permanent road design).

SUBGRADE COMPACTION

8-30. In fill sections, the subgrade is the top layer of the embankment, which is compacted to the required density and brought to the desired grade and section. For subgrades, plastic soils should be compacted at moisture contents that are close to optimum. Moisture contents cannot always be carefully con-trolled during military construction, but certain practical limits must be recognized. Generally, plastic soils cannot be compacted satisfactorily at moisture contents more than 10 percent above or below optimum. Much better results are obtained if the moisture content is controlled to within 2 percent of optimum. For cohesionless soils, moisture control is not as important, but some sands tend to bulk at low moisture content. Compaction should not be attempted until this situation is corrected. Normally, cohesionless soils are compacted at moisture contents that approach 100 percent saturation.

8-31. In cut sections, particularly when flexible pavements are being built to carry heavy wheel loads, subgrade soils that gain strength with compaction should be compacted to the general requirements given earlier. This may make it necessary to remove the soil, replace it, and compact it in layers to obtain the required densities at greater depths. In most construction in the theater of operations, subgrade soil in cut sections should be scarified to a depth of about 6 inches and recompacted. This is commonly referred to as a scarify/compact in-place (SCIP) operation. This procedure is generally desirable in the interest of uniformity.

Expansive Clays

8-32. As indicated previously, soils that have a high clay content (particularly (CH), (MH), and (OH)) may expand in detrimental amounts if compacted to a high density at a low moisture content and then exposed to water. Such soils are not desirable as subgrades and are difficult to compact. If they have to be used, they must be compacted to the maximum density obtainable using the MMC that will result in a minimum amount of swelling. Swelling soils, if placed at moisture contents less than the MMC, can be expected to swell more than 3 percent. Soil volume increases of up to 3 percent generally do not adversely affect theater-of-operations structures. This method requires detailed testing and careful control of compaction. In some cases, a base of sufficient thickness should be constructed to ensure against the harmful effects of expansion.

Clays and Organic Soils

8-33. Certain clay soils and organic soils lose strength when remolded. This is particularly true of some (CH) and (OH) soils. They have high strengths in their undisturbed condition, but scarifying, reworking, and compacting them in cut areas may reduce their shearing strengths, even though they are compacted to design densities. Because of these qualities, they should be removed from the construction site.

Silts

8-34. When some silts and very fine sands (predominantly (ML) and (SC) soils) are compacted in the presence of a high water table, they will pump water to the surface and become "quick", resulting in a loss of shearing strength. These soils cannot be properly compacted unless they are dried. If they can be compacted at the proper moisture content, their shearing resistance is reasonably high. Every effort should be made to lower the water table to reduce the potential of having too much water present. If trouble occurs with these soils in localized areas, the soils can be removed and replaced with more suitable ones. If removal, or drainage and later drying, cannot be accomplished, these soils should not be disturbed by attempting to compact them. Instead, they should be left in their natural state and additional cover material used to prevent the subgrade from being overstressed.

8-35. When these soils are encountered, their sensitivity may be detected by performing unconfined compression tests on the un- disturbed soil and on the remolded soil compacted to the design density at the design moisture content. If the undisturbed value is higher, do not attempt to compact the soil; manage construction operations to produce the least possible disturbance of the soil. Base the pavement design on the bearing value of the undisturbed soil.

BASE COMPACTION

8-36. Selected soils that are used in base construction must be compacted to the general requirements given earlier. The thickness of layers must be within limits that will ensure proper compaction. This limit is generally from 4 to 8 inches, depending on the material and the method of construction.

8-37. Smooth-wheeled or vibratory rollers are recommended for compacting hard, angular materials with a limited amount of fines or stone screenings. Pneumatic-tired rollers are recommended for softer materials that may break down (degrade) under a steel roller.

MAINTENANCE OF SOIL DENSITY

8-38. Soil densities obtained by compaction during construction may be changed during the life of the structure. Such considerations are of great concern to the engineer engaged in the construction of semipermanent installations, although they should be kept in mind during the construction of any facility to ensure satisfactory performance. The two principal factors that tend to change the soil density are—

- Climate.
- Traffic.

8-39. As far as embankments are concerned, normal embankments retain their degree of compaction unless subjected to unusual conditions and except in their outer portions, which are subjected to seasonal wetting and drying and frost action. Subgrades and bases are subject to more severe climatic changes and traffic than are embankments. Climatic changes may bring about seasonal or permanent changes in soil moisture and accompanying changes in density, which may distort the pavement surface. High-volume-change soils are particularly susceptible and should be compacted to meet conditions of minimum swelling and shrinkage. Granular soils retain much of their compaction under exposure to climatic conditions. Other soils may be somewhat affected, particularly in areas of severe seasonal changes, such as—

- Semiarid regions (where long, hot, dry periods may occur).
- Humid regions (where deep freezing occurs).

8-40. Frost action may change the density of a compacted soil, particularly if it is fine-grained. Heavy traffic, particularly for subgrades and bases of airfields, may bring about an increase in density over that obtained during construction. This increase in density may cause the rutting of a flexible pavement or the subsidence of a rigid pavement. The protection that a subgrade soil receives after construction is complete has an important effect on the permanence of compaction. The use of good shoulders, the maintenance of tight joints in a concrete pavement, and adequate drainage all contribute toward maintaining the degree of compaction achieved during construction.

SECTION III - CONSTRUCTION PROCEDURES

GENERAL CONSIDERATIONS

8-41. The general construction process of a rolled-earth embankment requires that the fill be built in relatively thin layers or "lifts," each of which is rolled until a satisfactory degree of compaction is obtained. The subgrade in a fill section is usually the top lift in the compacted fill, while the subgrade in a cut section is usually compacted in in-place soil. Soil bases are normally compacted to a high degree of density. Compaction requirements frequently stipulate a certain minimum density. For military construction, this is generally a specified minimum percentage of CE 55 maximum density for the soil concerned. The moisture content of the soil is maintained at or near optimum, within the practical limits of field construction operations (normally ±2 percent of the OMC). Principal types of equipment used in field compaction are sheepsfoot, smooth steel-wheeled, vibratory, and pneumatic-tired rollers.

SELECTION OF MATERIALS

8-42. Soils used in fills generally come from cut sections of the road or airfield concerned, provided that this material is suitable. If the material excavated from cut sections is not suitable, or if there is not enough of it, then some material is obtained from other sources. Except for highly organic soils, nearly any soil can be used in fills. However, some soils are more difficult to compact than others and some require flatter side slopes for stability. Certain soils require elaborate protective devices to maintain the fill in its original condition. When time is available, these considerations and others may make it advantageous to thoroughly investigate construction efforts, compaction characteristics, and shear strengths of soils to be used in major fills. Under expedient conditions, the military engineer must simply make the best possible use of the soils at hand.

8-43. In general terms, the coarse-grained soils of the USCS are desirable for fill construction, ranging from excellent to fair. The fine-grained soils are less desirable, being more difficult to compact and requiring more careful control of the construction process. Tables 5-2 and 5-3, pages 5-7 and 5-8, respectively contain more specific information concerning the suitability of these soils.

DUMPING AND SPREADING

8-44. Since most fills are built up of thin lifts to the desired height, the soil for each lift must be spread in a uniform layer of the desired thickness. In typical operations, the soil is brought in, dumped, and spread by scraper units. The scrapers must be adjusted carefully to accomplish this objective. Materials may also be brought in by trucks or wagons and dumped at properly spaced locations so that a uniform layer may be easily spread by blade graders or bulldozers. Working alone, bulldozers may form very short and shallow fills. End dumping of soil material to form a fill without compaction is rarely permitted in modern embankment construction except when a fill is being built over very weak soils, as in a swamp. The bottom layers may then be end dumped until sufficient material has been placed to allow hauling and compacting equipment to operate satisfactorily. The best thickness of the layer to be used with a given soil and a given equipment cannot be determined exactly in advance. It is best determined by trial during the early stages of rolling on a project. No lift, however, will have a thickness less than twice the diameter of the largest size particle in the lift. As stated previously, compacted lifts will normally range from 4 to 8 inches in depth (see table 8-3, page 8-15).

COMPACTION OF EMBANKMENTS

8-45. If the fill consists of cohesive or plastic soils, the embankment generally must be built up of uniform layers (usually 4 to 6 inches in compacted thickness), with the moisture content carefully controlled. Rolling should be done with the sheepsfoot or tamping-foot rollers. Bonding of a layer to the one placed on top of it is aided by the thin layer of loose material left on the surface of the rolled layer by the roller feet. Rubber-tired or smooth-wheeled rollers may be used to provide a smooth, dense, final surface. Rubber-tired construction equipment may provide supplemental compaction if it is properly routed over the area.

8-46. If the fill material is clean sand or sandy gravel, the moisture range at which compaction is possible is generally greater. Because of their rapid draining characteristics, these soils may be compacted effectively at or above OMC. Vibratory equipment may be used. Soils may be effectively compacted by combined saturation and the vibratory effects of crawler tractors, particularly when tractors are operated at fairly high speeds so that vibration is increased.

8-47. For adequate compaction, sands and gravels that have silt and clay fines require effective control of moisture. Certain soils of the (GM) and (SM) groups have especially great need for close control. Pneumatic-tired rollers are best for compacting these soils, although vibratory rollers may be used effectively.

8-48. Large rock is sometimes used in fills, particularly in the lower portion. In some cases, the entire fill may be composed of rock layers with the voids filled with smaller rocks or soil and only a cushion layer of soil for the subgrade. The thickness of such rock layers should not be more than 24 inches with the diameter of the largest rock fragment being not greater than 90 percent of the lift thickness. Compaction of this type of fill is difficult but may generally be done by vibration from the passage of tack-type equipment over the fill area or possibly 50-ton pneumatic-tired rollers.

8-49. Finishing in embankment construction includes all the operations necessary to complete the earthwork. Included among these operations are the trimming of the side and ditch slopes, where necessary, and the fine grading needed to bring the embankment section to final grade and cross section. Most of these are not separate operations performed after the completion of other operations but are carried along as the work progresses. The tool used most often in finishing operations is the motor grader, while scraper and dozer units may be used if the finish tolerances are not too strict. The provision of adequate drainage facilities is an essential part of the work at all stages of construction, temporary and final.

Table 8-3. Soil classification and compaction requirements (average)

Major Division	Symbol	Symbol Description	Value as a Base, Subbase, or Subgrade	Potential Frost Action	Sheepsfoot, Standard With Ballast (Towed by Dozer) Single Drum: 4 ft, Dual Drum: 8 ft — Compacted Lift Thickness (Inches)	Rolling Speed (mph)(vpm)	Number of Passes	Self-Propelled Vibratory Roller Rolling Width = 7 ft — Compacted Lift Thickness (Inches)	Rolling Speed (mph)(vpm)	Number of Passes	Compactor, High Speed, Tamping Foot, Self-Propelled, BOMAG Model Rolling Width = 5 ft (Not Recommended for Finish Grade) — Compacted Lift Thickness (Inches)	Rolling Speed (mph)(vpm)	Number of Passes
Coarse-grained soils with 50% or more larger than No 200 sieve opening — Gravel and/or gravelly soils	GW	Well-graded gravels or gravel-sand mixture with 5% or less amount of fines	Good to excellent for subbase and subgrade. Fair to good for base.	None to very slight	*	N/A	N/A	Best 18	4 mph 1,400 vpm or more	8	12	10	5
	GP	Poorly graded gravels or gravel sand mixture with little or no fines	Fair to good for all.	None to very slight	*	N/A	N/A	Best 18	4 mph 1,400 vpm or more	8	12	10	5
	GM	Silty gravel, gravel-sand-silt mixtures	Not suitable for base, 15% or less of fines with PI of 5 or less. 50% or less of fines for subbase and subgrade.	Slight to medium	*	N/A	N/A	12	4 mph 1,100 vpm	6	9	10	6
	GC	Clayey gravel, gravel-sand-clay mixture	Not suitable for base, 15% or less of fines with PI of 5 or less. Poor to good for subbase and subgrade.	Slight to medium	6	3	10	12	4 mph 700 vpm to none	6	9	8	7
Sand and/or sandy soils	SW	Well-graded sands or gravelly sand mixture with 5% or less amount of fines	Poor for base. Fair to good for subbase and subgrade.	None to very slight	*	N/A	N/A	Best 18	4 mph 1,400 vpm or more	8	12	10	5
	SP	Poorly graded sands or gravelly sand mixture with 5% or less amount of fines	Poor to not suitable for base. Poor to fair for subbase and subgrade.	None to very slight	*	N/A	N/A	Best 18	4 mph 1,400 vpm or more	8	12	10	5
	SM	Silty sands, sand-silt mixture	Not suitable for base. Poor to good for subbase and subgrade.	Slight to high	*	N/A	N/A	12	4 mph 1,100 vpm	6	9	10	6
	SC	Clayey sands, sand-clay mixture	Not suitable for base. Poor to fair for subbase and subgrade.	Slight to high	Best 6	3	10	12	3 mph 700 vpm to none	7	9	8	6
Fine-grained soils with more than 50% smaller than No 200 sieve opening — Silt and clays with LL < 50	ML	Inorganic silt, silty fine sands	Not suitable for base or subbase. Poor to fair for subgrade.	Medium to very high	6	3	10	8	3 mph 700 vpm to none	7	6	8	5
	CL	Inorganic clay of low to medium plasticity, lean clays	Not suitable for base or subbase. Poor to fair for subgrade.	Medium to high	Best 6	2	12	8	3 mph 700 vpm to none	7	6	4	5
	OL	Organic silt and organic silt-clay of low plasticity	Not suitable for base or subbase. Poor to very poor for subgrade.	Medium to high	6	2	12	*	N/A	N/A	6	4	5
Silt and clays with LL > 50	MH	Inorganic silt micaceous or diatomaceous silty soil	Not suitable for base or subbase. Poor to fair for subgrade.	Medium to very high	6	2	12	*	N/A	N/A	6	4	6
	CH	Inorganic clay of high plasticity, fatty clays	Not suitable for base or subbase. Poor to fair for subgrade.	Medium	Best 6	2	14	*	N/A	N/A	6	3	6
	OH	Organic clay of medium to high plasticity	Not suitable for base or subbase. Poor to very poor for subgrade.	Medium	6	2	14	*	N/A	N/A	6	3	6

* NOT recommended

Table 8-3. Soil classification and compaction requirements (average)

Vibratory Roller (Wheel Towed) Rolling Width = 4 ft			9-Wheel Pneumatic, Self-Propelled With Ballast 100 psi Rolling Width = 6 ft			50-Ton Pneumatic Compactor With Ballast (Wheel Towed) 100 psi Rolling Width = 7 ft			13-Wheel Pneumatic Compactor With Ballast (Wheel Towed) 100 psi Rolling Width = 7 ft		
Number of Passes	Rolling Speed (mph) (vpm)	Compacted Lift Thickness (Inches)	Number of Passes	Rolling Speed (mph)	Compacted Lift Thickness (Inches)	Number of Passes	Rolling Speed (mph)	Compacted Lift Thickness (Inches)	Number of Passes	Rolling Speed (mph)	Compacted Lift Thickness (Inches)
8	4 mph 1400 vpm or more	12	5	6	6	10	5	18	10	5	6
8	4 mph 1,400 vpm or more	12	6	6	6	10	5	18	10	5	6
8	4 mph 1,100 vpm	9	7	6	6	8	4	12	10	4	6
9	4 mph 700 vpm to none	9	7	5	6	8	3	12	10	4	6
8	4 mph 1,400 vpm or more	12	7	6	6	9	5	18	10	5	6
8	4 mph 1,100 vpm or more	12	7	6	6	9	5	18	10	5	6
6	4 mph 1,100 vpm	9	8	6	6	7	4	12	10	4	6
10	3 mph 700 vpm to none	9	8	5	6	7	3	12	12	3	6
10	3 mph 700 vpm to none	6	6	4	4	6	4	9	7	3	4
10	3 mph 700 vpm to none	6	6	4	4	6	3	9	7	3	4
N/A	N/A	*	6	4	4	6	3	9	7	3	4
N/A	N/A	*	6	4	4	6	3	9	8	3	4
N/A	N/A	*	6	3	4	7	3	9	9	2	4
N/A	N/A	*	6	3	4	7	3	9	9	2	4

DENSITY DETERMINATIONS

8-50. Density determinations are made in the field by measuring the wet weight of a known volume of compacted soil. The sample to be weighed is taken from a roughly cylindrical hole that is dug in the compacted layer. The volume of the hole may be determined by one of several methods, including the use of—

- Heavy oil of known specific gravity.
- Rubber balloon density apparatus.

- Calibrated sand.
- Nuclear densimeter.

8-51. When the wet weight and the volume are known, the unit wet weight may then be calculated, as described in FM 5-430.

8-52. In very arid regions, or when working with soils that lose strength when remolded, the adequacy of compaction should be judged by performing the in-place CBR test on the compacted soil of a subgrade or base. The CBR thus obtained can then be compared with the design CBR, provided that the design was based on CBR tests on unsoaked samples. If the design was based on soaked samples, the results of field in-place CBR tests must be correlated with the results of laboratory tests performed on undisturbed mold samples of the in-place soil subjected to soaking. Methods of determining the in-place CBR of a soil are described in TM 5-530.

FIELD CONTROL OF COMPACTION

8-53. As stated in previous paragraphs, specifications for adequate compaction of soil used in military construction generally require the attainment of a certain minimum density in field rolling. This requirement is most often stated in terms of a specified percentage range of CE 55 maximum density. With many soils, the close control of moisture content is necessary to achieve the stated density with the available equipment. Careful control of the entire compaction process is necessary if the required density is to be achieved with ease and economy. Control generally takes the form of field checks of moisture and density to—

- Determine if the specified density is being achieved.
- Control the rolling process.
- Permit adjustments in the field, as required.

8-54. The following discussion assumes that the laboratory compaction curve is available for the soil being compacted so that the maximum density and OMC are known. It is also assumed that laboratory-compacted soil and field-compacted soil are similar and that the required density can be achieved in the field with the equipment available.

DETERMINATION OF MOISTURE CONTENT

8-55. It may be necessary to check the moisture content of the soil during field rolling for two reasons. First, since the specified density is in terms of dry unit weight and the density measured directly in the field is generally the wet unit weight, the moisture content must be known so that the dry unit weight can be calculated. Second, the moisture content of some soils must be maintained close to optimum if satisfactory densities are to be obtained. Adjustment of the field moisture content can only be done if the moisture content is known. The determination of density and moisture content is often done in one overall test procedure; these determinations are described here separately for convenience.

Field Examination

8-56. Experienced engineers who have become familiar with the soils encountered on a particular project can frequently judge moisture content accurately by visual and manual examination. Friable or slightly plastic soils usually contain enough moisture at optimum to permit the forming of a strong cast by compressing it in the hand. As noted, some clay soils have OMCs that are close to their PLs; thus, a PL or "thread" test conducted in the field may be highly informative.

Field Drying

8-57. The moisture content of a soil is best and most accurately determined by drying the soil in an oven at a controlled temperature. Methods of determining the moisture content in this fashion are described in TM 5-530.

8-58. The moisture content of the soil may also be determined by air drying the soil in the sun. Frequent turning of the soil speeds up the drying process. From a practical standpoint, this method is generally too slow to be of much value in the control of field rolling.

8-59. Several quick methods may be used to determine approximate moisture contents under expedient conditions. For example, the sample may be placed in a frying pan and dried over a hot plate or a field stove. The temperature is difficult to control in this procedure, and organic materials may be burned, thus causing a slight to moderate error in the results. On large-scale projects where many samples are involved, this quick method may be used to speed up determinations by comparing the results obtained from this method with comparable results obtained by oven-drying.

8-60. Another quick method that may be useful is to mix the damp soil with enough denatured grain alcohol to form a slurry in a perforated metal cup, ignite the alcohol, and permit it to burn off. The alcohol method, if carefully done, produces results roughly equivalent to those obtained by careful laboratory drying. For best results, the process of saturating the soil with alcohol and burning it off completely should be repeated three times. This method is not reliable with clay soils. Safety measures must be observed when using this method. The burning must be done outside or in a well-ventilated room and at a safe distance from the alcohol supply and other flammable materials. The metal cup gets extremely hot, and it should be allowed to cool before handling.

"Speedy" Moisture-Content Test

8-61. The "speedy" moisture test kit provided with the soil test set provides a very rapid moisture- content determination and can be highly accurate if the test is performed properly. Care must be exercised to ensure that the reagent used has not lost its strength. The reagent must be very finely powdered (like portland cement) and must not have been exposed to water or high humidity before it is used. The specific test procedures are contained in the test set.

Nuclear Densimeter

8-62. This device provides real-time in-place moisture content and density of a soil. Accuracy is high if the test is performed properly and if the device has been calibrated with the specific material being tested. Operators must be certified, and proper safety precautions must be taken to ensure that the operator does not receive a medically significant dose of radiation during the operation of this device. There are stringent safety and monitoring procedures that must be followed. The method of determining the moisture content of a soil in this fashion is described in the operator's manual.

DETERMINATION OF WATER TO BE ADDED

8-63. If the moisture content of the soil is less than optimum, the amount of water to be added for efficient compaction is generally computed in gallons per square yards. The computation is based on the dry weight of soil contained in a compacted layer. For example, assume that the soil is to be placed in 6-inch, compacted layers at a dry weight of 120 pcf. The moisture content of the soil is determined to be 5 percent while the OMC is 12 percent. Assume that the strip to be compacted is 40 feet wide. Compute the amount of water that must be added per 100-foot station to bring the soil to optimum moisture. The following formula applies:

$$density\ desired\ (pcf)\ x\ \frac{W_{desired} - W_{actual}}{100}$$

$$x\ Vol\ (cu\ ft) x\ \frac{1\ gal}{8.33\ lb}$$

8-64. Substituting in the above formula from the conditions given:

$$120\ pcf\ x\ \frac{12 - 5}{100}\ x\ 100\ ft\ x\ 40\ x\ 0.5\ ft$$

$$x\ \frac{1\ gal}{8.33\ lb} = 2{,}017\ gal/station$$

8-65. If either drying conditions or rain conditions exist at the time work is in progress, it may be advisable to either add to or reduce this quantity by up to 10 percent.

COMPACTION EQUIPMENT

8-66. Equipment normally available to the military engineer for the compaction of soils includes the following types of rollers:

- Pneumatic-tired.
- Sheepsfoot.
- Tamping-foot.
- Smooth steel-wheeled.
- Vibratory.

PNEUMATIC-TIRED ROLLER

8-67. These heavy pneumatic-tired rollers are designed so that the weight can be varied to apply the desired compactive effort. Rollers with capacities up to 50 tons usually have two rows of wheels, each with four wheels and tires designed for 90 psi inflation. They can be obtained with tires designed for inflation pressures up to 150 psi. As a rule, the higher the tire pressure the greater the contact pressures and, consequently, the greater the compactive effort obtained. Information available from projects indicates that large rubber-tired compactors are capable of compacting clay layers effectively up to about 6 inches compacted depth and coarse granular or sand layers slightly deeper. Often it is used especially for final compaction (proof rolling) of the upper 6 inches of subgrade, for subbases, and for base courses. These rollers are very good for obtaining a high degree of compaction. When a large rubber-tired roller is to be used, care should be exercised to ensure that the moisture content of cohesive materials is low enough so that excessive pore pressures do not occur. Weaving or springing of the soil under the roller indicates that pore pressures are developing.

8-68. Since this roller does not aerate the soil as much as the sheepsfoot, the moisture content at the start of compaction should be approximately the optimum. In a soil that has the proper moisture content and lift thickness, tire contact pressure and the number of passes are the important variables affecting the degree of compaction obtained by rubber-tired rollers. Generally, the tire contact pressure can be assumed to be approximately equal to the inflation pressure.

8-69. Variants of the pneumatic-tired roller include the pneumatic roller and the self- propelled pneumatic-tired roller.

PNEUMATIC ROLLER

8-70. As used in this manual, the term "pneumatic roller" applies to a small rubber-tired roller, usually a "wobble wheel." The pneumatic roller is suitable for granular materials; however, it is not recommended for fine-grained clay soils except as necessary for sealing the surface after a sheepsfoot roller has "walked out." It compacts from the top down and is used for finishing all types of materials, following immediately behind the blade and water truck.

SELF-PROPELLED, PNEUMATIC-TIRED ROLLER

8-71. The self-propelled, pneumatic-tired roller has nine wheels (see figure 8-6, page 8-20). It is very maneuverable, making it excellent for use in confined spaces. It compacts from the top down. Like the towed models, the self-propelled, pneumatic-tired roller can be used for compaction of most soil materials. It is also suitable for the initial compaction of bituminous pavement.

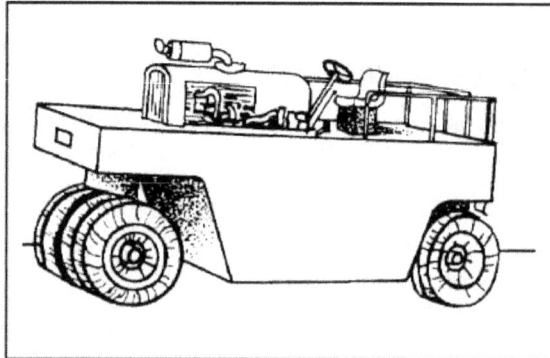

Figure 8-6. Self-propelled, pneumatic-tired roller

8-72. For a given number of passes of a rubber-tired roller, higher densities are obtained with the higher tire pressures. However, caution and good judgment must be used and the tire pressure adjusted in the field depending on the nature of the soil being compacted. For compaction to occur under a rubber-tired roller, permanent deformation has to occur. If more than slight pumping or spring occurs under the tires, the roller weight and tire pressure are too high and should be lowered immediately. Continued rolling under these conditions causes a decrease in strength even though a slight increase in density may occur. For any given tire pressure, the degree of compaction increases with additional passes, although the increase may be negligible after six to eight passes.

SHEEPSFOOT ROLLER

8-73. This roller compacts all fine-grained materials, including materials that will break down or degrade under the roller feet, but it will not compact cohesionless granular materials. The number of passes necessary for this type of roller to obtain the required densities must be determined for each type of soil encountered. The roller compacts from the bottom up and is used especially for plastic materials. The lift thickness for sheepsfoot rollers is limited to 6 inches in compacted depth. Penetration of the roller feet must be obtained at the start of rolling operations. This roller "walks out" as it completes its compactive effort, leaving the top 1 to 2 inches uncompacted.

8-74. The roller may tend to "walk out" before proper compaction is obtained. To prevent this, the soil may be scarified lightly behind the roller during the first two or three passes, and additional weight may be added to the roller.

8-75. A uniform density can usually be obtained throughout the full depth of the lift if the material is loose and workable enough to allow the roller feet to penetrate the layer on the initial passes. This produces compaction from the bottom up; therefore, material that becomes compacted by the wheels of equipment during pulverizing, wetting, blending, and mixing should be thoroughly loosened before compaction operations are begun. This also ensures uniformity of the mixture. The same amount of rolling generally produces increased densities as the depth of the lift is decreased. If the required densities are not being obtained, it is often necessary to change to a thinner lift to ensure that the specified density is obtained.

8-76. In a soil that has the proper moisture content and lift thickness, foot contact pressure and the number of passes are the important variables affecting the degree of compaction obtained by sheepsfoot rollers. The minimum foot contact pressure for proper compaction is 250 psi. Most available sheepsfoot rollers are equipped with feet having a contact area of 5 to 8 square inches. The foot pressure can be changed by varying the weight of the roller (varying the amount of ballast in the drum), or in special cases, by welding larger plates onto the faces of the feet. For the most efficient operation of the roller, the contact pressure should be close to the maximum at which the roller will "walk out" satisfactorily, as indicated in figure 8-7.

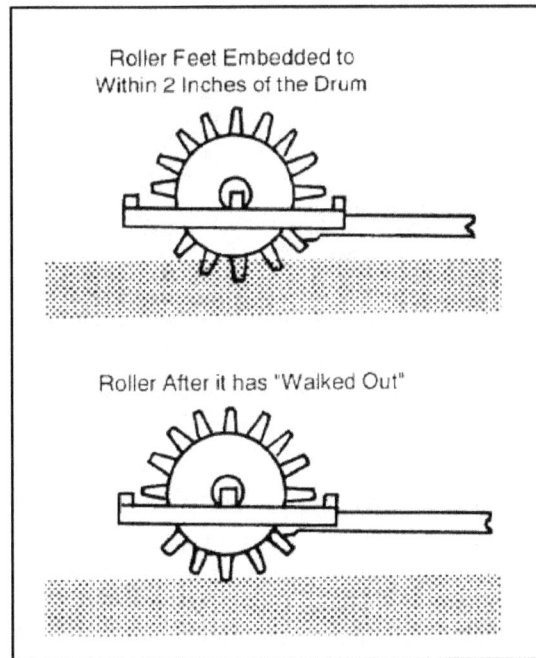

Roller Feet Embedded to
Within 2 Inches of the Drum

Roller After it has "Walked Out"

Figure 8-7. Compaction by a sheepsfoot roller

8-77. The desirable foot contact pressure varies for different soils, depending on the bearing capacity of the soil; therefore, the proper adjustments have to be made in the field based on observations of the roller. If the feet of the roller tend to "walk out" too quickly (for example, after two passes), then bridging may occur and the bottom of the lift does not get sufficient compaction. This indicates that the roller is too light or the feet too large, and the weight should be increased. However, if the roller shows no tendency to "walk out" within the required number of passes, then the indications are that the roller is too heavy and the pressure on the roller feet is exceeding the bearing capacity of the soil. After making the proper adjustments in foot pressure (by changing roller size), the only other variable is the repetition of passes. Tests have shown that density increases progressively with an increase in the number of passes.

TAMPING-FOOT ROLLER

8-78. A tamping-foot roller is a modification of the sheepsfoot roller. The tamping feet are trapezoidal pads attached to a drum. Tamping-foot rollers are normally self-propelled, and the drum may be capable of vibrating. The tamping-foot roller is suitable for use with a wide range of soil types.

STEEL-WHEELED ROLLER

8-79. The steel-wheeled roller is much less versatile than the pneumatic roller. Although extensively used, it is normally operated in conjunction with one of the other three types of compaction rollers. It is used for compacting granular materials in thin lifts. Probably its most effective use in subgrade work is in the final finish of a surface, following immediately behind the blade, forming a dense and watertight surface. Figure 8-8, page 8-22, shows a two-axle tandem (5- to 8-ton) roller.

Figure 8-8. Two axle, tandem steel-wheeled roller

SELF-PROPELLED, SMOOTH-DRUM VIBRATORY ROLLER

8-80. The self-propelled, smooth-drum vibratory roller compacts with a vibratory action that rearranges the soil particles into a denser mass (see figure 8-9). The best results are obtained on cohesionless sands and gravels. Vibratory rollers are relatively light but develop high dynamic force through an eccentric weight arrangement. Compaction efficiency is impacted by the ground speed of the roller and the frequency and amplitude of the vibrating drum.

Figure 8-9. Self-propelled, smooth-drum vibratory roller

OTHER EQUIPMENT

8-81. Other construction equipment may be useful in certain instances, particularly crawler-type tractor units and loaded hauling units, including rubber-tired scrapers. Crawler tractors are practical compacting units, especially for rock and cohesionless gravels and sands. The material should be spread in thin layers (about 3 or 4 inches thick) and is usually compacted by vibration.

COMPACTOR SELECTION

8-82. Table 8-3, page 8-15, gives information concerning compaction equipment and compactive efforts recommended for use with each of the groups of the USCS.

8-83. Normally, there is more than one type of compactor suitable for use on a project's type(s) of soil. When selecting a compactor, use the following criteria:

- Availability.
- Efficiency.

AVAILABILITY

8-84. Ascertain the types of compactors that are available and operationally ready. On major construction projects or when deployed, it may be necessary to lease compaction equipment. The rationale for leasing compaction equipment is based on the role it plays in determining overall project duration and construction quality. Uncompacted lifts cannot be built on until they are compacted. Substituting less efficient types of compaction equipment decreases productivity and may reduce project quality if desired dry densities are not achieved.

EFFICIENCY

8-85. Decide how many passes of each type of compactor are required to achieve the specified desired dry density. Determining the most efficient compactor is best done on a test strip. A test strip is an area that is located adjacent to the project and used to evaluate compactors and construction procedures. The compactive effort of each type of compactor can be determined on the test strip and plotted graphically. Figure 8-10 compares the following types of compactors:

- Vibratory (vibrating drum) roller.
- Tamping-foot roller.
- Pneumatic-tired roller.

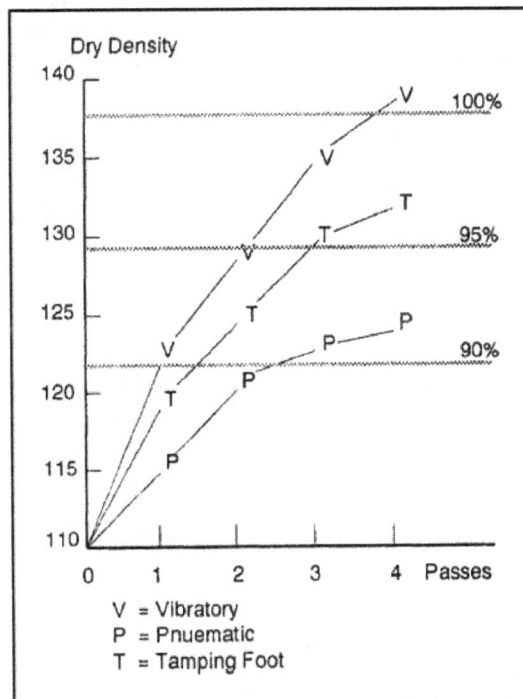

Figure 8-10. Use of test strip data to determine compactor efficiency

8-86. In this example, a dry density of 129 to 137 pcf is desired. The vibrating roller was the most efficient, achieving densities within the specified density range in three passes. The tamping foot compactor also compacted the soil to the desired density in three passes. However, the density achieved (130 pcf) is so close to the lower limit of the desired density range that any variation in the soil may cause the achieved density to drop below 129 pcf. The pneumatic-tired roller was the least efficient and did not densify the soil material to densities within the specified density range.

8-87. Once the type(s) of compactor is selected, optimum lift thicknesses can be determined. Table 8-3, page 8-15, provides information on average optimum lift thicknesses, but this information must be verified. Again, the test strip is a way to determine optimum lift thickness without interfering with other operations occurring on the actual project.

8-88. In actual operation, it is likely that more than one type of compactor will be operating on the project to maintain peak productivity and to continue operations when the primary compactors require maintenance or repair. Test-strip data helps to maintain control of project quality while providing the flexibility to allow construction at maximum productivity.

SECTION IV - QUALITY CONTROL

PURPOSE

8-89. Poor construction procedures can invalidate a good pavement or embankment design. Therefore, quality control of construction procedures is as important to the final product as is proper design. The purpose of quality control is to ensure that the soil is being placed at the proper density and moisture content to provide adequate bearing strength (CBR) in the fill. This is accomplished by taking samples or testing at each stage of construction. The test results are compared to limiting values or specifications, and the compaction should be accepted or reworked based on the results of the density and moisture content tests. A quality- control plan should be developed for each project to ensure that high standards are achieved. For permanent construction, statistical quality-control plans provide the most reliable check on the quality of compaction.

QUALITY-CONTROL PLAN

8-90. Generally, a quality-control plan consists of breaking the total job down into lots with each lot consisting of "X" units of work. Each lot is considered a separate job, and each job will be accepted or rejected depending on the test results representing this lot. By handling the control procedure in this way, the project engineer is able to determine the quality of the job on a lot-by-lot basis. This benefits the engineering construction unit and project engineer by identifying the lots that will be accepted and the lots that will be rejected. As this type of information is accumulated from lot to lot, a better picture of the quality of the entire project is obtained.

8-91. The following essential items should be considered in a quality-control plan:

- Lot size.
- Random sampling.
- Test tolerance.
- Penalty system.

LOT SIZE

8-92. There are two methods of defining a lot size (unit of work). A lot size may be defined as an operational time period or as a quantity of production. One advantage that the quantity-of-production method has over the operational-time-period method is that the engineering construction unit will probably have plant and equipment breakdowns and other problems that would require that production be stopped for certain periods of time. This halt in production could cause difficulties in recording production time. On the other hand, there are always records that would show the amount of materials that have been produced.

Therefore, the better way to describe a lot is to specify that a lot will be expressed in units of quantity of production. By using this method, each lot will contain the same amount of materials, establishing each one with the same relative importance. Factors such as the size of the job and the operational capacity usually govern the size of a production lot. Typical lot sizes are 2,000 square yards for subbase construction and 1,200 square yards for stabilized subgrade construction. To statistically evaluate a lot, at least four samples should be obtained and tested properly.

RANDOM SAMPLING

8-93. For a statistical analysis to be acceptable, the data used for this analysis must be obtained from random sampling. Random sampling means that every sample within the lot has an equal chance of being selected. There are two common types of random sampling. One type consists of dividing the lot into a number of equal size sublots; one random sample is then taken from each of the sublots. The second method consists of taking the random samples from the entire lot. The sublot method has one big advantage, especially when testing during production, in that the time between testing is spaced somewhat; when taking random samples from the lot, all tests might occur within a short time. The sublot method is recommended when taking random samples. It is also recommended that all tests be conducted on samples obtained from in-place material. By conducting tests in this manner, obtaining additional samples for testing would not be a problem.

TEST TOLERANCE

8-94. A specification tolerance for test results should be developed for various tests with consideration given to a tolerance that could be met in the field and a tolerance narrow enough so that the quality of the finished product is satisfactory. For instance, the specifications for a base course would usually state that the material must be compacted to at least 100 percent CE 55 maximum density. However, because of natural variation in material, the 100 percent requirement cannot always be met. Field data indicates that the average density is 95 percent and the standard deviation is 3.5. Therefore, it appears that the specification should require 95 percent density and a standard deviation of 3.5, although there is a good possibility that the material will further densify under traffic.

PENALTY SYSTEM

8-95. After the project is completed, the job should be rated based on the results of the statistical quality-control plan for that project. A satisfactory job, meeting all of the specification tolerances, should be considered 100 percent satisfactory. On the other hand, those jobs that are not 100 percent satisfactory should be rated as such. Any job that is completely unsatisfactory should be removed and reconstructed satisfactorily.

THEATER-OF-OPERATIONS QUALITY CONTROL

8-96. In the theater of operations, quality control is usually simplified to a set pattern. This is not as reliable as statistical testing but is adequate for the temporary nature of theater-of-operations construction. There is no way to ensure that all areas of a project are checked; however, guidelines for planning quality control are as follows:

- Use a "test strip" to determine the approximate number of passes needed to attain proper densities.
- Test every lift as soon as compaction is completed.
- Test every roller lane.
- Test obvious weak spots.
- Test roads and airfields every 250 linear feet, staggering tests about the centerline.
- Test parking lots and storage areas every 250 square yards.
- Test trenches every 50 linear feet.
- Remove all oversized materials.

- Remove any pockets of organic or unsuitable soil material.
- Increase the distance between tests as construction progresses, if initial checks are satisfactory.

CORRECTIVE ACTIONS

8-97. When the density and/or moisture of a soil does not meet specifications, corrective action must be taken. The appropriate corrective action depends on the specific problem situation. There are four fundamental problem situations:

- Overcompaction.
- Undercompaction.
- Too wet.
- Too dry.

8-98. It is possible to have a situation where one or more of these problems occur at the same time, such as when the soil is too dry and also undercompacted. The specification block that was plotted on the moisture density curve (CE 55) is an excellent tool for determining if a problem exists and what the problem is.

OVERCOMPACTION

8-99. Overcompaction occurs when the material is densified in excess of the specified density range. An overcompacted material may be stronger than required, which indicates—

8-100.

- Wasted construction effort (but not requiring corrective action to the material).
- Sheared material (which no longer meets the design CBR criteria).

8-101. In the latter case, scarify the overcompacted lift and recompact to the specified density. Laboratory analysis of overcompacted soils (to include CBR analysis) is required before a corrective action decision can be made.

UNDERCOMPACTION

8-102. Undercompaction may indicate—

- A missed roller pass.
- A change in soil type.
- Insufficient roller weight.
- A change in operating frequency or amplitude (if vibratory rollers are in use).
- A defective roller drum.
- The use of an improper type of compaction equipment.

8-103. Corrective action is based on a sequential approach. Initially, apply additional compactive effort to the problem area. If undercompacting is a frequent problem or develops a frequent pattern, look beyond a missed roller pass as the cause of the problem.

TOO WET

8-104. Soils that are too wet when compacted are susceptible to shearing and strength loss. Corrective action for a soil compacted too wet is to—

- Scarify.
- Aerate.
- Retest the moisture content.
- Recompact, if moisture content is within the specified range.
- Retest for both moisture and density.

Too Dry

8-105. Soils that are too dry when compacted do not achieve the specified degree of densification as do properly moistened soils. Corrective action for a soil compacted too dry is to—

- Scarify.
- Add water.
- Mix thoroughly.
- Retest the moisture content.
- Recompact, if moisture content is within the specified range.
- Retest for both moisture and density.

This page intentionally left blank.

Chapter 9

Soil Stabilization for Roads and Airfields

Soil stabilization is the alteration of one or more soil properties, by mechanical or chemical means, to create an improved soil material possessing the desired engineering properties. Soils may be stabilized to increase strength and durability or to prevent erosion and dust generation. Regardless of the purpose for stabilization, the desired result is the creation of a soil material or soil system that will remain in place under the design use conditions for the design life of the project.

Engineers are responsible for selecting or specifying the correct stabilizing method, technique, and quantity of material required. This chapter is aimed at helping to make the correct decisions. Many of the procedures outlined are not precise, but they will "get you in the ball park." Soils vary throughout the world, and the engineering properties of soils are equally variable. The key to success in soil stabilization is soil testing. The method of soil stabilization selected should be verified in the laboratory before construction and preferably before specifying or ordering materials.

SECTION I - METHODS OF STABILIZATION

BASIC CONSIDERATIONS

9-1. Deciding to stabilize existing soil material in the theater of operations requires an assessment of the mission, enemy, terrain, troops (and equipment), and time available (METT-T).

- Mission. What type of facility is to be constructed—road, airfield, or building foundation? How long will the facility be used (design life)?
- Enemy. Is the enemy interdicting lines of communications? If so, how will it impact on your ability to haul stabilizing admixtures delivered to your construction site?
- Terrain. Assess the effect of terrain on the project during the construction phase and over the design life of the facility. Is soil erosion likely? If so, what impact will it have? Is there a slope that is likely to become unstable?
- Troops (and equipment). Do you have or can you get equipment needed to perform the stabilization operation?
- Time available. Does the tactical situation permit the time required to stabilize the soil and allow the stabilized soil to cure (if necessary)?

9-2. There are numerous methods by which soils can be stabilized; however, all methods fall into two broad categories. They are—

- Mechanical stabilization.
- Chemical admixture stabilization.

9-3. Some stabilization techniques use a combination of these two methods. Mechanical stabilization relies on physical processes to stabilize the soil, either altering the physical composition of the soil (soil blending) or placing a barrier in or on the soil to obtain the desired effect (such as establishing a sod cover to prevent dust generation). Chemical stabilization relies on the use of an admixture to alter the chemical properties of the soil to achieve the desired effect (such as using lime to reduce a soil's plasticity).

9-4.   Classify the soil material using the USCS. When a soil testing kit is unavailable, classify the soil using the field identification methodology. Mechanical stabilization through soil blending is the most economical and expedient method of altering the existing material. When soil blending is not feasible or does not produce a satisfactory soil material, geotextiles or chemical admixture stabilization should be considered. If chemical admixture stabilization is being considered, determine what chemical admixtures are available for use and any special equipment or training required to successfully incorporate the admixture.

MECHANICAL STABILIZATION

9-5.   Mechanical stabilization produces by compaction an interlocking of soil-aggregate particles. The grading of the soil-aggregate mixture must be such that a dense mass is produced when it is compacted. Mechanical stabilization can be accomplished by uniformly mixing the material and then compacting the mixture. As an alternative, additional fines or aggregates may be blended before compaction to form a uniform, well-graded, dense soil-aggregate mixture after compaction. The choice of methods should be based on the gradation of the material. In some instances, geotextiles can be used to improve a soil's engineering characteristics (see chapter 11).

9-6.   The three essentials for obtaining a properly stabilized soil mixture are—

- Proper gradation.
- A satisfactory binder soil.
- Proper control of the mixture content.

9-7.   To obtain uniform bearing capacity, uniform mixture and blending of all materials is essential. The mixture will normally be compacted at or near OMC to obtain satisfactory densities.

9-8.   The primary function of the portion of a mechanically stabilized soil mixture that is retained on a Number 200 sieve is to contribute internal friction. Practically all materials of a granular nature that do not soften when wet or pulverize under traffic can be used; however, the best aggregates are those that are made up of hard, durable, angular particles. The gradation of this portion of the mixture is important, as the most suitable aggregates generally are well-graded from coarse to fine. Well-graded mixtures are preferred because of their greater stability when compacted and because they can be compacted more easily. They also have greater increases in stability with corresponding increases in density. Satisfactory materials for this use include—

- Crushed stone.
- Crushed and uncrushed gravel.
- Sand.
- Crushed slag.

9-9.   Many other locally available materials have been successfully used, including disintegrated granite, talus rock, mine tailings, caliche, coral, limerock, tuff, shell, slinkers, cinders, and iron ore. When local materials are used, proper gradation requirements cannot always be met.

Note. If conditions are encountered in which the gradation obtained by blending local materials is either finer or coarser than the specified gradation, the size requirements of the finer fractions should be satisfied and the gradation of the coarser sizes should be neglected.

9-10. The portion of the soil that passes a Number 200 sieve functions as filler for the rest of the mixture and supplies cohesion. This aids in the retention of stability during dry weather. The swelling of clay material serves somewhat to retard the penetration of moisture during wet weather. Clay or dust from rock-crushing operations are commonly used as binders. The nature and amount of this finer material must be carefully controlled, since too much of it results in an unacceptable change in volume with change in moisture content and other undesirable properties. The properties of the soil binder are usually controlled by controlling the plasticity characteristics, as evidenced by the LL and PI. These tests are performed on the portion of the material that passes a Number 40 sieve. The amount of fines is controlled by limiting the amount of material that may pass a Number 200 sieve. When the stabilized soil is to be subjected to frost action, this factor must be kept in mind when designing the soil mixture.

USES

9-11. Mechanical soil stabilization may be used in preparing soils to function as—

- Subgrades.
- Bases.
- Surfaces.

9-12. Several commonly encountered situations may be visualized to indicate the usefulness of this method. One of these situations occurs when the surface soil is a loose sand that is incapable of providing support for wheeled vehicles, particularly in dry weather. If suitable binder soil is available in the area, it may be brought in and mixed in the proper proportions with the existing sand to provide an expedient all-weather surface for light traffic. This would be a sand-clay road. This also may be done in some cases to provide a "working platform" during construction operations. A somewhat similar situation may occur in areas where natural gravels suitable for the production of a well-graded sand-aggregate material are not readily available. Crushed stone, slag, or other materials may then be stabilized by the addition of suitable clay binder to produce a satisfactory base or surface. A common method of mechanically stabilizing an existing clay soil is to add gravel, sand, or other granular materials. The objectives here are to—

- Increase the drainability of the soil.
- Increase stability.
- Reduce volume changes.
- Control the undesirable effects associated with clays.

OBJECTIVE

9-13. The objective of mechanical stabilization is to blend available soils so that, when properly compacted, they give the desired stability. In certain areas, for example, the natural soil at a selected location may have low load-bearing strength because of an excess of clay, silt, or fine sand. Within a reasonable distance, suitable granular materials may occur that may be blended with the existing soils to markedly improve the soil at a much lower cost in manpower and materials than is involved in applying imported surfacing.

9-14. The mechanical stabilization of soils in military construction is very important. The engineer needs to be aware of the possibilities of this type of construction and to understand the principles of soil action previously presented. The engineer must fully investigate the possibilities of using locally available materials.

LIMITATIONS

9-15. Without minimizing the importance of mechanical stabilization, the limitations of this method should also be realized. The principles of mechanical stabilization have frequently been misused, particularly in areas where frost action is a factor in the design. For example, clay has been added to "stabilize" soils, when in reality all that was needed was adequate compaction to provide a strong, easily drained base that would not be susceptible to detrimental frost action. An understanding of the densification that can be achieved by modern compaction equipment should prevent a mistake of this sort. Somewhat similarly, poor trafficability of a soil during construction because of lack of fines should not necessarily provide an excuse for mixing in clay binder. The problem may possibly be solved by applying a thin surface treatment or using some other expedient method.

SOIL BASE REQUIREMENTS

9-16. Grading requirements relative to mechanically stabilized soil mixtures that serve as base courses are given in Table 7-3 of TM 5-330/Air Force Manual (AFM) 86-3, Volume II. Experience in civil highway construction indicates that best results are obtained with this type of mixture if the fraction passing the Number 200 sieve is not greater than two-thirds of the fraction passing the Number 40 sieve. The size of the largest particles should not exceed two-thirds of the thickness of the layer in which they are incorporated. The mixture should be well-graded from coarse to fine.

9-17. A basic requirement of soil mixtures that are to be used as base courses is that the PI should not exceed 5. Under certain circumstances, this requirement may be relaxed if a satisfactory bearing ratio is developed. Experience also indicates that under ideal circumstances the LL should not exceed 25. These requirements may be relaxed in theater-of-operations construction. The requirements may be lowered to a LL of 35 and a PI of 10 for fully operational airfields. For emergency and minimally operational airfields, the requirements may be lowered to a LL of 45 and a PI of 15, when drainage is good.

SOIL SURFACE REQUIREMENTS

9-18. Grading requirements for mechanically stabilized soils that are to be used directly as surfaces, usually under emergency conditions, are generally the same as those indicated in table 7-3 of TM 5-330/AFM 86-3, Volume II. Preference should be given to mixtures that have a minimum aggregate size equal to 1 inch or perhaps 1½ inches. Experience indicates that particles larger than this tend to work themselves to the surface over a period of time under traffic. Somewhat more fine soil is desirable in a mixture that is to serve as a surface, as compared with one for a base. This allows the surface to be more resistant to the abrasive effects of traffic and penetration of precipitation. To some extent, moisture lost by evaporation can be replaced by capillarity.

9-19. Emergency airfields that have surfaces of this type require a mixture with a PI between 5 and 10. Experience indicates that road surfaces of this type should be between 4 and 9. The surface should be made as tight as possible, and good surface drainage should be provided. For best results, the PI of a stabilized soil that is to function first as a wearing surface and then as a base, with a bituminous surface being provided at a later date, should be held within very narrow limits. Consideration relative to compaction, bearing value, and frost action are as important for surfaces of this type as for bases.

PROPORTIONING

9-20. Mixtures of this type are difficult to design and build satisfactorily without laboratory control. A rough estimate of the proper proportions of available soils in the field is possible and depends on manual and visual inspection. For example, suppose that a loose sand is the existing subgrade soil and it is desired to add silty clay from a nearby borrow source to achieve a stabilized mixture. Each soil should be moistened to the point where it is moist, but not wet; in a wet soil, the moisture can be seen as a shiny film on the surface. What is desired is a mixture that feels gritty and in which the sand grains can be seen. Also, when the soils are combined in the proper proportion, a cast formed by squeezing the moist soil mixture in the hand will not be either too strong or too weak; it should just be able to withstand normal handling without breaking. Several trial mixtures should be made until this consistency is obtained. The proportion of each of the two soils should be carefully noted. If gravel is available, this may be added, although there is no real rule of thumb to tell how much should be added. It is better to have too much gravel than too little.

Use of Local Materials

9-21. The essence of mechanical soil stabilization is the use of locally available materials. Desirable requirements for bases and surfaces of this type were given previously. It is possible, especially under emergency conditions, that mixtures of local materials will give satisfactory service, even though they do not meet the stated requirements. Many stabilized mixtures have been made using shell, coral, soft limestone, cinders, marl, and other materials listed earlier. Reliance must be placed on—

- Experience.
- An understanding of soil action.
- The qualities that are desired in the finished product.
- Other factors of local importance in proportioning such mixtures in the field.

Blending

9-22. It is assumed in this discussion that an existing subgrade soil is to be stabilized by adding a suitable borrow soil to produce a base course mixture that meets the specified requirements. The mechanical analysis and limits of the existing soil will usually be available for the results of the subgrade soil survey (see chapter 3). Similar information is necessary concerning the borrow soil. The problem is to determine the proportions of these two materials that should be used to produce a satisfactory mixture. In some cases, more than two soils must be blended to produce a suitable mixture. However, this situation is to be avoided when possible because of the difficulties frequently encountered in getting a uniform blend of more than two local materials. Trial combinations are usually made on the basis of the mechanical analysis of the soil concerned. In other words, calculations are made to determine the gradation of the combined materials and the proportion of each component adjusted so that the gradation of the combination falls within specified limits. The PI of the selected combination is then determined and compared with the specification. If this value is satisfactory, then the blend may be assumed to be satisfactory, provided that the desired bearing value is attained. If the plasticity characteristics of the first combination are not within the specified limits, additional trials must be made. The proportions finally selected then may be used in the field construction process.

Numerical Proportioning

9-23. The process of proportioning will now be illustrated by a numerical example (see table 9-1, page 9-6). Two materials are available, material B in the roadbed and material A from a nearby borrow source. The mechanical analysis of each of these materials is given, together with the LL and PI of each. The desired grading of the combination is also shown, together with the desired plasticity characteristics.

Specified Gradation

9-24. Proportioning of trial combinations may be done arithmetically or graphically. The first step in using either the graphical or arithmetical method is to determine the gradation requirements. Gradation requirements for base course, subcourse, and select material are found in table 7-1, page 7-17, TM 5-330/AFM 86-3, Volume II. In the examples in figure 9-1, page 9-6, and figure 9-2, page 9-8, a base course material with a maximum aggregate size of 1 inch has been specified. In the graphical method, the gradation requirements are plotted to the outside of the right axis. In the arithmetical method, they are plotted in the column labelled "Specs." Then the gradations of the soils to be blended are recorded. The graphical method has the limitation of only being capable of blending two soils, whereas the arithmetical method can be expanded to blend as many soils as required. At this point, the proportioning methods are distinctive enough to require separate discussion.

Graphical Proportioning

9-25. The actual gradations of soil materials A and B are plotted along the left and right axes of the graph, respectively. As shown in figure 9-1, material A has 92 percent passing the ¾-inch sieve while material B has 72 percent passing the same sieve. Once plotted, a line is drawn across the graph, connecting the percent passing of material A with the percent passing of material B for each sieve size.

Note. Since both materials A and B had 100 percent passing the 1-inch sieve, it was omitted from the graph and will not affect the results.

9-26. Mark the point where the upper and lower limits of the gradation requirements intersect the line for each sieve size. In figure 9-1, the allowable percent passing the Number 4 sieve ranges from 35 to 65 percent passing. The point along the Number 4 line at which 65 percent passing intersects represents 82 percent material A and 18 percent material B. The 35 percent passing intersects the Number 4 line at 19 percent material A and 81 percent material B. The acceptable ranges of material A to be blended with material B is the widest range that meets the gradation requirements for all sieve sizes. The shaded area of the chart represents the combinations of the two materials that will meet the specified gradation requirements. The boundary on the left represents the combination of 44 percent material A and 56 percent material B. The position of this line is fixed by the upper limit of the requirement relating to the material passing the Number 200 sieve (15 percent). The boundary on the right represents the combination of 21 percent material A and 79 percent material B. This line is established by the lower limit of the requirement relative to the fraction passing the Number 40 sieve (15 percent). Any mixture falling within these limits satisfies the gradation requirements. For purposes of illustration, assume that a combination of 30 percent material A and 70 percent material B is selected for a trial mixture. A similar diagram can be prepared for any two soils.

Table 9-1. Numerical example of proportioning

Mechanical Analysis			
Sieve Designation	Percent Passing, by Weight		
	Material A	Material B	Desired
1-inch	100	100	100
¾ inch	92	72	70-100
⅜ inch	83	45	50-80
Number 4	75	27	35-65
Number 10	67	15	20-50
Number 40	52	—	15-30
Number 200	33	1	5-15
Plasticity Characteristics			
Liquid limit	32	12	≤28
Plasticity index	9	0	≤6

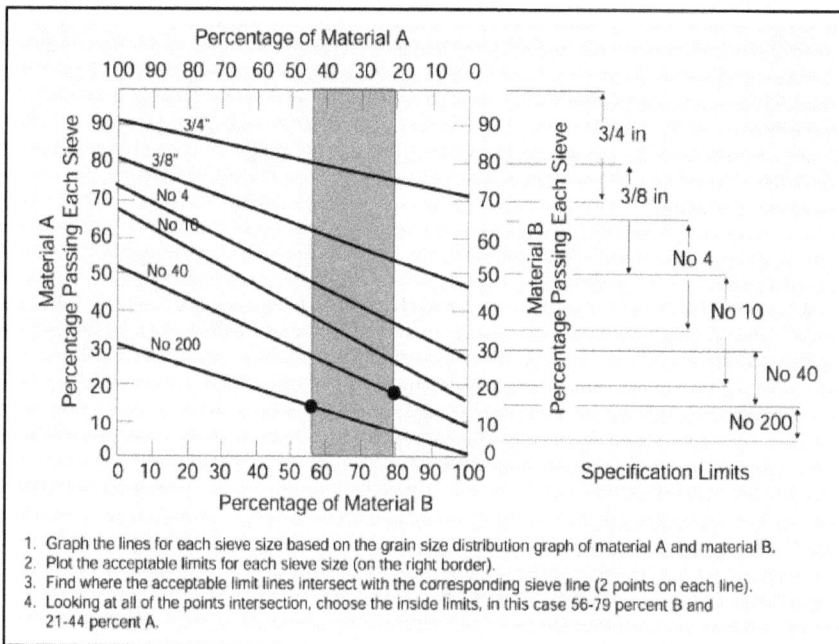

1. Graph the lines for each sieve size based on the grain size distribution graph of material A and material B.
2. Plot the acceptable limits for each sieve size (on the right border).
3. Find where the acceptable limit lines intersect with the corresponding sieve line (2 points on each line).
4. Looking at all of the points intersection, choose the inside limits, in this case 56-79 percent B and 21-44 percent A.

Figure 9-1. Graphical method of proportioning two soils to meet gradation requirements

Arithmetical Proportioning

9-27. Record the actual gradation of soils A and B in their respective columns (Columns 1 and 2, figure 9-2, page 9-8). Average the gradation limits and record in the column labelled "S". For example, the allowable range for percent passing a 3/8-inch sieve in a 1-inch minus base course is 50 to 80 percent. The average, 50+80/2, is 65 percent. As shown in figure 9-2, S for 3/8 inch is 65. Next, determine the absolute value of S-A and S-B for each sieve size and record in the columns labelled "|S-A|" and "|S-B|", respectively. Sum columns |S-A| and |S-B|. To determine the percent of soil A in the final mix, use the formula—

In the example in figure 9-2:

$$\%A = \frac{\Sigma B}{\Sigma A + \Sigma B} x\ 100$$

$$\Sigma A = 134$$

$$\Sigma B = 103$$

thus:

$$\frac{103}{134 + 103} = \frac{103}{237} x\ 100\% = 43.5\%$$

9-28. The percent of soil B in the final mix can be determined by the formula:

$$\%B = \frac{\Sigma A}{\Sigma A + \Sigma B} x\ 100$$

or

$$100\% - \%A = \%B$$

Note. If three or more soils are to be blended, the formula would be—

$$\%C = \frac{\Sigma A + \Sigma B + \Sigma n}{\Sigma A + \Sigma B + \Sigma C + \Sigma n} x\ 100$$

9-29. This formula can be further expanded as necessary.

9-30. Multiply the percent passing each sieve for soil A by the percentage of soil A in the final mix; record the information in column 4 (see figure 9-2). Repeat the procedure for soil B and record the information in column 5 (see figure 9-2). Complete the arithmetical procedure by adding columns 4 and 5 to obtain the percent passing each sieve in the blended soil.

9-31. Both the graphical and arithmetical methods have advantages and disadvantages. The graphical method eliminates the need for precise blending under field conditions and the methodology requires less effort to use. Its drawback becomes very complex when blending more than two soils. The arithmetical method allows for more precise blending, such as mixing at a batch plant, and it can be readily expanded to accommodate the blending of three or more soils. It has the drawback in that precise blending is often unattainable under field conditions. This reduces the quality assurance of the performance of the blended soil material.

Sieve Size	Percent Passing			$\bar{S}$	$	\bar{S}\text{-}A	$	$	\bar{S}\text{-}B	$	0.435 A	0.565 B	Percent Passing Final Mix
	Soil A	Soil B	Specs S						
	(1)	(2)		(3)	(3-1)	(3-2)	(4)	(5)	(4+5)
1-in	100	100	100	100	0	0	43.5	56.5	100
3/4-in	92	72	70-100	85	7	13	40.0	40.7	80.7
3/8-in	83	45	50-80	65	18	20	36.1	25.4	61.5
No 4	75	45	35-65	50	25	23	32.6	15.3	47.9
No 10	67	15	25-50	35	32	20	29.1	8.5	37.6
No 40	52	5	15-30	23	29	18	22.6	2.8	25.4
No 200	33	1	5-15	10	+23	+9	14.4	0.6	15.0

$$\sum A = 134 \qquad \sum B = 103$$

Percent Soil A in final mix =

$$percent\ A = \frac{\sum B}{\sum A + \sum B} \times 100 \qquad\qquad \frac{103}{(134+103)} \times 100 = 43.5\ percent$$

Percent Soil B in final mix =

$$percent\ B = \frac{\sum A}{\sum A + \sum B} \times 100 \qquad\qquad \frac{134}{(134+103)} \times 100 = 56.5\ percent$$

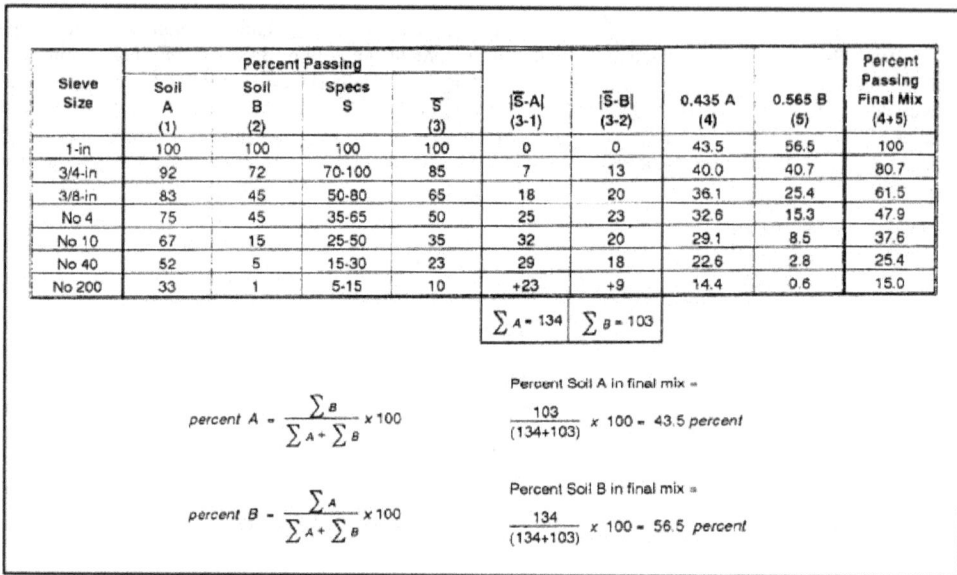

Figure 9-2. Arithmetical method of proportioning soils to meet gradation requirements

Plasticity Requirements

9-32. A method of determining the PI and LL of the combined soils serves as a method to indicate if the proposed trial mixture is satisfactory, pending the performance of laboratory tests. This may be done either arithmetically or graphically. A graphical method of obtaining these approximate values is shown in figure 9-3. The values shown in figure 9-3 require additional explanation, as follows. Consider 500 pounds of the mixture tentatively selected (30 percent as material A and 70 percent as material B). Of this 500 pounds, 150 pounds are material A and 350 pounds material B. Within the 150 pounds of material A, there are 150 (0.52) = 78 pounds of material passing the Number 40 sieve. Within the 350 pounds of material B, there are 150 (0.05) = 17.5 pounds of material passing the Number 40 sieve. The total amount of material passing the Number 40 sieve in the 500 pounds of blend = 78 + 17.5 = 95.5 pounds. The percentage of this material that has a PI of 9 (material A) is (78/95.5) 100 = 82. As shown in figure 9-3, the approximate PI of the mixture of 30 percent material A and 70 percent material B is 7.4 percent. By similar reasoning, the approximate LL of the blend is 28.4 percent. These values are somewhat higher than permissible under the specification. An increase in the amount of material B will somewhat reduce the PI and LL of the combination.

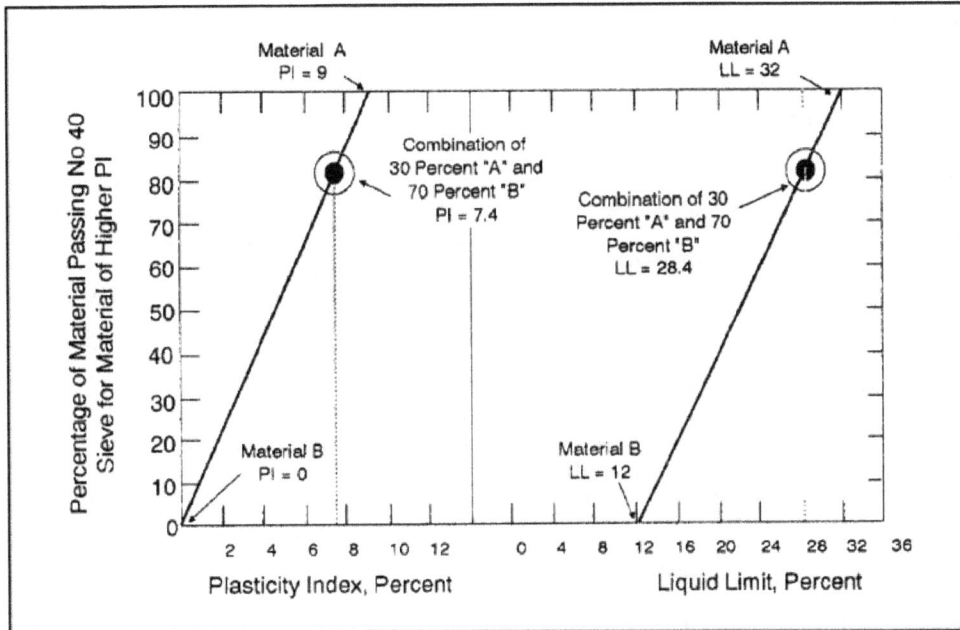

Figure 9-3. Graphical method of estimating plasticity characteristics of a combination two soils

Field Proportioning

9-33. In the field, the materials used in a mechanically stabilized soil mixture probably will be proportioned by loose volume. Assume that a mixture incorporates 75 percent of the existing subgrade soil, while 25 percent will be brought in from a nearby borrow source. The goal is to construct a layer that has a compacted thickness of 6 inches. It is estimated that a loose thickness of 8 inches will be required to form the 6-inch compacted layer. A more exact relationship can be established in the field as construction proceeds. Of the 8 inches loose thickness, 75 percent (or 0.75(8) = 6 inches) will be the existing soil. The remainder of the mix will be mixed thoroughly to a depth of 8 inches and compacted by rolling. The proportions may be more accurately controlled by weight, if weight measurements can be made under existing conditions.

WATERPROOFING

9-34. The ability of an airfield or road to sustain operations depends on the bearing strength of the soil. Although an unsurfaced facility may possess the required strength when initially constructed, exposure to water can result in a loss of strength due to the detrimental effect of traffic operations. Fine-grained soils or granular materials that contain an excessive amount of fines generally are more sensitive to water changes than coarse-grained soils. Surface water also may contribute to the development of dust by eroding or loosening material from the ground surface that can become dust during dry weather conditions.

Sources of Water

9-35. Water may enter a soil either by the percolation of precipitation or ponded surface water, by capillary action of underlying groundwater, by a rise in the water-table level, or by condensation of water vapor and accumulation of moisture under a vapor-impermeable surface. As a general rule, an existing groundwater table at shallow depths creates a low load-bearing strength and must be avoided wherever possible. Methods to protect against moisture ingress from sources other than the ground surface will not be considered here. In most instances, the problem of surface water can be lessened considerably by following the proper procedures for—

- Grading.
- Compaction.
- Drainage.

Objectives of Waterproofers

9-36. The objective of a soil-surface waterproofer is to protect a soil against attack by water and thus preserve its in-place or as-constructed strength during wet-weather operations. The use of soil waterproofers generally is limited to traffic areas. In some instances, soil waterproofers may be used to prevent excessive softening of areas, such as shoulders or overruns, normally considered nontraffic or limited traffic areas.

9-37. Also, soil waterproofers may prevent soil erosion resulting from surface water runoff. As in the case of dust palliatives, a thin or shallow-depth soil waterproofing treatment loses its effectiveness when damaged by excessive rutting and thus can be used efficiently only in areas that are initially firm. Many soil waterproofers also function well as dust palliatives; therefore, a single material might be considered as a treatment in areas where the climate results in both wet and dry soil surface conditions. Geotextiles are the primary means of waterproofing soils when grading, compaction, and drainage practices are insufficient. Use of geotextiles is discussed in detail in chapter 11.

CHEMICAL ADMIXTURE STABILIZATION

9-38. Chemical admixtures are often used to stabilize soils when mechanical methods of stabilization are inadequate and replacing an undesirable soil with a desirable soil is not possible or is too costly. Over 90 percent of all chemical admixture stabilization projects use—

- Cement.
- Lime.
- Fly ash.
- Bituminous materials.

9-39. Other stabilizing chemical admixtures are available, but they are not discussed in this manual because they are unlikely to be available in the theater of operations.

WARNING

Chemical admixtures may contain hazardous materials. Consult appendix C to determine the necessary safety precautions for the selected admixture.

9-40. When selecting a stabilizer additive, the factors that must be considered are the—

- Type of soil to be stabilized.
- Purpose for which the stabilized layer will be used.
- Type of soil quality improvement desired.
- Required strength and durability of the stabilized layer.
- Cost and environmental conditions.

9-41. Table 9-2 lists stabilization methods most suitable for specific applications. To determine the stabilizing agent(s) most suited to a particular soil, use the gradation triangle in figure 9-4, page 9-12, to find the area that corresponds to the gravel, sand, and fine content of the soil. For example, soil D has the following characteristics:

- With 95 percent passing the Number 4 sieve, the PI is 14.
- With 14 percent passing the Number 200 sieve, the LL is 21.

9-42. Therefore the soil is 5 percent gravel, 81 percent sand, and 14 percent fines. Figure 9-4 shows this soil in Area 1C.

Table 9-2. Stabilization methods most suitable for specific applications

Purpose	Soil Type	Method
Subgrade Stabilization Improves load-carrying and stress-distribution characteristics	Fine-grained	SA, SC, MB, C
	Coarse-grained	SA, SC, MB, C
	Clays of low PI	C, SC, CMS, LMS, SL
	Clays of low PI	SL, LMS
Reduces frost susceptibility	Fine-grained	CMS, SA, SC, LF
	Clays of low PI	CMS, SC, SL, LMS
Improves waterproofing and runoff	Clays of low PI	CMS, SA, LMS, SL
Controls shrinkage and swell	Clays of low PI	CMS, SC, C, LMS, SL
	Clays of low PI	SL
Reduces resiliency	Clays of low PI	SL, LMS
	Elastic silts or clays	SC, CMS
Base Course Stabilization Improves substandard materials	Fine-grained	SC, SA, LF, MB
	Clays of low PI	SC, SL
Improves load-carrying and stress-distribution characteristics	Coarse-grained	SA, SC, MB, LF
	Fine-grained	SC, SA, LF, MB
Reduces pumping	Fine-grained	SC, SA, LF, MB, membranes
Dust Palliative	Fine-grained	CMS, SA, oil or bituminous surface spray, APSB
	Plastic soils	CMS, SL, LMS, APSB, DCA 70

Legend:
The methods of treatment are—
APSB = Asphalt penetration surface binder LMS = Lime-modified soil
C = Compaction MB = Mechanical blending
CMS = Cement-modified soil SA = Soil-asphalt
DCA 70 = Polyvinyl acetate emulsion SC = Soil-cement
LF = Lime-fly ash SL = Soil-lime

9-43. Table 9-3, page 9-12, shows that the stabilizing agents recommended for Area 1C soils include bituminous material, portland cement, lime, and lime-cement-fly ash. In this example, bituminous agents cannot be used because of the restriction on PI, but any of the other agents can be used if available.

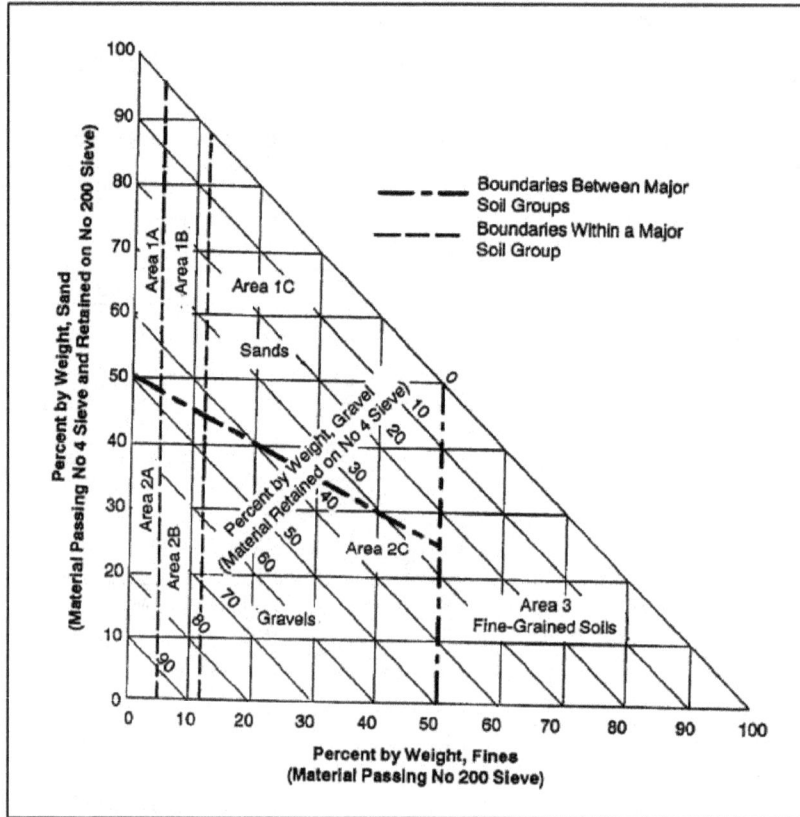

Figure 9-4. Gradation triangle for use in selecting a stabilizing additive

Table 9-3. Guide for selecting a stabilizing additive.

Area	Soils Class	Type of Stabilizing Additive Recommended	Restriction on LL and PI of Soil	Restriction on Percent Passing No 200 Sieve	Remarks
1A	SW or SP	(1) Bituminous (2) Portland cement (3) Lime-cement-fly ash	 PI not to exceed 25		
1B	SW-SM or SP-SM or SW-SC or SP-SC	(1) Bituminous (2) Portland cement (3) Lime (4) Lime-cement-fly ash	PI not to exceed 10 PI not to exceed 30 PI not to exceed 12 PI not to exceed 25	PI 30 or less PI 12 or greater	
1C	SM or SC or SM-SC	(1) Bituminous (2) Portland cement (3) Lime (4) Lime-cement-fly ash	PI not to exceed 10 —* PI not less than 12 PI not to exceed 25	Not to exceed 30 percent by weight	

Table 9-3. Guide for selecting a stabilizing additive.

Area	Soils Class	Type of Stabilizing Additive Recommended	Restriction on LL and PI of Soil	Restriction on Percent Passing No 200 Sieve	Remarks
2A	GW or GP	(1) Bituminous			Well-graded material only
		(2) Portland cement			Material should contain at least 45 percent by weight of material passing No 4 sieve
		(3) Lime-cement-fly ash	PI not to exceed 25		
2B	GW-GM or	(1) Bituminous	PI not to exceed 10		Well-graded material only
	GP-GM or	(2) Portland cement	PI not to exceed 30		Material should contain at least 45 percent by weight of material passing No 4 sieve
	GW-GC or CP-GC				
		(3) Lime	PI not less than 12		
		(4) Lime-cement-fly ash	PI not to exceed 25		
2C	GH or GC or GM-GC	(1) Bituminous	PI not to exceed 10	Not to exceed 30 percent by weight	Well-graded material only
		(2) Portland cement	—*		Material should contain at least 45 percent by weight of material passing No 4 sieve
		(3) Lime	PI not less than 12		
		(4) Lime-cement-fly ash	PI not to exceed 25		
3	CH or CL	(1) Portland cement	LL less than 40 and		Organic and strongly acid soils falling within this area are not susceptible to stabilization by ordinary means
	or MH or ML or OH or OL or HL-CL		PI less than 20		
		(2) Lime	PI not less than 12		

$* \ PI \leq 20 + \dfrac{50-percent\ passing\ No\ 200\ sieve}{4}$

CEMENT

9-44. Cement can be used as an effective stabilizer for a wide range of materials. In general, however, the soil should have a PI less than 30. For coarse-grained soils, the percent passing the Number 4 sieve should be greater than 45 percent.

9-45. If the soil temperature is less than 40 degrees Fahrenheit and is not expected to increase for one month, chemical reactions will not occur rapidly. The strength gain of the cement-soil mixture will be minimal. If these environmental conditions are anticipated, the cement may be expected to act as a soil modifier, and another stabilizer might be considered for use. Soil-cement mixtures should be scheduled for construction so that sufficient durability will be gained to resist any freeze-thaw cycles expected.

9-46. Portland cement can be used either to modify and improve the quality of the soil or to transform the soil into a cemented mass, which significantly increases its strength and durability. The amount of cement additive depends on whether the soil is to be modified or stabilized. The only limitation to the amount of cement to be used to stabilize or modify a soil pertains to the treatment of the base courses to be used in flexible pavement systems. When a cement-treated base course for Air Force pavements is to be surfaced with asphaltic concrete, the percent of cement by weight is limited to 4 percent.

Modification

9-47. The amount of cement required to improve the quality of the soil through modification is determined by the trial-and-error approach. To reduce the PI of the soil, successive samples of soil-cement mixtures must be prepared at different treatment levels and the PI of each mixture determined.

9-48. The minimum cement content that yields the desired PI is selected, but since it was determined based on the minus 40 fraction of the material, this value must be adjusted to find the design cement content based on total sample weight expressed as—

$$A = 100Bc$$

where—

A = design cement content, percent of total weight of soil
B = percent passing Number 40 sieve, expressed as a decimal
c = percent of cement required to obtain the desired PI of minus Number 40 material, expressed as a decimal

9-49. If the objective of modification is to improve the gradation of granular soil through the addition of fines, the analysis should be conducted on samples at various treatment levels to determine the minimum acceptable cement content. To determine the cement content to reduce the swell potential of fine-grained plastic soils, mold several samples at various cement contents and soak the specimens along with untreated specimens for four days. The lowest cement content that eliminates the swell potential or reduces the swell characteristics to the minimum becomes the design cement content. The cement content determined to accomplish soil modification should be checked to see if it provides an unconfined compressive strength great enough to qualify for a reduced thickness design according to criteria established for soil stabilization (see tables 9-4 and 9-5).

9-50. Cement-modified soil may be used in frost areas also. In addition to the procedures for the mixture design described above, cured specimens should be subjected to the 12 freeze-thaw cycles test (omit wire brush portion) or other applicable freeze-thaw procedures. This should be followed by a frost-susceptibility test, determined after freeze- thaw cycling, and should meet the requirements set forth for the base course. If cement- modified soil is used as the subgrade, its frost susceptibility (determined after freeze-thaw cycling) should be used as the basis of the pavement thickness design if the reduced subgrade-strength design method is applied.

Table 9-4. Minimum unconfined compressive strengths for cement, lime, and combined lime-cement-fly ash stabilized soils

Stabilized Soil Layer	Minimum Unconfined Compressive Strength, psi[a]		
	Flexible Pavement		Rigid Pavement, All
	Army and Air Force	Navy	
Base coarse,	7.50	750	500
Subbase coarse, select material or subgrade	250	300 (cement) 150 (lime)	200

[a] Unconfined compressive strength determined at seven days for cement stabilization and 28 days for lime or lime-cement-fly ash stabilization.

Table 9-5. Durability requirements

| Type of Soil Stabilized | Maximum Allowable Weight Loss After 12 Wet-Dry Freeze-Thaw Cycles, Percent of Initial Specimen Weight | |
	Army and Air Force	Navy
Granular, PI <10	11	14
Granular, PI >10	8	14
Silt	8	14
Clays	6	14

Stabilization

9-51. The following procedure is recommended for determining the design cement content for cement-stabilized soils:

Step 1. Determine the classification and gradation of the untreated soil. The soil must meet the gradation requirements shown in table 9-6, page 9-16, before it can be used in a reduced thickness design (multilayer design).

Step 2. Select an estimated cement content from table 9-7, page 9-16, using the soil classification.

Step 3. Using the estimated cement content, determine the compaction curve of the soil-cement mixture.

Step 4. If the estimated cement content from step 2 varies by more than ±2 percent from the value in tables 9-8 or 9-9, page 9-17, conduct additional compaction tests, varying the cement content, until the value from table 9-8 or 9-9 is within 2 percent of that used for the moisture-density test.

Note. Figure 9-5, page 9-18, is used in conjunction with table 9-9. The group index is obtained from figure 9-5 and used to enter table 9-9.

Step 5. Prepare samples of the soil-cement mixture for unconfined compression and durability tests at the dry density and at the cement content determined in step 4. Also prepare samples at cement contents 2 percent above and 2 percent below that determined in step 4. The samples should be prepared according to TM 5-530 except that when more than 35 percent of the material is retained on the Number 4 sieve, a CBR mold should be used to prepare the specimens. Cure the specimens for seven days in a humid room before testing. Test three specimens using the unconfined compression test and subject three specimens to durability tests. These tests should be either wet-dry tests for pavements located in nonfrost areas or freeze-thaw tests for pavements located in frost areas.

Step 6. Compare the results of the unconfined compressive strength and durability tests with the requirements shown in tables 9-4 and 9-5. The lowest cement content that meets the required unconfined compressive strength requirement and demonstrates the required durability is the design content. If the mixture should meet the durability requirements but not the strength requirements, the mixture is considered to be a modified soil.

9-52. Theater-of-operations construction requires that the engineer make maximum use of the locally available construction materials. However, locally available materials may not lend themselves to classification under the USCS method. The average cement requirements of common locally available construction materials is shown in table 9-10, page 9-19.

Table 9-6. Gradation requirements

Type Course	Sieve Size	Percent Passing	
		Army and Air Force	Navy
Base	22	100	—
	1 1/2 in	70-100	—
	1 in	45-100	100
	3/4 in	—	90-100
	1/2 in	30-90	—
	Number 4	20-70	40-70
	Number 10	15-60	—
	Number 30	—	12-40
	Number 40	5-40	—
	Number 200	0-20	3-15
Subbase	3 in	100	100
	Number 4	—	45-100
	Number 10	—	36-60
	Number 1000	—	3-20
	Number 2000	0-25	0-3

Table 9-7. Estimated cement requirements for various soil types

Soil Classification	Initial Estimated Cement Requirement, Percent Dry Weight
GW, SW	5
GP, SW-SM, SW-SC	6
GW-GM, GW-GC	
GM, SM, GC, SC, SP-SM, SP-SC, GP-GM, GP-GC, SM-SC, GM-GC	7
SP, CL, ML, ML-CL	10
MH-OH	11
CH	10

Table 9-8. Average cement requirements for granular and sandy soils

Material Retained on No 4 Sieve, Percent	Material Smaller Than 0.05 mm, Per cent	Cement Content, Percent by Weight Maximum Dry Density, lb/cu ft (Treated Material)					
		116-120	121-126	127-131	132-137	138-142	143 or more
0-14	0-19	10	9	8	7	6	5
	20-39	9	8	7	7	5	5
	40-50	11	10	9	8	6	5
15-29	0-19	10	9	8	6	5	5
	20-39	9	8	7	6	6	5
	40-50	12	10	9	8	7	6
30-45	0-19	10	8	7	6	5	5
	20-39	11	9	8	7	6	5
	40-50	12	11	10	9	8	6

Note. Base course goes to 70 percent retained on the No 4 sieve.

Table 9-9. Average cement requirements for silty and clayey soils

Group Index	Material Between 0.05 and 0.005 mm, Percent	Cement Content, Percent by Weight Maximum Dry density, lb/cu ft (Treated Material)						
		99-104	105-109	110-115	116-120	121-126	127-131	132 or more
0-3	0-19	12	11	10	8	8	7	7
	20-39	12	11	10	9	8	8	7
	40-59	13	12	11	9	9	8	8
	60 or more	—	—	—	—	—	—	—
3-7	0-19	13	12	11	9	8	7	7
	20-39	13	12	11	10	0	8	8
	40-59	14	13	12	10	10	9	8
	60 or more	15	14	12	11	10	9	9
7 11	0-19	14	13	11	10	9	8	8
	20-39	15	14	11	10	9	9	9
	40-59	16	14	12	11	10	10	9
	60 or more	17	15	13	11	10	10	10
11-15	0-19	15	14	13	12	11	9	9
	20-39	16	15	13	12	11	10	10
	40-59	17	16	14	12	12	11	10
	60 or more	18	16	14	13	12	11	11
15-20	0-19	17	16	14	13	12	11	10
	20-39	18	17	15	14	13	11	11
	40-59	19	18	15	14	14	12	12
	60 or more	20	19	16	15	14	13	12

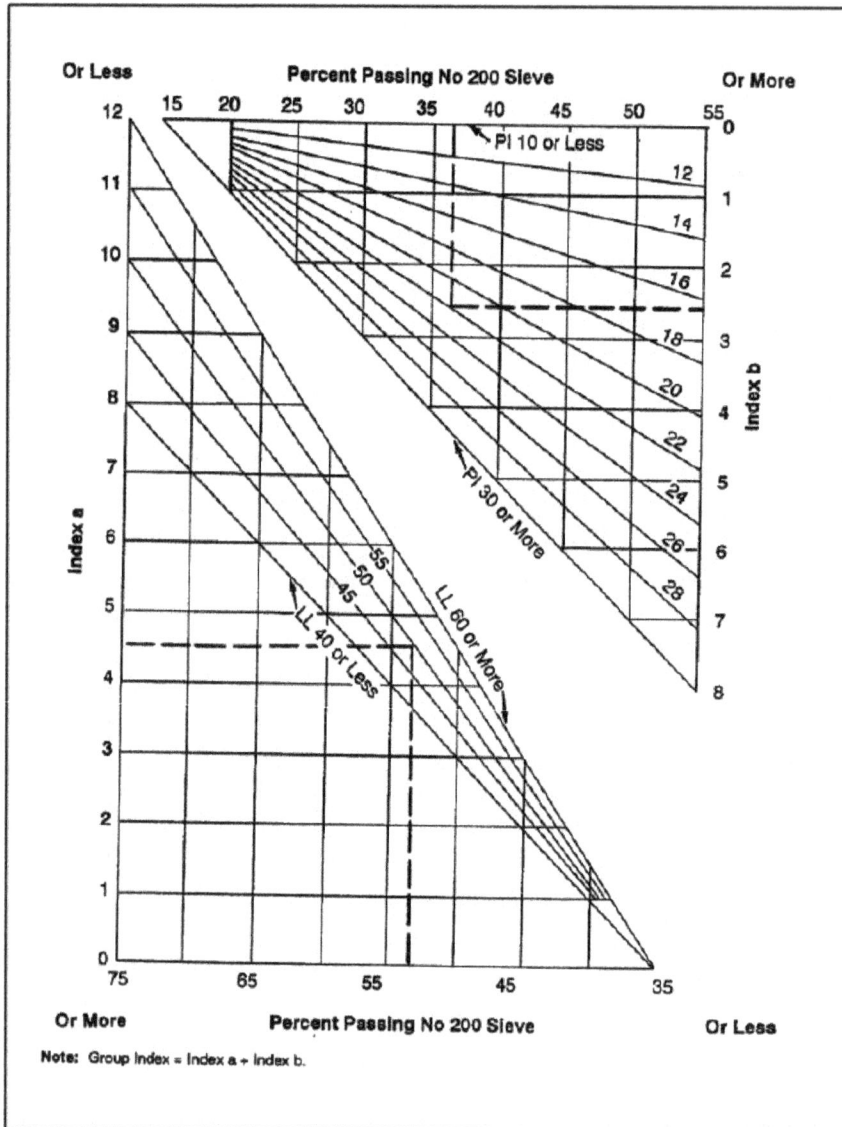

Figure 9-5. Group index for determining average cement requirements

Table 9-10. Average cement requirements of miscellaneous materials

Type of Miscellaneous Material	Estimated Cement Content and That Used in Moisture-Density Test		Cement Contents for Wet-Dry and Freeze-Thaw Tests, Percent by Weight
	Percent by Volume	Percent by Volume	
Shell soils	8	7	5-7-9
Limestone screenings	75	3-5-7	
Red dog	9	8	6-8-10
Shale or disintegrated shale	11	10	8-10-12
Caliche	8	7	5-7-9
Cinders	8	8	6-8-10
Chert	9	8	6-8-10
Chat	8	7	5-7-9
Marl	11	11	9-11-13
Scoria (containing material retained on the No 4 sieve)	12	11	9-11-13
Scoria (not containing material retained on the No 4 sieve)	8	7	5-7-9
Air-cooled slag	9	7	5-7-9
Water-cooled slag	10	12	10-12-14

LIME

9-53. Experience has shown that lime reacts with medium-, moderately fine-, and fine-grained soils to produce decreased plasticity, increased workability and strength, and reduced swell. Soils classified according to the USCS as (CH), (CL), (MH), (ML), (SC), (SM), (GC), (GM), (SW-SC), (SP-SC), (SM-SC), (GW-GC), (GP-GC), and (GM-GC) should be considered as potentially capable of being stabilized with lime.

9-54. If the soil temperature is less than 60 degrees Fahrenheit and is not expected to increase for one month, chemical reactions will not occur rapidly. Thus, the strength gain of the lime-soil mixture will be minimal. If these environmental conditions are expected, the lime may be expected to act as a soil modifier. A possible alternative stabilizer might be considered for use. Lime-soil mixtures should be scheduled for construction so that sufficient durability is gained to resist any freeze-thaw cycles expected.

9-55. If heavy vehicles are allowed on the lime-stabilized soil before a 10- to 14-day curing period, pavement damage can be expected. Lime gains strength slowly and requires about 14 days in hot weather and 28 days in cool weather to gain significant strength. Unsurfaced lime-stabilized soils abrade rapidly under traffic, so bituminous surface treatment is recommended to prevent surface deterioration.

9-56. Lime can be used either to modify some of the physical properties and thereby improve the quality of a soil or to transform the soil into a stabilized mass, which increases its strength and durability. The amount of lime additive depends on whether the soil is to be modified or stabilized. The lime to be used may be either hydrated or quicklime, although most stabilization is done using hydrated lime. The reason is that quicklime is highly caustic and dangerous to use. The design lime contents determined from the criteria presented herein are for hydrated lime. As a guide, the lime contents determined herein for hydrated lime should be reduced by 25 percent to determine a design content for quicklime.

Modification

9-57. The amount of lime required to improve the quality of a soil is determined through the same trial-and-error process used for cement-modified soils.

Stabilization

9-58. To take advantage of the thickness reduction criteria, the lime-stabilized soil must meet the unconfined compressive strengths and durability requirements shown in tables 9-4 and 9-5, pages 9-14 and 9-15, respectively.

9-59. When lime is added to a soil, a combination of reactions begins to take place immediately. These reactions are nearly complete within one hour, although substantial strength gain is not reflected for some time. The reactions result in a change in both the chemical composition and the physical properties. Most lime has a pH of about 12.4 when placed in a water solution. Therefore, the pH is a good indicator of the desirable lime content of a soil-lime mixture. The reaction that takes place when lime is introduced to a soil generally causes a significant change in the plasticity of the soil, so the changes in the PL and the LL also become indicators of the desired lime content. Two methods for determination of the initial design lime content are presented in the following steps:

Step 1. The preferred method is to prepare several mixtures at different lime-treatment levels and determine the pH of each mixture after one hour. The lowest lime content producing the highest pH of the soil-lime mixtures is the initial design lime content. Procedures for conducting a pH test on lime-soil mixtures are presented in TM 5-530. In frost areas, specimens must be subjected to the freeze-thaw test as discussed in step 2 below. An alternate method of deter- mining an initial design lime content is shown in figure 9-6. Specific values required to use this figure are the PI and the percent of material passing the Number 40 sieve. These properties are determined from the PL and the gradation test on the untreated soil for expedient construction; use the amount of stabilizer determined from the pH test or figure 9-6.

Step 2. After estimating the initial lime content, conduct a compaction test with the lime-soil mixture. The test should follow the same procedures for soil-cement except the mixture should cure no less than one hour and no more than two hours in a sealed container before molding. Compaction will be accomplished in five layers using 55 blows of a 10-pound hammer having an 18-inch drop (CF 55). The moisture density should be determined at lime con- tents equal to design plus 2 percent and design plus 4 percent for the preferred method at design $\pm$ 2 percent for the alternate method. In frost areas, cured specimens should be subjected to the 12 freeze-thaw cycles (omit wire brush portion) or other applicable freeze-thaw procedures, followed by frost susceptibility determinations in standard laboratory freezing tests. For lime-stabilized or lime-modified soil used in lower layers of the base course, the frost susceptibility (deter- mined after freeze-thaw cycling) should meet the requirements for the base course. If lime-stabilized or lime- modified soil is used as the subgrade, its frost susceptibility (determined after freeze-thaw cycling) should be the basis of the pavement thickness design if the reduced subgrade strength design method is applied.

Step 3. Uniformed compression tests should be performed at the design percent of maximum density on three specimens for each lime content tested. The design value would then be the minimum lime content yielding the required strength. Procedures for the preparation of lime-soil specimens are similar to those used for cement-stabilized soils with two exceptions: after mixing, the lime-soil mixture should be allowed to mellow for not less than one hour nor more than two hours; after compaction, each specimen should be wrapped securely to prevent moisture loss and should be cured in a constant-temperature chamber at 73 degrees Fahrenheit $\pm$2 degrees Fahrenheit for 28 days. Procedures for conducting unconfined compression tests are similar to those used for soil-cement specimens except that in lieu of moist curing, the lime-soil specimens should remain securely wrapped until testing.

Step 4. Compare the results of the unconfined compressive tests with the criteria in table 9-4. The design lime content must be the lowest lime content of specimens meeting the strength criteria indicated.

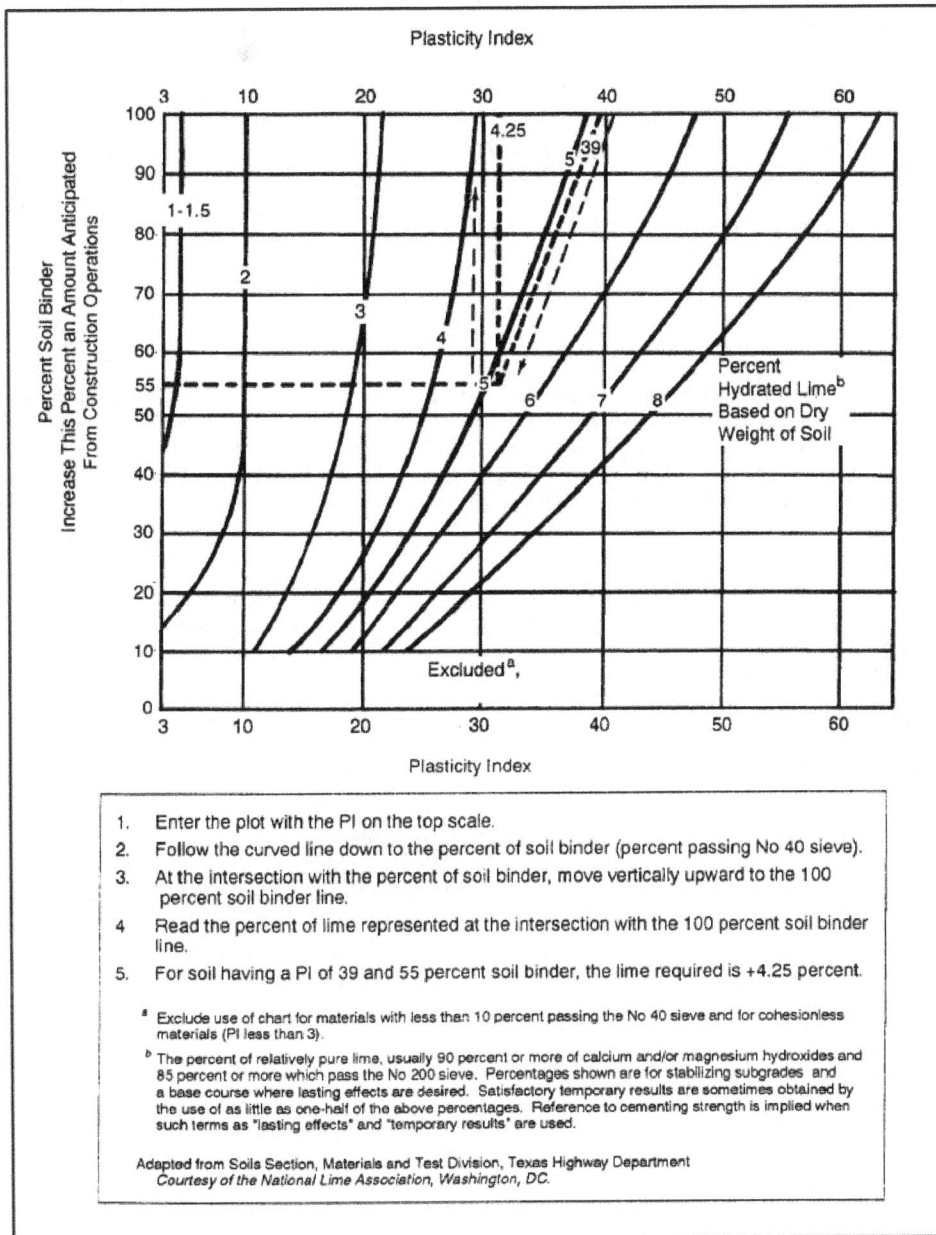

Figure 9-6. Alternate method of determining initial design lime content

Other Additives

9-60. Lime may be used as a preliminary additive to reduce the PI or alter gradation of a soil before adding the primary stabilizing agent (such as bitumen or cement). If this is the case, then the design lime content is the minimum treatment level that will achieve the desired results. For nonplastic and low-PI materials in which lime alone generally is not satisfactory for stabilization, fly ash may be added to produce the necessary reaction.

FLY ASH

9-61. Fly ash is a pozzolanic material that consists mainly of silicon and aluminum compounds that, when mixed with lime and water, forms a hardened cementitious mass capable of obtaining high compression strengths. Fly ash is a by-product of coal-fired, electric power-generation facilities. The liming quality of fly ash is highly dependent on the type of coal used in power generation. Fly ash is categorized into two broad classes by its calcium oxide (CaO) content. They are—

- Class C.
- Class F.

Class C

9-62. This class of fly ash has a high CaO content (12 percent or more) and originates from subbituminous and lignite (soft) coal. Fly ash from lignite has the highest CaO content, often exceeding 30 percent. This type can be used as a stand-alone stabilizing agent. The strength characteristics of Class C fly ash having a CaO less than 25 percent can be improved by adding lime. Further discussion of fly ash properties and a listing of geographic locations where fly ash is likely to be found are in appendix B.

Class F

9-63. This class of fly ash has a low CaO content (less than 10 percent) and originates from anthracite and bituminous coal. Class F fly ash has an insufficient CaO content for the pozzolanic reaction to occur. It is not effective as a stabilizing agent by itself; however, when mixed with either lime or lime and cement, the fly ash mixture becomes an effective stabilizing agent.

Lime Fly Ash Mixtures

9-64. LF mixtures can contain either Class C or Class F fly ash. The LF design process is a four-part process that requires laboratory analysis to determine the optimum fines content and lime-to-fly-ash ratio.

Step 1. Determine the optimum fines content. This is the percentage of fly ash that results in the maximum density of the soil mix. Do this by conducting a series of moisture-density tests using different percentages of fly ash and then determining the mix level that yields maximum density. The initial fly ash content should be about 10 percent based on the weight of the total mix. Prepare test samples at increasing increments (2 percent) of fly ash, up to 20 percent. The design fines content should be 2 percent above the optimum fines content. For example, if 14 percent fly ash yields the maximum density, the design fines content would be 16 percent. The moisture density relation would be based on the 16 percent mixture.

Step 2. Determine the rates of lime to fly ash. Using the design fines con- tent and the OMC determined in step 1, prepare triplicate test samples at LF ratios of 1:3, 1:4, and 1:5. Cure all test samples in sealed containers for seven days at 100 degrees Fahrenheit.

Step 3. Evaluate the test samples for unconfined compressive strength. If frost is a consideration, subject a set of test samples to 12 cycles of freeze-thaw durability tests (refer to FM 5-530 for actual test procedures).

Step 4. Determine the design LF ratio. Compare the results of the unconfined strength test and freeze-thaw durability tests with the minimum requirements found in tables 9-4 and 9-5, pages 9-14 and 9-15, respectively. The LF ratio with the lowest lime content that meets the required unconfined compressive strength and demonstrates the required durability is the design LF content. The treated material must also meet frost susceptibility requirements as indicated in Special Report 83-27. If the mixture meets the durability requirements but not the strength requirements, it is considered to be a modified soil. If neither strength nor durability criteria are met, a different LF content may be selected and the testing procedure repeated.

Lime-Cement-Fly Ash (LCF) Mixtures

9-65. The design methodology for determining the LCF ratio for deliberate construction is the same as for LF except cement is added in step 2 at the ratio of 1 to 2 percent of the design fines content. Cement may be used in place of or in addition to lime; however, the design fines content should be maintained.

9-66. When expedient construction is required, use an initial mix proportion of 1 percent portland cement, 4 percent lime, 16 per-cent fly ash, and 79 percent soil. Minimum unconfined strength requirements (see table 9-4) must be met. If test specimens do not meet strength requirements, add cement in ½ percent increments until strength is adequate. In frost-susceptible areas, durability requirements must also be satisfied (see table 9-5).

9-67. As with cement-stabilized base course materials, LCF mixtures containing more than 4 percent cement cannot be used as base course material under Air Force airfield pavements.

BITUMINOUS MATERIALS

9-68. Types of bituminous-stabilized soils are—

- Soil bitumen. A cohesive soil system made water-resistant by admixture.
- Sand bitumen. A system in which sand is cemented together by bituminous material.
- Oiled earth. An earth-road system made resistant to water absorption and abrasion by means of a sprayed application of slow- or medium-curing liquid asphalt.
- Bitumen-waterproofed, mechanically stabilized soil. A system in which two or more soil materials are blended to produce a good gradation of particles from coarse to fine. Comparatively small amounts of bitumen are needed, and the soil is compacted.
- Bitumen-lime blend. A system in which small percentages of lime are blended with fine-grained soils to facilitate the penetration and mixing of bitumens into the soil.

Soil Gradation

9-69. The recommended soil gradations for subgrade materials and base or subbase course materials are shown in table 9-11 and table 9-12, page 9-24, respectively. Mechanical stabilization may be required to bring soil to proper gradation.

Table 9-11. Recommended gradations for bituminous-stabilized subgrade materials

Sieve Size	Percent Passing
3-in	100
Number 4	50 – 100
Number 30	38 – 100
Number 200	2 - 30

Table 9-12. Recommended gradations for bituminous-stabilized base and subbase materials

Sieve Size	1½ in Max	1 in Max	¾ in Max	½ in Max
1 ½-in	100	—	—	—
1-in	84 ± 9	100	—	—
¾-in	76 ± 9	83 ± 9	100	—
½-in	66 ± 9	73 ± 9	82 ± 9	100
⅜-in	59 ± 9	64 ± 9	72 ± 9	83 ± 9
No 4	45 ± 9	48 ± 9	54± 9	62 ± 9
No 8	35 ± 9	37 ± 9	41 ± 9	47 ± 9
No 16	27 ± 9	28 ± 9	32 ± 9	36 ± 9
No 30	20 ± 9	21 ± 9	24 ± 9	28 ± 9
No 50	14 ± 7	16 ± 7	17 ± 7	20 ± 7
No 100	9 ± 5	11 ± 5	12 ± 5	14 ± 5
No 200	5 ± 2	5 ± 2	5 ± 2	5 ± 2

Types of Bitumen

9-70. Bituminous stabilization is generally accomplished using—

- Asphalt cement.
- Cutback asphalt.
- Asphalt emulsions.

9-71. The type of bitumen to be used depends on the type of soil to be stabilized, the method of construction, and the weather conditions. In frost areas, the use of tar as a binder should be avoided because of its high-temperature susceptibility. Asphalts are affected to a lesser extent by temperature changes, but a grade of asphalt suitable to the prevailing climate should be selected. Generally the most satisfactory results are obtained when the most viscous liquid asphalt that can be readily mixed into the soil is used. For higher quality mixes in which a central plant is used, viscosity-grade asphalt cements should be used. Much bituminous stabilization is performed in place with the bitumen being applied directly on the soil or soil-aggregate system. The mixing and compaction operations are conducted immediately thereafter. For this type of construction, liquid asphalts (cutbacks and emulsions) are used. Emulsions are preferred over cutbacks because of energy constraints and pollution control effects. The specific type and grade of bitumen depends on the characteristics of the aggregate, the type of construction equipment, and the climatic conditions. Table 9-13 lists the types of bituminous materials for use with soils having different gradations.

Table 9-13. Bituminous materials for use with soils of different gradations

Material	Grade
Open-Graded Aggregate	
Rapid- and medium-curing liquid asphalts	RC-250, RC-800, and MC-3000
Medium-setting asphalt emulsion	MS-2 and CMS-2
Well-Graded Aggregate With Little or No Material Passing the No 200 Sieve	
Rapid- and medium-curing liquid asphalts	RC-250, RC-800, MC-250, and MC-800
Slow-curing liquid asphalts	SC-250 and SC-800
Medium-setting and slow-setting asphalt emulsions	MS-2, CMS-2, S-1, and CSS-1
Aggregate With a Considerable Percentage of Fine Aggregate and Material Passing the No 200 Sieve	
Medium-curing liquid asphalts	MC-250 and MC-800
Slow-curing liquid asphalts	SC-250 and SC-800
Slow-setting asphalt emulsions	SS-1, SS-1h, CSS-1, and CSS-1h
Medium-setting asphalt emulsions	MS-2 and CMS-2
The simplest type of bituminous stabilization is the application of liquid asphalt to the surface of an unbound aggregate road. For this type of operation, the slow- and medium-curing liquid asphalts SC-70, SC-250, MC-70, and MC-250 are used.	
Legend: RC rapid curing SS slow setting MC medium curing CMS cement-modified soil MS medium setting CSS cationic slow setting SC slow curing	

Mix Design

9-72. Guidance for the design of bituminous-stabilized base and subbase courses is contained in TM 5-822-8. For subgrade stabilization, the following equation may be used for estimating the preliminary quantity of cutback asphalt to be selected:

$$P = \frac{0.2_{(a)} + 0.07_{(b)} + 1.5_c + 0.20_{(d)}}{(100 - S)} X\ 100$$

where—

P = percent of cutback asphalt by weight of dry aggregate
a = percent of mineral aggregate retained on Number 50 sieve
b = percent of mineral aggregate passing Number 50 and retained on Number 100 sieve
c = percent of mineral aggregate passing Number 100 and retained on Number 200 sieve
d = percent of mineral aggregate passing Number 200 sieve
S = percent solvent

9-73. The preliminary quantity of emulsified asphalt to be used in stabilizing subgrades can be determined from table 9-14, page 9-26. Either cationic or anionic emulsions can be used. To ascertain which type of emulsion is preferred, first determine the general type of aggregate. If the aggregate contains a high content of silica, as shown in figure 9-7, page 9-26, a cationic emulsion should be used (see figure 9-8, page 9-27). If the aggregate is a carbonate rock (limestone, for example), an anionic emulsion should be used.

Table 9-14. Emulsified asphalt requirements

Percent Passing No 200 Sieve	Pounds of Emulsified Asphalt Per 100 lbs of Dry Aggregate at Percent Passing No10 Sieve					
	≤50	60	70	80	90	100
0	6.0	6.3	6.5	6.7	7.0	7.2
2	6.3	6.5	6.7	7.0	7.2	7.5
4	6.5	6.7	7.0	7.2	7.5	7.7
6	6.7	7.0	7.2	7.5	7.7	7.9
8	7.0	7.2	7.5	7.7	7.9	8.2
10	7.2	7.5	7.7	7.9	8.2	8.4
12	7.5	7.7	7.9	8.2	8.4	8.6
14	7.2	7.5	7.7	7.9	8.2	8.4
16	7.0	7.2	7.5	7.7	7.9	8.2
18	6.7	7.0	7.2	7.5	7.7	7.9
20	6.5	6.7	7.0	7.2	7.5	7.7
22	6.3	6.5	6.7	7.0	7.2	7.5
24	6.0	6.3	6.5	6.7	7.0	7.2
25	6.2	6.4	6.6	6.9	7.1	7.3

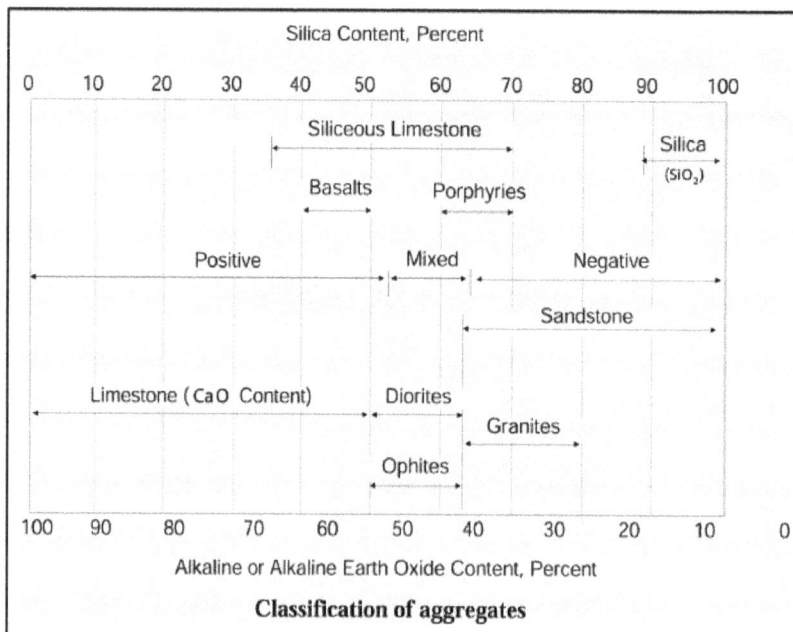

Figure 9-7. Classification of aggregate

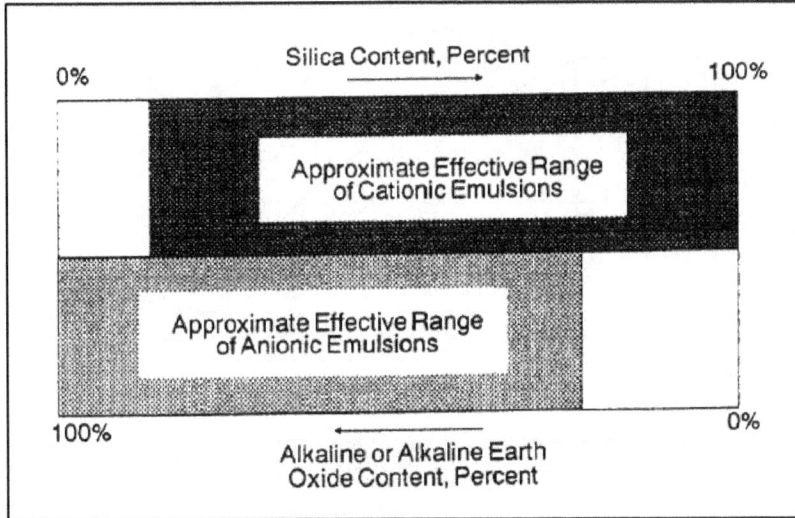

Figure 9-8. Approximate effective range of cationic and aniomic emulsion of various types of asphalt

9-74. Figure 9-9 and figure 9-10, page 9-28, can be used to find the mix design for asphalt cement. These preliminary quantities are used for expedient construction. The final design content of asphalt should be selected based on the results of the Marshall stability test procedure. The minimum Marshall stability recommended for subgrades is 500 pounds; for base courses, 750 pounds is recommended. If a soil does not show increased stability when reasonable amounts of bituminous materials are added, the gradation of the soil should be modified or another type of bituminous material should be used. Poorly graded materials may be improved by adding suitable fines containing considerable material passing a Number 200 sieve. The amount of bitumen required for a given soil increases with an increase in percentage of the finer sizes.

Climatic Zones	Asphalt Grade Penetration
Arctic	100 to 120
Temperate	85 to 100
Tropic	60 to 70
Desert	40 to 50

Figure 9-9. Determination of asphalt grade for expedient construction

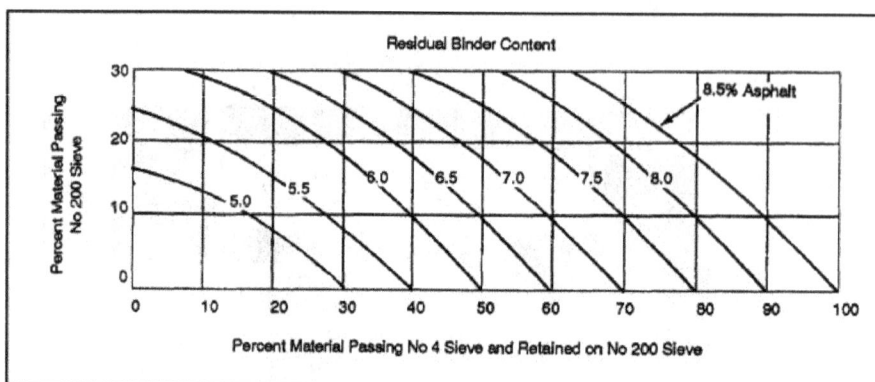

Figure 9-10. Selection of asphalt cement content

SECTION II - DESIGN CONCEPTS

STRUCTURAL CATEGORIES

9-75. Procedures are presented for determining design thicknesses for two structural categories of pavement. They are—

- Single-layer.
- Multilayer.

9-76. Typical examples of these pavements are indicated in figure 9-11.

9-77. A typical single-layer pavement is a stabilized soil structure on a natural subgrade. The stabilized layer may be mixed in place or premixed and later placed over the existing subgrade. A waterproofing surface such as membrane or a single bituminous surface (SBST) or a double bituminous surface treatment (DBST) may also be provided. A multilayer structure typically consists of at least two layers, such as a base and a wearing course, or three layers, such as a subbase, a base, and a wearing course. A thin waterproofing course may also be used on these structures. Single-layer and multilayer pavement design procedures are presented for all categories of roads and for certain categories of airfields as indicated in table 9-15, page 9-30.

9-78. Both single-layer and multilayer pavement structures may be constructed under either the expedient or nonexpedient concept. Different structural designs are provided to allow the design engineer wider latitude of choice. However, single-layer structures are often associated with expedient construction rather than nonexpedient construction, and multilayers are nonexpedient and permanent. Certain considerations should be studied to determine whether to use a single-layer or multilayer design under either concept.

9-79. The overall concept of design as described herein can be defined in four basic determinations as indicated in table 9-16, page 9-30.

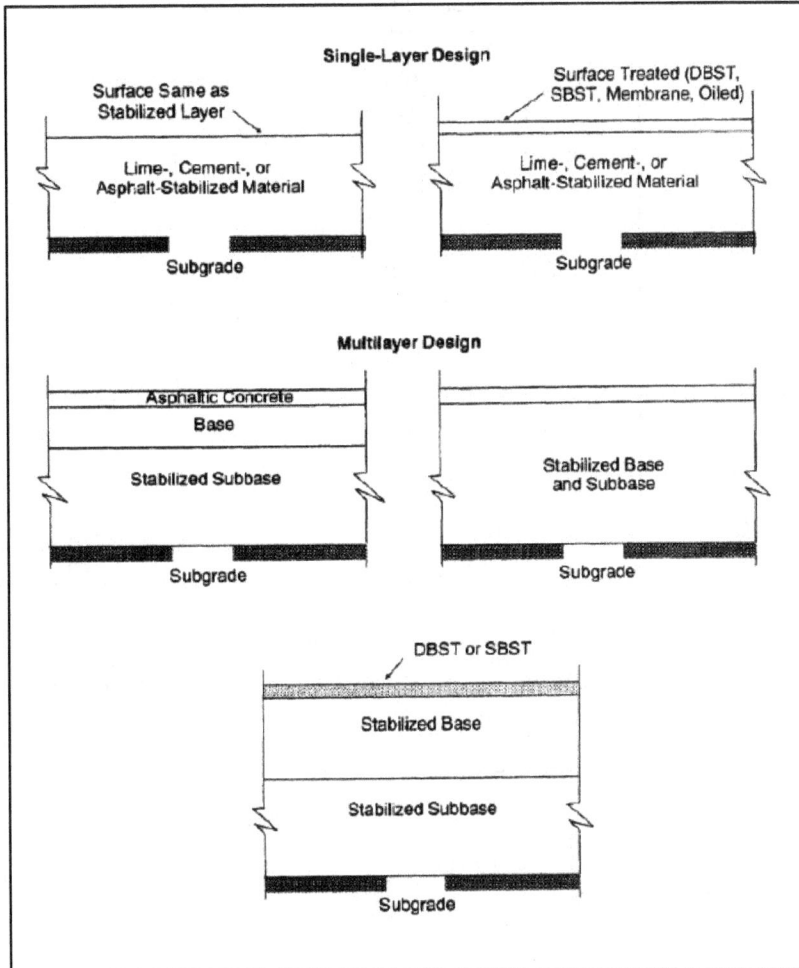

Figure 9-11. Typical sections for single-layer and multilayer design

Table 9-15. Thickness design procedures by airfield category

Airfield Category			Thickness Design Procedures	
Area	Length	Usage	Single-Layer	Multilayer
Close battle	2,000'	Load	Yes	No
	2,500'	Liaison	Yes	No
		Surveillance	Yes	No
		Load	Yes	No
	3,000'	Load	Yes	No
Rear	3,500'	Liaison	Yes	No
		Surveillance	Yes	Yes
		Load	Yes	Yes
	6,000'	Load	Yes	Yes
		Tactical	Yes	Yes
COMMZ	6,000'	Load	Yes	Yes
	10,000'	Load	Yes	Yes
	8,000'	Tactical	Yes	Yes

Table 9-16. Design determinations

Determination	Basis for Making Determination
Whether to use expedient or nonexpedient construction	Mission objective, construction time, proposed facility use
Whether to construct a single-layer or multilayer facility	Availability of material, manpower, equipment, time
Type and quantify of stabilizer to be used	Soil type and properties
Design thickness of pavement	Procedure described under "Thickness Design Procedures"

STABILIZED PAVEMENT DESIGN PROCEDURE

9-80. To use different stabilized materials effectively in transportation facilities, the design procedure must incorporate the advantages of the higher quality materials. These advantages are usually reflected in better performance of the structures and a reduction in total thicknesses required. From a standpoint of soil stabilization (not modification), recent comparisons of behavior based on type and quality of material have shown that stabilization provides definite structural benefits. Design results for airfield and road classifications are presented to provide guidance to the designer in determining thickness requirements when using stabilized soil elements. The design thickness also provides the planner the option of comparing the costs of available types of pavement construction, thereby providing the best structure for the situation.

9-81. The design procedure primarily incorporates the soil stabilizers to allow a reduction of thickness from the conventional flexible pavement-design thicknesses. These thickness reductions depend on the proper consideration of the following variables:

- Load.
- Tire pressure.
- Design life.
- Soil properties.
- Soil strength.
- Stabilizer type.
- Environmental conditions.
- Other factors.

9-82. The design curves for theater-of-operations airfields and roads are given for single-layer and multilayer pavements later in this section.

9-83. In the final analysis, the choice of the admixture to be used depends on the economics and availability of the materials involved. The first decision that should be made is whether stabilization should be attempted at all. In some cases, it may be economical merely to increase the compaction requirements or, as a minimum, to resort to increased pavement thickness. If locally available borderline or unacceptable materials are encountered, definite consideration should be given to upgrading an otherwise unacceptable soil by stabilization.

9-84. The rapid method of mix design should be indicative of the type and percentage of stabilizer required and the required design thickness. This procedure is meant to be a first-step type of approach and is by no means conclusive. Better laboratory tests are needed to evaluate strength and durability and should be performed in specific cases where time allows. Estimated time requirements for conducting tests on stabilized material are presented in table 9-17. Even when stabilized materials are used, proper construction techniques and control practices are mandatory.

Table 9-17. Estimated time required for test procedures

Type of Construction	Stabilizing Agent	Time Required*
Expedient	Lime	None
	LCF	None
	Cement	None
	Bitumen	None
Non expedient	Lime	30 days
	LCF	30 days
	Cement	6-9 days
	Bitumen	1 day
* These criteria do not include time required for gradation and classification tests on the untreated soil.		

THICKNESS DESIGN PROCEDURES

9-85. The first paragraphs of this section give the design engineer information concerning soil stabilization for construction of theater-of-operations roads and airfields. The information includes procedures for determining a soil's suitability for stabilization and a means of determining the appropriate type and amount of stabilizer to be used. The final objective in this total systematic approach is to determine the required design thicknesses. Depending on the type of facility and the AI or the CBR of the unstabilized subgrade, the design procedure presented in this section allows determination of the required thickness of an overlying structure that must be constructed for each anticipated facility.

9-86. This basic structural design problem may have certain conventional overriding factors, such as frost action, that influence this required thickness. The decision to stabilize or not may be based on factors other than structural factors, such as economy, availability of stabilizer, and time. It must be realized that soil stabilization is not a cure for all military engineering problems. Proper use of this manual as a guide allows, in some cases, reductions in required thicknesses. The primary benefit in soil stabilization is that it can provide a means of accomplishing or facilitating construction in situations in which environmental factors or lack of suitable materials could preclude or seriously hamper work progress. Through the proper use of stabilization, marginal soils can often be transformed into acceptable construction materials. In many instances, the quantity of materials required can be reduced and economic advantages gained if the cost of chemical stabilization can be offset by a savings in material transportation costs.

9-87. The structural benefits of soil stabilization, shown by increased load-carrying capability, are generally known. In addition, increased strength and durability also occur with stabilization.

9-88. Generally, lesser amounts of stabilizers may be used for increasing the degree of workability of a soil without effectively increasing structural characteristics. Also, greater percentages may be used for increasing strength at the risk of being uneconomical or less durable. Some of the information presented is intended for use as guidance only and should not supersede specific trial-proven methods or laboratory testing when either exists.

9-89. Primary considerations in determining thickness design are those that involve the decision to construct a single-layer or multilayer facility, as discussed earlier. The method chosen depends on the type of construction. All permanent construction and most multilayer designs should use the reduced thickness design procedure. Usually the single layer is of expedient design.

ROADS

9-90. Specific procedures for determining total and/or layer thicknesses for roads are discussed below. The more expedient methods are shown first, followed by more elaborate procedures. Road classification is based on equivalent number 18-kip, single-axle, dual-wheel applications. Table 9-18 lists the classes of roads.

Table 9-18. Road classifications

Classification	Equivalent Number 18-kip, Single-Axle, Dual-Wheel Applications
A	1.7×10^8
B	1.7×10^8
C	1×10^8
D	4.7×10^7
E	2×10^7
F	Not considered for stabilization concepts

SINGLE-LAYER

9-91. For each category of roads (Classes A through E), a single design curve is presented that applies to all types of stabilization (see figures 9-12 through 9-15, pages 9-33 through 9-34). These curves indicate the total pavement thickness required on an unstabilized subgrade over a range of subgrade strength values. It should be noted that each curve terminates above a certain subgrade CBR. This is because design strength criteria for unsurfaced roads indicate that a natural soil of this appropriate strength could sustain the traffic volume required of this category of facility without chemical stabilization. The following flow diagram indicates the use of these design procedures:

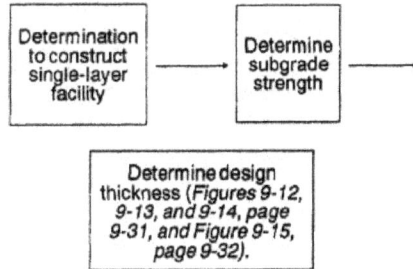

9-92. On a single-layer road, a thin wearing course may be advisable to provide waterproofing and to offset the effects of tire abrasion.

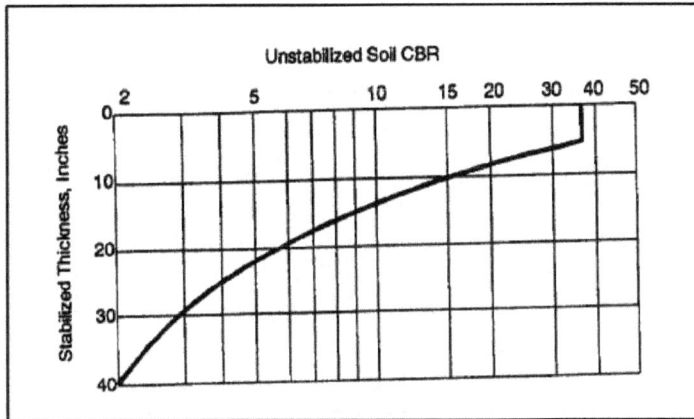

Figure 9-12. Design curve for Class A and Class B single-layer roads using stabilized soils

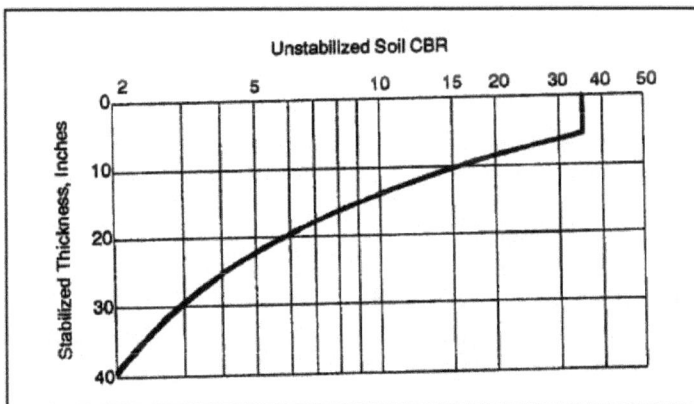

Figure 9-13. Design curve for Class C single layer roads using stabilized soils

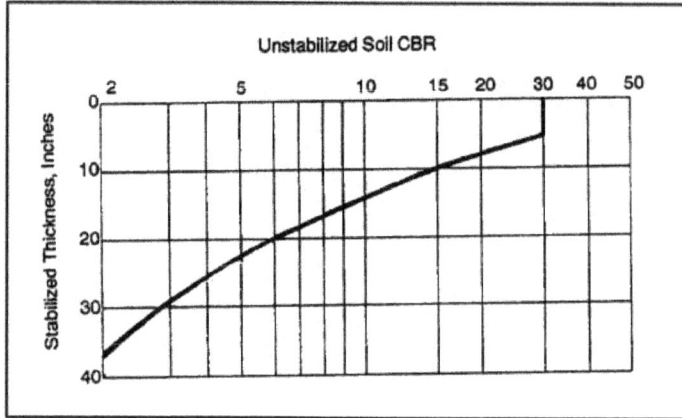

Figure 9-14. Design curve for Class D single-layer roads using stabilized soils

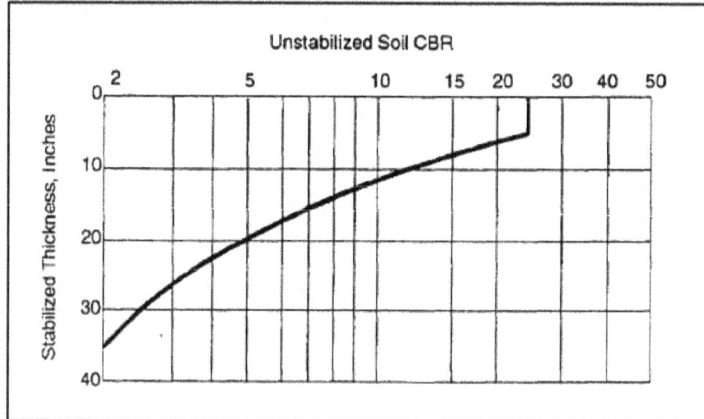

Figure 9-15. Design curve for Class E single-layer roads using stabilized soils

MULTILAYER

9-93. For each road category, four design curves are shown (see figure 9-16 and figures 9-17 through 9-19, pages 9-36 and 9-37). These curves indicate the total thickness required for pavements incorporating one of the following combinations of soil and stabilizer:

- Lime and fine-grained soils.
- Asphalt and coarse-grained soils.
- Portland cement and coarse-grained soils.

9-94. Coarse- and fine-grained soils are defined according to the USCS. The curves presented in figures 9-16 through 9-19 are applicable over a range of subgrade CBR values.

9-95. Individual layer thickness can be accomplished using table 9-19, page 9-37. This table indicates minimum base and wearing course thickness requirements for road Classes A through E. Minimum surface course thickness requirements are indicated for a base course with a strength of 50 to 100 CBR. If a stabilized soil layer is used as a subbase, the design base thickness is the total thickness minus the combined thickness of base and wearing courses. If a stabilized layer is used as a base course over an untreated subgrade, the design base thickness is the total thickness minus the wearing course thickness. The following flow diagram shows these procedures:

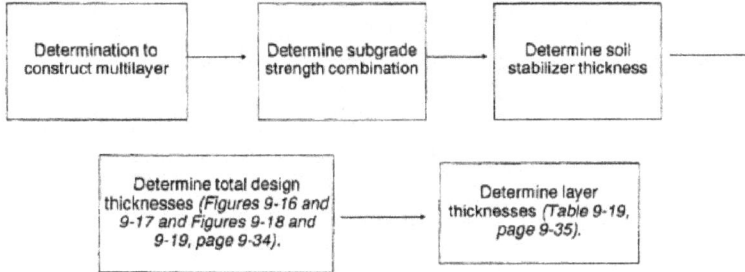

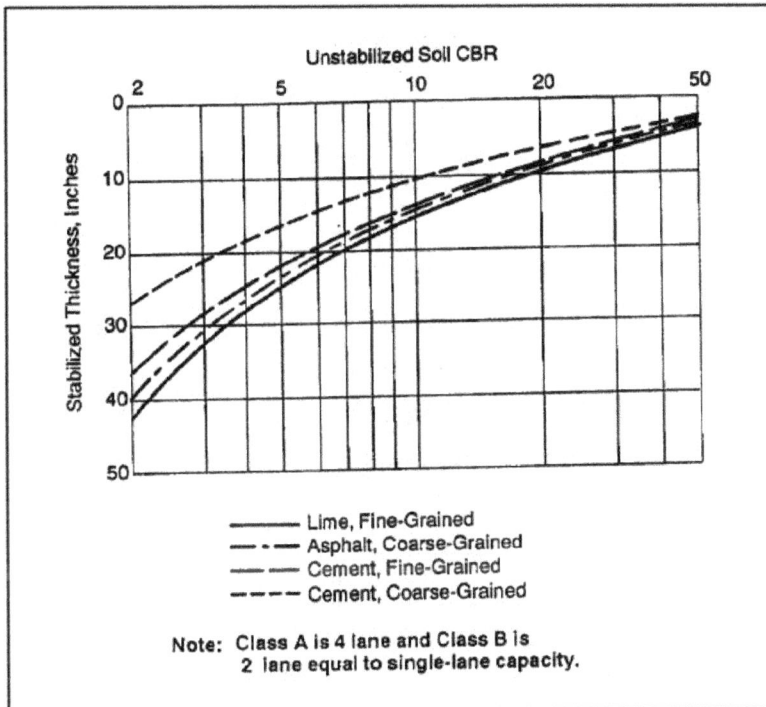

Figure 9-16. Design curve for Class A and Class B multilayer roads using stabilized soils

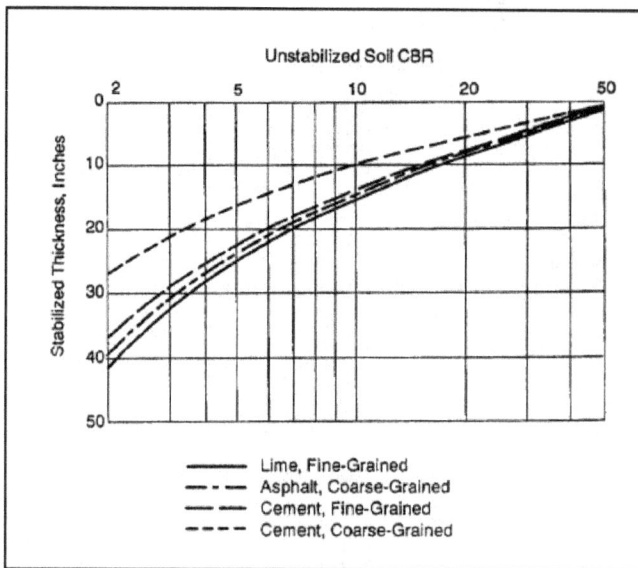

Figure 9-17. Design curve for Class C multilayer roads using stabilized soils

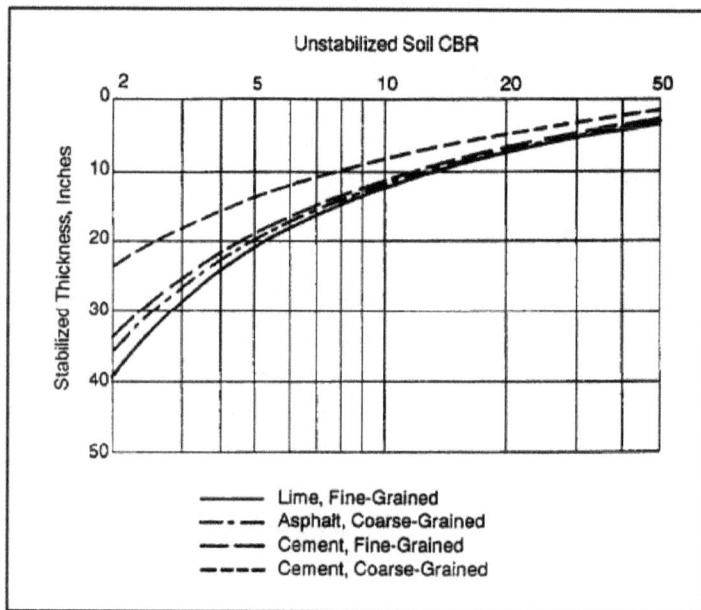

Figure 9-18. Design curve for Class D multilayer roads using stabilized soils

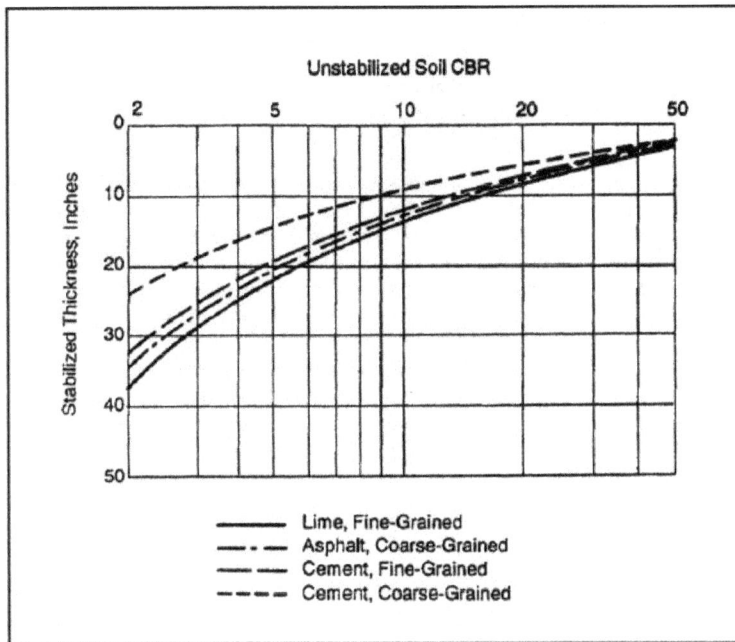

Figure 9-19. Design curve for Class E multilayer roads using stabilize soils

Table 9-19. Recommended minimum thickness of pavement and base coarse for roads in the theater-of-operations

Equivalent 18,000-lb Dual-wheel Load Operations	100 CBR Base			80 CBR Base			50 CBR Base		
	Pavement (in)	Base (in)	Total (in)	Pavement (in)	Base (in)	Total (in)	Pavement (in)	Base (in)	Total (in)
3 x10³ or less	SBST[a]	4	4 ½	MBST	4	5	1½	4	5 ½
3 x10³ -1.5x10⁴	SBST	4	4 ½	MBST	4	5	2	4	6
1.5 x10⁴ - 7x10⁴	MBST[b]	4	5	1 ½	4	5½	2½	4	6½
7 x10⁴ - 7x10⁵	MBST	4	5	1 ½	4	5½	3	4	7
7 x10⁵ - 7x10⁴	1½	4	5 ½	2	4	6	3½	4	7½
7 x10⁶ - 7x10⁷	1½	4	5 ½	2 ½	4	6 ½	4	4	8
7 x10⁷ - 7x10⁸	2	4	6	3 ½	4	7 ½	4½	4	8½
7 x10⁸ - 2x10⁹	3	4	7	3 ½	4	7 ½	5	4	9

[a] Single bituminous surface treatment (SBST).
[b] Multiple bituminous surface treatment (MBST).

9-96. Reduced thickness design factors, (see table 9-20, page 9-38, and figure 9-20, page 9-38) should be applied to conventional design thickness when designing for permanent and nonexpedient road and airfield design. The use of stabilized soil layers within a flexible pavement provides the opportunity to reduce the overall thickness of pavement structure required to support a given load. To design a pavement containing stabilized soil layers requires the application of equivalency factors to a layer or layers of a conventionally designed pavement. To qualify for application of equivalency factors, the stabilized layer must meet appropriate strength and durability requirements set forth in *TM 5-822-4/AFM 88-7*, Chapter 4. An equivalency factor represents the number of inches of a conventional base or subbase that can be replaced by 1 inch of stabilized material. Equivalency factors are determined from—

- Table 9-20 for bituminous stabilized materials.
- Figure 9-20 for materials stabilized with cement, lime, or a combination of fly ash mixed with cement or lime.

9-97. Selection of an equivalency factor from the tabulation depends on the classification of the soil to be stabilized. Selection of an equivalency factor from figure 9-20, requires that the unconfined compressive strength, determined according to ASTM D1633, is known. Equivalency factors are determined from figure 9-20, for subbase materials only. The relationship established between a base and a subbase is 2:1. Therefore, to determine an equivalency factor for a stabilized base course, divide the subbase factor from figure 9-20, by 2. See TM 5-330/AFM 86-3, Volume II for conventional design procedures.

Table 9-20. Reduced thickness criteria for permanent and nonexpedient road and airfield design

Material	Equivalency Factors	
	Base	Subbase
All bituminous concrete	1.15	2.30
GW, GP GM, GC	1.00	2.00
SW, SP, SM, SC	a	1.50
a Not used for base coarse material		

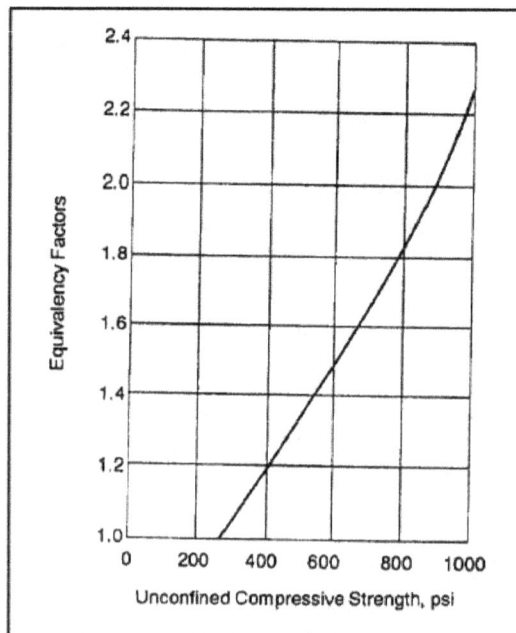

Figure 9-20. Equivalency factors for soils stabilized with cement, lime, or cement and lime mixed with fly ash

AIRFIELDS

9-98. Specific procedures for determining the total and/or layer thicknesses for airfields are discussed in the following paragraphs. The more expedient methods are shown first, followed by more elaborate procedures. Airfields are categorized by their position on the battlefield, the runway length, and the controlling aircraft. Table 9-21 lists aircraft categories.

Table 9-21. Airfield categories

Airfields				
Facility Identification Airfield Area Categories	Controlling Aircraft	Gross Weight (kips)	Takeoffs and Landings Cycles	Design Curve (Figure Number)
Close Battle Area				
2,000	C-130	125.0 175.0	85	9-21
Liaison 2,500'	O-1	2.5	380	9-21
Surveillance 2,500'	OV-1	15.5	480	9-21
3,000'	C-130	125.0 175.0	420	9-21
Rear Area				
Liaison 3,500'	O-1	2.5	1,900	9-22
Surveillance, 3,500'	OV-1	15.5	2,400	9-22
3,500'	C-130	125.0 175.0	2,100	9-22
6,000'	C-141	250.0 320.0	2,200	9-23
Tactical				
Rear area 6,000'	F-4C	59.0	7,100	9-24
COMMZ 8,000'	F-4C	59.0	35,500	9-25
	F-111B	111.0	53,000	9-25
COMMZ				
Liaison 3,000'	OV-1	15.5	12,000	9-26
6,000'	C-130	125.0 175.0	10,000	9-27
10,000'	C-141	200.0 300.0	10,000	9-27
Semipermanent	C-5A	570.0 770.0	700	9-28
	C-5A	550.0 650.0 750.0	3,500	9-28

SINGLE-LAYER

9-99. Design curves for single-layer airfield construction are in figures 9-21 through 9-28, pages 9-40 through 9-46. In these figures the controlling aircraft and design life in cycles (one cycle is one takeoff and one landing) are indicated for each airfield category. The design curves are applicable for all types of stabilization over a range of subgrade strengths up to a maximum above which stabilization would generally be unwarranted if the indicated material subgrade strength could be maintained. Design curves are presented for typical theater-of-operations gross weights for the controlling aircraft category. For a single-layer facility, a thin wearing course may provide waterproofing or minimize abrasion resulting from aircraft tires. The following flow diagram indicates these procedures:

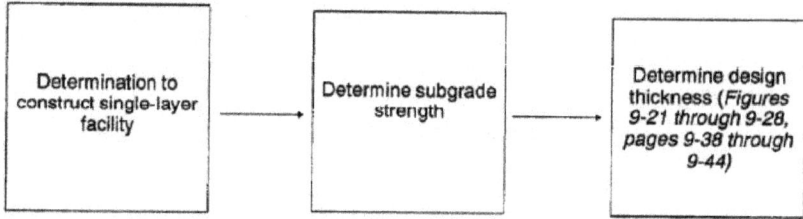

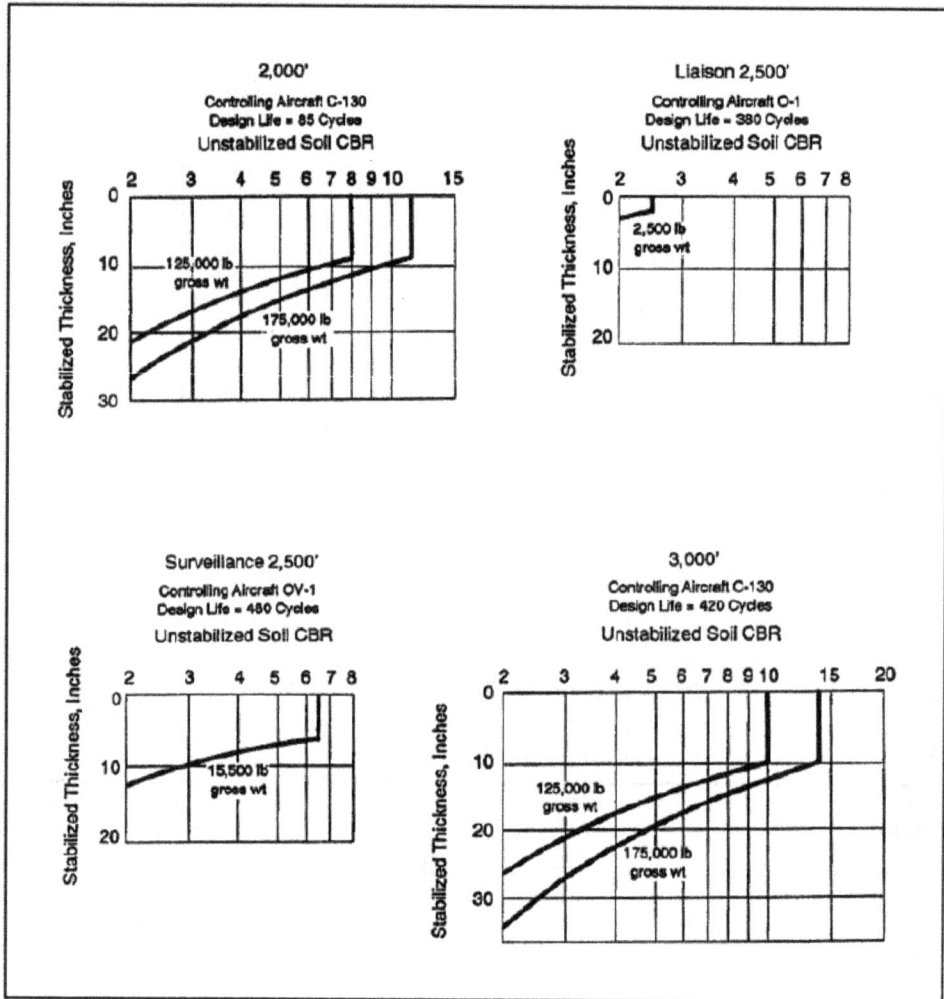

Figure 9-21. Design curves for single layer airfields using stabilized soils in close battle areas

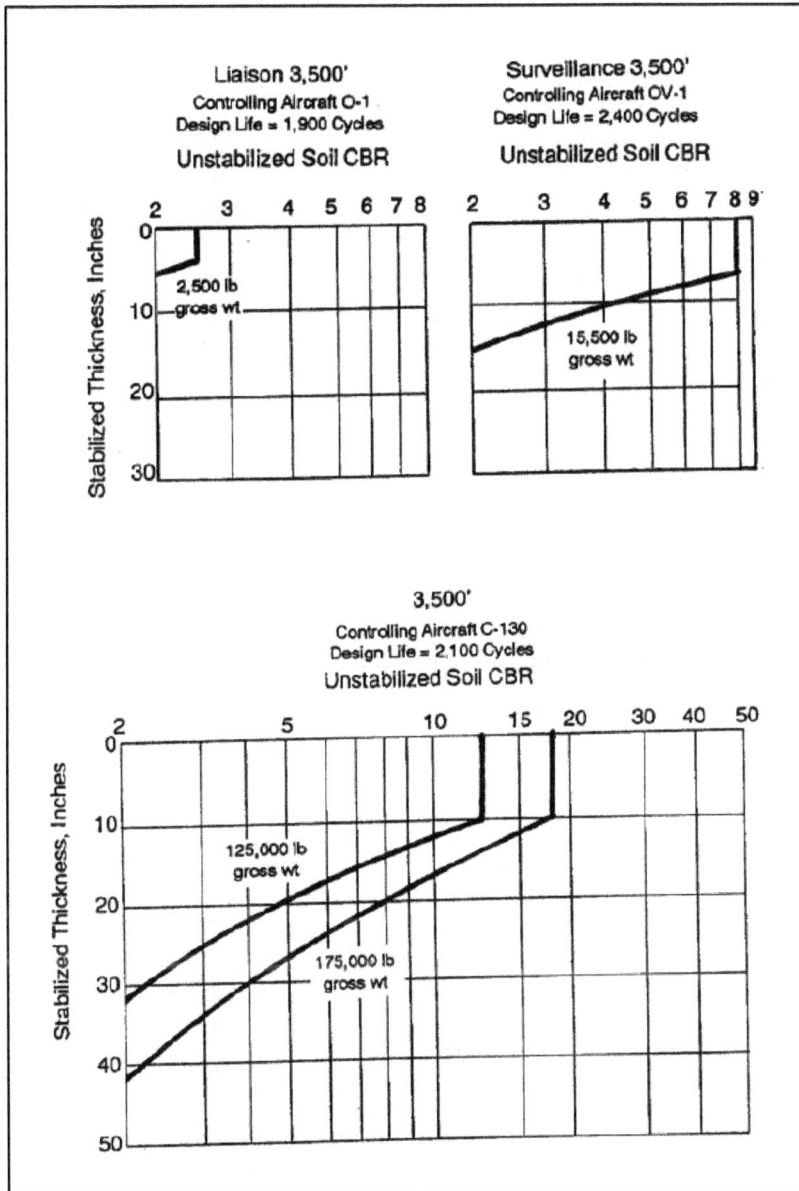

Figure 9-22. Design curves for single-layer airfields using stabilized soils in rear areas

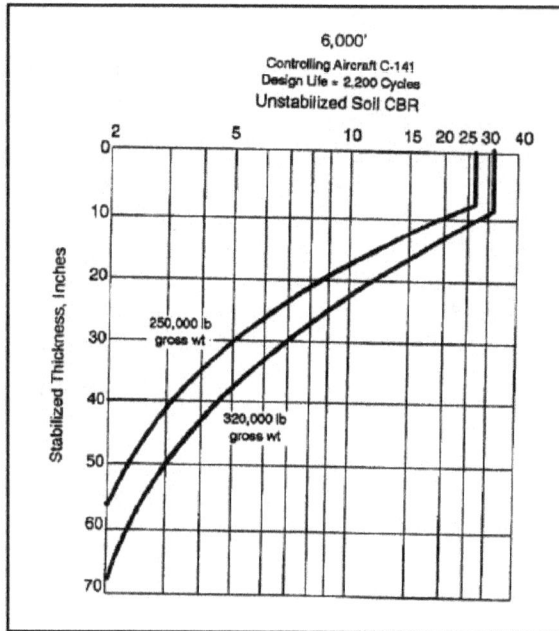

Figure 9-23. Design curves for single-layer airfields using stabilized soils in rear area 6,000'

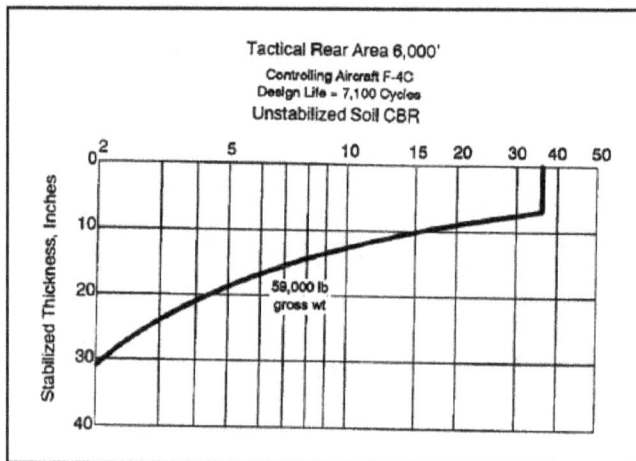

Figure 9-24. Design curves for single-layer airfields using stabilized soils in tactical rear area

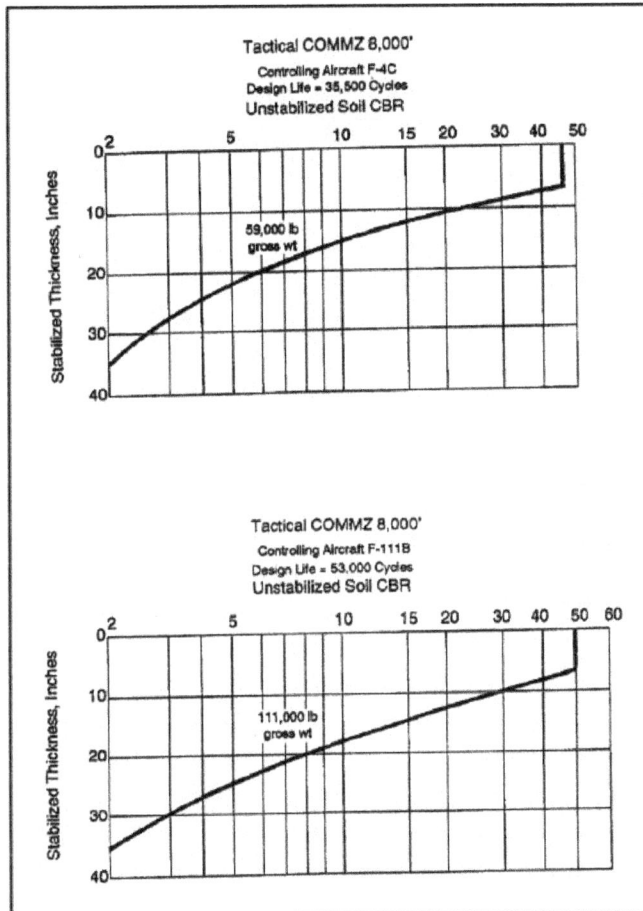

Figure 9-25. Design curves for single-layer airfields using stabilized
soils in tactical COMMZ area

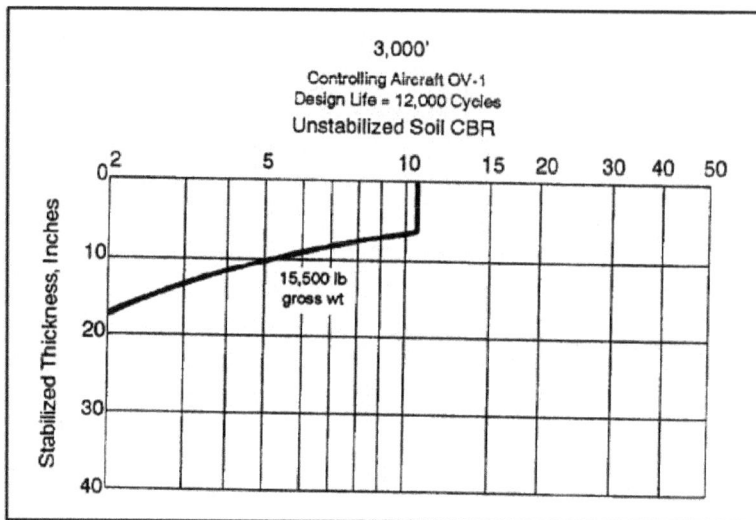

Figure 9-26. Design curves for single-layer airfields using stabilized soils in liaison COMMZ airfields

MULTILAYER

9-100. In the design of multilayer airfields, it is first necessary to determine the total design thickness based on conventional flexible pavement criteria. Then an appropriate reduction factor is applied for the particular soil-stabilizer combination anticipated for use. Determinations of individual layer thickness finalizes the design. Conventional flexible pavement design curves and procedures may be found in TM 5-330/AFM 86-3, Volume II. After total thickness has been determined, a reduction factor is applied (see table 9-22 or 9-23, page 9-47). Individual layer thicknesses can be determined using table 9-24, page 9-48, and procedures indicated for multilayer roads. The following flow diagram indicates these design procedures:

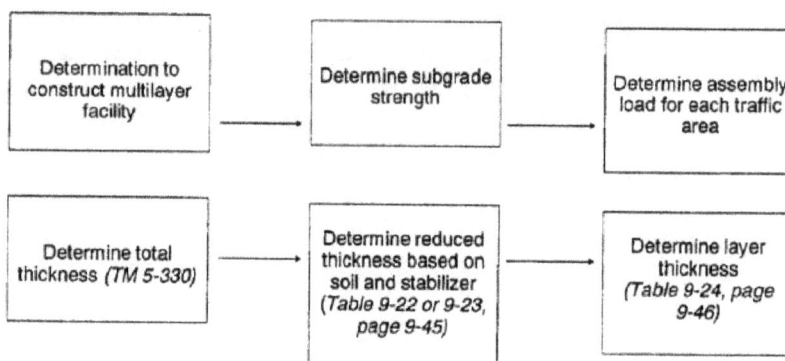

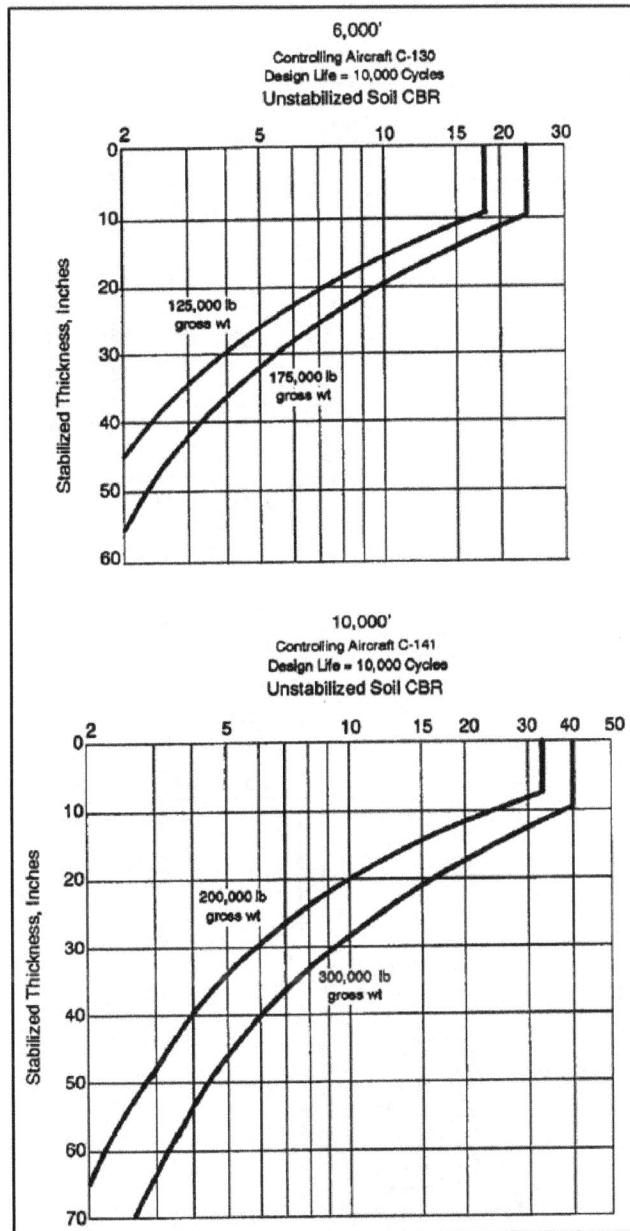

Figure 9-27. Design curves for single-layer airfields using stabilized soils in COMMZ

Semipermanent

Controlling Aircraft C-5A
Design Life = 700 Cycles
Unstabilized Soil CBR

Semipermanent

Controlling Aircraft C-5A
Design Life = 3,500 Cycles
Unstabilized Soil CBR

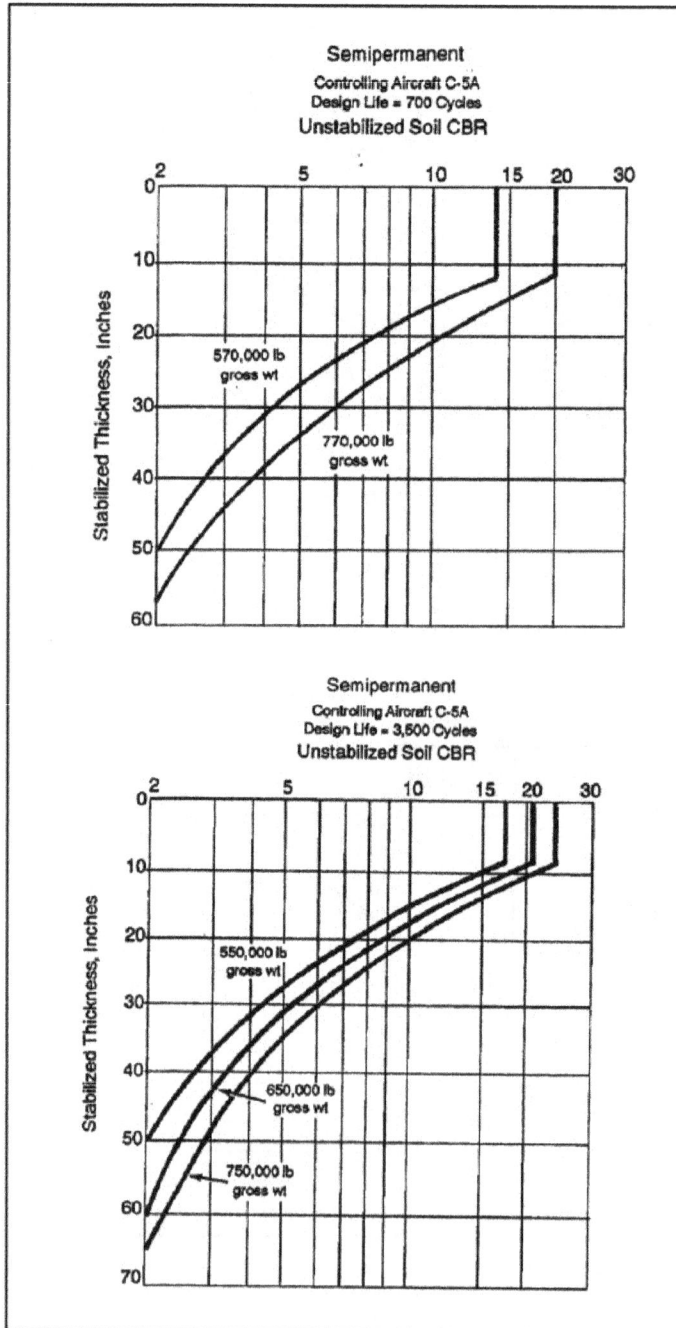

Figure 9-28. Design curves for single-layer airfields using stabilized soils in semi-permanent COMMZ airfields

Table 9-22. Thickness reduction factors for Navy design

Stabilized Material	Equivalency Factors
1 inch of lime-stabilized subbase may be substituted for	1.2 inches of unstabilized subbase coarse
1 inch of cement-stabilized subbase may be substituted for	1.2 inches of unstabilized subbase coarse
1 inch of cement-stabilized base may be substituted for	1.5 inches of unstabilized base coarse
1 inch of bituminous base may be substituted for	1.5 inches of unstabilized base coarse

Table 9-23. Equivalency factors for Air Force bases and Army airfields

Material	Equivalency factors	
	Base	Subbase
Unbound crushed stone	1.00	2.00
Unbound aggregate	a	1.00
Asphalt-stabilized		
All-bituminous concrete	1.15	2.30
GW, GP, GM, GC, SW, SP, SM, SC	a	2.00
Cement-stabilized		
GW, GP SW, SP	1.15[b]	2.30
GC, GM	1.00[b]	2.00
ML, MH, CL, CH	a	1.70
SC, SM	a	1.50
Lime-stabilized		
ML,MH, CL, CH	a	1.00
SC, SM, GC, GM	a	1.10
Lime-cement fly-ash stabilized		
ML, MH, CL, CH	a	1.30
SC, SM, GC, GM	a	1.40

[a] Not used as a base coarse
[b] For Air force bases, cement is limited to 4 percent by weight or less

EXAMPLES OF DESIGN

9-101. Use of the design criteria in this section can best be illustrated by examples of typical design situations.

EXAMPLE 1

9-102. The mission is to construct an airfield for the logistical support of an infantry division and certain nondivision artillery units. The facility must sustain approximately 210 takeoffs and landings of C-130 aircraft, operating at 150,000 pounds gross weight, along with operations of smaller aircraft. Because of unsatisfactory soil strength requirements and availability of chemical stabilizing agents, stabilization is to be considered. The facility is also considered an expedient single-layer design.

9-103. A site reconnaissance and a few soil samples at the proposed site indicate the following:

- The natural strength is 8 CBR.
- It has a PI of 15.
- It has a LL of 30.
- Twenty percent passes a Number 200 sieve.
- Thirty percent is retained on a Number 4 sieve.
- The classification is (SC).

9-104. Using this information, a determination can be made from figure 9-4, page 9-12, and table 9-3, page 9-12, that the proper agent is cement, lime, or fly ash. The soil-lime pH test indicates that a lime content of 3 percent is required to produce a pH of 12.4. Since the soil classified as an (SC), an estimated cement content of 7 percent is selected from table 9-7, page 9-16. The fly ash ratio is 4 percent lime, 1 percent cement, 16 percent fly ash, and 79 percent soil. The characteristics of all additives are then reviewed, and because of predicted cool weather conditions, cement stabilization is chosen.

9-105. The design thickness is then determined. The facility will be designed as a close battle area 3,000' airfield designed for 420 cycles of a C-130 aircraft. To determine the design thickness, Figure 9-1, page 9-6, is used. For a subgrade strength of 8 CBR and interpolating between the 125,000- and 175,000-pound curves, the required design thickness is 13½ inches.

Table 9-24. Recommended minimum thickness of pavements and bases for airfields

Aircraft	89 CBR Base (in)						100 CBR Base (in)					
	Pavement			Base			Pavement			Base		
	E[a]	M[a]	F[a]	E	M	F	E	M	F	E	M	F
Bomber												
B-52H	3 ½	4½	5½	6	6	6	2½	3½	4½	6	6	6
Cargo												
C-5A	2	2	2	6	6	6	2	2	2	6	6	6
C-7A	MBST	MBST	MBST	6	6	6	MBST	MBST	MBST	6	6	6
C-123B	MBST	1½	1½	6	6	6	MBST	MBST	2	6	6	6
C-124C	2½	2½	3	6	6	6	1½	2	2½	6	6	6
C-130E	2	2½	3	6	6	6	1½	2	2½	6	6	6
HC-130H	2	2½	3	6	6	6	1½	2	2½	6	6	6
C-135A	2	2½	3	6	6	6	2	2	2	6	6	6
KC-135A	2	3	3½	6	6	6	2	2	2½	6	6	6
C-141B	2½	3	3½	6	6	6	2	2	2½	6	6	6
Fighter												
F-100F	2	2	2½	6	6	6	2	2	2	6	6	6
F-104A	2	2	2	6	6	6	2	2	2	6	6	6
F-104C	2	2	2	6	6	6	2	2	2	6	6	6
F-11A	—	—	—	—	—	—	—	—	—	—	—	—
F-4C	2½	3	3½	6	6	6	2	2 ½	3	6	6	6
F-5A	2	2	2	6	6	6	2	2	2	6	6	6
[a] E = emergency; M = Minimum; F = Full.												

EXAMPLE 2

9-106. The mission is to provide a rear area 6,000' airfield facility for C-5A aircraft operating at 320,000 pounds gross weight. Time and materials indicate that a multilayer facility can be constructed using nonexpedient methods. A site reconnaissance indicates the following:

- The natural strength is 5 CBR.
- It has a PI of 5.
- It has a LL of 35.
- Fifteen percent passes the Number 200 sieve.
- Sixty percent is retained on the Number 4 sieve.
- The classification is (GM).

9-107. Chemical stabilization is considered for the upper subgrade, but a supply of 100-CBR base course material is available. An asphaltic concrete wearing course will be used. Since the soil classified as (GM), either bituminous, fly ash, or cement stabilization is appropriate (see figure 9-4 and table 9-3). Because of the lack of adequate quantities of cement and fly ash, bituminous stabilization will be tried.

9-108. The material is termed "sand-gravel bitumen." Table 9-13, page 9-25, recommends either asphalt cutbacks or emulsions (considerable materials passing the Number 200 sieve); since cutback asphalt is available, it will be used. It is anticipated that the in-place temperature of the sand will be about 100 degrees Fahrenheit. (From table 9-13, it can then be determined that the grade of cutback to be used is MC-800. From the equation given on page 9-25 and the gradation curve (not shown for the example), a preliminary design content of 6.7 percent asphalt is determined.) Design specimens are then molded and tested using the procedures indicated in TM 5-530. Comparing the test results with the criteria given previously (a minimum of 500 pounds), it can be determined that the upper subgrade can be stabilized with cutback asphalt. An optimum asphalt content of 6.5 percent is indicated.

9-109. The design thickness is then determined. (Only procedures for determining design thickness of a Type A runway area will be indicated.) Since the airfield is a rear-area 6,000' facility with the C-141 as the controlling aircraft category and is a multilayer design, TM 5-330, figure D-36 is used. A subgrade strength of 5 CBR and a design thickness of 45 inches is required for a conventional pavement. Since soil stabilization is involved, reduced thickness design is allowed. Table 9-20, page 9-38, shows that the equalizing factor for an asphalt-stabilized subbase of a (GM) soil is 2.00. Therefore, the required thickness for the pavement, including the surface and base course, is 22.5 inches.

9-110. To determine individual layer thicknesses, use table 9-24. For a rear-area, 6,000' airfield with the C-141 as the controlling aircraft and a 100 CBR base-course strength, the minimum surface-course and base course thicknesses are 2½ and 6 inches, respectively. Thus, the individual layer thickness would be as follows: surface course, 2½ inches; 100-CBR base course, 6 inches; and stabilized upper subgrade, 14.0 inches. Another viable solution would be to stabilize the base course also.

EXAMPLE 3

9-111. The mission is to quickly provide an expedient road of geometrical classification between two organizational units. Only the in-place material can be stabilized. The preliminary site investigation indicates the following:

- The natural strength is 15 CBR.
- It has a PI of 15.
- It has a LL of 30.
- Sixty percent passes a Number 40 sieve.
- Fifty-five percent passes a Number 200 sieve.
- The classification is (CL).

9-112. Expedient design procedures indicate that lime or cement stabilization is feasible (see table 9-3, page 9-12, and figure 9-4 page 9-12). Only lime is readily available. Design procedures for a single-layer

Class E road will be used. Figure 9-6, page 9-21, is used to determine the initial design lime content. Since the soil has a PI of 15 and 60 percent passing the number 40 sieve, an estimated lime content of 2½ percent is selected. Figure 9-15, page 9-34, indicates that the required design thickness of stabilized material is 9 inches.

EXAMPLE 4

9-113. The mission is to construct a road between two rear-area units. Time and material conditions allow nonexpedient procedures. A hot-mix plant is available so that an asphaltic concrete wearing course can be applied. However, the upper portion of the in-place material must be upgraded to provide a suitable base course. Geometric criteria indicate that a Class D multilayer road is required. A soil survey reveals the following with respect to the in-place material:

- The average soil strength is 7 CBR.
- It has a PI of 9.
- It has a LL of 25.
- Thirty-seven percent passes the Number 200 sieve.
- Forty-five percent passes the Number 4 sieve (25 percent is smaller than 0.05 mm, and 5 percent is smaller than 0.005 mm).
- The classification is (GC).

9-114. Following procedures in figure 9-4 and table 9-3 it is determined that cement or lime-cement-fly ash stabilization will work with this soil; however, fly ash is not available. The soil-cement laboratory test (see TM 5-530) is run. Test results indicate that a cement content of 6 percent is required. Figure 9-18, page 9-36, indicates that a total pavement thickness of 12 inches is required above the 7-CBR subgrade for a cement-stabilized, coarse-grained soil. A minimum base-course strength of 70 CBR is assumed. Table 9-18, page 9-32, indicates that a Class D road is designed for 18-kip equivalent loads and a CBR of 50; a 4-inch asphaltic cement pavement and a 10-inch cement-stabilized base are required.

EXAMPLE 5

9-115. The mission is to provide an expedient tactical support area airfield for the operation of approximately 7,000 cycles of F-4C traffic. The single-layer design is selected. A site reconnaissance reveals the following:

- The natural strength is 4 CBR.
- It has a PI of 12.
- Eleven percent passes a Number 200 sieve.
- Twenty percent retained on a Number 4 sieve.
- Organic material occurs as a trace in the soil samples.

9-116. Climatological data indicate a trend for subfreezing weather, and full traffic must be applied immediately upon completion. Ordinarily, based on information from figure 9-4 and table 9-3, either cement, lime, or fly ash stabilization would be the appropriate agent for this situation and the soil would classify as an (SW-SM) borderline. With the constraints on curing times, soil stabilization would not be the appropriate method of construction. Another means, possibly landing mats, must be considered for the successful completion of the mission.

EXAMPLE 6

9-117. The mission is to provide an expedient Class E road between two organizational task forces. The single-layer design is selected. The preliminary site investigation for a portion of the road indicates a natural soil strength of 30 CBR. The design curve for this road classification, shows that a 30-CBR soil is adequate for the intended traffic and that it does not require any stabilization (see figure 9-15). Therefore, no soil sampling or testing is necessary. A problem area may later arise from a reduction of strength, that is, a large volume of rainfall or a dust problem on this particular road.

THEATRE-OF-OPERATIONS AIRFIELD CONSIDERATIONS

9-118. In the theater of operations, the lack of trained personnel, specialized equipment, or time often eliminates consideration of many laboratory procedures. The CBR and special stabilization tests in particular will not be considered for these reasons. As a result, other methods for determining design pavement thicknesses have been developed using the AI (see TM 5-330/AFM 86-3, Volume II). This system is purely expedient and should not replace laboratory testing and reduced thickness design procedures.

FUNCTIONS OF SOIL STABILIZATION

9-119. As previously discussed, the three primary functions of stabilization are—

- Strength improvement.
- Dust control.
- Waterproofing.

9-120. Use of table 9-25 allows the engineer to evaluate the soil stabilization functions as they relate to different types of theater-of-operations airfields. It is possible to easily see the uses of stabilization for the traffic or nontraffic areas of airfields. This table, developed from table 9-26, page 9-53, shows the possible functional considerations for situations where either no landing mat, a light-duty mat, or a medium-duty mat may be employed. (Landing mats are discussed in TM 5-330/AFM 86-3, Volume II and TM 5-337.) As an example of the use of this table, consider the construction of the "heavy lift in the support area."

Table 9-25. Stabilization functions pertinent to theater-of-operations airfields

| Airfield Type | Possible Functions of Stabilization for Indicated Areas | | | | | |
| | Traffic Areas [a] | | | Nontraffic Areas [b] | | |
	Strength Improvement	Dust Control	Soil Water-proofing	Strength Improvement	Dust Control	Soil Water proofing
Close Battle Area 2,000' No mat	X	X	X		X	
With LM					X	
2,500' Liaison Surveillance		X	X		X	
No mat	X	X	X		X	
With LM					X	
3,000' No mat	X	X	X		X	
With LM					X	
Rear Area Liaison 3,500' Surveillance 3,500*		X	X		X	
No mat		X	X		X	X
With LM					X	X
3,500' No mat	X	X	X	X	X	X
With LM	X			(X)	(X)	(X)
Wiih MM				(X)	(X)	(X)
6,000' No mat	X	X	X	X	X	X
With LM	X			(X)	(X)	(X)

Table 9-25. Stabilization functions pertinent to theater-of-operations airfields

Airfield Type	Possible Functions of Stabilization for Indicated Areas					
	Traffic Areas [a]			Nontraffic Areas [b]		
	Strength Improvement	Dust Control	Soil Water-proofing	Strength Improvement	Dust Control	Soil Water proofing
Wiih MM	X			(X)	(X)	(X)
Tactical Rear area 6.000' No mat				X	X	X
With LM	X			(X)	(X)	(X)
With MM				(X)	(X)	(X)
COMMZ 8,000' No mat				X	X	X
With LM	X			(X)	(X)	(X)
With MM	X			(X)	(X)	(X)
COMMZ Liaison 3,000' No mat	X	X	X		X	X
With LM					X	X
6,000' No mat	X	X	X	X	X	X
With LM	X			(X)	(X)	(X)
With MM	X			(X)	(X)	(X)
10,000' No mat				X	X	X
With LM	X			(X)	(X)	(X)
With MM	X			(X)	(X)	(X)

Notes. 1. References to the use or no use of mat for a particular airfield apply only to heir use in traffic areas.
2. X = functions for which stabilization may be considered; blank space = no function for stabiliza ion, (X) = function will exist only if landing mat is not used in nontraffic areas.
[a] Traffic areas include runways, taxiways, and aprons.
[b] Nontraffic areas include overruns, shoulders, and peripheral zones that receive little or no traffic.

9-121. Referring to the traffic areas, a certain minimum strength is required for unsurfaced-soil operations (that is, without a landing mat) or if either the light duty mat (LM) or the medium duty mat (MM) is used. If the existing soil strength is not adequate, stabilization for strength improvement may be considered either to sustain unsurfaced operations or to be a necessary base for the landing mat. Further, if no mat is used, stabilization might be needed only to provide dust control and/or soil waterproofing. If a landing mat is used, however, the functions of dust control and soil waterproofing would be satisfied and stabilization need not be considered in any event. Possible stabilization functions for nontraffic areas have been shown in a similar manner. For certain airfields, such as the "light lift in the battle area," no function for strength improvement in either traffic or nontraffic areas is indicated. Such airfields have an AI requirement of 5 or more unsurfaced operations (see table 9-26). Site selection should be exercised in most instances to avoid areas of less than a 5 AI. For certain airfields, such as the "tactical in the support area," a landing mat or improved surfacing always will be provided. Therefore a "no mat" situation pertains only to the nontraffic areas.

Table 9-26. Basic airfield expedient surfacing requirements

Airfield Type	Typical Sector	Anticipated Service Life	Possible Using Aircraft US Type	Runway, Taxiway, and Apron Surfacing Requirements for Airfield Index of						Overrun Area and Shoulder Surfacing Requirements for Airfield Index of					
				5-6	6-8	8-10	10-12	12-15	>15	5-6	6-8	8-10	10-12	12-15	>15
Close Battle Area 2,000'	Brigade base	1 week	C-130[a] C-123	I (4c)	U	U	U	U	U	U (4c)	U	U	U	U	U
2.500' Liaison	Division or brigade base	1-4 weeks	O-1[a]	U (3c)	U	U	U	U	U	U (2c)	U	U	U	U	U
Surveillance	Division base		OV-1[a]	LM (3c)(5c)	LM	U	U	U	U	U (4c)(3c)	U	U	U	U	U
3.000'	Division base		C-130[a] C-7A	LM (5c)	LM	U	U	U	U	U (4c)	U	U	U	U	U
Rear Area Liaison 3,500'	Corps area	1-6 months	O-1[a]	U (3c)	U	U	U	U	U	U (2c)	U	U	U	U	U
Surveillance 3,500'	Corps area		OV-1[a]	LM (3c)	LM	U	U	U	U	U (5c)	U	U	U	U	U
3,500'	Corps area or FASCOM		C-130a	MM (4c)	MM	LM	U	U	U	LM (3c)	U				
			C-7A								U				
6,000'	Corps area or FASCOM		C-141[a]			MM	LM	LM	U	LM	LM	LM	U	U	U
				I	MM (6c)					(3c)					
Tactical Rear area 6,000'	Corps area or FASCOM	1-6 months	F-4C[a]	MM	MM	MM	LM	LM	LM	LM	LM	LM	LM	U	U
COMMZ 8.000'		6-24 months	F-4C[a]	I	MM (7c)	MM	MM	LM	LM	LM (4c)	LM	LM	LM	U	U
COMMZ Liaison 3000*	COMMZ[b]	6-24 months	OV-1[a]	LM (5c)	LM	LM	U	U	U	U (5c)	U	U	U	U	U
6,000'			C-130[a]	I	MM (6c)	MM	LM	LM	U	LM (4c)	LM	U	U	U	U
10.000'			C-141[a]	I	I		MM	LM	LM	MM (3c)	LM	LM	LM	U	U
			C-135[a]	I	I	MM (8c)									

Note. U = unsurfaced soil with or without membrane.   MM = medium duty mat,   LM = light duty mat, and
I = subgrade airfield index must be increased to that required for medium duty mat

[a] Particular aircraft that is critical in load and/or ground run, from which area requirements, geometries, and expedient surfacing requirements were developed.

[b] Communications zone.

[c] Minimum Airfield Index.

DESIGN REQUIREMENTS FOR STRENGTH IMPROVEMENT

9-122.   Where stabilization for strength improvement is considered, certain basic design requirements, in terms of strength and thickness of a stabilized soil layer on a given subgrade, must be met. The strength and thickness requirements vary depending on the operational traffic parameters and the strength of the soil directly beneath the stabilized soil layer. Since the traffic parameters are known for each airfield type, a minimum strength requirement for the stabilized soil layer can be specified for each airfield based on unsurfaced-soil criteria. For any given subgrade condition, the thickness of a minimum-strength, stabilized soil layer necessary to prevent overstress of the subgrade also can be determined. Table 9-27 gives design requirements for traffic and nontraffic areas of different airfield types for which stabilization may be used for strength improvement. As seen, the minimum-strength requirement in terms of AI is a function only of the applied traffic for a particular airfield and is independent of the subgrade strength. However, the thickness is a direct function of the underlying subgrade strength.

9-123. Proper evaluation of the subgrade is essential for establishing thickness requirements. In evaluating the subgrade for stabilization purposes, a representative AI strength profile must be established to a depth that would preclude the possibility of overstress in the underlying subgrade. This depth varies depending on the—

- Airfield.
- Pattern of the profile itself.
- Manner of stabilization.

9-124. In this regard, the thickness data given in table 9-27 can be used also to provide guidance in establishing an adequate strength profile. Generally, a profile to a depth of 24 inches is sufficient to indicate the strength profile pattern. However, if a decrease in strength is suspected in greater depths, the strength profile should be obtained to no less than the thickness indicated in table 9-27 under the 5-6 subgrade AI column for the appropriate airfield.

9-125. The use of table 9-27 to establish the design requirements for soil stabilization is best illustrated by the following example: Assume that a rear area 3,500' airfield is to be constructed and that a subgrade AI evaluation has been made from which a representative profile to a sufficient depth can be established. One of three general design cases can be considered de-pending on the shape of the strength profile:

- The first case considers constant strength with depth; therefore, the required thickness is read directly from table 9-27 under the appropriate subgrade AI column. Thus, in the example, if a subgrade AI of 8 is measured, the required thickness of a stabilized soil layer if no landing mat were used would be 18 inches. The required minimum strength of this stabilized soil layer is an AI of 15. If the light landing mat were used, a 6-inch-thick layer with a minimum AI of 10 would be required as a base overlying the subgrade AI of 8.
- The second case considers an increase in strength with depth; therefore, the required thickness of stabilization may be considerably less than indicated in the table. For this example, assume that the AI increases with depth as shown in figure 9-29, page 9-56. A stabilized layer can be provided either by building up a compacted base on top of the existing ground surface or by treating the in-place soil. Because of this, each situation represents a somewhat different design problem.

9-126. An in-place treatment is analogous to replacing the existing soil to some depth with an improved quality material. Where strength increases with depth, the point at which thickness is compatible with the strength at that particular point must be determined. This point can be determined graphically simply by superimposing a plot of the thickness design requirements versus subgrade AI (see table 9-27) directly on the strength profile plot. This procedure is shown in figure 9-29. The depth at which the two plots intersect is the design thickness requirement for a stabilized-soil layer. In the example, a thickness of 9.5 inches (or say 10 inches) is required.

Table 9-27. Design requirements for strength improvement

Pertinent Airfield Type	Traffic Areas							NonTraffic Areas						
	Minimum Stabilization Strength Required [a]	Thickness of Soil Stabilization Required (in) for Subgrade Airfield Index of						Minimum Stabilization Strength Required [a]	Thickness of Soil Stabilization Required (in) for Subgrade Airfield Index of					
		5-6	6-8	8-10	10-12	12-15	>15		5-6	6-8	8-10	10-12	12-15	>15
Close Battle Area														
2,000'														
No mat	6	11												
2,500*														
Surveillance														
No mat	8	8	7											
No mat	8	15	12											
Rear Area														
3,500'														
No mat	10	22	18	12				6	10					
With LM	8	10	6											
6,000'														
No mat	15	40	29	19	14	10		9	10	9	7			
With LM	10	28	18	8										
With MM	6	18												
Tactical														
Rear area 6,000'														
No mat								11	7	6	5	5		
With LM	10	8	6	5										
COMMZ 8,000'														
No mat								13	10	8	6	5		
With LM	12	12	9	5	5									
With MM	8	7	5											
COMMZ														
Liaison 3,000'														
No mat	10	12	10	8										
6,000														
No mat	15	30	26	18	14	10		8	12	10				
With LM	10	18	14	6										
With MM	6	8												
10,000'														
No mat								12	14	12	10	9		
With LM	12	42	33	16	5			6	5					
With MM	8	33	22											

Note. Practical limitations generally will preclude the consideration of chemical stabilization methods to develop an improved quality layer in excess of 16 in thick

[a] Airfield Index
LM = Light duty mat
MM = Medium duty mat

9-127. If a compacted base of a select borrow soil is used to provide a stronger layer on the subgrade shown in figure 9-29 the thickness must again be consistent with the strength at some depth below the surface of the placed base-course layer. Since the base-course layer itself will be constructed to a minimum AI of 15, the weakest point under the placed base will be at the surface of the existing ground, or in this instance an AI of 8. Using this value, table 9-27 gives a thickness of 18 inches of base course. Compaction of the existing ground would be beneficial in terms of thickness requirements if it would increase the critical subgrade strength to a higher value. If, for example, the minimum AI of the existing ground could be increased from 8 to 12, the thickness of base required would be reduced to 10 inches (see table 9-27, page 9-55).

- The third case considers a decrease in strength with depth. The strength profile shown in figure 9-30 indicates a crust of firm material over a significantly weaker zone of soil beneath. In this example, the importance of proper analysis of subgrade conditions is stressed. If strength data were obtained to less than 30 inches, the adequacy of the design could not be fully determined.

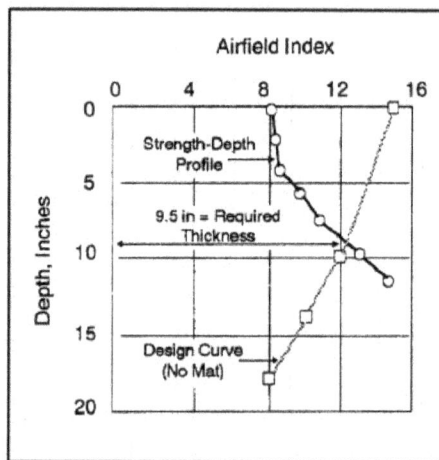

Figure 9-29. Thickness design procedure for subgrades that increase in strength with depth

9-128. Consider again an in-place stabilization process. Although the strength profile and design curve intersect initially at a shallow depth (about 3 inches) (see figure 9-30), the strength profile does not remain to the right of the design curve. This indicates that the design requirement has been satisfied. The second and final intersection occurs at 24 inches. Since there is no indication of a further decrease in strength with depth, a thickness of 24 inches is therefore required.

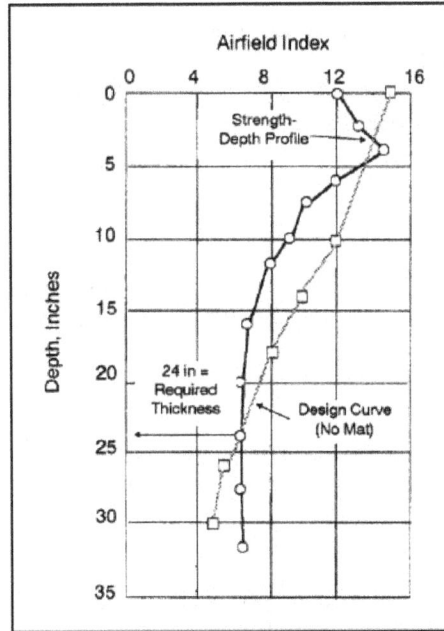

Figure 9-30. Thickness design procedure for subgrades that decrease in strength with depth

9-129. In the case of a compacted base placed on a subgrade that decreases in strength with depth, the procedure for determining the design thickness is more difficult. The design thickness can be determined by comparing the strength-depth profile with the design curve. If the measured AI at any given depth is less than the minimum requirement shown by the design curve, a sufficient thickness of improved quality soil must be placed on the existing ground surface to prevent overstress at that depth. However, the thickness of base necessary must be such that the requirements will be met at all depths. To satisfy this condition, the required thickness must be equal to the maximum difference, which will occur at a particular strength value, between the depth indicated by the design curve and the depth from the strength-depth profile. In the example shown in figure 9-30, this maximum difference occurs at an AI of 12. The difference is 10 inches, which is the required thickness for an improved quality base.

9-130. The same procedures described for a decrease in strength with depth can be used to derive the strength and thickness requirements for a base course under either an LM or MM. The thickness design requirements given herein are for stabilized soil layers having a minimum strength property to meet the particular airfield traffic need. Although the strength actually achieved may well exceed the minimum requirement, no consideration should be given to reducing the design thickness as given in table 9-27 or as developed by the stated procedures.

SECTION III - DUST CONTROL

EFFECTS OF DUST

9-131. Dust can be a major problem during combat (and training) operations. Dust negatively impacts morale, maintenance, and safety. Experience during Operation Desert Shield/Storm suggests that dust was a major contributor to vehicle accidents. It also accelerated wear and tear on vehicles and aircraft components.

9-132. Dust is simply airborne soil particles. As a general rule, dust consists predominantly of soil that has a particle size finer than 0.074 mm (that is, passing a Number 200 sieve).

9-133. The presence of dust can have significant adverse effects on the overall efficiency of aircraft by—

- Increasing downtime and maintenance requirements.
- Shortening engine life.
- Reducing visibility.
- Affecting the health and morale of personnel.

9-134. In addition, dust clouds can aid the enemy by revealing positions and the scope of operations.

DUST FORMATION

9-135. The presence of a relative amount of dust-size particles in a soil surface does not necessarily indicate a dust problem nor the severity of dust that will result in various situations. Several factors contribute to the generation, severity, and perpetuity of dust from a potential ground source. These include—

- Overall gradation.
- Moisture content.
- Density and smoothness of the ground surface.
- Presence of salts or organic matter, vegetation, and wind velocity and direction.
- Air humidity.

9-136. When conditions of soil and environment are favorable, the position of an external force to a ground surface generates dust that exists in the form of clouds of various density, size, and height above the ground. In the case of aircraft, dust may be generated as a result of erosion by propeller wash, engine exhaust blast, jet-blast impingement, and the draft of moving aircraft. Further, the kneading and abrading action of tires can loosen particles from the ground surface that may become airborne.

9-137. On unsurfaced roads, the source of dust may be the roadway surface. Vehicle traffic breaks down soil structure or abrades gravel base courses, creating fine-grained particles that readily become airborne when trafficked.

DUST PALLIATIVES

9-138. The primary objective of a dust palliative is to prevent soil particles from becoming airborne. Dust palliatives may be required for control of dust on nontraffic or traffic areas or both. If a prefabricated landing mat, membrane, or conventional pavement surfacing is used in the traffic areas of an airfield, the use of dust palliatives would be limited to nontraffic areas. For nontraffic areas, a palliative is needed that can resist the maximum intensity of air blast impingement by an aircraft or the prevailing winds. Where dust palliatives provide the necessary resistance against air impingement, they may be totally unsuitable as wearing surfaces. An important factor limiting the applicability of a dust palliative in traffic areas is the extent of surface rutting that will occur under traffic. If the bearing capacity allows the soil surface to rut under traffic, the effectiveness of a shallow-depth palliative treatment could be destroyed rapidly by breakup and subsequent stripping from the ground surface. Some palliatives tolerate deformations better than others, but normally ruts inches deep result in the virtual destruction of any thin layer or shallow depth penetration dust palliative treatment.

9-139. The success of a dust-control program depends on the engineer's ability to match a dust palliative to a specific set of factors affecting dust generation. These factors include—

- Intensity of area use.
- Topography.
- Soil type.
- Soil surface features.
- Climate.

INTENSITY OF AREA USE

9-140. Areas requiring dust-control treatments should be divided into traffic areas based on the expected amount of traffic. The three classes of traffic areas are—

- Nontraffic.
- Occasional traffic.
- Traffic.

Nontraffic Areas

9-141. These areas require treatment to withstand air-blast effects from wind or aircraft operations and are not subjected to traffic of any kind. Typical nontraffic areas include—

- Graded construction areas.
- Denuded areas around the periphery of completed construction projects.
- Areas bordering airfield or heliport complexes.
- Protective petroleum, oil, and lubricant (POL) dikes.
- Magazine embankments or ammunition storage barricades.
- Bunkers and revetments.
- Cantonment, warehouse, storage, and housing areas, excluding walkways and roadways.
- Unimproved grounds.
- Areas experiencing wind-borne sand.

Occasional-Traffic Areas

9-142. Besides resisting helicopter rotor downwash, aircraft propwash, and air blast from jet engines, these areas are also subjected to occasional traffic by vehicles, aircraft, or personnel. Vehicle traffic is limited to occasional, non-channelized traffic. Typical occasional-traffic areas include the following:

- Shoulders and overruns of airfields.
- Shoulders hover lanes, and peripheral areas of heliports and helipads.
- Nontraffic areas where occasional traffic becomes necessary.

Traffic Areas

9-143. Areas subjected to regular channelized traffic by vehicles, aircraft, or personnel. Properly treated traffic areas resist the effects of air blasts from fixed- or rotary-wing aircraft. Typical traffic areas include:

- Roadways and vehicle parking areas.
- Walkways.
- Open storage areas.
- Construction sites.
- Runways, taxiways, shoulders, overruns, and parking areas of airfields.
- Hover lanes and landing and parking pads of heliports.
- Tank trails.

TOPOGRAPHY

9-144. Dust palliatives for controlling dust on flat and hillside areas are based on the expected traffic, but the specific palliative selected may be affected by the slope. For example, a liquid palliative may tend to run off rather than penetrate hillside soils, which degrades the palliative's performance.

9-145. Divide the area to be treated into flat and hillside areas. Flat is defined as an average ground slope of 5 percent or less, while hillside refers to an average ground slope greater than 5 percent. Particular areas can be given special attention, if required.

SOIL TYPE

9-146. Soil type is one of the key features used to determine which method and material should be used for dust control. Soils to be treated for dust control are placed into five general descriptive groupings based on the USCS. They are—

- Silts or clays (high LL) (types (CH), (OH), and (MH)).
- Silts or clays (low LL) (types (ML), (CL), (ML-CL), and (OL)).
- Sands or gravels (with fines) (types (SM), (SC), (SM-SC), (GM), (GC), (GM-GC), and (GW-GM)).
- Sands (with little or no fines) (types (SW-SM), (SP), and (SW)).
- Gravels (with little or no fines) (types (GP) and (GW)).

SOIL SURFACE FEATURES

9-147. Soil surface features refer to both the state of compaction and the degree of soil saturation in the area to be treated. Loose surface conditions are suitable for treatment in nontraffic or occasional areas only. Firm surface conditions are suitable for treatment under any traffic condition.

Loose and Dry or Slightly Damp Soil

9-148. The surface consists of a blanket (¼ to 2 inches thick) of unbound or uncompacted soil, overlying a relatively firm subgrade and ranging in moisture content from dry to slightly damp.

Loose and Wet or Scurry Soil

9-149. The surface condition consists of a blanket (¼ to 2 inches thick) of unbound or uncompacted soil, overlying a soft to firm subgrade and ranging in moisture content from wet to slurry consistency. Soil in this state cannot be treated until it is dried to either a dry or slightly damp state.

Firm and Dry or Slightly Damp Soil

9-150. The surface condition consists of less than a -inch-thick layer of loose soil, ranging in moisture content from dry to slightly damp and overlying a bound or compacted firm soil subgrade.

Firm and Wet Soil

9-151. The surface condition resembles that of the previous category. This soil must be dried to either a dry or slightly damp state before it can be treated.

CLIMATE

9-152. Climatic conditions influence the storage life, placement, curing, and aging of dust palliatives. The service life of a dust palliative may vary with the season of the year. For example, salt solutions become ineffective during the dry season when the relative humidity drops below 30 percent.

DUST-CONTROL METHODS

9-153. The four general dust-control-treatment methods commonly used are—

- Agronomic.
- Surface penetrant.
- Admix.
- Surface blanket.

AGRONOMIC

9-154. This method consists of establishing, promoting, or preserving vegetative cover to prevent or reduce dust generation from exposed soil surfaces. Vegetative cover is often considered the most satisfactory form of dust palliative. It is aesthetically pleasing, durable, economical, and considered to be permanent. Some agronomic approaches to dust control are suitable for theater-of-operations requirements. Planning construction to minimize disturbance to the existing vegetative cover will produce good dust-palliative results later.

9-155. Agronomic practices include the use of—

- Grasses.
- Shelter belts.
- Rough tillage.

9-156. Grounds maintenance management and fertilizing will help promote the development of a solid ground cover. Agronomic methods are best suited for nontraffic and occasional-traffic areas; they are not normally used in traffic areas.

Grasses

9-157. Seeding, sprigging, or sodding grasses should be considered near theater-of-operations facilities that have a projected useful life exceeding 6 months. Combining mulch with seed promotes quicker establishment of the grass by retaining moisture in the soil. Mulching materials include straw, hay, paper, or brush. When mulches are spread over the ground, they protect the soil from wind and water erosion. Mulches are effective in preventing dust generation only when they are properly anchored. Anchoring can be accomplished by disking or by applying rapid curing (RC) bituminous cutbacks or rapid setting (RS) asphalt emulsions. Mulch is undesirable around airports and heliports since it may be ingested into jet engines, resulting in catastrophic engine failure.

Shelter Belts

9-158. They are barriers formed by hedges, shrubs, or trees that are high and dense enough to significantly reduce wind velocities on the leeward side. Their placement should be at right angles to the prevailing winds. While a detailed discussion of shelter-belt planning is beyond the scope of this manual, shelter belts should be considered for use on military installations and near forward landing strips (FLS) constructed for contingency purposes in austere environments (such as those constructed in Central America).

Rough Tillage

9-159. This method consists of using a chisel, a lister, or turning plows to till strips across nontraffic areas. Rough tillage works best with cohesive soils that form clods. It is not effective in cohesionless soils and, if used, may contribute to increased dust generation.

SURFACE PENETRANT

9-160. The surface penetration method involves applying a liquid dust palliative directly to the soil surface by spraying or sprinkling and allowing the palliative to penetrate the surface. The effectiveness of this method depends on the depth of penetration of the dust palliative (a function of palliative viscosity and soil permeability). Using water to prewet the soil that is to be treated enhances penetration of the palliative.

9-161. Surface penetrants are useful under all traffic conditions; however, they are only effective on prepared areas (for example, on unsurfaced gravel roads). Dust palliatives that penetrate the soil surface include—

- Bitumens.
- Resins.

- Salts.
- Water.

Bitumens

9-162.   Conventional types of bituminous materials that may be used for dust palliatives include—

- Cutback asphalts.
- Emulsified asphalts.
- Road tars.
- Asphaltic penetrative soil binder (APSB).

9-163.   These materials can be used to treat both traffic and nontraffic areas. All bituminous materials do not cure at the same rate. This fact may be of importance when they are being considered for use in traffic areas. Also, bituminous materials are sensitive to weather extremes. Usually bituminous materials impart some waterproofing to the treated area that remains effective as long as the treatment remains intact (for example, as placed or as applied). Bituminous materials should not be placed in the rain or when rain is threatening.

9-164.   A cutback asphalt (cutbacks) is a blend of an asphalt cement and a petroleum solvent. These cutbacks are classified as RC, medium curing (MC), and slow curing (SC), depending on the type of solvent used and its rate of evaporation. Each cutback is further graded by its viscosity. The RC and SC grades of 70 and 250, respectively, and MC grades of 30, 70, and 250 are generally used. Regardless of classification or grade, the best results are obtained by preheating the cutback. Spraying temperatures usually range from 120 to 300 degrees Fahrenheit. The actual range for a particular cutback is much narrower and should be requested from the supplier at the time of purchase. The user is cautioned that some cutbacks must be heated above their flash point for spraying purposes; therefore, no smoking or open flames should be permitted during the application or the curing of the cutback. The MC-30 grade can be sprayed without being heated if the temperature of the asphalt is 80 degrees Fahrenheit or above. A slightly moist soil surface assists penetration. The curing time for cutbacks varies with the type. Under favorable ground temperature and weather conditions, RC cures in 1 hour, MC in 3 to 6 hours, and SC in 1 to 3 days. In selecting the material for use, local environmental protection regulations must be considered.

9-165.   Asphalt emulsions (emulsions) are a blend of asphalt, water, and an emulsifying agent. They are available either as anionic or cationic emulsions. The application of emulsions at ambient temperatures of 80 degrees Fahrenheit or above gives the best results. Satisfactory results may be obtained below this temperature, especially if the application is made in the morning to permit the warming effects of the afternoon sun to aid in curing. Emulsions should not be placed at temperatures below 50 degrees Fahrenheit. Emulsions placed at temperatures below freezing will freeze, producing a substandard product. For best results in a freezing environment, emulsions should be heated to between 75 and 130 degrees Fahrenheit. The temperature of the material should never exceed the upper heating limit of 185 degrees Fahrenheit because the asphalt and water will separate (break), resulting in material damage. Emulsions generally cure in about 8 hours. The slow setting (SS) anionic emulsions of grades SS-1 and SS-1h may be diluted with 1 to 5 or more parts water to one part emulsified asphalt by volume before using. As a general rule, an application of 3 parts water to 1 part emulsion solution is satisfactory. The slow-setting cationic emulsions or grades cationic slow setting (CSS)-1 and CSS-1h are easiest to use without dilution. If dilution is desired, the water used must be free of any impurities, minerals, or salts that might cause separation (breaking) of the emulsion within the distribution equipment.

9-166.   Road tars (RTs) (tars) are viscous liquids obtained by distillation of crude tars obtained from coal. Tars derived from other basic materials are also available but are not normally used as soil treatments. Tars are graded by viscosity and are available in grades ranging from 1 to 12. They are also available in the road tar cutback (RTCB) form of viscosity grades 5 and 6 and in the emulsified form. Tar emulsions are difficult to prepare and handle. The low-viscosity grades RT-1 and RT-2 and the RTCB grades can be applied at temperatures as low as 60 degrees Fahrenheit without heating. The tar cutbacks generally have better penetrating characteristics than asphalts and normally cure in a few hours. Tars produce excellent surfaces, but curing proceeds very slowly. Several days or even weeks may be required to obtain a completely cured layer. Tars are susceptible to temperature changes and may soften in hot weather or become brittle in cold weather.

9-167. APSB, a commercial product, is a special liquid asphalt composed of a high penetration grade of asphalt and a solvent blend of kerosene and naphtha. It is similar in character to a standard low-viscosity, medium-curing liquid asphalt, but it differs in many specific properties. The APSB is suitable for application to soils that are relatively impervious to conventional liquid asphalts and emulsion systems. Silts and moderately plastic clays (to a PI of 15) can be treated effectively. Curing time for the APSB is 6 to 12 hours under favorable ground temperature and weather conditions. On high-plasticity solids (with a PI greater than 15), the material remains on the surface as an asphalt film that is tacky at a ground temperature of approximately 100 degrees Fahrenheit and above. The APSB must be heated to a temperature between 130 to 150 degrees Fahrenheit to permit spraying with an asphalt distributor.

Resins

9-168. These dust palliatives may be used as either surface penetrants or surface blankets. They have a tendency to either penetrate the surface or form a thin surface film depending on the type of resin used, the soil type, and the soil condition. The materials are normally applicable to nontraffic areas and occasional-traffic areas where rutting will not occur. They are not recommended for use with silts and clays.

9-169. Resin-petroleum-water emulsions are quite stable and highly resistant to weathering. A feature of this type of dust palliative is that the soil remains readily permeable to water after it is treated. This type of product is principally manufactured under the trade name Coherex. Application rates range from 0.33 to 0.5 gallon per square yard. The material may be diluted for spraying using 4 parts water to 1 part concentrate. This material is primarily suited for dry sandy soils; it provides unsuitable results when used on silty and clayey soils.

9-170. Lignin is a by-product of the manufacture of wood pulp. It is soluble in water and therefore readily penetrates the soil. Its solubility also makes it susceptible to leaching from the soil; thus, application is repeated as necessary after rainfall. Lignin is readily available in the continental United States and certain other sections of the world. It is useful in areas where dust control is desirable for short periods of time; it is not recommended for use where durability is an important factor. The recommended application rate is 1 gallon per square yard of a resinous solution of 8 percent solid lignin sulphite.

9-171. Concrete curing compounds can be used to penetrate sands that contain little or no silts or clays. This material should be limited to areas with no traffic. The high cost of this material is partly offset by the low application rate required (0.1 to 0.2 gallon per square yard). Standard asphalt pressure distributors can be used to apply the resin; however, the conventional spray nozzles should be replaced with nozzles with smaller openings to achieve a uniform distribution at the low application rate.

Salts

9-172. Salts in water emulsions have been used with varying success as dust palliatives. Dry calcium chloride ($CaCl_2$) is deliquescent and is effective when the relative humidity is about 30 percent or greater. A soil treated with calcium chloride retains more moisture than the untreated soil under comparable drying conditions. Its use is limited to occasional-traffic areas. Sodium chloride (NaCl) achieves some dust control by retaining moisture and also by some cementing from salt crystallization. Both calcium chloride and sodium chloride are soluble in water and are readily leached from the soil surface; thus, frequent maintenance is required. Continued applications of salt solutions can ultimately build up a thin, crusted surface that will be fairly hard and free of dust. Most salts are corrosive to metal and should not be stored in the vehicle used for application. Magnesium chloride ($MgCl_2$) controls dust on gravel roads with tracked-vehicle traffic. Best results can be expected in areas with occasional rainfall or where the humidity is above 30 percent. The dust palliative selected and the quantity used should not exceed local environmental protection regulations.

Water

9-173. As a commonly used (but very temporary) measure for allaying dust, a soil surface can be sprinkled with water. As long as the ground surface remains moist or damp, soil particles resist becoming airborne. Depending on the soil and climate, frequent treatment may be required. Water should not be applied to clay soil surfaces in such quantity that puddles forms since a muddy or slippery surface may result where the soil remains wet.

ADMIX

9-174. The admix method involves blending the dust palliative with the soil to produce a uniform mixture. This method requires more time and equipment than either the penetration or surface blanket methods, but it has the benefit of increasing soil strength.

9-175. Normally, a minimum treatment depth of 4 inches is effective for traffic areas and 3 inches for other areas. The admixture can be mixed in place or off site. Typical admixture dust palliatives include—

- Portland cement.
- Hydrated lime.
- Bituminous materials.

In-Place Admixing

9-176. In-place admixing is the blending of the soil and a dust palliative on the site. The surface soil is loosened (if necessary) to a depth slightly greater than the desired thickness of the treated layer. The dust palliative is added and blended with the loosened surface soil, and the mixture is compacted. Powders may be spread by hand or with a mechanical spreader; liquids should be applied with an asphalt distributor. Mixing equipment that can be used includes—

- Rotary tillers.
- Rotary pulverizer-mixers.
- Graders.
- Scarifiers.
- Disk harrows.
- Plows.

9-177. Admixing and/or blending should continue until a uniform color of soil and dust palliative mixture, both horizontally and vertically, is achieved. The most effective compaction equipment that can be used is a sheepsfoot or rubber-tired rollers. The procedure for in-place admixing closely resembles the soil stabilization procedure for changing soil characteristics and soil strength used in road construction. For dust control on a nontraffic area, adequate compaction can be achieved by trafficking the entire surface with a 5-ton dual-wheel truck. For all other traffic situations, the procedure should follow TM 5-822-4. This procedure is time-consuming and requires the use of more equipment than the other three. Following placement, admixing, and compaction, a minimum of seven days is required for curing.

9-178. Two cementing-type powders (portland cement and hydrated lime) are primarily used to improve the strength of soils. However, when they are admixed with soils in relatively small quantities (2 to 5 percent by dry soil weight), the modified soil is resistant to dusting. Portland cement is generally suited to all soil types, if uniform mixing can be achieved, whereas hydrated lime is applicable only to soils containing a high percentage of clay. The compacted soil surface should be kept moist for a minimum of 7 days before allowing traffic on it.

9-179. Bituminous materials are more versatile than cementing materials in providing adequate dust control and waterproofing of the soil. Cutbacks, emulsion asphalts, and road tars can all be used successfully. The quantity of residual bituminous material used should range from 2 to 3 percent of dry soil weight (for soils having less than 30 percent passing the Number 200 sieve) to 6 to 8 percent (for soils having more than 30 percent fine-grained soils passing the Number 200 sieve). The presence of mica in a soil is detrimental to the effectiveness of a soil-bituminous material admixture. There are no simple guides or shortcuts for designing mixtures of soil and bituminous materials. The maximum effectiveness of soil-bituminous material admixtures can usually be achieved if the soil characteristics are within the following limits:

- The PI is ≤ 10.
- The amount of material passing the Number 200 sieve is ≤ 30 percent by weight.

9-180.   This data and additional construction data can be found in TM 5-822-4. Traffic should be detoured around the treated area until the soil-bituminous material admixture cures.

9-181.   Cutback asphalt provides a dust-free, waterproof surface when admixed into soil to depths of 3 inches or more on a firm subgrade. More satisfactory results are obtained if the cutback asphalt is preheated before using it. Soils should be fairly dry when cutback asphalts are admixed. When using SC or MC types of cutback asphalt, aerate the soil-asphalt mixture to allow the volatiles to evaporate.

9-182.   Emulsified asphalts are admixed with a conditioned soil that allows the emulsion to break before compaction. A properly conditioned soil should have a soil moisture content not to exceed 5 percent in soils having less than 30 percent passing the Number 200 sieve. Emulsified asphalts, particularly the cationics (CSS-1 or CSS-lb), are very sensitive to the surface charge of the aggregate or soil. When they are used improperly, the emulsion may break prematurely or after some delay. The slow-setting anionic emulsions of grades SS-1 and SS-1h are less sensitive.

9-183.   Road tars with RT and RTCB grades can be used as admixtures in the same manner as other bituminous materials. Road tar admixtures are susceptible to temperature changes and may soften in hot weather or become brittle in cold weather.

Off-Site Admixing

9-184.   Off-site admixing is generally used where in-place admixing is not desirable and/or soil from another source provides a more satisfactory treated surface. Off-site admixing may be accomplished with a stationary mixing plant or by windrow-mixing with graders in a central working area. Processing the soil and dust palliative through a central plant produces a more uniform mixture than in-place admixing. The major disadvantage of off-site operations is having to transport and spread the mixed material.

SURFACE BLANKET

9-185.   The principle of the surface blanket method is to place a "blanket" cover over the soil surface to control dust. The three types of materials used to form the blanket are—

- Minerals (aggregates).
- Synthetics (prefabricated membranes and meshes).
- Liquids (bituminous or polyvinyl acetate liquids).

9-186.   These materials may be used alone or in the combinations discussed later.

9-187.   The type of treatment used dictates the equipment required. However, in all cases, standard construction equipment can be used effectively to place any of the blanket materials. Mechanized equipment should be used wherever possible to assure uniformity of treatment.

9-188.   The surface blanket method is applicable to nontraffic, occasional-traffic, and traffic areas. Aggregate, prefabricated membrane, and mesh treatments are easy to place and can withstand considerable rutting. The other surface blanket methods only withstand considerable rutting. Once a surface blanket treatment is torn or otherwise compromised and the soil exposed, subsequent traffic or air blasts increase the damage to the torn surface blanket and produce dust from the exposed soil. Repairs (maintenance) should begin as soon as possible to protect the material in place and keep the dust controlled.

Minerals (Aggregates)

9-189.   Aggregate is appropriate in arid areas where vegetative cover cannot be effectively established. It is effective as a dust palliative on nontraffic and occasional-traffic areas. The maximum recommended aggregate size is 2 inches; except for airfields and heliports. To prevent the aggregate from being picked up by the prop (propeller) wash, rotor wash, or air blast, 4-inch aggregate is recommended (see table 9-28, page 9-66).

Table 9-28. Recommended aggregate gradation for dust control on airfields and heliports

Sieve Designation		Percent Passing
100.0 mm	4.0 in	100
90.0 mm	3.5 in	90 - 100
63.0 mm	2.5 in	25 - 60
37.5 mm	1.5 in	0 - 15
19.0 mm	¾ in	0 - 5

Prefabricated Membrane

9-190. Membrane used to surface an area controls dust and even acts as a surface course or riding surface for traffic that does not rut the soil. When subjected to traffic, the membrane can be expected to last approximately 5 years. Minor repairs can be made easily. For optimum anchorage, the membrane should be extended into 2-foot-deep ditches at each edge of the covered area; then it should be staked in place and the ditches backfilled. Further details on the use and installation of prefabricated membranes can be obtained from TM 5-330/AFM 86-3, Volume II.

Prefabricated Mesh

9-191. Heavy, woven jute mesh, such as commonly used in conjunction with grass seed operations, can be used for dust control of nontraffic areas. The mesh should be secured to the soil by burying the edges in trenches and by using large U-shaped staples that are driven flush with the soil surface. A minimum overlap of 3 inches should be used in joining rolls of mesh; covered soil should be sprayed with a bituminous material. Trial applications are recommended at each site and should be adjusted to suit each job situation.

Bituminous Liquid

9-192. Single- or double-bituminous surface treatments can be used to control dust on most soils. A medium-curing liquid asphalt is ordinarily used to prime the soil before placing the surface treatment. Fine-grained soils are generally primed with MC-30 and coarse-grained soils with MC-70. After the prime coat cures, a bituminous material is uniformly applied, and gravel, slag, or stone aggregate is spread over the treated area at approximately 25 pounds per square yard. The types of bituminous materials, aggregate gradations, application rates, and methods of placing surface treatments are described in TM 5-822-8/AFM 88-6, Chapter 9. Single-or double-bituminous surface treatments should not be used where turf is to be established.

Polyvinyl Acetate (DCA 1295) (Without Reinforcement)

9-193. DCA 1295 has a slight odor and an appearance similar to latex paint. The material is diluted 3 parts DCA 1295 to 1 part water and cures in 2 to 4 hours under ideal conditions of moderate to high temperature and low relative humidity. A clear, flexible film forms on the treated surface. DCA 1295 can be sprayed with a conventional asphalt distributor provided modifications are made to the pump to permit external lubrications. The DCA 1295 can be used alone or over a fiberglass reinforcement. Adding fiberglass does not affect the basic application procedures or the curing characteristics of the DCA 1295. This material is suitable for use on nontraffic, occasional-traffic, and traffic areas. It is also effective when sprayed over grass seed to protect the soil until grass occurs. Uniform soil coverage is enhanced by sprinkling (prewetting) the surface with water.

Polyvinyl Acetate (DCA 1295) (With Reinforcement)

9-194. A fiberglass scrim material is recommended for use with the DCA 1295 when a reinforcement is desired. Fiberglass scrim increases the expected life of the dust-control film by reducing the expansion and contraction effects of weather extremes. The scrim material should be composed of fiberglass threads with a plain weave pattern of 10 by 10 (ten threads per inch in the warp direction and 10 threads per inch in the fill direction) and a greige finish. It should weigh approximately 1.6 ounces per square yard. Using scrim material does not create any health or safety hazards, and special storage facilities are not required. Scrim materials can be applied under any climatic conditions suitable for dispensing the DCA 1295. (Under special conditions, continuous strands of fiberglass may be chopped into ½-inch-long segments and blown over the area to be protected.) The best method of placement is for the fiberglass scrim material to be placed immediately after prewetting with water, followed by the DCA 1295.

Polypropylene-Asphalt Membrane

9-195. The polypropylene-asphalt membrane is recommended for use in all traffic areas. It has considerable durability and withstands rutting up to approximately 2 inches in depth. This system is a combination of a polypropylene fabric sprayed with an asphalt emulsion. Normally a cationic emulsion is used; however, anionic emulsions have also been used successfully. Several types of polypropylene fabric are commercially available.

9-196. This treatment consists of the following steps:

- Place a layer of asphalt (0.33 to 0.50 gallon per square yard) on the ground, and cover this with a layer of polypropylene fabric.
- Place 0.33 gallon per square yard of asphalt on top of the polypropylene.
- Apply a sand-blotter course.

9-197. This system does not require any rolling or further treatment and can be trafficked immediately.

9-198. Care should be taken during construction operations to ensure adequate longitudinal and transverse laps where two pieces of polypropylene fabric are joined. Longitudinal joints should be lapped a minimum of 12 inches. On a superelevated section, the lap should be laid so the top lap end is facing downhill to help prevent water intrusion under the membrane. On a transverse joint, the minimum overlap should be at least 24 inches. Additional emulsion should be on the top side of the bottom lap to provide enough emulsion to adhere to and waterproof the top lap. Figure 9-31, page 9-68, illustrates this process on tangential sections. Applying polypropylene on roadway curves requires cutting and placing the fabric as shown in figure 9-32, page 9-68. The joints in curved areas should be overlapped a minimum of 24 inches.

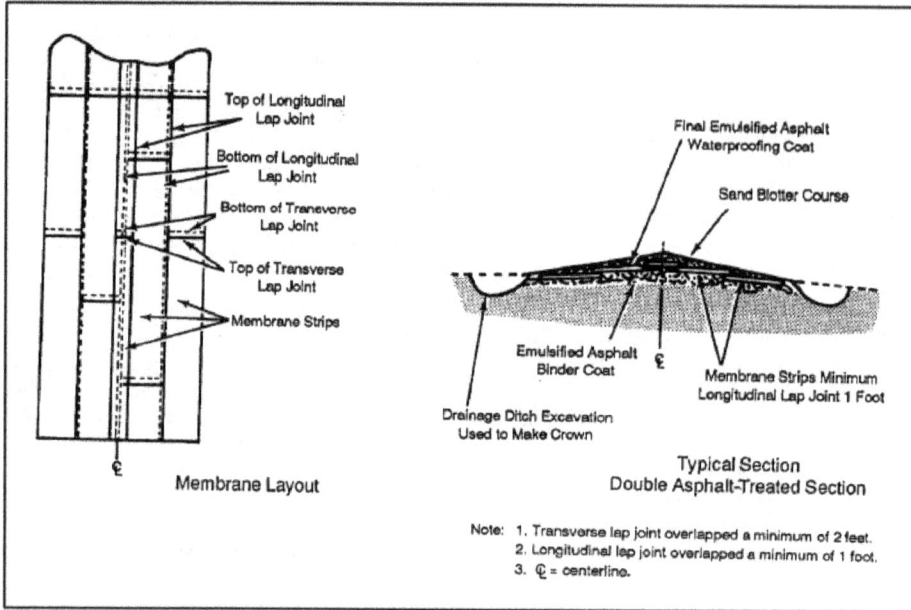

Figure 9-31. Polypropylene membrane layout for tangential sections

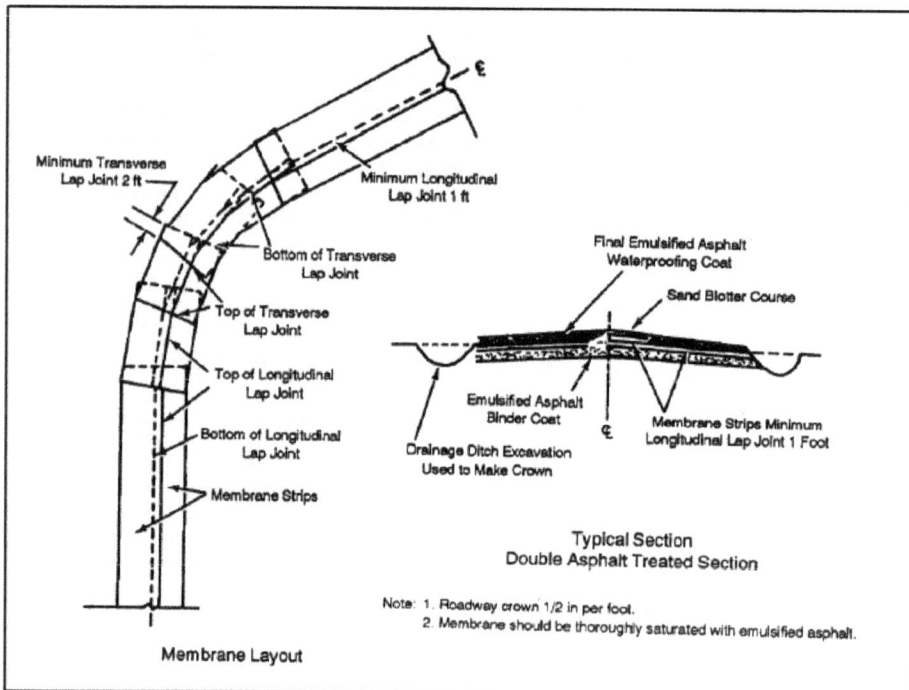

Figure 9-32. Polypropylene membrane layout for curved sections

SELECTION OF DUST PALLIATIVES

9-199. There are many dust palliatives that are effective over a wide range of soils and climatic conditions. Engineering judgment and material availability play key roles in determining the specific dust palliative to select. Tables 9-29 through 9-32, pages 9-69 through 9-73, were developed from evaluation of their actual performance to assist in the selection process. The dust palliatives and dust control methods are not listed in any order of effectiveness.

9-200. Where no dust palliative is listed for a particular dust control method, none was found to be effective under those conditions. For example, for the agronomic method, a dust palliative is not recommended for a loose, sandy soil with no binder, nor is a dust palliative recommended for the surface penetration of a firm, clay soil (see table 9-29, page 9-70, and table 9-30, page 9-71). Also, the agronomic method of dust control is not recommended for any traffic area (see table 9-31, page 9-72).

9-201. In table 9-32, page 9-73, numbers representing dust palliatives are listed in numerical order and separated by the dust- control method. This table includes the suggested rates of application for each dust-palliative; for instance, gallon per square yard for liquid spray on applications or gallon per square yard per inch for liquid (or pound per square yard per inch for powders) admix applications.

Table 9-29. Dust palliative numbers for dust control in nontraffic areas

Nontraffic Area — Flat*

Soil type or condition		Dust Control Method			
Condition	Soil type	Agronomic	Surface Penetrant	Admix	Surface Blanket
Loose and Dry (or Slightly Damp)	Silts or Clays (High Liquid Limit)	1, 2, 3, 4	5, 8, 11, 25	31, 37, 40	43, 46, 47, 48, 51, 54, 60
Loose and Wet (or Slurry)	Silts or Clays (Low Liquid Limit)	1, 2, 3, 4	5, 8, 15, 20, 22, 25	28, 34, 37, 40	43, 46, 47, 48, 51, 54, 60
Loose and Wet (or Slurry)	Sands and Gravels (with Fines)	1, 2, 3, 4	5, 8, 15, 20, 22, 25	28, 34, 37, 40	43, 46, 47, 48, 51, 54, 60
Loose and Wet (or Slurry)	Sands (with Little or No Fines)		5, 9, 12, 16, 18, 21, 22, 25	30, 34, 37, 40	43, 46, 47, 48, 51, 54, 60
Firm and Dry (or Slightly Damp)	Silts or Clays (High Liquid Limit)	1, 2, 3, 4		31, 37, 40	43, 46, 47, 51, 54, 57
Firm and Dry (or Slightly Damp)	Silts or Clays (Low Liquid Limit)	1, 2, 3, 4	8, 12, 15, 22, 25	28, 34, 37, 40	43, 46, 47, 51, 54, 57
Firm and Dry (or Slightly Damp)	Sands or Gravels (with Fines)	1, 2, 3, 4	8, 12, 15, 22, 25	28, 34, 37, 40	43, 46, 47, 51, 54, 57
Firm and Wet	Gravels (with Little or No Fines)	No Treatment Necessary			

Note. Numbers refer to palliative numbers listed in table 9-32, pages 9-68 through 9-70.
* Hillside applications for liquid dust palliatives should be reduced by half and then repeated, if necessary, to avoid runoff/waste.

Table 9-30. Dust palliative numbers for dust control in occasional-traffic areas

Occasional-traffic Area (Up to 80 mph Airblast) — Flat*

Soil type or condition		Loose and Dry (or Slightly Damp)		Loose and Wet (or Slurry)		Firm and Dry (or Slightly Damp)			Firm and Wet
		Silts or Clays (High Liquid Limit)	Silts or Clays (Low Liquid Limit)	Sands and Gravels (with Fines)	Sands (with Little or No Fines)	Silts or Clays (High Liquid Limit)	Silts or Clays (Low Liquid Limit)	Sands or Gravels (with Fines)	Gravels (with Little or No Fines)
Dust Control Method	Agronomic	1	1	1		1	1	1	No Treatment Necessary
	Surface Penetrant	6, 10, 13, 17, 26	6, 10, 13, 17, 26	6, 10, 13, 17, 23, 26	6, 10, 13, 17, 19, 26		6, 10, 13, 17, 23, 26	6, 10, 13, 17, 23, 26	
	Admix	32, 35, 38, 41		29, 32, 35, 38, 41	29, 35, 38, 41	32, 35, 38, 41	29, 32, 35, 38, 41	29, 32, 35, 38, 41	
	Surface Blanket	43, 45, 49, 52, 55, 61		43, 45, 49, 52, 55, 61	43, 45, 49, 52, 55, 61	43, 45, 49, 52, 55, 58	43, 45, 49, 52, 55, 58	43, 45, 49, 52, 55, 58	

Note. Numbers refer to palliative numbers listed in table 9-32, pages 9-68 through 9-70.
* Hillside applications for liquid dust palliatives should be reduced by half and then repeated, if necessary, to avoid runoff/waste.

Table 9-31. Dust palliative numbers for dust control in traffic areas

Traffic Area

Flat*

Loose and Wet (or Slurry)
- Silts or Clays (Low Liquid Limit)
- Sands and Gravels (with Fines)

Loose and Dry (or Slightly Damp)
- Silts or Clays (High Liquid Limit)
- Sands (with Little or No Fines)

Firm and Dry (or Slightly Damp)
- Silts or Clays (High Liquid Limit)
- Silts or Clays (Low Liquid Limit)
- Sands or Gravels (with Fines)

Firm and Wet
- Gravels (with Little or No Fines) — No Treatment Necessary

**

Dust Control Method	Silts or Clays (High Liquid Limit)	Silts or Clays (Low Liquid Limit) / Sands or Gravels (with Fines)
Surface Penetrant	33 36 39 42	7 10 14 17 24 27
Admix	44 50 53 56 62	30 38 39 42
Surface Blanket		44 50 56 59

Soil type or condition

Dust Control Method

Note. Numbers refer to palliative numbers listed in table 9-32, pages 9-68 through 9-70.
* Hillside applications for liquid dust palliatives should be reduced by half and then repeated, if necessary, to avoid runoff/waste.

Table 9-32. Dust palliative electives

Palliative Number	Material[a]	Rate of Application[b]	Estimated[c] Service Life
Agronomic Method			
1	Vegetative	See TM 5-830-21/AFM 88-17, Chapter 2, and AR 420-74	5 years to permanent
2	Mulch	See TM 5-830-21AFM 88-17, Chapter 2, and AR 420-74	6-12
3	Shelter belt	As determined by trial	3-5 years to permanent
4	Rough tillage	Each 25 to 100 feet	1-4
Surface-Penetrant Method			
	Bituminous materials		
	Cutback asphalt		
5	SC, MC, RC; grades 30-250	0.33	4-6
6	SC, MC, RC; grades 30-250	0.50	1-3
7	SC, MC, RC; grades 30-250	0.50	1
	Emulsified asphalt		
8	SS or CSS	0.33	4-6
9	SS or CSS	0.50	1-3
10	SS or CSS	0.50	1
	Road tar and road tar cutback		
11	RT grades 1-6, RTCB grades 5-6	0.33	2-4
12	↓	0.50	5-7
13		0.50	2-4
14		0.50	1-2
15	APSB	0.33	5-8
16	APSB	0.50	5-8
17	APSB	0.50	1-4
	Resinous materials		
18	Resin in water emulsion	0.50	3-9
19	Resin in water emulsion	0.50	1-3
20	Lignin (8% solids)	0.50	1-3
21	Concrete curing compound	0.33	1-3
	Brine materials		
22	Salt in water emulsion	0.33	10-14
23	Salt in water emulsion	0.50	8-12
24	Sail in water emulsion	0.67	6-12
25	Water	0.25	1 hour
26	Water	0.33	1 hour
27	Water	0.50	1 hour

[a].Users must ensure that materials comply with existing Environmental Protection Agency (EPA) regulations for the intended use.
[b] Rate of application in gallons per square yard unless o herwise noted.
[c] Estimated service life in months unless otherwise noted.

Table 9-32. Dust palliative electives

Palliative Number	Material[a]	Rate of Application[b]	Estimated[c] Service Life
	Admix Method*		
	Cementing materials		
28	Portland cement	1.5 lb per sq yd per in	4-6
29	Portland cement	2.5 lb per sq yd per in	4-6
30	Portland cement	4.0 lb per sq yd per in	4-6
31	Hydrated lime	1.5 lb per sq yd per in	4-6
32	Hydrated lime	2.5 lb per sq yd per in	4-6
33	Hydrated lime	4.0 lb per sq yd per in	4-6
	Bituminous materials		
	Cutback asphalt		
34	SC, MC, RC; grades 70-250	0.15 gal per sq yd per in	4-6
35	SC, MC, RC; grades 70-250	0.25 gal per sq yd per in	4-6
36	SC. MC, RC; grades 70-250	0.40 gal per sq yd per in	4-6
	Emulsified asphalt		
37	SS or CSS	0.10 gal per sq yd per in	4-6
38	SS or CSS	0.30 gal per sq yd per in	4-6
39	SS or CSS	0.50 gal per sq yd per in	4-6
	Road tar and road tar cutback		
40	RT grades 1-6; RTCB grades 5-6	0.15 gal per sq yd per in	4-6
41	RT grades 1-6; RTCB grades 5-6	0.25 gal per sq yd per in	4-6
42	RT grades 1-6; RTCB grades 5-6	0.40 gal per sq yd per in	4-6
	Surface-Blanket Method		
43	Aggregate	2 in thick	2-3 yr
44	Prefabricated membrane	1 layer	3-6
45	Prefabricated membrane	1 layer	6-9
46	Prefabricated membrane	1 layer	4-5 yr
47	Fabricated mesh	1 layer	9-12
48	Bituminous surface treatment	0.15 prime; 0.25-0.35 cover	1-2 yr
49	Bituminous surface treatment	0.25 prime; 0.25-0.35 cover	4-6
50	Bituminous surface treatment	0.40 prime: 0.25-0.35 cover	4-6
51	DCA 1295 diluted 3 parts concentrate to 1 part water	0.33	8-12
52	DCA 1295 diluted 3 Darts concentrate to 1 part water	0.50	4-8
53	DCA 1295 diluted 3 parts concentrate to 1 part water	0.67	3-4
54	DCA 1295 diluted 3 parts concentrate to 1 part water with fiberglass reinforcing	0.33	8-16
55	DCA 1295 diluted 3 parts concentrate to 1 part water with fiberglass reinforcing	0.50	4-12

* Suggested minimum thickness, 4 inches (see *TM 5-331A and TM 5-822-4*).

Table 9-32. Dust palliative electives

Palliative Number	Materials[a]	Rate of Application[b]	Estimated[c] Service Life
Surface Blanket Method			
56	DCA 1295 diluted 3 parts concentrate to 1 part water with f berglass reinforcing	0.67	3-6
57	Polypropylene-asphalt membrane	0.67	4-6
58	Polypropylene-asphalt membrane	0.67	8-12
59	Polypropylene-asphalt membrane	0.67	1-2 yr
60	Polypropylene-asphalt membrane	0.83	4-6
61	Polypropylene-asphalt membrane	0.83	8-12
62	Polypropylene-asphalt membrane	0.83	1-2 yr

a Users must ensure that materials comply with existing Environmental Protection Agency (EPA) regulations for the intended use.
b Rate of applica ion in gallons per square yard unless otherwise noted.
c Estimated service life in months unless otherwise noted.

APPLICATION RATES

9-202. The application rates should be considered estimates, as stated above. Unfortunately, the admix method and some surface-blanket methods represent a full commitment. Should failure occur after selection and placement, the only recourse is to completely retreat the failed area, which is a lengthy and involved process. However, should failure occur on a section treated with a liquid dust palliative, retreatment of the failed area is relatively simple, involving only a distributor and operator. A second application is encouraged as soon as it is determined that the initial application rate is not achieving the desired results.

PLACEMENT

9-203. No treatment is suggested for areas containing large dense vegetation and/or large debris. Loose soil in a wet or slurry condition and firm soil that is wet should not be treated. Dust problems should not exist in any of these areas; however, if the areas are known dust producers when dry, they should be dried or conditioned and then treated.

DILUTION

9-204. Several dilution ratios are mentioned for some liquid dust palliatives. The ratios are presented as volume of concentrate to volume of water and should be viewed as a necessary procedure before a particular liquid can be sprayed. The water is a necessary vehicle to get the dust palliative on the ground. The stated application rate is for the dust palliative only. When high dilution ratios are required to spray a dust palliative, extra care should be taken to prevent the mixture from flowing into adjacent areas where treatment may be unnecessary and/or into drainage ditches. Two or more applications may be necessary to achieve the desired application rate. Considerable time can be saved by first determining the minimum dilution that permits a dust palliative to be sprayed.

PREWETTING

9-205. All liquid dust palliatives present a better finished product when they are sprayed over an area that has been prewet with water. The actual amount of water used in prewetting an area varies but usually ranges from 0.03 to 0.15 gallons per square yard. The water should not be allowed to pond on the surface, and all exposed soil should be completely dampened. The performance of brine materials is enhanced by increasing the amount of water to two to three times the usual recommendation. However, the water should not be allowed to pond, and the fine-sized particles should not be washed away.

CURING

9-206. Most liquid dust palliatives require a curing period. DCA 1295 dries on the soil surface to form a clear film. The curing time is around 4 hours but may vary with weather conditions. Brine materials do not require a curing period, making them immediately available to traffic. Bituminous materials may be ready to accept traffic as soon as the material temperature drops to the ambient temperature.

DUST CONTROL ON ROADS AND CANTONMENT AREAS

9-207. Controlling dust on roads and in and around cantonment areas is important in maintaining health, morale, safety, and speed of movement. Table 9-33 lists several dust palliatives suitable for controlling dust on roads and in cantonment areas, the equipment required to apply the palliative, the level of training required, and the life expectancy of the dust palliative.

Table 9-33. Roads and cantonment area treatments

Treatment	Material Avail-ability	Equipment Required	Experience Required	Expected Life (Months)
Sand grid/surface	S	1, 2, 3, 5, 6, 7, 8	ET	6 - 12
Crushed aggregate with the following treatment:	A	1, 2, 4, 5, 6,10	ET	—
/Brine solutions				
Calcium chloride	S	6 or 7	ET	1 - 2
Magnesium chloride	S	6 or 7	ET	1 - 2
/Crude oil	A	7	ET	3 - 6
/Diesel	A	7	ET	1 - 2
/Waste oil	A	7	ET	2 - 4
/Asphalt cutback				
RC	A	4, 7, 9	E	12 - 24
MC	A	4, 7, 9	E	12 - 24
SC	A	4, 7, 9	E	12 - 24
/Asphalt emulsion				
SS	A/S	4, 7, 9	E	12 - 24
/Asphalt concrete	A	1, 8, 9,11,13	E	36 - 72
/Asphalt surface treatment	A	1, 7, 9,12	E	24 - 60

Material, experience, and life codes:

A - Available
S - Shipping required
ET- Easily trained
E - Experience required

Equipment codes:

1 - Dump truck
2 - Scoop loader
3 - Fork lift
4 - Road grader
5 - Tractor (dozer)
6 - Water truck
7 - Bituminous distributor

8 - Vibratory drum compactor
9 - Rubber- ired compactor
10- Rock crushers
11- Mixing plant
12- Aggregate spreader
13- Paver

DUST CONTROL FOR HELIPORTS

9-208. Dust control for heliports is essential for safety reasons. Because of the nature of heliborne operations, many aircraft are likely to be arriving or departing simultaneously. Obscuration of the airfield due to dust reduces air traffic controllers' ability to control flight operations and represents a significant safety hazard. Adequate dust control for the heliport is essential for safe and efficient flight operations. Figure 9-33 illustrates heliport areas requiring dust control and lists the dimensions of the areas to be treated based on aircraft type. Table 9-34, page 9-78, lists dust palliatives suitable for use around helipads and aircraft maintenance areas.

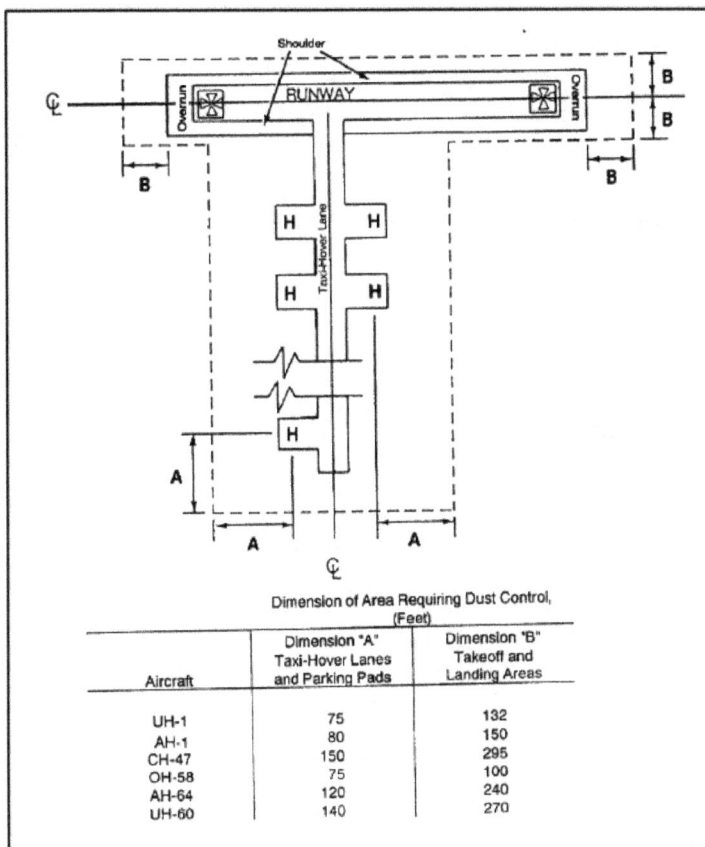

Figure 9-33. Dust control effort required for heliports

| | Dimension of Area Requiring Dust Control, (Feet) | |
Aircraft	Dimension "A" Taxi-Hover Lanes and Parking Pads	Dimension "B" Takeoff and Landing Areas
UH-1	75	132
AH-1	80	150
CH-47	150	295
OH-58	75	100
AH-64	120	240
UH-60	140	270

Table 9-34. Helipad/helicopter maintenance area treatments

Treatment	Material Availability	Equipment Required	Experience Required	Expected Life (Months)
Landing mat	S	3	ET	UL
MO-MAT	S	3	ET	6 - 12
T17 membrane	S	3	ET	6*
Crushed stone (open graded)	A	1, 2, 4, 5,10	E	6 - 24
Sand grid/surface	S	1, 2, 3, 4, 6, 7, 8	ET	6 - 18
Crushed aggregate with the following treatment:	A	1, 2, 4, 5, 6, 10	ET	—
/Crude oil	A	7	ET	6
/Diesel	A	7	ET	2
/Waste oil	A	7	ET	4
/Asphalt cutback				
RC	A	4, 7, 9	E	24 - 36
MC	A	4, 7, 9	E	24 - 36
SC	A	4, 7, 9	E	24 - 36
/Asphalt emulsion				
SS	A/S	4, 7, 9	E	24 - 36
/Asphalt concrete	A	1, 8, 9, 11,13	E	36 - 72
/Asphalt surface treatment	A	1, 7, 9, 12	E	24 - 60

* Performance depends on soil strength beneath membrane

Material, experience, and life codes:
A - Available
S - Shipping required
ET - Easily trained
E - Experience required
UL - Unlimited

Equipment codes:
1 - Dump truck
2 - Scoop loader
3 - Forklift
4 - Road grader
5 - Tractor (dozer)
6 - Water truck
7 - Bituminous distributor

8 - Vibratory drum compactor
9 - Rubber-tired compactor
10 - Rock crushers;
11 - Mixing plant
12 - Aggregate spreader
13 - Paver

CONTROL OF SAND

9-209. Operation Desert Storm highlighted the problems of stabilizing airborne and migrating sand. Airborne sand reduces the life expectancy of mechanical parts exposed to its abrasive effects. From a construction standpoint, migrating sand poses a significant engineering problem—how to prevent dune formation on facilities.

9-210. There are many ways to control migrating sand and prevent sand-dune formation on roads, airfields, and structures. There are certain advantages and disadvantages in each one. The following methods for the stabilization and/or destruction of wind-borne sand dunes are the most effective:

- Fencing.
- Paneling.
- Bituminous materials.
- Vegetative treatment.
- Mechanical removal.
- Trenching.
- Water.
- Blanket covers.
- Salt solutions.

9-211. These methods may be used singularly or in combination.

FENCING

9-212.   This method of control employs flexible, portable, inexpensive fences to destroy the symmetry of a dune formation. The fence does not need to be a solid surface and may even have 50 percent openings, as in snow fencing. Any material, such as wood slats, slender poles, stalks, or perforated plastic sheets, bound together in any manner and attached to vertical or horizontal supports is adequate. Rolled bundles that can be transported easily are practical. Prefabricated fencing is desirable because it can be erected quickly and economically. Because the wind tends to underscore and undermine the base of any obstacle in its flow path, the fence should be installed about 1 foot above ground level. To maintain the effectiveness of the fencing system, a second fence should be installed on top of the first fence on the crest of the sand accumulation. The entire windward surface of the dune should be stabilized with a dust-control material, such as bituminous material, before erecting the first fence. The old fences should not be removed during or after the addition of new fences. Figure 9-34 shows a cross section of a stabilized dune with porous fencing. As long as the fences are in place, the sand remains trapped. If the fences are removed, the sand soon moves downwind, forming an advancing dune. The proper spacing and number of fences required to protect a specific area can only be determined by trial and observation. Figure 9-35 illustrates a three-fence method of control. If the supply of new sand to the dune is eliminated, migration accelerates and dune volume decreases. As the dune migrates, it may move great distances downwind before it completely dissipates. An upwind fence may be installed to cut off the new sand supply if the object to be protected is far downwind of the dune. This distance usually should be at least four times the width of the dune.

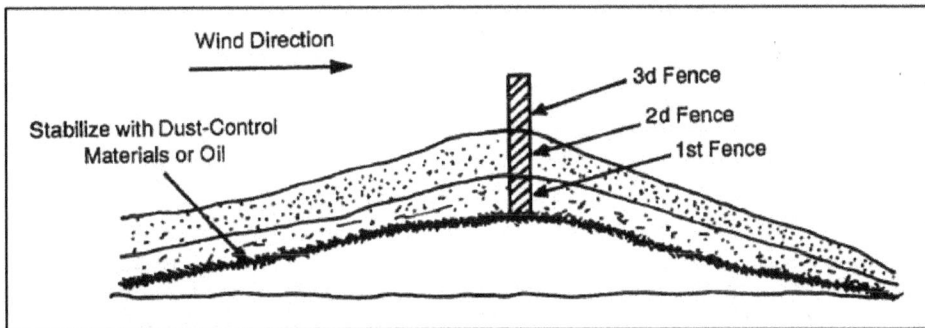

Figure 9-34. Cross section of dune showing initial and subsequent fences

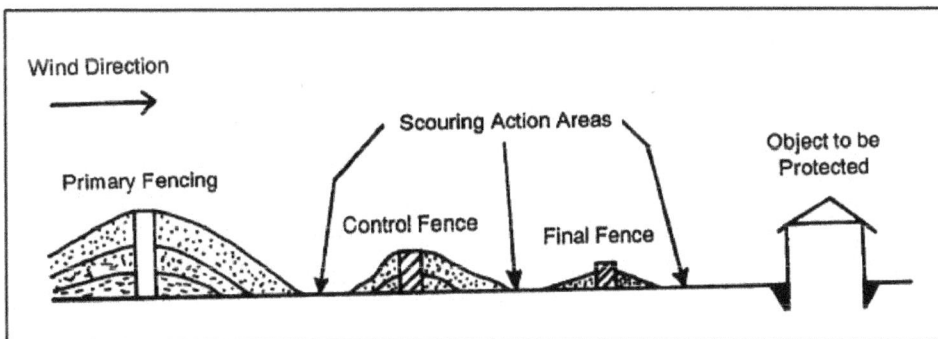

Figure 9-35. Three fences installed to control dune formation

Paneling

9-213. Solid barrier fences of metal, wood, plastic, or masonry can be used to stop or divert sand movement. To stop sand, the barriers should be constructed perpendicular to the wind direction. To divert sand, the panels should be placed obliquely or nearly parallel to the wind. They may be a single-slant or V-shaped pattern (see figure 9-36). When first erected, paneling appears to give excellent protection. However, panels are not self-cleaning, and the initial accumulations must be promptly removed by mechanical means. If the accumulation is not removed, sand begins to flow over and around the barrier and soon submerges the object to be protected. Mechanical removal is costly and endless. This method of control is unsatisfactory because of the inefficiency and expense. It should be employed only in conjunction with a more permanent control, such as plantings, fencing, or dust palliatives. Equally good protection at less cost is achieved with the fencing method.

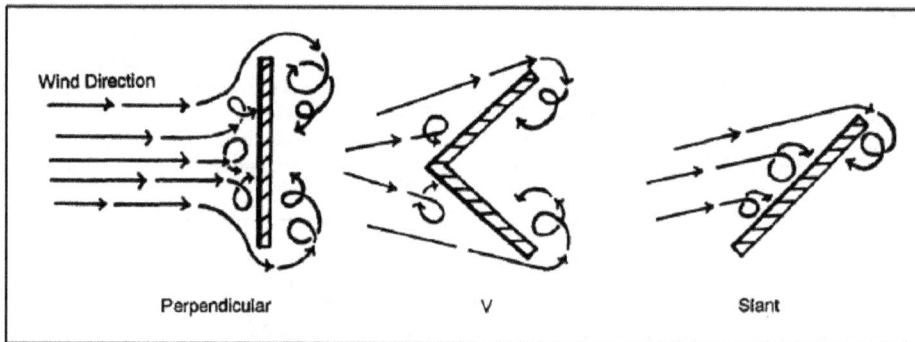

Figure 9-36. Three types of solid fencing or paneling for control of dune formation

Bituminous Materials

9-214. Destroying dune symmetry by spraying bituminous materials at either the center or the ends of the dune is an inexpensive and practical method of sand control. Petroleum resin emulsions and asphalt emulsions are effective. The desired stickiness of the sand is obtained by diluting 1 part petroleum resin emulsion with 4 parts waters and spraying at the rate of ½ gallon per square yard. Generally, the object to be protected should be downwind a distance of at least twice the tip-to-tip width of the dune. The center portion of a barchan dune can be left untreated, or it can be treated and unstabilized portions allowed to reduce in size by wasting. Figure 9-37 shows destruction of a typical barchan dune and stabilization depending on the area treated.

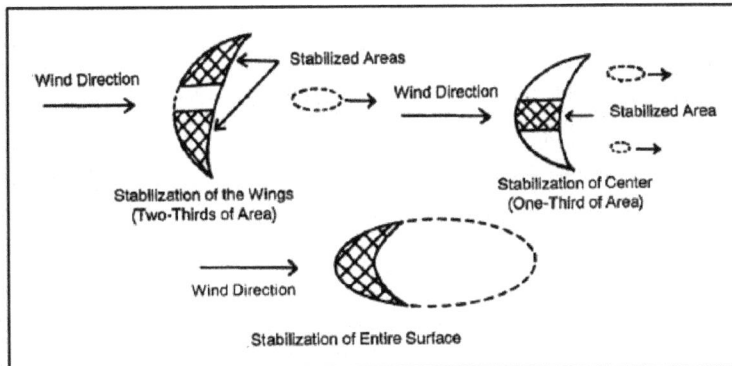

Figure 9-37. Schematic of dune destruction or stabilization by selective treatment

VEGETATIVE TREATMENT

9-215. Establishing a vegetative cover is an excellent method of sand stabilization. The vegetation to be established must often be drought resistant and adapted to the climate and soil. Most vegetative treatments are effective only if the supply of new sand is cut off. An upwind and water, fertilizers, and mulch are used liberally. To prevent the engulfment of vegetation, the upwind boundaries are protected by fences or dikes, and the seed may be protected by using mulch sprayed with a bituminous material. Seed on slopes may be anchored by mulch or matting. Oats and other cereal grasses may be planted as a fast-growing companion crop to provide protection while slower-growing perennial vegetation becomes established. Usually the procedure is to plant clonal plantings, then shrubs (as an intermediate step), followed by long-lived trees. There are numerous suitable vegetative treatments for use in different environments. The actual type of vegetation selected should be chosen by qualified individuals familiar with the type of vegetation that thrives in the affected area. Stabilization by planting has the advantages of permanence and environmental enhancement wherever water can be provided for growth.

MECHANICAL REMOVAL

9-216. In small areas, sand may be removed by heavy equipment. Conveyor belts and power-driven wind machines are not recommended because of their complexity and expense. Mechanical removal may be employed only after some other method has been used to prevent the accumulation of more deposits. Except for its use in conjunction with another method of control, the mechanical removal of sand is not practical or economical.

TRENCHING

9-217. A trench may be cut either transversely or longitudinally across a dune to destroy its symmetry. If the trench is maintained, the dune will be destroyed by wastage. This method has been used successfully in the Arizona Highway Program in the Yuma Desert, but it is expensive and requires constant inspection and maintenance.

WATER

9-218. Water may be applied to sand surfaces to prevent sand movement. It is widely used and an excellent temporary treatment. Water is required for establishing vegetative covers. Two major disadvantages of this method are the need for frequent reapplication and the need for an adequate and convenient source.

BLANKET COVERS

9-219.   Any material that forms a semipermanent cover and is immovable by the wind serves to control dust. Solid covers, though expensive, provide excellent protection and can be used over small areas. This method of sand control accommodates pedestrian traffic as well as a minimum amount of vehicular traffic. Blanket covers may be made from bituminous or concrete pavements, prefabricated landing mats, membranes, aggregates, seashells, and saltwater solutions. After placement of any of these materials, a spray application of bituminous material may be required to prevent blanket decomposition and subsequent dust.

SALT SOLUTIONS

9-220.   Water saturated with sodium chloride or other salts can be applied to sand dunes to control dust. Rainfall leaches salts from the soil in time. During periods of no rainfall and low humidity (below approximately 30 percent), water may have to be added to the treated area at a rate of 0.10 to 0.20 gallon per square yard to activate the salt solution.

SECTION IV - CONSTRUCTION PROCEDURES

MECHANICAL SOIL STABILIZATION

9-221.   This section provides a list of construction procedures, using mechanical stabilization methods, which will be useful to the engineer in the theater of operations.

ON-SITE BLENDING

9-222.   On-site blending involves the following steps:

Preparation

- Shape the area to crown and grade.
- Scarify, pulverize, and adjust the moisture content of the soil, if necessary.
- Reshape the area to crown and grade.

Addition of Imported Soil Materials

9-223.   Use one of the following methods:

- Distribute evenly by means of an improved stone spreader.
- Use spreader boxes behind dump trucks.
- Tailgate each measured truck, loading to cover a certain length.
- Dump in equally spaced piles, then form into windrows with a motor grader before spreading.

Mixing

- Add water, if required, to obtain a moisture content of about 2 percent above optimum and mix with either a rotary mixer, pulvimixer, blade, scarifier, or disk.
- Continue mixing until the soil and aggregate particles are in a uniform, well-graded mass.
- Blade to crown and grade, if needed.

Compaction

- Compact to specifications determined by the results of a CE 55 Proctor test performed on the blended soil material.
- Select the appropriate type(s) of compaction equipment, based on the gradation characteristics of the blended soil.

OFF-SITE BLENDING

9-224. Off-site blending involves the following steps:

Preparation

9-225. Shape area to crown and grade.

Addition of Blended Soil Materials

- Spread blended material evenly, using one of the methods discussed for on-site blending.
- Determine the moisture content of the placed, blended material. Adjust the moisture content, if necessary.

LIME STABILIZATION

9-226. Lime stabilization involves the following steps:

Preparation

- Shape the surface to crown and grade.
- Scarify to the specified depth.
- Partially pulverize the soil.

Spreading

9-227. Select one of the following methods; use about ½ of the total lime required.

- Spot the paper bags of lime on the runway, empty the bags, and level the lime by raking or dragging.
- Apply bulk lime from self-unloading trucks (bulk trucks) or dump trucks with spreaders.
- Apply the lime by slurry (1 ton of lime to 500 gallons of water). The slurry can be mixed in a central plant or in a tank truck and distributed by standard water or asphalt tank trucks with or without pressure.

Preliminary Mixing, Watering, and Curing

- Mix the lime and soil (pulverize soil to less than a 2-inch particle size exclusive of any gravel or stone).
- Add water.

CAUTION

The amount of water needs to be increased by approximately 2 percent for lime stabilization purposes.

- Mix the lime, water, and soil using rotary mixers (or blades).
- Shape the lime-treated layer to the approximate section.
- Compact lightly to minimize evaporation loss, lime carbonation, or excessive wetting from heavy rains.
- Cure lime-soil mixture for zero to 48 hours to permit the lime and water to break down any clay clods. For extremely plastic clays, the curing period may be extended to 7 days.

Final Mixing and Pulverization

- Add the remaining lime by the appropriate method.

- Continue the mixing and pulverization until all of the clods are broken down to pass a 1-inch screen and at least 60 percent of the material will pass a Number 4 sieve.
- Add water, if necessary, during the mixing and pulverization process.

Compaction

- Begin compaction immediately after the final mixing.
- Use pneumatic-tired or sheepsfoot rollers.

Final Curing

- Let cure for 3 to 7 days.
- Keep the surface moist by periodically applying an asphaltic membrane or water.

CEMENT STABILIZATION

9-228.   Cement stabilization involves the following steps:

Preparation

- Shape the surface to crown and grade.
- Scarify, pulverize, and prewet the soil, if necessary.
- Reshape the surface to crown and grade.

Spreading

9-229.   Use one of the following methods:

- Spot the bags of cement on the runway, empty the bags, and level the cement by raking or dragging.
- Apply bulk cement from self-unloading trucks (bulk trucks) or dump trucks with spreaders.

Mixing

- Add water and mix in place with a rotary mixer.
- Perform by processing in 6- to 8-foot-wide passes (the width of the mixer) or by mixing in a windrow with either a rotary mixer or motor grader.

Compaction

- Begin compaction immediately after the final mixing (no more than 1 hour should pass between mixing and compaction), otherwise cement may hydrate before compaction is completed.
- Use pneumatic-tired and sheepsfoot rollers. Finish the surface with steel-wheeled rollers.

Curing

9-230.   Use one of the following methods:

- Prevent excessive moisture loss by applying a bituminous material at a rate of approximately 0.15 to 0.30 gallon per square yard.
- Cover the cement with about 2 inches of soil or thoroughly wetted straw.

FLY-ASH STABILIZATION

9-231.   The following construction procedures for stabilizing soils apply to fly ash, lime-fly ash mixtures, and lime-cement-fly ash mixtures:

Preparation

- Shape the surface to crown and grade.

- Scarify and pulverize the soil, if necessary.
- Reshape the surface to crown and grade.

Spreading

9-232.   Use one of the following methods:

- Spot the bags of fly ash on the road or airfield; empty the bags into individual piles; and distribute the fly ash evenly across the surface with a rake or harrow.
- Apply fly ash or a fly ash mixture in bulk from self-unloading trucks (bulk trucks) or dump trucks with spreaders.

Mixing

- Begin mixing operations within 30 minutes of spreading the fly ash.
- Mix the soil and fly ash thoroughly by using a rotary mixer, by windrowing with a motor grader, or by using a disk harrow.
- Continue to mix until the mixture appears uniform in color.

Compaction

- Add water to bring the soil moisture content to 2 percent above the OMC.
- Begin compaction immediately following final mixing. Compaction must be completed within 2 hours of mixing.
- Minimum compactive effort for soils treated with fly ash is 95 percent of the maximum dry density of the mixed material.
- Reshape to crown and grade; then finish compaction with steel-wheeled rollers.

Curing

9-233.   After the fly ash treated lifts have been finished, protect the surface from drying to allow the soil material to cure for not less than 3 days. This may be accomplished by—

- Applying water regularly throughout the curing period.
- Covering the amended soil with a 2-inch layer of soil or thoroughly wetted straw.
- Applying a bituminous material at the rate of approximately 0.15 to 0.30 gallon per square yard.

BITUMINOUS STABILIZATION

9-234.   In-place stabilization using bituminous materials can be performed with a traveling plant mixer, a rotary-type mixer, or a blade. The methods for using these mixers are outlined below:

Traveling Plant Mixer

- Shape and compact the roadbed on which the mixed material is to be placed. A prime coat should be applied on the roadbed and allowed to cure. Excess asphalt from the prime coat should be blotted with a light application of dry sand.
- Haul aggregate to the job and wind-rowed by hauling trucks, a spreader box, or a blade.
- Add asphalt to the windrow by an asphalt distributor truck or added within the traveling plant mixer.
- Use one of the several types of single- or multiple-pass shaft mixers that are available.
- Work the material until about 50 percent of the volatiles have escaped. A blade is often used for this operation.
- Spread the aggregate to a uniform grade and cross section.
- Compact.

Rotary Mixer

- Prepare the roadbed as explained above for the traveling plant mixer.
- Spread the aggregate to a uniform grade and cross section.
- Add asphalt in increments of about 0.5 gallon per square yards and mix. Asphalt can be added within the mixer or with an asphalt distributor truck.
- Mix the aggregate by one or more passes of the mixer.
- Make one or more passes of the mixer after each addition of asphalt.
- Maintain the surface to the grade and cross section by using a blade during the mixing operation.
- Aerate the mixture.

Blade Mixing

- Prepare the roadbed as explained above for the traveling plant mixer.
- Place the material in a windrow.
- Apply asphalt to the flattened windrow with a distributor truck. A multiple application of asphalt could be used.
- Mix thoroughly with a blade.
- Aerate the mixture.
- Move the mixed windrow to one side of the roadway.
- Spread the mixture to the proper grade and crown.
- Compact the mixture.

CENTRAL PLANT CONSTRUCTION METHODS

9-235. Although central plant mixing is desirable in terms of the overall quality of the stabilized soil, it is not often used for theater-of-operations construction. Stabilization with asphalt cement, however, must be accomplished with a central hot-mix plant.

9-236. The construction methods for central plant mixing that are common for use with all types of stabilizers are given below:

Storing

- Prepare storage areas for soils and aggregates.
- Prepare storage area for stabilizer.
- Prepare storage area for water.

Mixing

- Prepare the area to receive the materials.
- Prepare the mixing areas.

Hauling

9-237. Use trucks.

Placing

9-238. Use a spreader box or bottom-dump truck, followed by a blade to spread to a uniform thickness.

Compacting

9-239. Use a steel-wheeled, pneumatic-tired, or sheepsfoot roller, depending on the material.

Curing

9-240. Provide an asphaltic membrane for cement-stabilized soil and an asphaltic membrane or water for lime-stabilized soils.

SURFACE WATERPROOFING

9-241. Surface waterproofing involves the following steps:

Preparation

- Shape the area to crown and grade.
- Remove all deleterious materials, such as stumps, roots, turf, and sharp-edged soil-aggregate particles.

Mixing

- Adjust the water content to about optimum to 2 percent below optimum, and mix with a traveling mixer, pulvimixer, blade, scarifier, or plow.
- Blade to the crown and grade.

Compaction

- Begin compaction after mixing.
- Use pneumatic-tired or sheepsfoot rollers.

Membrane Placement

- Grade the area to the crown and grade and cut anchor ditches.
- Use a motor grader.
- Roll the area with a steel-wheeled roller or a lightweight, pneumatic-tired roller.
- Place a neoprene-coated nylon fabric or a polypropylene-asphalt membrane on the area.

This page intentionally left blank.

Chapter 10

Slope Stabilization

This chapter pertains to the design of earth slopes as it relates to road construction. It particularly concerns slope stability and which slopes should be used under average conditions in cuts and embankments. Some of the subjects covered are geologic features that affect slope stability, soil mechanics, indicators of unstable slopes, types of slope failures, and slope stabilization.

Road failures can exert a tremendous impact on mission success. It is vital that personnel engaged in road-building activities be aware of the basic principles of slope stability. They must understand how these principles are applied to construct stable roads through various geologic materials with specific conditions of slope and soil.

Basic slope stability is illustrated by a description of the balance of forces that exist in undisturbed slopes, how these forces change as loads are applied, and how groundwater affects slope stability and causes road failure.

GEOLOGIC FEATURES

10-1. There are certain geologic features that have a profound effect on slope stability and that can consequently affect road construction in an area. Many of these geologic features can be observed in the field and may also be identified on topographic maps and aerial photographs. In some cases, the presence of these features may be located by comparing geologic and topographic maps. The following paragraphs describe geologic features that have a significant effect on slope stability and the techniques that may be used to identify them:

FAULTS

10-2. The geologic uplift that accompanies mountain building is evident in the mountainous regions throughout the world. Stresses built up in layers of rock by the warping that accompanies uplift is usually relieved by fracturing. These fractures may extend for great distances both laterally and vertically and are known as faults. Often the material on one side of the fault is displaced vertically relative to the other side; sometimes igneous material or serpentine may be intruded into faults. Faults are the focal point for stress relief and for intrusions of igneous rock and serpentine; therefore, fault zones usually contain rock that is fractured, crushed, or partly metamorphosed. It is extremely important to recognize that fault zones are zones of geologic weakness and, as such, are critical in road location. Faults often leave topographic clues to their location. An effort should be made to identify any faults in the vicinity of a proposed road location.

10-3. The location of these fault zones is established by looking for—

- Saddles, or low sections in ridges, that are aligned in the same general direction from one drainage to another.
- Streams that appear to deviate from the general direction of the nearby streams.

10-4. Notice that the proposed locations of the fault zones on the topographic map follow saddles and drainages in reasonably straight lines.

10-5. Aerial photographs should be carefully examined for possible fault zones when neither geologic maps nor topographic maps offer any clues. An important feature of a fault zone slide that may be detected from aerial photography is the slick, shiny surface caused by the intense heat developed by friction on sliding surfaces within the fault zone.

10-6. Field personnel should be alert for on-the-ground evidence of faulting when neither geologic maps nor topographic maps provide definite clues to the location of faults.

BEDDING PLANE SLOPE

10-7. There are many locations where sedimentary or metamorphic rocks have been warped or tilted and the bedding planes may be steeply sloped. If a road is planned for such an area, it is important to determine the slope of the bedding planes relative to the ground slope. In areas where bedding planes are approximately parallel to the slope of the sidehill, road excavation may remove enough support to allow large chunks of rock and soil to slide into the road (see figure 10-1). If preliminary surveys reveal that these conditions exist, then the route may need to be changed to the opposite side of the drainage area or ridge where the bedding planes slope into the hillside.

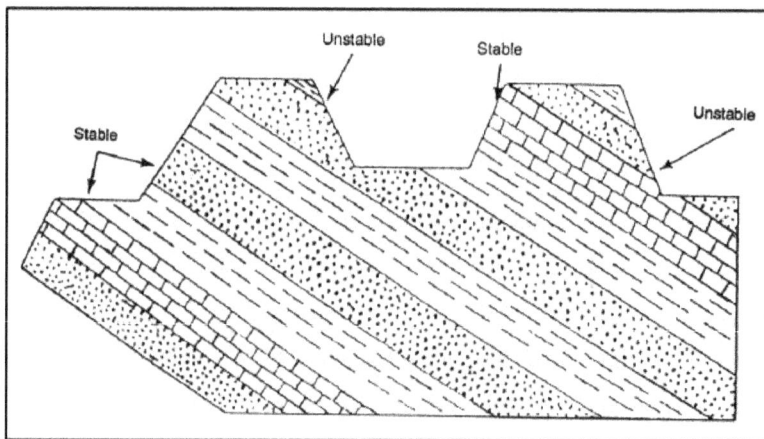

Figure 10-1. Slope of bedding planes

SOIL MECHANICS

10-8. The two factors that have the greatest effect on slope stability are—

- Slope gradient.
- Groundwater.

10-9. Generally, the greater the slope gradient and the more groundwater present, the less stable will be a given slope regardless of the geologic material or the soil type. It is absolutely essential that engineers engaged in locating, designing, constructing, and maintaining roads understand why slope gradient and groundwater are so important to slope stability.

SLOPE GRADIENT

10-10. The effects of slope gradient on slope stability can be understood by discussing the stability of pure, dry sand. Slope stability in sand depends entirely on frictional resistance to sliding. Frictional resistance to sliding, in turn, depends on—

- The slope gradient that affects the portion of the weight of an object that rests on the surface.
- The coefficient of friction.

Normal Force

10-11. The fraction of the weight of an object that rests on a surface is known as the normal force (N) because it acts normal to, or perpendicular to, the surface. The normal force changes as the slope of the surface changes.

10-12. The upper curve in figure 10-2 shows how the gradient of a surface changes the normal force of a 100-pound block resting on the surface. When the slope gradient is zero, the entire weight of the 100-pound block rests on the surface and the normal force is 100 pounds. When the surface is vertical, there is no weight on the surface and the normal force is zero. The coefficient of friction converts the normal force to frictional resistance to sliding (F). An average value for the coefficient of friction for sand is about 0.7. This means that the force required to slide a block of sand along a surface is equal to 0.7 times the normal force.

10-13. The lower curve in figure 10-2 shows how the frictional resistance to sliding changes with slope gradient. The lower curve was developed by multiplying the values of points on the upper curve by 0.7. Therefore, when the slope gradient is zero, the normal force is 100 pounds, and 70 pounds of force is required to slide the block along the surface. When the slope gradient is 100 percent, the normal force is 71 pounds (from point 3 on the upper curve), and 50 pounds (or 71 pounds x 0.7) is required to slide the block along the surface (from point 3 on the lower curve).

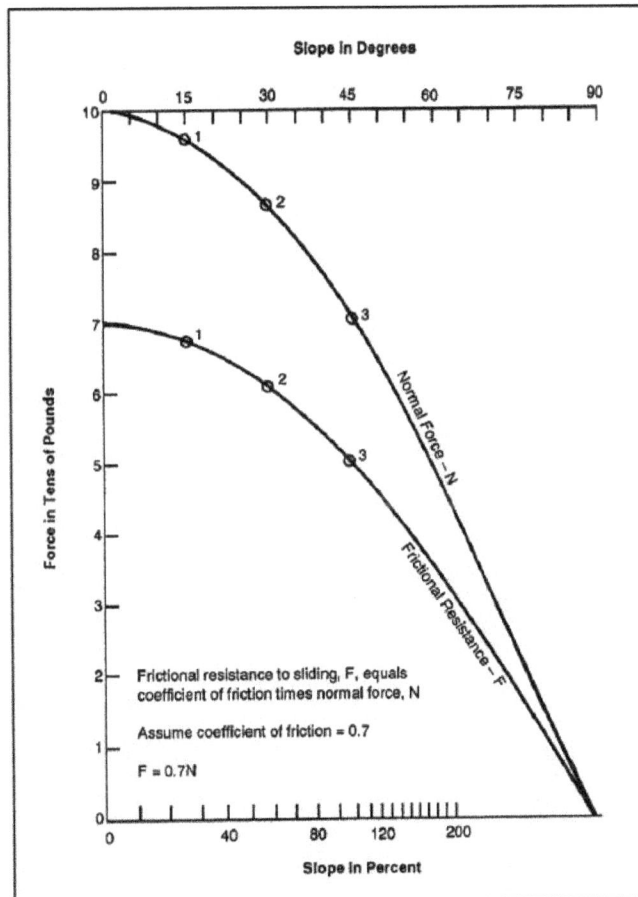

Figure 10-2. Normal force

Downslope Force

10-14. The portion of the weight that acts downslope provides some of the force to overcome frictional resistance to sliding. The downslope force, sometimes known as the driving force, also depends on the slope gradient and increases as the gradient increases (see figure 10-3). Obviously, when the slope gradient becomes steep enough, the driving force exceeds the frictional resistance to sliding and the block slides.

10-15. Figure 10-4 shows the curve for frictional resistance to sliding (from figure 10-2, page 10-3) superimposed on the curve of the driving force (from figure 10-3). These two curves intersect at 70 percent (35 degree) slope gradient. In this example, this means that for slope gradients less than 70 percent, the frictional resistance to sliding is greater than the downslope component of the weight of the block and the block remains in place on the surface. For slope gradients greater than 70 percent, the block slides because the driving force is greater than the frictional resistance to sliding. This discussion has been confined to the case of pure, dry sand, a case which is seldom found in soils, but the principles of the effects of slope gradient and frictional resistance to sliding apply to any dry soil.

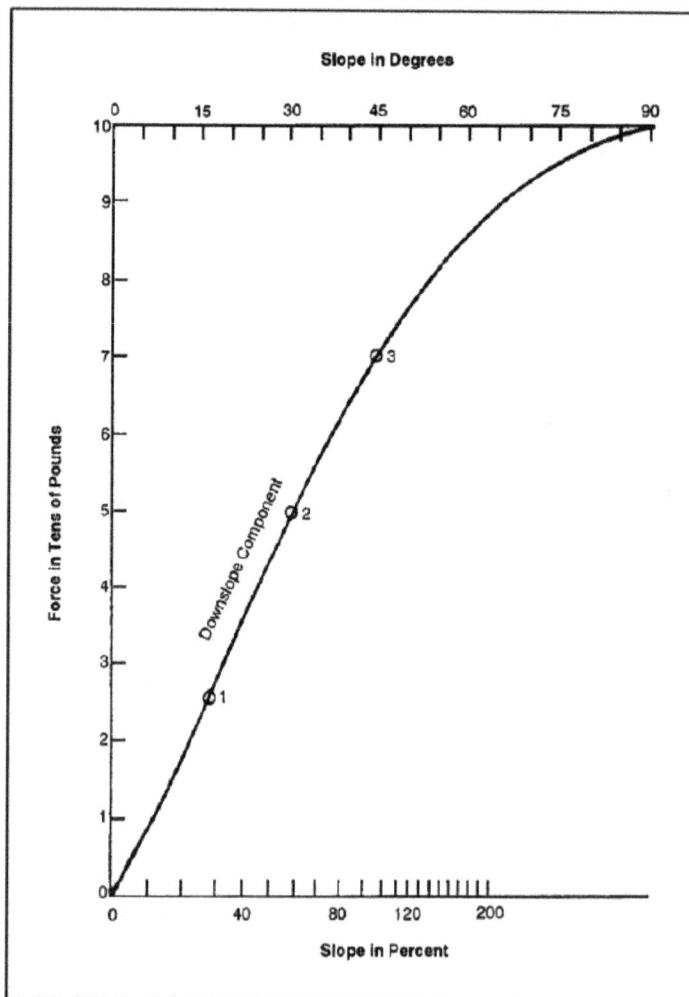

Figure 10-3. Downslope or driving force

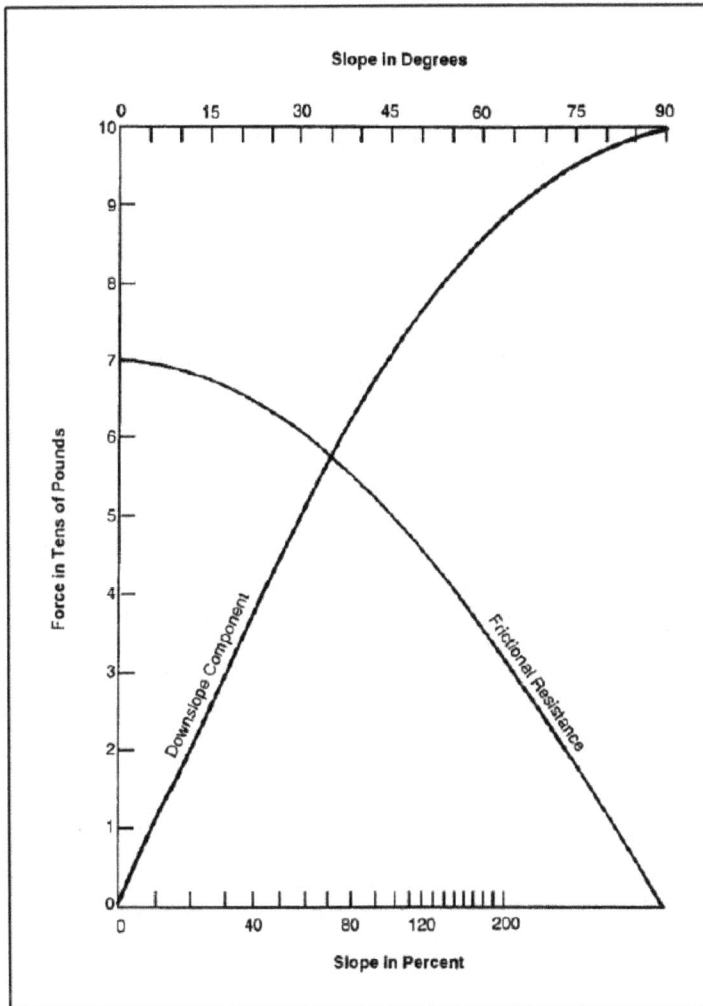

Figure 10-4. Frictional resistance to sliding

Shear Strength

10-16. A block of uniform soil fails, or slides, by shearing. That is, one portion of the block moves past another portion in a parallel direction. The surface along which this shearing action takes place is called the shear plane, or the plane of failure. The resistance to shearing is often referred to as shear strength. Pure sand develops shear strength by frictional resistance to sliding; however, pure clay is a sticky substance that develops shear strength because the individual particles are cohesive. The presence of clay in soils increases the shear strength of the soil over that of a pure sand because of the cohesive nature of the clay.

10-17. A dry clay has considerable shear strength as demonstrated by the great force required to crush a clod with the fingers. However, as a dry clay absorbs water, its shear strength decreases because water films separate the clay particles and reduce its cohesive strength. The structure of the clay particle determines how much water will be absorbed and, consequently, how much the shear strength will decrease upon saturation. There are some clays, such as illite and kaolinite, that provide relative stability to soils even when saturated. However, a saturated montmorillonite clay causes a significant decrease in slope stability. Saturated illite and kaolinite clays have about 44 percent of their total volume occupied by water compared to about 97 percent for a saturated montmorillonite clay. This explains why montmorillonite clay has such a high shrink-swell potential (large change in volume from wet to dry) and saturated clays of this type have a low shear strength. Thus, the type of clay in a soil has a significant effect on slope stability.

10-18. Granitoid rocks tend to weather to sandy soils as the weathering process destroys the grain-to-grain contact that holds the mineral crystals together. If these soils remain in place long enough, they eventually develop a significant amount of clay. If erosion removes the weathered material at a rapid rate, the resulting soil is coarse-textured and behaves as a sand for purposes of slope stability analysis. Many soils with a significant clay content have developed from granitoid material. They have greater shear strength and support steeper cut faces than a granitoid-derived soil with little clay.

10-19. The relative stability among soils depends on a comparison of their shear strength and the downslope component of the weight of the soil. For two soils developed from the same geologic material, the soil with the higher percentage of illite or kaolinite clay has greater shear strength than a soil with a significant amount of montmorillonite clay.

GROUNDWATER

10-20. A common observation is that a hillslope or the side slopes of a drainageway may be perfectly stable during the summer but may slide after the winter rains begin. This seasonal change in stability is due mainly to the change in the amount of water in the pores of the soil. The effect of groundwater on slope stability can best be understood by again considering the block of pure sand. Frictional resistance to sliding in dry sand is developed as the product of the coefficient of friction and the normal force acting on the surface of the failure plane. A closeup view of this situation shows that the individual sand grains are interlocked, or jammed together, by the weight of the sand. The greater the force that causes this interlocking of sand grains, the greater is the ability to resist the shear force that is caused by the downslope component of the soil weight. As groundwater rises in the sand, the water reduces the normal force because of the buoyant force exerted on each sand grain as it becomes submerged.

UPLIFT FORCE

10-21. The uplift force of the groundwater reduces the interlocking force on the soil particles, which reduces the frictional resistance to sliding. The uplift force of groundwater is equal to 62.4 pounds per foot of water in the soil. The effective normal force is equal to the weight of the soil resting on the surface minus the uplift force of the groundwater.

10-22. The following example illustrates the calculation of the effective normal force.

- If 100 pounds of sand rests on a horizontal surface and contains 3 inches (or 0.25 foot) of groundwater, then the effective normal force is 100 - (62.4 x 0.25) = 100 - 15.6 or 84.4 pounds. This shows how groundwater reduces frictional resistance to sliding.
- The frictional resistance to sliding with this groundwater condition is 84.4 x 0.7 (average coefficient of friction for sand), which equals 59.1 pounds.
- As a comparison, for 100 pounds of dry sand on a horizontal surface the frictional resistance to sliding is 100 x 0.7 or 70 pounds.

10-23. The following examples further emphasize how the presence of groundwater can decrease slope stability by reducing the frictional resistance to sliding:

- First, a layer of dry sand 5 feet thick is assumed to weigh 100 pounds per foot of depth. The downslope component of the dry weight and the frictional resistance to sliding for dry sand was calculated for various slope gradients as in figure 10-2, page 10-3, and figure 10-3, page 10-4.
- Next, 6 inches of groundwater is assumed to be present and the frictional resistance to sliding is recalculated, taking into account the uplift force of the groundwater. The results of these calculations are shown in figure 10-5.

Note. In a dry condition, sliding occurs when the slope gradient exceeds 70 percent. With 6 inches of groundwater, the soil slides when the slope gradient exceeds 65 percent.

- For a comparison, assume a dry sand layer only 2 feet thick that weighs 100 pounds per foot of depth. Again, assume 6 inches of groundwater and recalculate the downslope component of the soil weight and the frictional resistance to sliding with and without the groundwater (see figure 10-6, page 10-8).

Note. With 6 inches of groundwater, this thin layer of soil slides when the slope gradient exceeds 58 percent.

10-24. These examples demonstrate that the thinner soil mantle has a greater potential for sliding under the same groundwater conditions than a thicker soil mantle. The 6 inches of groundwater is a greater proportion of the total soil thickness for the 2-foot soil than for the 5-foot soil, and the ratio of uplift force to the frictional resistance to sliding is greater for the 2-foot soil. A pure sand was used in these examples for the sake of simplicity, but the principles still apply to soils that contain varying amounts of silt and clay together with sand.

10-25. Although adding soil may decrease the effect of uplift force on the frictional resistance to sliding, it is dangerous to conclude that slopes can be made stable solely through this approach. The added soil reduces uplift force, but it may increase another factor that in turn decreases frictional resistance, resulting in a slope failure. Decreasing the uplift force of water can be best achieved through a properly designed groundwater control system.

SEEPAGE FORCE

10-26. There is still another way that ground-water contributes to slope instability, and that is the seepage force of groundwater as it moves downslope. The seepage force is the drag force that moving water exerts on each individual soil particle in its path. Therefore, the seepage force contributes to the driving force that tends to move masses of soil downslope. The concept of the seepage force may be visualized by noting how easily portions of a coarse-textured soil may be dislodged from a road cut bank when the soil is conducting a relatively high volume of groundwater.

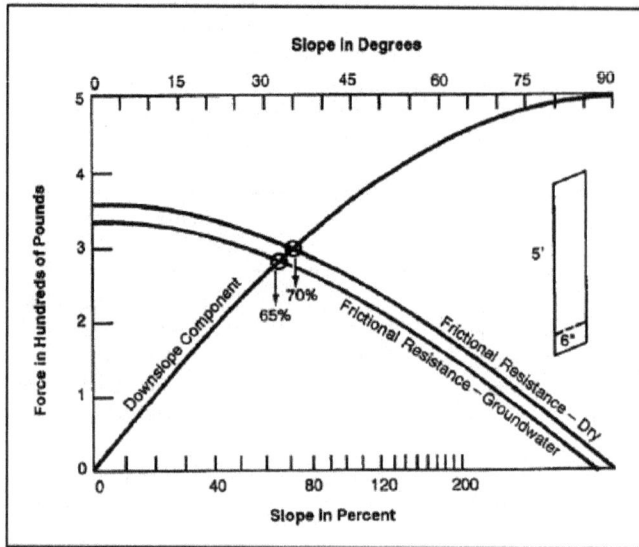

Figure 10-5. Frictional resistance to sliding with uplift force of groundwater

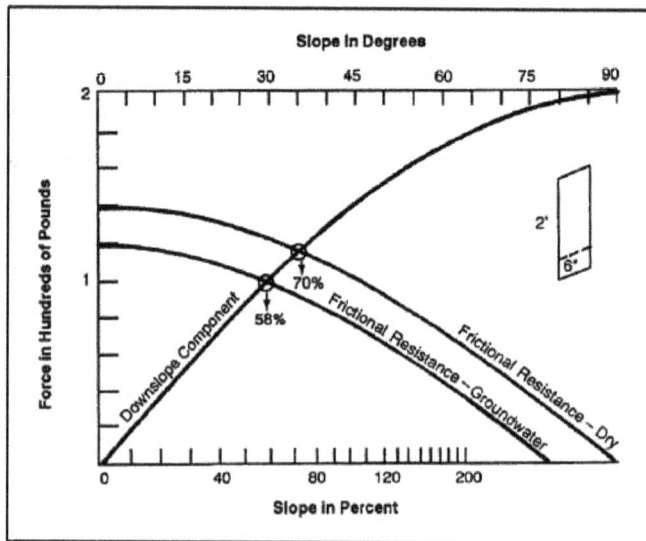

Figure 10-6. Frictional resistance to sliding with and without groundwater

SLOPE FAILURE

10-27.   Slope failure includes all mass soil movements on—

- Man-made slopes (such as road cuts and fills).
- Natural slopes (in clear-cut areas or undisturbed forest).

10-28. A classification of slope failure is useful because it provides a common terminology, and it offers clues to the type of slope stability problem that is likely to be encountered. Types of slope and road failures are remarkably consistent with soils, geologic material, and topography. For example, fast-moving debris avalanches or slides develop in shallow, coarse-textured soils on steep hillsides; large, rotational slumps occur in deep, saturated soils on gentle to moderate slopes.

ROCKFALLS AND ROCKSLIDES

10-29. Rockfalls and rockslides usually originate in bedded sediments, such as massive sandstone, where the beds are undercut by stream erosion or road excavation. Stability is maintained by the—

- Competence of the rock.
- Frictional resistance to sliding along the bedding planes.

10-30. These factors are particularly important where the bedding planes dip downslope toward a road or stream. Rockslides occur suddenly, slide with great speed, and sometimes extend entirely across the valley bottom. Slide debris consists of fractured rock and may include some exceptionally large blocks. Road locations through areas with a potential for rockslides should be examined by specialists who can evaluate the competence of the rock and determine the dip of the bedding planes.

DEBRIS AVALANCHES AND DEBRIS FLOWS

10-31. These two closely related types of slope failure usually originate on shallow soils that are relatively low in clay content on slopes over 65 percent. In southeast Alaska, the US Forest Service has found that debris avalanches develop on slopes greater than 65 percent on shallow, gravelly soils and that this type of slope failure is especially frequent on slopes over 75 percent.

10-32. Debris avalanches are the rapid downslope flowage of masses of loose soil, rock, and forest debris containing varying amounts of water. They are like shallow landslides resulting from frictional failure along a slip surface that is essentially parallel to the topographic surface, formed where the accumulated stresses exceed the resistance to shear. The detached soil mantle slides downslope above an impermeable boundary within the loose debris or at the unweathered bedrock surface and forms a disarranged deposit at the base. Downslope, a debris avalanche frequently becomes a debris flow because of substantial increases in water content. They are caused most frequently when a sudden influx of water reduces the shear strength of earth material on a steep slope, and they typically follow heavy rainfall.

10-33. There are two situations where these types of slope failure occur in areas with shallow soil, steep slopes, and heavy seasonal rainfall.

10-34. The first situation is an area where stream development and geologic erosion have formed high ridges with long slopes and steep, V-shaped drainages, usually in bedded sedimentary rock. The gradient of many of these streams increases sharply from the main stream to the ridge; erosion has created headwalls in the upper reaches. The bowl-shaped headwall region is often the junction for two or more intermittent stream channels that begin at the ridgetop. This leads to a quick rise in groundwater levels during seasonal rains. Past debris avalanches may have scoured round-bottom chutes, or troughs, into the relatively hard bedrock. The headwall region may be covered with only a shallow soil mantle of precarious stability, and it may show exposed bedrock, which is often dark with groundwater seepage.

10-35. The second situation with a high potential for debris avalanches and flows is where excavated material is sidecast onto slopes greater than 65 percent. The sidecast material next to the slope maintains stability by frictional resistance to sliding and by mechanical support from brush and stumps. As more material is sidecast, the brush and stumps are buried and stability is maintained solely by frictional resistance to sliding. Since there is very little bonding of this material to the underlying rock, the entire slope is said to be overloaded. It is quite common for new road fills on steep overloaded slopes to fail when the seasonal rains saturate this loose, unconsolidated material, causing debris avalanches and flows. Under these circumstances, the road fill, together with a portion of the underlying natural slope, may form the debris avalanche (see figure 10-7, page 10-10).

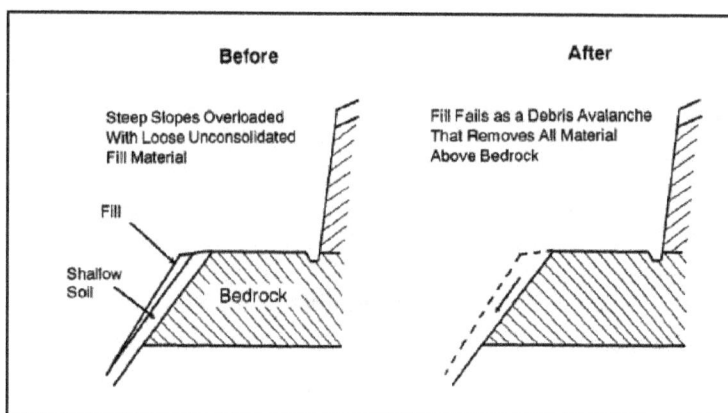

Figure 10-7. Debris avalanche

10-36. Debris avalanches and debris flows occur suddenly, often with little advance warning. There is practically nothing that can be done to stabilize a slope that shows signs of an impending debris avalanche. The best possible technique to use to prevent these types of slope failures is to avoid—

- Areas with a high potential for debris avalanches.
- Overloading steep slopes with excessive sidecast.

10-37. Engineers should learn the vegetative and soil indicators of this type of unstable terrain, especially for those areas with high seasonal groundwater levels.

10-38. If unstable terrain must be crossed by roads, then radical changes in road grade and road width may be required to minimize site disturbance. Excavated material may need to be hauled away to keep overloading of unstable slopes to an absolute minimum. The location of safe disposal sites for this material may be a serious problem in steep terrain with sharp ridges. Site selection will require just as much attention to the principles of slope stability as to the location and construction of the remainder of the road.

SLUMPS AND EARTHFLOWS

10-39. Slumps and earthflows usually occur in deep, moderately fine- or fine-textured soils that contain a significant amount of silt and/or clay. In this case, shear strength is a combination of cohesive shear strength and frictional resistance to sliding. As noted earlier, groundwater not only reduces frictional resistance to shear, but it also sharply reduces cohesive shear strength. Slumps are slope failures where one or more blocks of soil have failed on a hemispherical, or bowl-shaped, slip surface. They may show varying amounts of backward rotation into the hill in addition to downslope movement (see figure 10-8). The lower part of a typical slump is displaced upward and outward like a bulbous toe. The rotation of the slump block usually leaves a depression at the base of the main scarp. If this depression fills with water during the rainy season, then this feature is known as sag pond. Another feature of large slumps is the "hummocky" terrain, composed of many depressions and uneven ground that is the result of continued earthflow after the original slump. Some areas that are underlain by particularly incompetent material, deeply weathered and subject to heavy winter rainfall, show a characteristically hummocky appearance over the entire landscape. This jumbled and rumpled appearance of the land is known as melange terrain.

Figure 10-8. Backward rotation of a slump block

10-40. Depressions and sag ponds allow winter rains to enter the groundwater reservoir, reduce the stability of the toe of the slump, and promote further downslope movement of the entire mass. The mature timber that usually covers old slumps often contains "jackstrawed," or "crazy," trees that lean at many different angles within the stand. This indicates unstable soils and actively moving slopes (see figure 10-9).

Figure 10-9. Jackstrawed trees

10-41. There are several factors affecting slumps that need to be examined in detail to understand how to prevent or remedy this type of slope failure. The block of soil that is subject to slumping can be considered to be resting on a potential failure surface of hemispherical shape (see figure 10-10, page 10-12). The block is most stable when its center of gravity is at its lowest position on this failure surface. When the block fails, its center of gravity is shifted to a lower, more stable position as a result of the failure. Added weight, such as a road fill, at the head of a slump shifts the center of gravity of the block to a higher, more unstable position and tends to increase the potential for rotation. Similarly, removing weight from the toe of the slump, as in excavating for a road, also shifts the center of gravity of the block to a higher position on the failure surface. Therefore, loading the head of a slump and/or unloading the toe will increase the potential for further slumping on short slopes (see figure 10-11). The chance of slumping can be reduced by shifting the center of gravity of a potential slump block to a lower position by following the rule: Unload the head and load the toe.

10-42. If it is absolutely necessary to locate a road through terrain with a potential for slumping, there are several techniques that may be considered to help prevent slumps and earthflows. They are—

- Improve the surface drainage.
- Lower the groundwater level.
- Use rock riprap, or buttresses, to provide support.
- Install an interceptor drain.
- Compact fills.

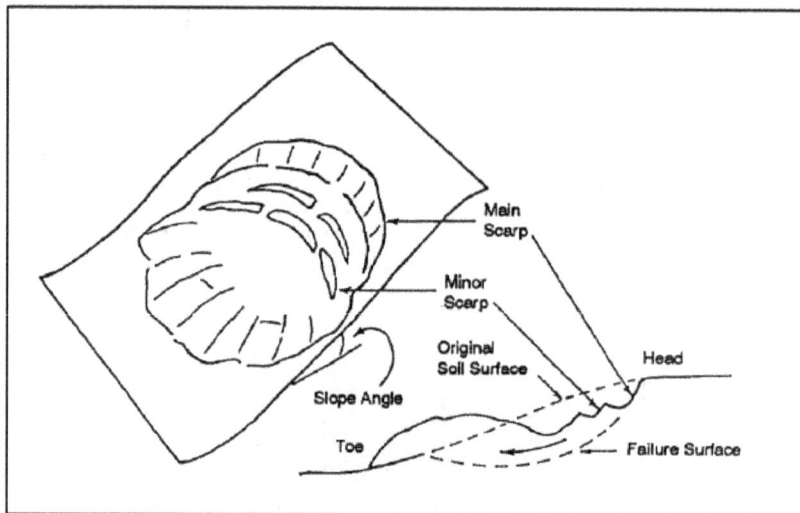

Figure 10-10. Structural features of a slump

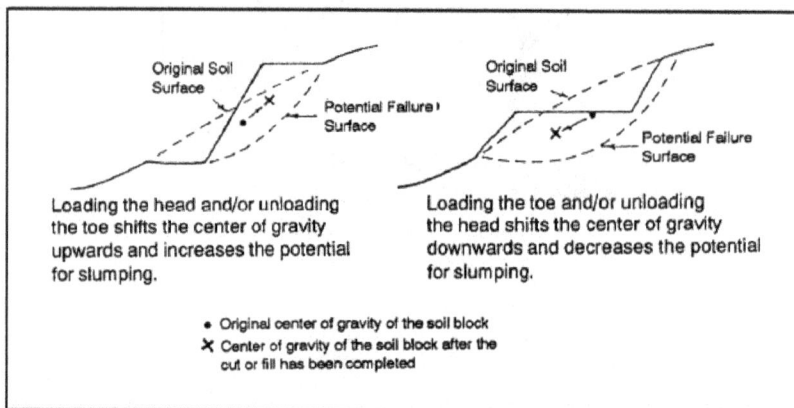

Figure 10-11. Road construction across short slopes

Surface Drainage

10-43. Improving the surface drainage is one of the least expensive and most effective techniques, but it is often overlooked. Sag ponds and depressions can be connected to the nearest stream channel with ditches excavated by a bulldozer or a grader. Figure 10-12, shows the theoretical effect on the groundwater reservoir of a surface drainage project. Improved drainage removes surface water quickly, lowers the groundwater level, and helps stabilize the slump.

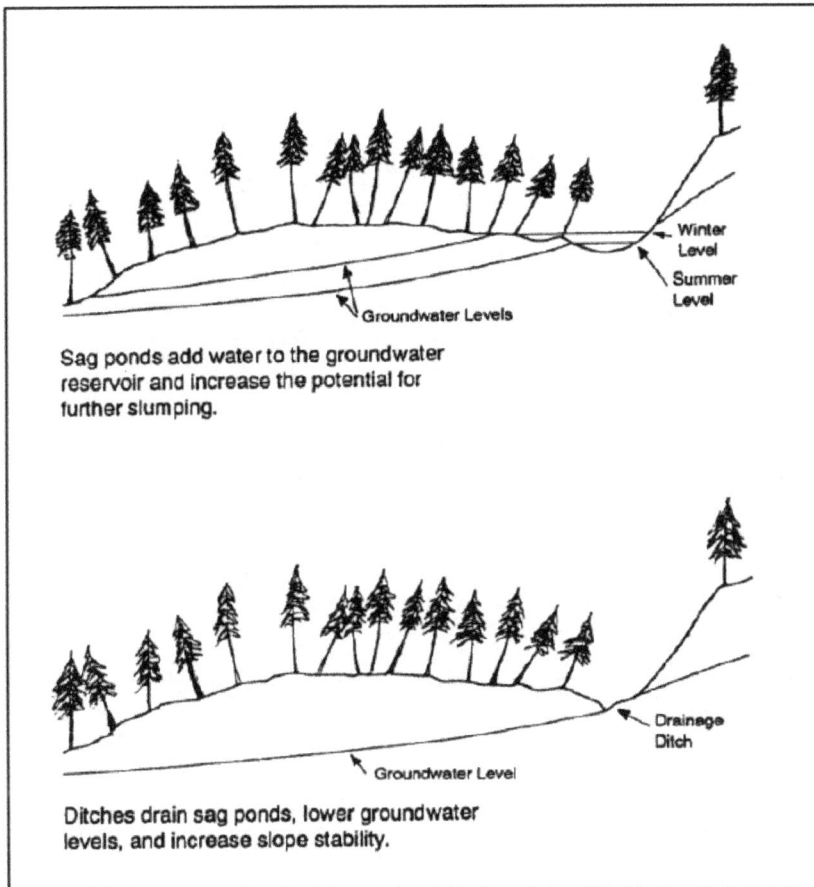

Figure 10-12. Increasing slope stability with surface drainage

Groundwater Level

10-44. Lower the ground-water level by means of a perforated pipe that is augered into the slope at a slight upward angle. These drains are usually installed in road cutbanks to stabilize areas above an existing road or below roads to stabilize fills. Installing perforated pipe is relatively expensive, and there is a risk that slight shifts in the slump mass may render the pipe ineffective. In addition, periodic cleaning of these pipes is necessary to prevent blockage by algae, soil, or iron deposits.

Rock Riprap, or Buttresses

10-45. Stabilize existing slumps and prevent potential slumps by using rock riprap, or buttresses, to provide support for road cuts or fills (see figure 10-13). Heavy rock riprap replaces the stabilizing weight that is by excavation during road construction (see figure 10-11, page 10-12). Another feature of riprap is that it is porous and allows groundwater to drain out of the slump material while providing support for the cut slope.

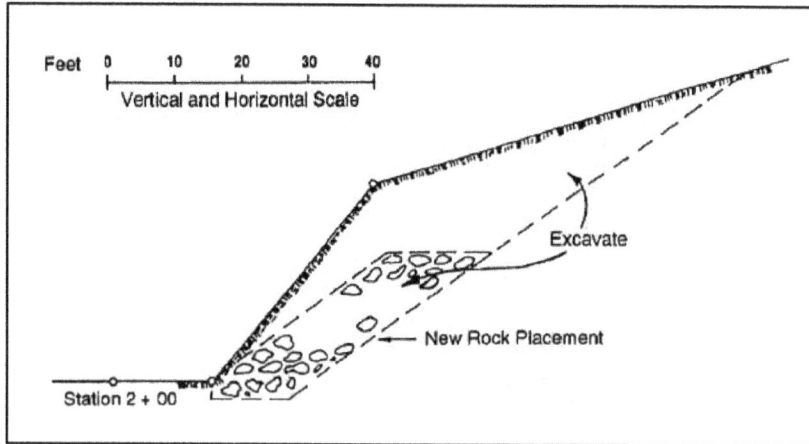

Figure 10-13. Using rock riprap to provide support for road cuts or fills

Interceptor Drain

10-46. Install an interceptor drain to collect groundwater that is moving laterally downslope and under the road, saturating the road fill. A backhoe can be used to install interceptor drains in the ditch along an existing road. Figure 10-14 shows a sample installation.

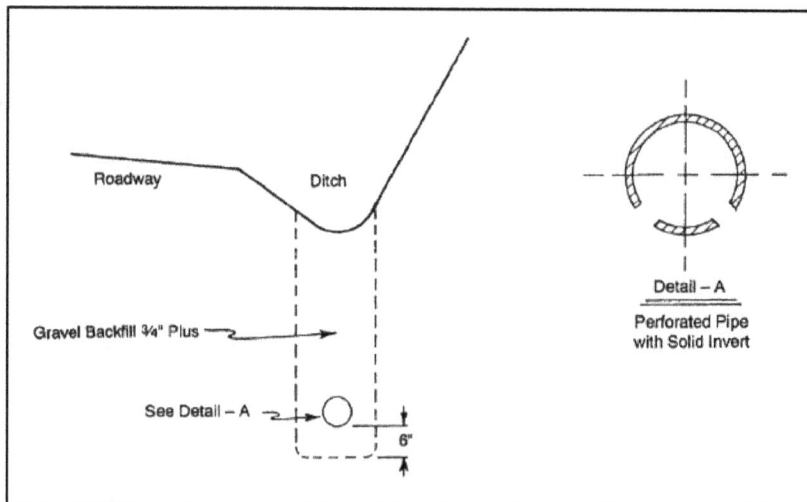

Figure 10-14. Installing interceptor drains along an existing road

Fills

10-47. Compact fills to reduce the risk of road failure when crossing small drainages. Compaction increases the density of the material, reduces the pore space, and thereby reduces the adverse effect of groundwater. The foundation material under the proposed fill should be evaluated as part of the design process to determine if this material will support a compacted fill without failure. Roads may often be built across gentle slopes of incompetent material with a high groundwater table by overexcavating the material, placing a thick blanket of coarse material, then building the road on the blanket (see figure 10-15). The coarse rock blanket distributes the weight of the roadway over a larger area and provides better drainage for groundwater under the road.

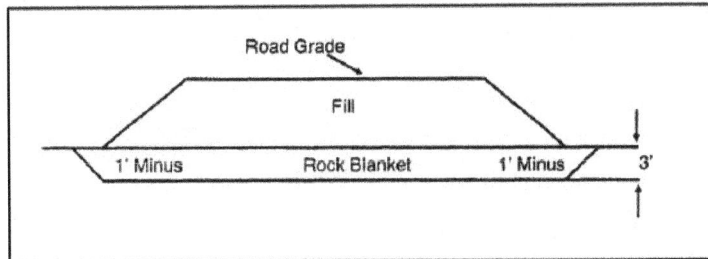

Figure 10-15. Building a road on a blanket

SOIL CREEP

10-48. Many of these slope failures may be preceded and followed by soil creep, a relatively slow-moving type of slope failure. Soil creep may be a continuous movement on the order of less than 1 foot per decade. The indicators of soil creep may be subtle, but you must be aware of the significance of this type of slope failure. Soil creep, at any moment, may be immeasurable; however, when the effect is cumulated over many years, it can create stresses within the soil mantle that may approach the limit of frictional resistance to sliding and/or the cohesive shear strength along a potential shear failure surface.

10-49. Soil creep is particularly treacherous in conjunction with debris avalanches. The balance between stability and failure may be approached gradually over a number of years until only a heavy seasonal rain or a minor disturbance is necessary to trigger a catastrophic slope failure. Soil creep also builds up stresses in potential slumps so that even moderate rainfall may start a slow earthflow on a portion of the slope. Depending on the particular conditions, minor movement may temporarily relieve these stresses and create sags or bulges in the slope; or it may slightly steepen the slope and increase the potential for a major slump during the next heavy rain. The point to remember is that soil creep is the process that slowly changes the balance of forces on slopes. Even though an area may be stable enough to withstand high seasonal groundwater levels this year, it may not be able to 5 years from now.

STABLE SLOPE-CONSTRUCTION IN BEDDED SEDIMENTS

10-50. Construction of stable roads requires not only a basic understanding of regional geology and soil mechanics but also specific, detailed information on the characteristics of soils, groundwater, and geology where the road is to be built. This following paragraphs present techniques on road location and the soil and vegetative indicators of slope instability and high groundwater levels.

10-51. Bedded sediments vary from soft siltstone to hard, massive sandstone. These different geologic materials, together with geologic processes and the effect of climate acting over long periods of time, determine slope gradient, soil, and the rate of erosion. These factors also determine the particular type of slope stability problem that is likely to be encountered. There are four slope stability problems associated with distinctive sites within the bedded sediments. They are—

- Sandstone - Type I.
- Sandstone - Type II.

- Deeply weathered siltstone.
- Sandstone adjoining ridges of igneous rock.

SANDSTONE - TYPE I

10-52. Type I sites are characterized by sharp ridges with steep slopes that may show a uniform gradient from near the ridgetop to the valley bottom. The landscape is sharply dissected by numerous stream channels that may become extremely steep as they approach the ridgetop. Headwalls (bowl-shaped areas with slope gradients often 100 percent or greater) may be present in the upper reaches of the drainage. The headwall is usually the junction for several intermittent streams that can cause sharp rises in the groundwater levels in the soil mantle during winter storms. It is quite common to note groundwater seepage on exposed bedrock in the headwall even during the summer. Figure 10-16 shows a block diagram that illustrates the features of Type I sites. This area was taken from the topographic map of the upper Smith River in Oregon (figure 10-17).

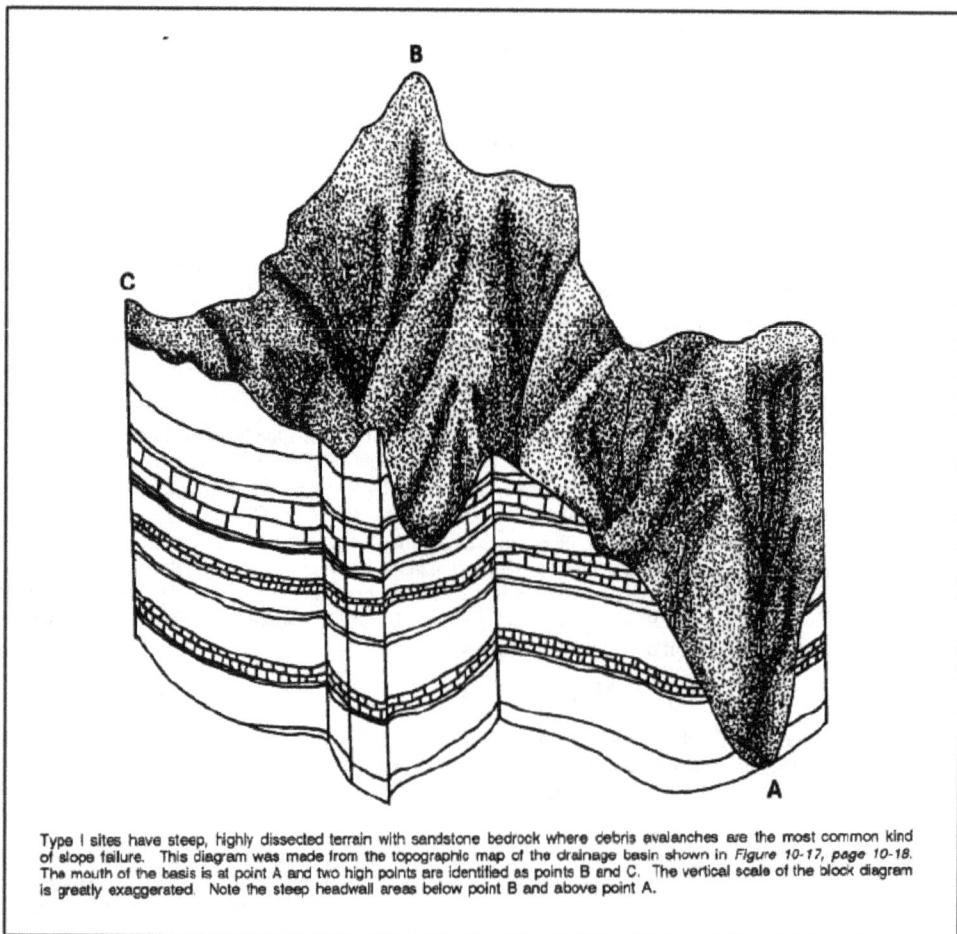

Type I sites have steep, highly dissected terrain with sandstone bedrock where debris avalanches are the most common kind of slope failure. This diagram was made from the topographic map of the drainage basin shown in *Figure 10-17, page 10-18*. The mouth of the basis is at point A and two high points are identified as points B and C. The vertical scale of the block diagram is greatly exaggerated. Note the steep headwall areas below point B and above point A.

Figure 10-16. Block diagram of type I site

10-53. The soils on the most critical portions of Type I sites are coarse-textured and shallow (less than 20 inches to bedrock). These soils are considered to be unstable on slopes greater than 80 percent. In areas where groundwater is present, these soils are considered to be unstable on slopes considerably less than 80 percent.

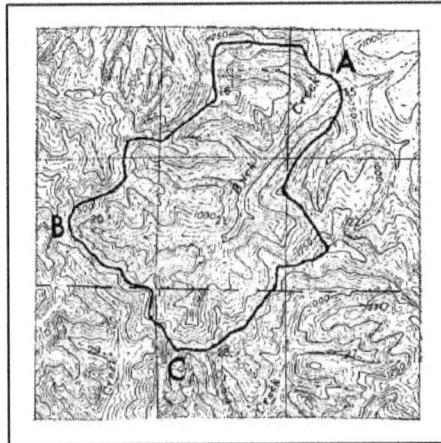

Figure 10-17. Topographic map of type I site

10-54. Debris avalanches and debris flows are the most common slope failure on Type I sites, and the headwall region is the most likely point of origin for these failures. Road construction through headwalls causes unavoidable sidecast. The probability is high for even minimum amounts of sidecast to overload slopes with marginal stability and to cause these slopes to fail. Observers often comment on the stability of full bench roads built through headwalls without realizing that debris avalanches may have occurred during construction before any traffic moved over the road.

Unstable-Slope Indicators

10-55. There are certain indicators of unstable slopes in Type I sites that may be used during road location. They are—

- Pistol-butted trees (see figure 10-18, page 10-18). Sliding soil or debris or active soil creep caused these trees to tip downslope while they were small. As the tree grew, the top regained a vertical posture. Pistol-butted trees are a good indicator of slope instability for areas where rain is the major component of winter precipitation; however, deep, heavy snow packs at high elevations may also cause this same deformation.

Figure 10-18. Pistol-butted trees

- Tipped trees (see figure 10-19). These trees have a sharp angle in the stem. This indicates that the tree grew straight for a number of years until a small shift in the soil mantle tipped the tree. The angled stem is the result of the recovery of vertical growth.

Figure 10-19. Tipped trees

- Tension cracks (see figure 10-20). Soil creep builds up stresses in the soil mantle that are sometimes relieved by tension cracks. These features may be hidden by vegetation, but they definitely indicate active soil movement.

Figure 10-20. Tension cracks

Road-Location Techniques

10-56. Techniques for proper road location on Type I sites include the following:

- Avoid headwall regions. Ridgetop locations are preferred rather than crossing through headwall regions.
- Roll the road grade. Avoid headwalls or other unstable areas by rolling the road grade. Short, steep pitches of adverse and favorable grade may be included.

Construction Techniques

10-57. Consider the following techniques when roads must be constructed across long, steep slopes or above headwall regions where sidecast must be held to a minimum:

- Reduce the road width. This may require a small tractor with a more narrow blade (for example, a D6) for construction. A U-shaped blade results in less sidecast than a straight blade, possibly because of better control of loose material.
- Control blasting techniques. These techniques may be used to reduce overbreakage of rock and reduce the amount of fractured material that is thrown out of the road right-of-way and into stream channels.
- Remove material. Hauling excavated material away from the steepest slopes may be necessary to avoid overloading the lower slopes.
- Select safe disposal sites. Disposal sites for excavated material should be chosen with care to avoid overloading a natural bench or spur ridge, causing slope failure (see figure 10-21, page 10-20). The closest safe disposal site may be a long distance from the construction site but the additional hauling costs must be weighed against the damage caused by failure of a closer disposal site with a higher probability of failure.

Figure 10-21. Safe disposal site

- Fill saddles. Narrow saddles may be used to hold excess material by first excavating bench roads below and on each side of the saddle. The saddle may then be flattened and the loose material that rolls downslope will be caught by the benches. Excavated material may be compacted on the flattened ridge to build up the grade.

- Choose the correct culvert size. A culvert should carry the maximum estimated flow volume for the design storm.

- Protect slopes. Culverts must not discharge drainage water onto the base of a fill slope. Culverts should either be designed to carry water on the natural grade at the bottom of the fill or downspouts or half-round culverts should be used to conduct water from the end of shorter culverts down the fill slope to the natural channel.

SANDSTONE - TYPE II

10-58. Type II sites have slopes with gradients that range from less than 10 percent up to 70 or 80 percent. The longer slopes may be broken by benches and have rounder ridges, fewer drainages, and gentler slope gradients than those on Type I sites. Headwalls are rare, and small patches of exposed bedrock are only occasionally found on the steeper slopes.

10-59. The soils on the gentle slopes developed over many centuries and are deep (often greater than 40 inches), with a clay content as high as 50 to 70 percent. The soils on the steeper slopes may be as deep as 40 inches, but the bedrock is fractured and weathered so there is a gradual transition from the soil into the massive bedrock. It is these factors of deeper soils, higher clay content, gentler slopes, and a gradual transition to bedrock that makes this terrain more stable than the terrain on Type I sites.

10-60. The factors that characterize Type II sites also cause this terrain to have more slope failures due to slumps and earthflows. The most unstable portions of Type II sites are the steep, concave slopes at the heads of drainages, the edge of benches, or the locations where groundwater tends to accumulate. Road failures frequently involve poorly consolidated or poorly drained road fills and embankments greater than 12 to 15 feet on any of the red clayey soils. Soil creep also creates tension within the clayey soils at the convex ridge nose where slope gradients may be only 50 percent. Excavation for roads at these points of sharp slope convexity sometimes causes failure of the embankment.

Unstable-Slope Indicators

10-61. Vegetative indicators of unstable portions of the landscape include mature trees that tip or lean as a result of minor earthflow or soil creep on the steeper slopes. Tipped or leaning trees may also be found on poorly drained soils adjacent to the stream channels. Actively moving slopes may show tension cracks, particularly on the steeper slopes.

10-62. There are several good indicators that may be used to determine the height that groundwater may rise in the soil and roughly how long during the year that the soil remains saturated. Iron compounds within the soil profile oxidize and turn rusty red or bright orange and give the soil a mottled appearance when the groundwater rises and falls intermittently during the winter. The depth below the soil surface where these mottles first occur indicates the average maximum height that this fluctuating water table rises in the soil. At locations where the water table remains for long periods during the year, the iron compounds are chemically reduced and give the soil profile a gray or bluish-gray appearance. The occurrence of these gleyed soils indicates a soil that is saturated for much of the year. Occasionally, mottles may appear above a gleyed subsoil, which indicates a seasonally fluctuating water table above a subsoil that is subject to prolonged saturation. Engineers should be aware of the significance of mottled and gleyed soils that are exposed during road construction. These indicators give clues to the need for drainage or extra attention concerning the suitability of a subsoil for foundation material.

Road-Location Techniques

10-63. Techniques for locating stable roads on Type II sites include the following:

- Avoid steep concave basins. Do not locate roads through these areas where stability is questionable, as indicated by vegetation and topography. Ridgetop locations are preferred.
- Choose stable benches. Benches may offer an opportunity for location of roads and landings, but these benches should be examined carefully to see that they are supported by rock and are not ancient, weathered slumps with marginal stability.
- Avoid cracked soil. Avoid locating a road around convex ridge noses or below the edge of benches where tension cracks or catsteps indicate a high probability for embankment failure.

Construction Techniques

10-64. Certain design and construction practices should be considered when building roads in this terrain.

- Avoid overconstruction. If it is necessary to build a road across steep drainages, avoid overconstruction and haul excess material away to avoid overloading the slopes.
- Avoid high cut embankments. An engineer or soil scientist may be able to suggest a maximum height at the ditch line for the particular soil and situation. A rule-of-thumb estimate for maximum height for a cut bank in deep, clayey soils is 12 to 15 feet.
- Pay special attention to fills. Fills of clayey material over steep stream crossings may fail if the material is not compacted and if groundwater saturates the base of the fill. Fill failures in this wet, clayey material on steep slopes tend to move initially as a slump, then may change to a mudflow down the drainage. To avoid this, compact the fill to accepted engineering standards, paying special attention to proper lift thicknesses, moisture content, and foundation conditions. Also, design drainage features, where necessary, to control ground-water in the base of a fill. For example, consider either a perforated pipe encased in a crushed rock filter or a blanket of crushed rock under the entire fill.

DEEPLY WEATHERED SILTSTONE

10-65. This stability problem originates in siltstone that is basically incompetent and easily weathered. Slumps and earthflows, both large and small, are very common when this material is subjected to heavy winter rainfall. The landscape may exhibit a benchy or hummocky appearance. Slopes with gradients as low as 24 percent may be considered unstable in deeply weathered siltstone with abundant water.

Unstable-Slope Indicators

10-66. Vegetative and topographic indicators of slope instability are numerous. Large patches of plants associated with set soils indicate high ground- water levels and impeded drainage. Conifers may tip or lean due to earthflow or soil creep. Slumps cause numerous benches, some of which show sag ponds. Blocks of soil may sag and leave large cracks, which gradually fill in with debris and living vegetation. The sharp contours of these features soften in time until the cracks appear as "blind drainages" or sections of stream channel that are blocked at both ends. The cracks collect water, keep the groundwater reservoir charged, and contribute to active soil movement.

Road-Location Techniques

10-67. Techniques for locating stable roads through terrain that has been derived from deeply weathered siltstone include the following:

- Check for indicators of groundwater. Avoid locating roads through areas where groundwater levels are high and where slope stability is likely to be at its worst. Such locations may be indicated by hydrophytes, tipped or leaning trees, and mottled or gleyed soils.
- Consider ridges. Ridgetop locations may be best because groundwater drainage is better there. Also, the underlying rock may be harder and may provide more stable roadbuilding material than weathered siltstone.
- Ensure adequate reconnaissance. Take pains to scout the terrain away from the proposed road location, using aerial photos and ground reconnaissance to be sure that the line does not run through or under an ancient slump that may become unstable due to the road construction.

Construction Techniques

10-68. Special road design and construction techniques for this type of terrain may include the following:

- Drainage ditches. Every effort should be made to improve drainage, both surface and subsurface, since ground-water is the major factor contributing to slope instability for this material. Sag ponds and bogs may be drained with ditches excavated by tractor or with ditching dynamite.
- Culverts. Extra culverts should be used to prevent water from ponding above the road and saturating the road prism and adjacent slopes.
- Road ditches. They should be carefully graded to provide plenty of fall to keep water moving. A special effort should be made to keep ditches and culverts clean following construction.

SANDSTONE ADJOINING RIDGES OF IGNEOUS ROCK

10-69. This slope stability problem in bedded sediments is caused by remnants of sandstone adjoining ridges of igneous rock. As a general rule, any contact zone between sedimentary material and igneous material is likely to have slope stability problems.

Unstable-Slope Indicators

10-70. The igneous rock may have caused fracturing and partial metamorphosis of the sedimentary rock at the time of intrusion. Also, water is usually abundant at the contact zone because the igneous material is relatively impermeable compared to the sediments; therefore, the sedimentary rock may be deeply weathered.

Road-Location Techniques

10-71. Special road location techniques for this type of slope stability problem include the following:

- Pay attention to the contact zone. Examine the terrain carefully on the ground and on aerial photos to determine if the mass of sandstone is large or small relative to the igneous rock mass. If the sandstone is in the form of a relatively large spur ridge, then the contact zone deserves special attention. The contact zone should be crossed as high as possible where groundwater accumulation is at a minimum. Elsewhere on the ridge of sandstone, the stability problems are the same as for Type I or Type II sandstone.
- Consider an alternative location. If the remnant of sandstone is relatively small, such as a ridge nose, then the entire mass of sandstone may be creeping rapidly enough to be considered unstable and the road should be located above this material in the more stable igneous rock.

Construction Techniques

10-72.   Design and construction techniques to be considered are as follows:

- Avoid high embankments. The sedimentary rock in the contact zone is likely to be fractured and may be somewhat metamorphosed as a result of the intrusion of igneous rock. In addition, the accumulation of ground-water is likely to have caused extensive weathering of this material. The road cut height at the ditch line should be kept as low as possible through this zone. Support by rock riprap may be necessary if the cut embankment must be high.
- Ensure good drainage. It is good a practice to put a culvert at the contact zone with good gradient on the ditches to keep the contact zone well drained. Other drainage measures, such as drain tile or perforated pipe, may be necessary.

This page intentionally left blank.

Chapter 11

Geotextiles

Other techniques are available for improving the condition of a soil besides mechanical blending and chemical stabilization. These techniques incorporate geotextiles in various pavement applications.

The term geotextile refers to any permeable textile used with foundation, soil, rock, earth, or any other geotechnical engineering-related material as an integral part of a human-made project, structure, or system. Geotextiles are commonly referred to as geofabrics, engineering fabrics, or just fabrics.

APPLICATIONS

11-1. Geotextiles serve four primary functions:

- Reinforcement.
- Separation.
- Drainage.
- Filtration.

11-2. In many situations, using these fabrics can replace soil, which saves time, materials, and equipment costs. In theater-of-operations horizontal construction, the primary concern is with separating and reinforcing low load- bearing soils to reduce construction time.

REINFORCEMENT

11-3. The design engineer attempts to reduce the thickness of a pavement structure whenever possible. Tests show that for low load-bearing soils (generally), the use of geofabrics can often decrease the amount of subbase and base course materials required. The fabric lends its tensile strength to the soil to increase the overall design strength. Figure 11-1, page 11-2, shows an example of this concept.

11-4. Swamps, peat bogs, and beach sands can also be quickly stabilized by the use of geofabrics. Tank trails have been successfully built across peat bogs using commercial geofabric.

SEPARATION

11-5. Construction across soft soils creates a dilemma for the engineer. The construction proceeds at a slow pace because much time is spent recovering equipment mired in muck and hauling large quantities of fill to provide adequate bearing strength. Traditionally, the following options may be considered:

- Bypass the area.
- Remove and replace the soil.
- Build directly on the soft soil.
- Stabilize mechanically or with an admixture.

Bypass

11-6. This course of action is often negated by the tactical situation or other physical boundaries.

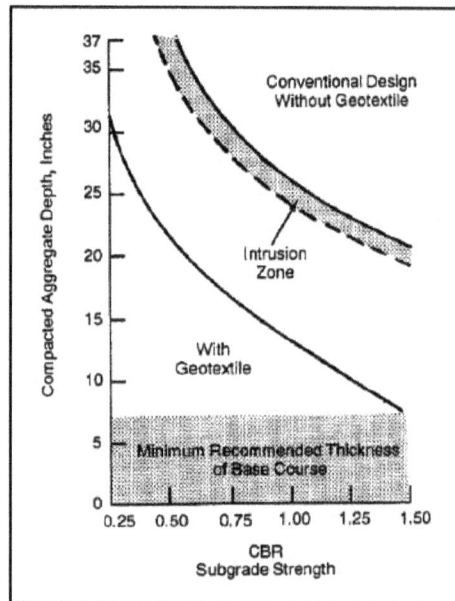

Figure 11-1. Comparison of aggregate depth requirements with and without a geotextile

Remove and Replace

11-7. Commonly referred to as "mucking," this option is sometimes a very difficult and time-consuming procedure. It can only be used if the area has good, stable soil underneath the poor soil. Furthermore, a suitable fill material must be found nearby.

Build On Directly

11-8. Base course construction material is often placed directly on the weak soil; however, the base course layer is usually very thick and the solution is temporary. A "pumping" action causes fines to intrude into the base course, which causes the base course to sink into the weak soil (see figure 11-2). As a result, the base course itself becomes weak. The remedy is to dump more material on the site.

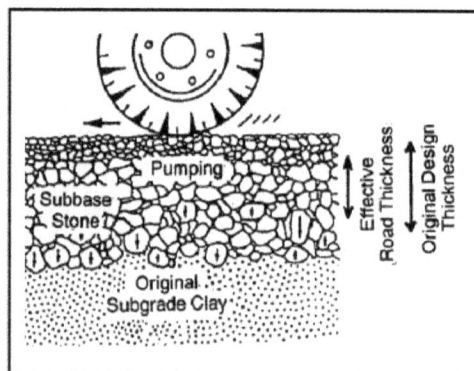

Figure 11-2. Effect of pumping action on a base course

Stabilize

11-9. As discussed in Chapter 10, stabilizing can be done mechanically or with an admixture could be used, but it may be very time consuming and costly.

11-10. For the last three options, the poor soil eventually intrudes into the base course, such as in a swampy area, or simply moves under the loads. By using geofabrics, the poor soils can be separated and confined to prevent intrusion or loss of soil (see figure 11-3).

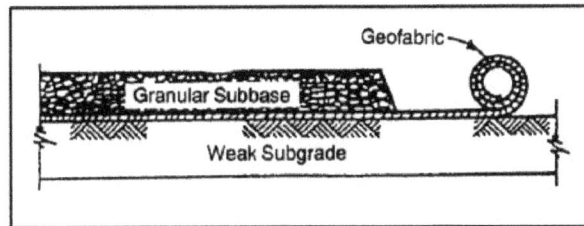

Figure 11-3. Separating a weak subgrade from a granular subbase with a geofabric

DRAINAGE

11-11. Geotextiles placed in situations where water is transmitted in the plane of their structure provide a drainage junction. Examples are geotextiles used as a substitute for granular material in trench drains, blanket drains, and drainage columns next to structures This woven fabric offers poor drainage characteristics; thick nonwoven fabrics have considerably more void space in their structure available for water transmission. A good drainage geotextile allows free water flow (but not soil loss) in the plane of the fabric.

FILTRATION

11-12. In filter applications, the geotextile is placed in contact with soil to be drained and allows water and any particles suspended in the water to flow out of the soil while preventing unsuspended soil particles from being carried away by the seepage. Filter fabrics are routinely used under riprap in coastal, river, and stream bank protection systems to prevent bank erosion. Another example of using a geotextile as a filter is a geotextile-lined drainage ditch along the edge of a road pavement.

UNPAVED AGGREGATE ROAD DESIGN

11-13. The widespread acceptance of geotextiles for use in engineering design has led to a proliferation of geotextile manufacturers and a multitude of geofabrics, each with different engineering characteristics. The design guidelines and methodology that follow will assist in selecting the right geofabric to meet construction requirements.

SITE RECONNAISSANCE

11-14. As with any construction project, a site reconnaissance provides the designer insight into the requirements and the problems that might be encountered during construction.

SUBGRADE SOIL TYPE AND STRENGTH

11-15. Identify the subgrade soil and determine its strength as outlined in chapter 9. If possible, determine the soil's shear strength (C) in psi. If you are unable to determine C, use the nomograph in figure 11-4, page 11-4, to convert the CBR value or Cone Index to C.

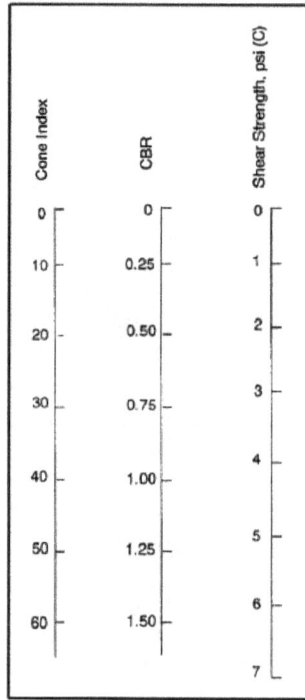

Figure 11-4. Determining the soil's shear strength by converting CBR value or cone index

SUBGRADE SOIL PERMISSIBLE LOAD

11-16. The amount of load that can be applied without causing the subgrade soil to fail is referred to as the permissible stress (S).

- Permissible subgrade stress without a geotextile:

    S = (2.8) C

- Permissible subgrade stress with a geotextile:

    S = (5.0) C

WHEEL LOAD, CONTACT PRESSURE, AND CONTACT AREA

11-17. Estimate wheel load, contact pressure, and contact area dimensions (see table 11-1). For the purpose of geotextile design, both single and dual wheels are represented as single-wheel loads (L) equal to one-half the axle load. The wheel load exerted by a single wheel is applied at a surface contact pressure (P) equal to the tire inflation pressure. Dual wheel loads apply a P equal to 75 percent of the tire inflation pressure. Tandem axles exert 20 percent more than their actual weight to the subgrade soil due to overlapping stress from the adjacent axle in the tandem set.

11-18. Estimate the area being loaded (B^2):

$$B^2 = \frac{L}{P}$$

$$B = \sqrt{B^2} = \text{length of one side of the square contact area}$$

Table 11-1. Vehicle input parameters

Vehicle Type (Choose Category Nearest the Actual Design Vehicles)	Axles S - Single T - Tandem	Wheels S - Single D - Dual	Axle Loads (lb)	Wheel Loads[1]* (lb)	Typical[2] Tire Inflation Pressure (psi)	Contact Pressure[3] P (psi)	Wheel Contact Area B[2] (in[2])	One Side of Square Contact Area B (in)
Highway Legal Vehicles								
Haul trucks[4] - F Axle	S	S	18,000	9,000	110	110	82	9 0
(stone, concrete) R Axle	T	D	18,000	10,800	110	83	130	11.4
Tractor trailer - F Axle	S	S	18,000	9,000	120	120	75	8.7
(18 wheeler) - R Axle	T	D	18,000	10,800	120	90	120	11.0
Off Highway Vehicles[5]								
35-ton trucks - F Axle	S	S	48,000	24,000	90	90	267	16.3
(CAT 769C) - R Axle	S	D	89,200	44,600	90	68	656	25.6
Wheel loader - F Axle	S	S	24,000	12,000	50	50	240	15.5
(CAT 910) -R Axle	S	S	10,000	5,000	50	50	100	10.0
Wheel loader - F Axle	S	S	37.000	18,500	60	60	308	17.6
(CAT 930) - R Axle	S	S	14,000	7,000	60	60	117	10.8
Wheel loader - F Axle	S	S	65,000	32,000	60	60	542	23.3
(CAT 966C) - R Axle	S	S	25,000	12,500	60	60	208	14.4
Wheel loader - F Axle	S	S	136,000	68,000	85	85	800	28.3
(CAT 988B)-R Axle	S	S	55,000	27,500	85	85	324	18.0
Wheel loader - F Axle	S	S	290,000	145,000	70	70	2071	45.5
(CAT 992) - R Axle	S	S	120,000	60,000	60	60	1000	31.6
Scraper - F Axle	S	S	88,600	44,300	80	80	554	23.5
(CAT 631D)-R Axle	S	S	75,400	37,700	75	75	503	22.4
Scraper - F Axle	S	S	120,000	60,000	85	85	706	26.6
(CAT 651B)-R Axle	S	S	110,800	55,400	80	80	692	26.3

Notes.
1. Wheel load is one-half the axle load and increased by 20% if the wheel is on a tandem axle.
2. Maximum tire inflation pressure is given for each class of vehicle. Using tires with lower inflation pressures would lower the contact pressures and allow for less thickness of he aggregate .structural section
3. Same as tire inflation pressure except that a factor of 0.75 times the inflation pressure must be used for all dual wheels.
4. Trucks used on- and off-highway use generally use lower inflation pressure tires requiring only 75 to 90 psi.
5. Manufacturers' specifications should be consulted for off-highway vehicles. Wide ranges of different inflation pressure tires are available for these vehicles.

AGGREGATE BASE THICKNESS

11-19. Assuming that wheel loads will be applied over a square area, use the Business theory of load distribution to determine the aggregate section thickness required to support the design load. The Boussinesq theory coefficientts are found in table 11-2, page 11-6.

- First, solve for X:
Without a geotextile:

$$X = \frac{S}{(4)P}$$

with a geotextile:

$$X\ geotextile = \frac{S\ geotextile}{(4)P}$$

- Using the calculated values of X and X geotextile, use table 11-2 to find the corresponding value of M and M geotextile.
- Then solve for the aggregate base thickness H and H geotextile.

- Without a geotextile:

$$H = \frac{B \text{ inches}}{(2)M}$$

- With a geotextile:

$$H \text{ geotextile} = \frac{B}{(2)M \text{ geotextile}}$$

Table 11-2. Boussinesq theory coefficients

If X =	Then M =
0.005	0.10
0.011	0.15
0.018	0.20
0.026	0.25
0.037	0.30
0.048	0.35
0.060	0.40
0.072	0.45
0.084	0.50
0.096	0.55
0.107	0.60
0.118	0.65
0.128	0.70
0.138	0.75
0.146	0.80
0.155	0.85
0.162	0.90
0.169	0.95
0.175	1.00
0.186	1.10
0.196	1.20
0.207	1.35
0.215	1.50
0.224	1.75
0.232	2.00
0.237	2.25
0.240	2.50
0.242	2.75
0.244	3.00
0.247	4.00
0.249	5.00
0.249	7.50
0.250	10.00
0.250	∞

11-20.   The difference between H and H geotextile is the aggregate savings due to the geotextile.

Aggregate Quality

11-21.   Adjust the aggregate section thickness for aggregate quality. The design method is based on the assumption that a good quality of aggregate (with a minimum CBR value of 80) is used. If a lower quality is used, the aggregate section thickness must be adjusted.

11-22.   Table 11-3, contains typical compacted strength properties of common structural materials. These values are approximations; use more specific data if it is available. Extract the appropriate thickness equivalent factor from table 11-3, then divide H by that factor to determine the adjusted aggregate section thickness.

Table 11-3. Compacted strength properties of common structural materials

Material	CBR Range	Thickness Equivalency Factor
Asphalt, concrete plant mix, high stability	>100	3.00
Crushed hard rock	80-100	1.00
Crushed medium-hard rock	60-80	0.85
Well-graded gravel	40-70	0.80
Shell	40-60	0.75
Sand-gravel mixtures	20-50	0.50
Soft rock	20-40	0.45
Clean sand	10-30	0.40
Lime-treated base1	>100	1.00-2.00
Cement-treated base1,2		
650 psi or more	>100	1.60
400 psi to 650 psi	>100	1.40
400 psi or less	>100	1.05

1 The strength of lime-treated and cement-treated bases depends on soil properties and construction procedures. Treated bases are also subject to long-term failure due to continuing chemical actions overtime.

2 Compressive strength at 7 days.

Note. The values listed above are general guidelines. More exact thickness equivalency factors can be determined by comparing the CBR of the available aggregate to he design CBR of 80. For example, an aggregate with a CBR of 55 would have an approximate thickness equivalency factor of 55/80 = 0.69.

Service Life

11-23.   Adjust the aggregate base thickness for the service life. The design method assumes that the pavement will be subjected to one thousand 18,000-pound equivalent vehicle passes. If you anticipate more than 1,000 equivalent passes, you will need to increase the design thickness by 30 percent and monitor the performance of the road.

11-24.   A second method of determining minimum required cover above a subgrade for wheeled vehicles with and without a geotextile requires fewer input parameters. Again, use figure 11-4, page 11-4, to correct CBR or cone index values to a C value. Determine the permissible stress (S) on the subgrade soil by multiplying C times 2.8 without a geotextile and 5.0 with a geotextile. Select the heaviest vehicle using the road and the design vehicle for each wheel load configuration: single, dual, or tandem. Using the appropriate graph (see figures 11-5, 11-6, or 11-7, pages 11-9 and 11-10) enter the graph at S. Round the design-vehicle wheel loads to the next higher wheel-load weight curve (for example, a dual wheel load of 10,500 pounds is rounded to 12,000 pounds (see figure 11-6). Determine the intersection between the appropriate wheel-load curve and S (with and without a geotextile) then read the minimum required thickness on the left axis. Use the greatest thickness values as the design thickness with and without a geotextile. Compare the cost of the material saved with the cost of the geotextile to determine if using the geotextile is cost effective.

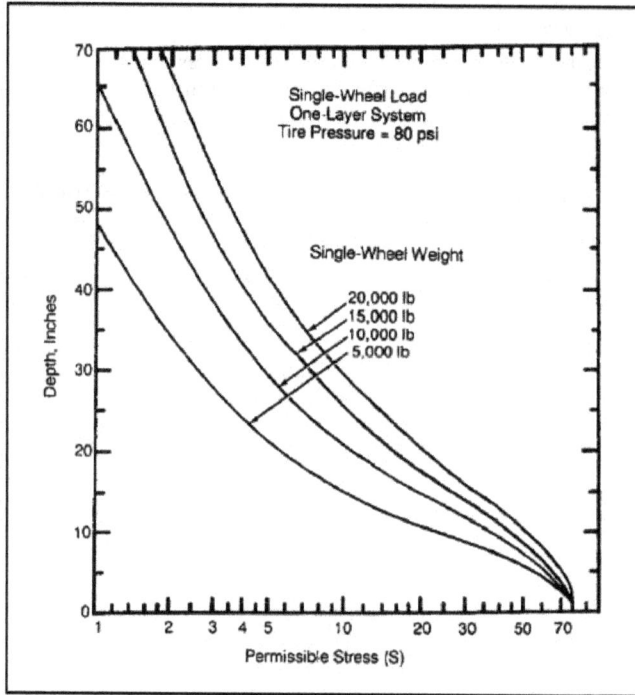

Figure 11-5. Thickness design curve for single-wheel load on gravel-surfaced pavements

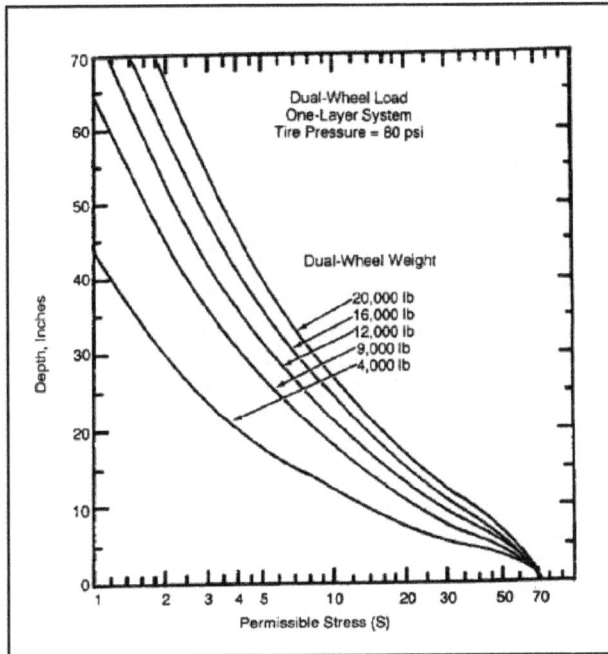

Figure 11-6. Thickness design curve for dual-wheel load on gravel-surfaced pavements

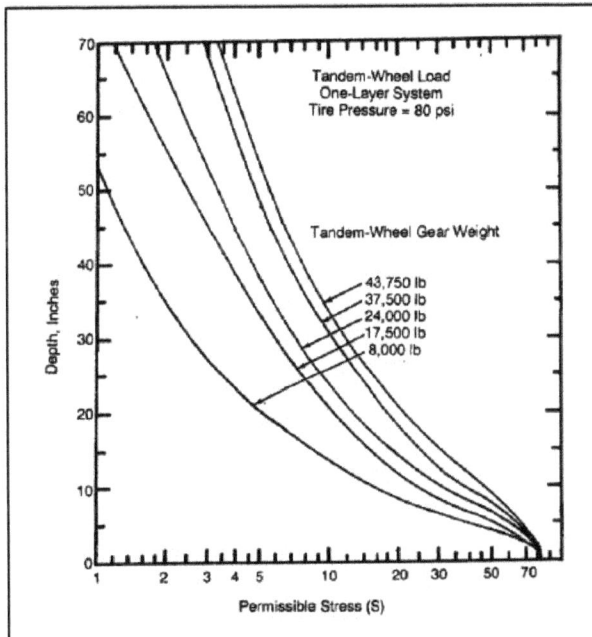

Figure 11-7. Thickness design curve for tandem-wheel load on gravel-surfaced pavements

SELECTING A GEOTEXTILE

11-25. Up to this point in the geotextile design process, we have been concerned with general design properties for designing unpaved aggregate roads. Now you must decide which geotextile fabric best meets your project requirements.

11-26. There are two major types of geotextile fabric: woven and nonwoven. Woven fabrics have filaments woven into a regular, usually rectangular, pattern with openings that are fairly evenly spaced and sized. Nonwoven fabrics have filaments connected in a method other than weaving, typically needle punching or head bonding at intersection points of the filaments. The pattern and the spacing and size of the openings are irregular in nonwoven fabrics. Woven fabrics are usually stronger than nonwoven fabrics of the same fabric weight. Woven geotextiles typically reach peak strength at between 5 and 25 percent strain. Nonwoven fabrics have a high elongation of 50 percent or more at maximum strength.

11-27. Table 11-4 provides information on important criteria and principle properties to consider when selecting or specifying a geotextile for a particular application. The type of equipment used to construct the road or airfield pavement structure on top of the geotextile must be considered. Equipment ground pressure (in psi) is an important factor in determining the geotextile fabric thickness. A thicker fabric is necessary to stand up to high equipment ground pressure (see table 11-5).

11-28. Once the required degree of geotextile survivability is determined, minimum specification requirements can be established based on ASTM standards (see table 11-6, page 11-12). After determining the set of testing standards the geotextile will be required to withstand to meet the use and construction requirements, either specify a geotextile for ordering or evaluate on-hand stock.

Table 11-4. Criteria and properties for geotextile evaluation

Criteria and Parameter	Property	Application			
		F	D	S	R
Design Requirements					
Mechanical strength					
Tensile strength	Wide-width strength	—	—	—	X
Tensile modulus	Wide-width modulus	—	—	—	X
Seam strength	Wide width	—	—	—	X
Tension creep	Creep	—	—	—	X
Soil-fabric friction	Friction angle	—	—	—	X
Hydraulic					
Flow capacity	Permeability,	X	X	X	X
	Transmissivity	—	X	—	—
Piping resistance	Apparent opening size(AOS)	X	—	X	X
Clogging resistance	Pommetry	X	—	X	X
	Gradient ratio	X	—	—	—
Constructability Requirements					
Tensile strength	Grab strength	X	X	X	X
Seam strength	Grab strength	X	X	X	X
Bursting resistance	Mullen burst	X	X	X	X
Puncture resistance	Red puncture	X	X	X	X
Tear resistance	Trapesoidal tear	X	X	X	X
F - Filtration D - Drainage S - Separation R - Reinforcement					

Table 11-5. Geotextile survivability for cover material and construction equipment

Cover Material	6-to 12- Inch Initial Lift Thickness		12- to Inch Initial Lift Thickness		18- to -24 Inch Initial Lift Thickness	>24-Inch Initial Lift Thickness
	Low Ground-Pressure Equipment <4 psi	Medium Ground-Pressure Equipment >4 psi, <8 psi	Medium Ground-Pressure Equipment > 4 psi, <8 psi	High Ground-Pressure Equipment >8 psi	High Ground-Pressure Equipment >8 psi	High Ground-Pressure Equipment >8 psi
Fine sand to ± 2-inch-diameter gravel, round to subangular	Low	Moderate	Low	Moderate	Low	Low
Coarse aggregate with diameter up to one-half	Moderate	High	Moderate	High	Moderate	Low

Table 11-5. Geotextile survivability for cover material and construction equipment

Cover Material	6-to 12- Inch Initial Lift Thickness		12- to Inch Initial Lift Thickness		18- to -24 Inch Initial Lift Thickness	>24-Inch Initial Lift Thickness
	Low Ground-Pressure Equipment <4 psi	Medium Ground-Pressure Equipment >4 psi, <8 psi	Medium Ground-Pressure Equipment > 4 psi, <8 psi	High Ground-Pressure Equipment >8 psi	High Ground-Pressure Equipment >8 psi	High Ground-Pressure Equipment >8 psi
proposed lift thickness, may be angular						
Some to most aggregate with diameter greater than one-half proposed lift thickness, angular and sharp-edged, few fines	High	Very high	High	Very high	High	Moderate

Notes.
1. For special construction techniques such as prerutting, increase geotextile survivability requirement one level.
2. Placement of excessive initial-cover material thickness may cause bearing failure of soft subgrades

Table 11-6. Minimum properties for geotextile survivability

Required Degree of Geotextile Survivability	Grab Strength[1] lb	Puncture Strength[2] lb	Burst Strength[3] psi	Trap Tear[4] lb
Very high	270	110	430	75
High	180	75	290	50
Moderate	130	40	210	40
Low	90	30	145	30

[1] ASTM D 4632

[2] ASTM D 4833

[3] ASTM D 3786

[4] ASTM D 4533, either principal direction

Note. All values represent minimum average roll values (for example, any roll in a lot should meet or exceed the minimum values in this table.) These values are normally 20 percent lower than manufacturer reported typical values.

ROADWAY CONSTRUCTION

11-29.   There is no singular way to construct with geofabrics. However, there are several applications and general guidelines that can be used.

Prepare the Site

11-30.   Clear, grub, and excavate the site to design grade, filling in ruts and surface irregularities deeper than 3 inches (see figure 11-8). Lightly compact the subgrade if the soil is CBR 1. The light compaction aids in locating unsuitable materials that may damage the fabric. Remove these materials when it is practical to do so.

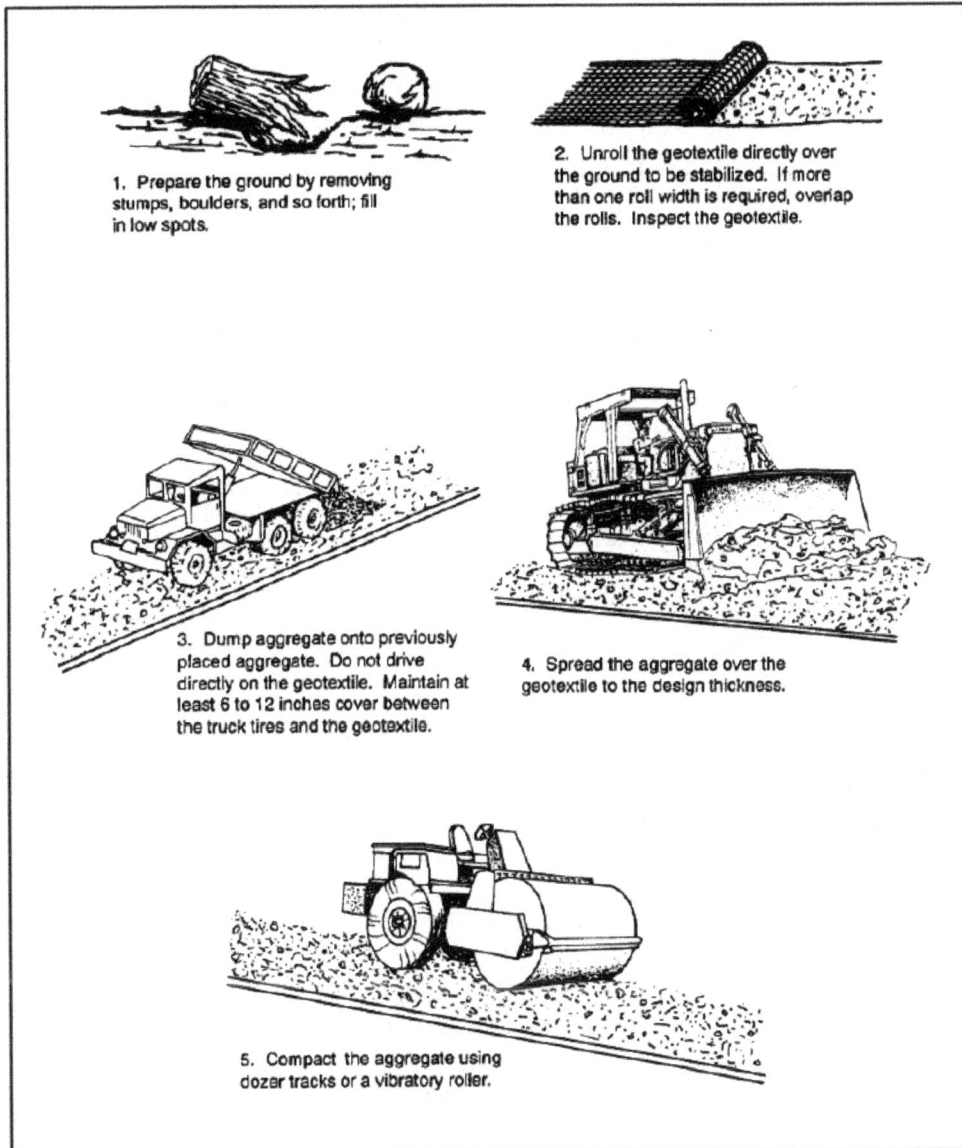

1. Prepare the ground by removing stumps, boulders, and so forth; fill in low spots.

2. Unroll the geotextile directly over the ground to be stabilized. If more than one roll width is required, overlap the rolls. Inspect the geotextile.

3. Dump aggregate onto previously placed aggregate. Do not drive directly on the geotextile. Maintain at least 6 to 12 inches cover between the truck tires and the geotextile.

4. Spread the aggregate over the geotextile to the design thickness.

5. Compact the aggregate using dozer tracks or a vibratory roller.

Figure 11-8. Construction sequence using geotextiles

11-31.   When constructing over extremely soft soils, such as peat bogs, the surface materials, such as the root mat, may be advantageous and should be disturbed as little as possible. Use sand or sawdust to cover protruding roots, stumps, or stalks to cushion the geotextile and reduce the potential for fabric puncture. Nonwoven geotextiles, with their high elongation properties, are preferred when the soil surface is uneven.

Lay the Fabric

11-32.   The fabric should be rolled out by hand, ahead of the backfilling and directly on the soil subgrade. The fabric is commonly, but not always, laid in the direction of the roadway. Where the subgrade cross section has large areas and leveling is not practical, the fabric may be cut and laid transverse to the roadway. Large wrinkles should be avoided. In the case of wide roads, multiple widths of fabric are laid to overlap. The lap length normally depends on the subgrade strength. Table 11-7 provides general guidelines for lap lengths.

Table 11-7. Recommended minimum overlap requirements

CBR	Minimum Overlap
>2	1 – 1.5 feet
1 – 2	2 – 3 feet
0.5 – 1	3 feet or sewn
< 0.5	Sewn
All roll ends	3 feet or sewn

Lay the Base

11-33.   If angular rock is to form the base, it is a common procedure to first place a protective layer of 6 to 8 inches of finer material. The base material is then dumped directly onto the previously spread load, pushed out over the fabric, and spread from the center using a bulldozer. It is critical that the vehicles not drive directly on the fabric nor puncture it. Small tracked bulldozers with a maximum ground pressure of 2 psi are commonly used. The blade is kept high to avoid driving rock down into the fabric. Finally, compaction and grading can be carried out with standard compaction equipment. If the installation has side drains, these are constructed after the pavement.

EARTH RETAINING WALLS

11-34.   As with road construction, there is no specific or preferred method for using geotextiles for retaining walls. Figure 11-9 shows one method that can be adapted to the specific needs of the engineer. The backfill material can be coarse-grained, fine-grained, or alternating layers of coarse- and fine-grained materials.

CONSTRUCTION ON SAND

11-35.   Construction on sand, such as a beach or a desert, presents a severe trafficability problem. The construction of an expedient road through this soil can be expedited by using a plastic geocell material called "sand grid". This material is in the Army's inventory stockage (National Stock Number (NSN) 5680-01-198-7955). The sand grid is a honeycomb-shaped geotextile measuring 20 feet long, 8 feet wide, and 8 inches deep when fully expanded (see figure 11-10, page 11-16).

11-36.   Sand grids are very useful when developing a beachhead for logistics-over-the-shore (LOTS) operations. Construction of sand-grid roadways proceeds rapidly. A squad-sized element augmented with a scoop loader, light bulldozers, and compaction equipment are all that is required to construct a sand-grid road.

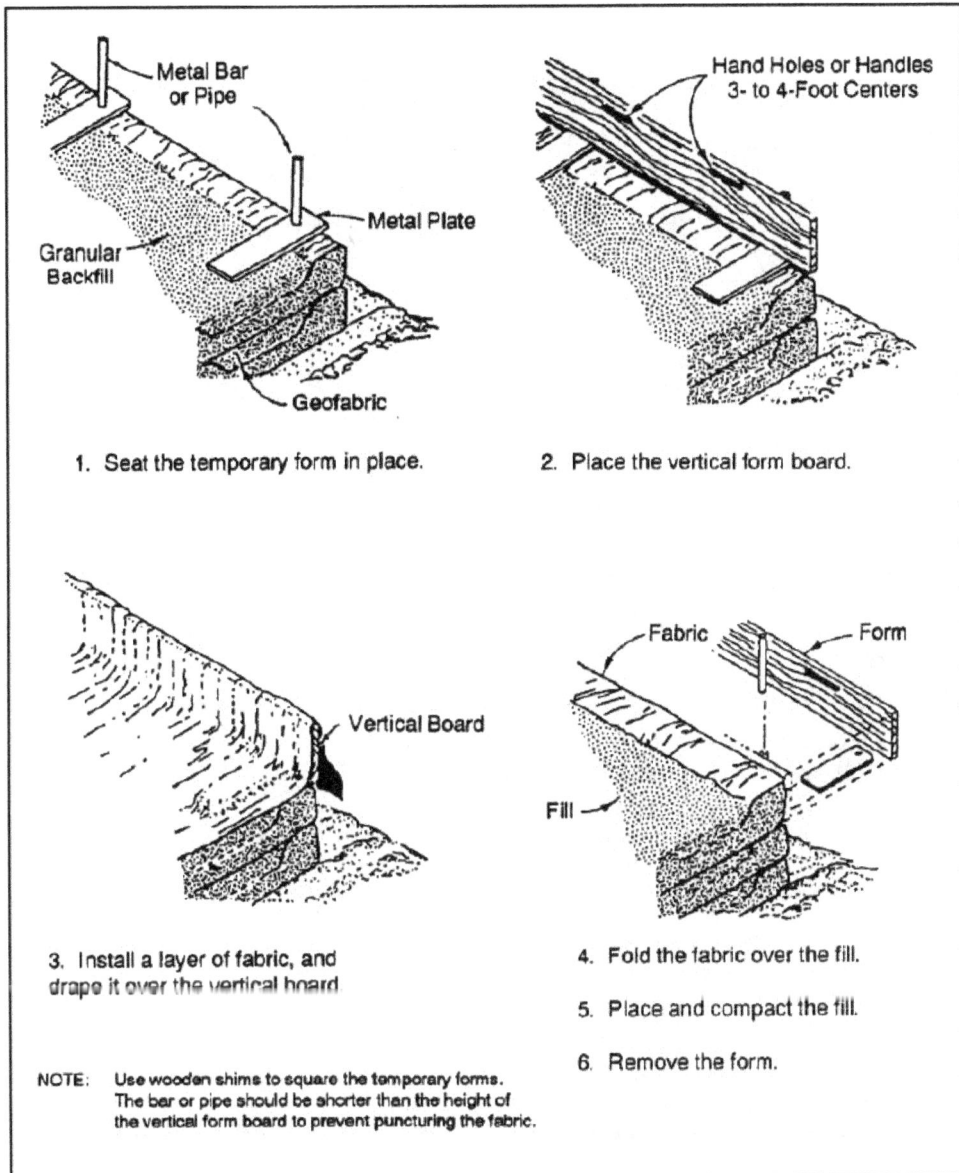

Figure 11-9. Constructing an earth retaining wall using geofabrics

1. Seat the temporary form in place.

2. Place the vertical form board.

3. Install a layer of fabric, and drape it over the vertical board

4. Fold the fabric over the fill.

5. Place and compact the fill.

6. Remove the form.

NOTE: Use wooden shims to square the temporary forms. The bar or pipe should be shorter than the height of the vertical form board to prevent puncturing the fabric.

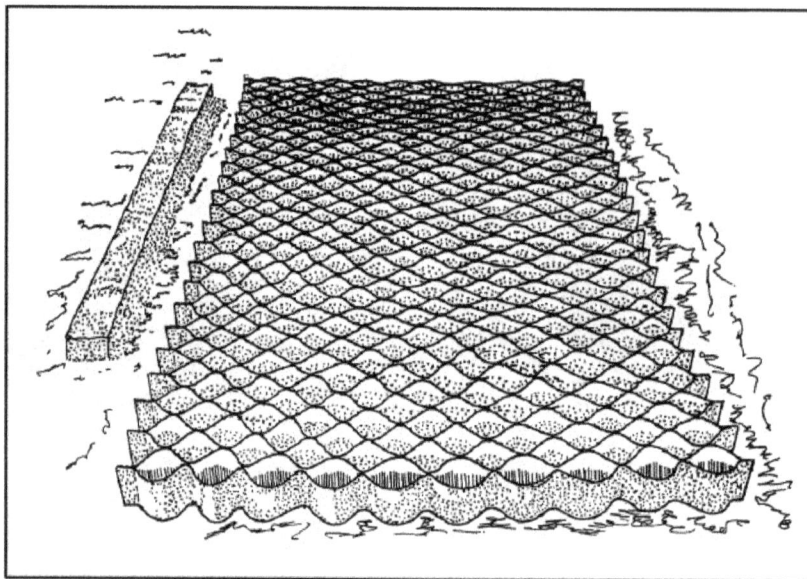

Figure 11-10. Sand grid

Procedures

11-37. Use the following construction procedures for a sand-grid road:

- Lay out the road. Establish a centerline that follows the course of the proposed road.
- Perform the earthwork necessary to level the roadway.
- Distribute folded sand grids along the roadway.
- Expand the sand-grid sections and secure them in place. Use shovels to fill the end cells and some of the side cells with sand.
- Use a scoop loader to fill the grids. They should be overfilled to allow for densification when compacted.
- Compact the sand, using compaction equipment.
- Use a scoop loader to back drag excess sand to the road shoulder area if asphalt or gravel surfacing is used.
- If available, apply an asphalt surface treatment over the filled sand grids to enhance their service life.

11-38. Sand grids perform well under wheeled-vehicle traffic. Tracked-vehicle traffic is very destructive to the sand-grid road. Asphalt surface treatments reduce sand-grid damage when a limited number of tracked vehicles must use the road.

Maintenance

11-39. Sand-grid roads are easily maintained. Entire damaged sections can be re- moved or the damaged portion can be cut and removed and a new grid fitted in its place.

11-40. Sand grids have many uses in the theater of operations other than just roads. They can also be used to construct bunkers, revetments, retaining walls, and a host of other expedient structures.

Chapter 12

Special Soil Problems

Misunderstanding soils and their properties can lead to construction errors that are costly in effort and material. The suitability of a soil for a particular use should be determined based on its engineering characteristics and not on visual inspection or apparent similarity to other soils. The considerations presented in the following paragraphs will help reduce the chances of selecting the wrong soil.

AGGREGATE BEHAVIOR

12-1. Gravel, broken stone, and crushed rock are commonly used to provide a stable layer, such as for a road base. These aggregate materials have large particles that are not separated or lubricated by moisture; thus, there is not a severe loss of strength with the addition of water. They remain stable even in the presence of rain or ponding within the layer. As a result, these soils can be compacted underwater, such as in the construction of a river ford.

SOIL-AGGREGATE MIXTURES

12-2. Construction problems can arise with certain mixtures of soil and aggregate. These include dirty gravel, pit-run broken stone from talus slopes, clays used in fills, and others. If there are enough coarse particles present to retain grain-to-grain contact in a soil-aggregate mixture, the fine material between these coarse particles often has only a minor effect. Strength often remains high even in the presence of water. On the other hand, mixtures of soil and aggregate in which the coarse particles are not continuously in contact with one another will lose strength with the addition of water as though no aggregate particles were present. High plasticity clays (PI > 15) in sand and gravel fills can expand on wetting to reduce grain-to-grain contact, so that strength is reduced and differential movement is increased. The change in texture from stable to unstable soil-aggregate mixtures can be rather subtle, often occurring between two locations in a single borrow pit. To assure a stable material, usually the mixture should contain no more than about 12 to 15 percent by weight of particles passing a Number 200 sieve; the finer fraction should ideally show low plasticity (PI < 5).

LATERITES AND LATERITIC SOILS

12-3. Laterites and lateritic soils form a group comprising a wide variety of red, brown, and yellow, fine-grained residual soils of light texture as well as nodular gravels and cemented soils. They may vary from a loose material to a massive rock. They are characterized by the presence of iron and aluminum oxides or hydroxides, particularly those of iron, which give the colors to the soils. For engineering purposes, the term "laterite" is confined to the coarse-grained vermicular concrete material, including massive laterite. The term "lateritic soils" refers to materials with lower concentrations of oxides.

12-4. Laterization is the removal of silicone through hydrolysis and oxidation that results in the formation of laterites and lateritic soils. The degree of laterization is estimated by the silica-sesquioxide (S-S) ratio ($SiO_2/(Fe_2O_3 + Al_2O_3)$). Soils are classed by the S-S ratios. The conclusion is—

- An S-S ratio of 1.33 = laterite.
- An S-S ratio of 1.33 to 20 = lateritic soil.
- An S-S ratio of 20 = nonlateritic, tropical soil.

LATERITES

12-5. Most laterites are encountered in an already hardened state. In some areas of the world, natural laterite deposits that have not been exposed to drying are soft with a clayey texture and mottled coloring, which may include red, yellow, brown, purple, and white. When the laterite is exposed to air or dried out by lowering the groundwater table, irreversible hardening often occurs, producing a material suitable for use as a building or road stone.

12-6. Frequently, laterite is gravel-sized, ranging from pea-sized gravel to 3 inches minus, although larger cemented masses are possible. A specific form of laterite rock, known as plinthite, is noteworthy for its potential use as a brick. In place, plinthite is soft enough to cut with a metal tool, but it hardens irreversibly when removed from the ground and dried.

LATERITIC SOILS

12-7. The lateritic soils behave more like fine-grained sands, gravels, and soft rocks. The laterite typically has a porous or vesicular appearance. Some particles of laterite tend to crush easily under impact, disintegrating into a soil material that may be plastic. Lateritic soils may be self-hardening when exposed to drying; or if they are not self-hardening, they may contain appreciable amounts of hardened laterite rock or laterite gravel.

LOCATION

12-8. Laterites and lateritic materials occur frequently throughout the tropics and subtropics. They tend to occur on level or gently sloping terrain that is subject to very little mechanical erosion. Laterite country is usually infertile. However, lateritic soils may develop on slopes in undulating topography (from residual soils), on alluvial soils that have been uplifted.

PROFILES

12-9. There are many variations of laterites and lateritic profiles, depending on factors such as the—

- Mode of soil formation.
- Cycles of weathering and erosion.
- Geologic history.
- Climate.

ENGINEERING CLASSIFICATION

12-10. The usual methods of soil classification, involving grain-size distribution and Atterberg limits, should be performed on laterites or suspected laterites that are anticipated for use as fill, base-course, or surface-course materials. Consideration should be given to the previously stated fact that some particles of laterite crush easily; therefore, the results obtained depend on such factors as the—

- Treatment of the sample.
- Amount of breakdown.
- Method of preparing the minus Number 40 sieve material.

12-11. The more the soil's structure is handled and disturbed, the finer the aggregates become in grading and the higher the Atterberg limit. While recognizing the disadvantage of these tests, it is still interesting to note the large spread and range of results for both laterites and lateritic soils.

COMPACTED SOIL CHARACTERISTICS

12-12. Particularly with some laterites, breakdown of the coarser particles will occur under laboratory test conditions. This breakdown is not necessarily similar to that occurring under conditions of field rolling; therefore, the prediction of density, shear, and CBR characteristics for remolded laboratory specimens is not reliable. Strength under field conditions tends to be much higher than is indicated by the laboratory tests, provided the material is not given excessive rolling. Field tests would be justified for supplying data necessary for pavement design. The laboratory CBR results for laterite show higher values after soaking than initially found after compaction, indicating a tendency for recementation. For lateritic soils, the presence of cementation in the grain structure also gives the same effect. In addition, the field strengths may be higher than those in the laboratory, although in a much less pronounced fashion. If the soil is handled with a minimum of disturbance and remolding, laboratory tests do furnish a usable basis for design; however, lesser disturbance and higher strengths may be achieved under field conditions.

PAVEMENT CONSTRUCTION

12-13. The laterized soils work well in pavement construction in the uses described in the following paragraphs, particularly when their special characteristics are carefully recognized. While the AASHTO and the Corps of Engineers classification systems and specifications are the basis for laterite material specification, experience by highway agencies in countries where laterites are found indicates that excellent performance from lateritic soils can be achieved by modifying the temperate zone specifications. The modified specifications are less exacting than either the AASHTO or the Corps of Engineers specifications, but they have proven satisfactory under tropical conditions.

Subgrade

12-14. The laterites, because of their structural strength, can be very suitable subgrades. Care should be taken to provide drainage and also to avoid particle breakdown from overcompaction. Subsurface investigation should be made with holes at relatively close spacing, since the deposits tend to be erratic in location and thickness. In the case of the lateritic soils, subgrade compaction is important because the leaching action associated with their formation tends to leave behind a loose structure. Drainage characteristics, however, are reduced when these soils are disturbed.

Base Course

12-15. The harder types of laterite should make good base courses. Some are even suitable for good quality airfield pavements. The softer laterites and the better lateritic soils should serve adequately for subbase layers. Although laterites are resistant to the effects of moisture, there is a need for good drainage to prevent softening and breakdown of the structure under repeated loadings. Base-course specifications for lateritic soils are based on the following soil use classifications:

- Class I (CBR 100).
- Class II (CBR 70 to 100).
- Class III (CBR 50 to 70).

12-16. Gradation requirements for laterite and laterite gravel soils used for base courses and subbases are listed in table 12-1. Atterberg limits and other test criteria for laterite base course materials are listed in table 12-2, page 12-4.

Table 12-1. Gradation requirements for laterite and laterite gravels

Sieve	Class I	Class II	Class III	Subbase
2 inch	100	100	100	—
1 ½ inch	82-100	82-100	90-100	100
¾ inch	51-100	31-100	69-100	85-100
⅜ inch	30-90	30-90	47-75	70-95
3/16 inch or No 4	19-73	19-73	36-59	55-76
No 8	8-51	8-51	24-43	40-57
No 30	4-31	4-31	20-35	35-48
No 200	0-15	0-15	15-25	25-42

Subbase

12-17. Subbase criteria are listed in table 12-1, page 12-3, and table 12-3. These criteria are less stringent than those found in TM 5-330, which limits subbase soil materials to a maximum of 15 percent fines, a LL of 25, and a PI of 5.

Table 12-2. Criteria for laterite base course materials

Criteria	Class I	Class II	Class III
CBR	100 min	80 min	50 min
LL	35 max	40 max	—
(LL) x (% Passing No 200)	600 max	900 max	1,250 max
PI	10 max	12 max	—
(PI) x (% Passing No 200)	200 max	400 max	600 max
Aggregate Crushing Value	<35	35-40	40-50
Los Angeles Abrasion	< 65%	< 65%	—

Table 12-3. Criteria for laterite subbase materials

CBR	Gradation	Maximum Values		
		PI	LL	No 200
≥ 20	See Table 12-1, page 12-3.	25	40	40

Surfacing

12-18. Laterite can provide a suitable low-grade wearing course when it can be compacted to give a dense, mechanically stable material; however, it tends to corrugate under road traffic and becomes dusty during dry weather. In wet weather, it scours and tends to clog the drainage system. To prevent corrugating, which is associated with loss of fines, a surface dressing may be used. Alternatively, as a temporary expedient, regular brushing helps. The lateritic soils, being weaker than the laterites, are not suitable for a wearing course. Their use for surfacing would be restricted to—

- Emergencies.
- Use under landing mats.
- Other limited purposes.

Stabilization

12-19. The laterite and lateritic soils can be effectively stabilized to improve their properties for particular uses. However, because of the wide range in lateritic soil characteristics, no one stabilizing agent has been found successful for all lateritic materials. Laboratory studies, or preferably field tests, must be performed to determine which stabilizing agent, in what quantity, performs adequately on a particular soil. Some that have been used successfully are—

- Cement.
- Asphalt.
- Lime.
- Mechanical stabilization.

12-20. Laterite and lateritic soils can still perform satisfactorily in a low-cost, unsurfaced road, even though the percent of fines is higher than is usual in the continental United States. This is believed to be due to the cementing action of the iron oxide content. Cement and asphalt work best with material of a lower fines content. When fines are quite plastic, adding lime reduces the plasticity to produce a stable material. Field trials make it possible to determine the most suitable compaction method, which gives sufficient density without destroying the granule structure and cementation. Generally, vibratory compaction is best. Laboratory test programs for tropical conditions do not ordinarily include freeze-thaw tests but should check the influence of wetting and drying. The finer-grained lateritic soils may contain enough active clay to swell and shrink, thus tending to destroy both the natural cementation and the effect of stabilization.

SLOPES

12-21.   The stiff natural structure of the laterites allows very steep or vertical cuts in the harder varieties, perhaps to depths of 15 to 20 feet. The softer laterites and the lateritic soils should be excavated on flatter than vertical slopes but appreciably steeper than would be indicated by the frictional characteristics of the remolded material. It is important to prevent access of water at the top of the slope.

CORAL

12-22.   "Coral" is a broad term applied to a wide variety of construction materials derived from the accumulation of skeletal residues of coral like marine plants and animals. It is found in various forms depending on the degree of exposure and weathering and may vary from a hard limestone like rock to a coarse sand. Coral develops in tropical ocean waters primarily in the form of coral reefs, but many of the South Pacific islands and atolls are comprised of large coral deposits. As a general rule, living colonies of coral are bright-colored, ranging from reds through yellows. Once these organisms die, they usually become either translucent or assume various shades of white, gray, and brown.

TYPES

12-23.   The three principal types of coral used in military construction are—

- Pitrun coral.
- Coral rock.
- Coral sand.

Pit-Run Coral

12-24.   Pit-run coral usually consists of fragmented coral in conjunction with sands and marine shells. At best, it classifies as a soft rock even in its most cemented form. The CBR values for this material may vary between 5 and 70. This material seldom shows any cohesive properties. In general, pit-run coral tends to be well graded, but densities above 120 pcf are seldom achieved.

Coral Rock

12-25.   Coral rock is commonly found in massive formations. The white type is very hard, while the gray type tends to be soft, brittle, and extremely porous. The CBR values for this material vary from 50 to 100. Densities above 120 pcf are common except in some of the soft rock.

Coral Sand

12-26.   Coral sand consists of decomposed coral rock that may be combined with washed and sorted beach sands. Generally, it classifies as a poorly graded sand and seldom will more than 20 percent of the soil particles pass a Number 200 sieve. The CBR values for this material vary between 15 and 50. Because of the lack of fines, compaction is difficult and densities above 120 pcf are uncommon.

SOURCES

12-27.   Coral may be obtained from—

- Construction site cuts.

- Quarries.
- Wet or dry borrow pits worked in benches.

12-28. Rooters should be used to loosen the softer deposits, and the loosened material should be moved with bulldozers to power shovels for loading into trucks or other transport equipment. Rooting and panning are preferable in soft coral pits or shallow lagoons. Where coral requires little loosening, draglines and carryall scrapers can be used. Occasionally, coral from fringing reefs and lagoons can be dug by draglines or shovels, piled as a causeway, and then trucked away progressively from the seaward end. Hard coral rock in cuts or aggregate quarries requires considerable blasting. In both pit and quarry operations, hard coral "heads" are often found embedded in the softer deposits, presenting a hazard to equipment and requiring blasting.

12-29. Blasting hard coral rock differs somewhat from ordinary rock quarrying since coral formations contain innumerable fissures in varying directions and many large voids. The porosity of the coral structure itself decreases blasting efficiency. Conventional use of low-percent dynamite in tamped holes produces the most satisfactory results.

12-30. Shaped charges are ineffective, especially when used underwater; cratering charges, although effective, are uneconomical. Usually holes 8 to 12 feet deep on 4- to 8-foot centers are required to get adequate blasting efficiency.

USES

12-31. Coral may be used as—

- Fills, subgrades, and base courses.
- Surfacing.
- Concrete aggregate.

Fills, Subgrades, and Base Courses

12-32. When properly placed, selected coral that is stripped from lagoon or beach floors or quarried from sidehills is excellent for fills, subgrades, and base courses.

Surfacing

12-33. White or nearly white coral with properly proportioned granular sizes compacted at OMC creates a concretelike surface. The wearing surface requires considerable care and heavy maintenance since coral breaks down and abrades easily under heavy traffic. Conversely, coral surfaces are extremely abrasive, and tire durability is greatly reduced.

Concrete Aggregate

12-34. Hard coral rock, when properly graded, is a good aggregate for concrete. Soft coral rock makes an inferior concrete, which is low in strength, difficult to place, and often of honeycomb structure.

CONSTRUCTION

12-35. Construction with coral presents some special problems, even though the uses are almost the same as for standard rock and soil materials. Since coral is derived from living organisms, its engineering characteristics are unique. Use the steps discussed in the following paragraphs to help minimize construction problems:

- Whenever coral is quarried from reefs or pits containing living coral deposits, allow the material to aerate and dry for a period of 6 days, if possible, but not less than 72 hours. Living coral organisms can remain alive in stockpiles for periods of up to 72 hours in the presence of water. If "live" coral is used in construction, the material exhibits high swell characteristics with accompanying loss of density and strength.

- Carefully control moisture content when constructing with coral. Increases of even 1 percent above OMC can cause reductions of 20 percent or more in densities achieved in certain types of coral. For best results, compaction should take place between the OMC and 2 percent below the OMC. Maintenance on coral roads is best performed when the coral is wet.
- When added to coral materials, salt water gives higher densities, with the same compactive effort, than fresh water. Use salt water in compaction whenever possible.
- Hard coral rock should not be used as a wearing surface. This rock tends to break with sharp edges when crushed and easily cuts pneumatic tires. When constructing with hard coral rock, use tracked equipment as much as possible.

DESERT SOILS

12-36. Deserts are very arid regions of the earth. Desert terrain varies widely as do the soils that compose the desert floor. The desert climate has a pronounced effect on the development of desert soils and greatly impacts on the engineering properties of these soils. Engineering methods effective for road or airfield construction in temperate or tropical regions of the world are often ineffective when applied to desert soils.

12-37. Because of wind erosion, desert soils tend to be of granular material, such as—

- Rock.
- Gravel.
- Sand.

12-38. Stabilization of the granular material is required to increase the bearing capacity of the soil. The options available for stabilizing desert soils are more limited. The primary means of stabilizing desert soils are—

- Soil blending.
- Geotextiles.
- Bituminous stabilization.

12-39. Chemical admixtures generally perform poorly under desert conditions because they cease hydration or "set up" too quickly and therefore do not gain adequate strength.

12-40. If a source of fines is located, the fines may be blended with the in-place soil, improving the engineering characteristics of the resultant soil. However, before using the fine material for blending, perform a complete soil analysis on the material. Also, consider unusual climatic conditions that may occur during the design life of the pavement structure. During Operation Desert Storm, many roads in Saudi Arabia were stabilized with fine material known as marl. The roads performed well until seasonal rains occurred; then they failed.

12-41. Geotextiles can be used alone, such as the sand grid, or in combination with bituminous surfacing. The latter is the most effective.

12-42. Bituminous treatment is the most effective method of stabilizing desert soils for temporary road and airfield use.

ARTIC AND SUBARTIC SOILS

12-43. Construction in arctic and subarctic soils in permafrost areas is more difficult than in temperate regions. The impervious nature of the underlying permafrost produces poor soil drainage conditions. Cuts cause changes to the subsurface thermal regime. Stability and drainage problems result when cuts are made. If the soil contains visible ice, or if when a sample of the frozen soil is thawed it becomes unstable, it is termed thaw-unstable. Cuts into these types of frozen soils should be avoided; however, if this is not possible, substantial and frequent maintenance will undoubtedly be required. Cuts can generally be made into dry frozen (for example, thaw-stable) soils without major problems. Thaw-stable soils are generally sandy or gravelly materials.

12-44. Road and runway design over frost-susceptible soils must consider both frost effects and permafrost conditions. Adjustments to the flexible pavement design may be required to counter the effects of frost and permafrost.

12-45. Depths of freeze and thaw are usually significantly altered by construction. Figure 12-1 illustrates the effect of clearing and stripping on the depth to permafrost after 5 years. The total depth of thaw is strongly influenced by the surface material and its characteristics (see table 12-4).

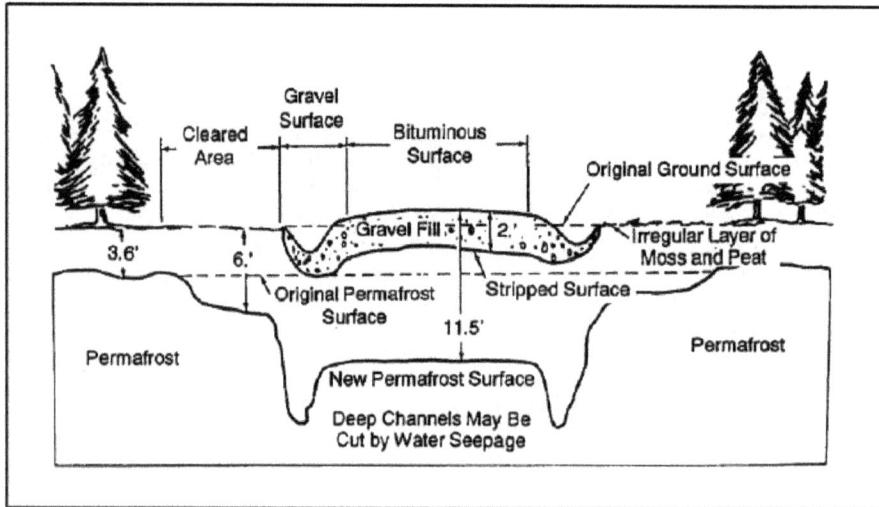

Figure 12-1. Maximum depth to permafrost below a road after 5 years in a subarctic region

Table 12-4. Measured depth of thaw below various surfaces in the subarctic after 5 years. (Fairbanks, Alaska, mean annual temperature 26 degrees Fahrenheit)

Type, of Surface	Color of Surface	Thickness of Pavement (ft)	Nature of Base Course	Thickness of Base Course (ft)	Approximate Elevation of Water Table	Observed Total Depth of Thaw (ft)	Depth of Thaw into Silt Subgrade (ft)
Gravel	Natural		Sand and gravel	4.0	Bottom of base course	8.0- 10.5	4.0-6.5
Concrete	Natural	0.5	Sand and gravel	4.0	Bottom of base	8.5-9.5	4.0-5.0
Concrete	Natural	0.5	Sand	4.0	course	8.5-9.5	4.0-5.0
Asphalt	Black	0.4	Sand and gravel	4.0	Bottom of base	8.5- 10.0	4.0-5.5
Trees, brush, grass, and moss	Natural vegetation				Surface	3.0-4.0	3.0-4,0
Grass and moss	Natural minus trees and brush				Surface	5.0 - 6.0	5.0-6.0
Grass without moss	Natural grass				Surface	8.0-9.0	8.0-9.0

12-46. Increasing the depth of the nonfrost-susceptible base course material can prevent subgrade thawing. The required base thickness may be determined from figure 12-2. The thawing index is determined locally by calculating the degree days of thawing that take place. Thawing degree days data is usually available from local weather stations or highway departments. If thawing degree days data is unavailable, they can be estimated from information contained in figures 12-3 through 12-5, pages 12-10 through 12-12.

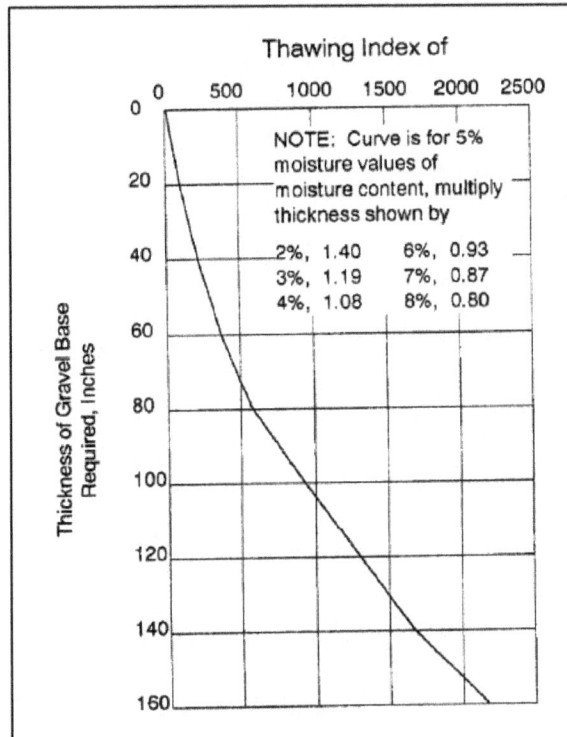

Figure 12-2. Thickness of base required to prevent thawing of subgrade

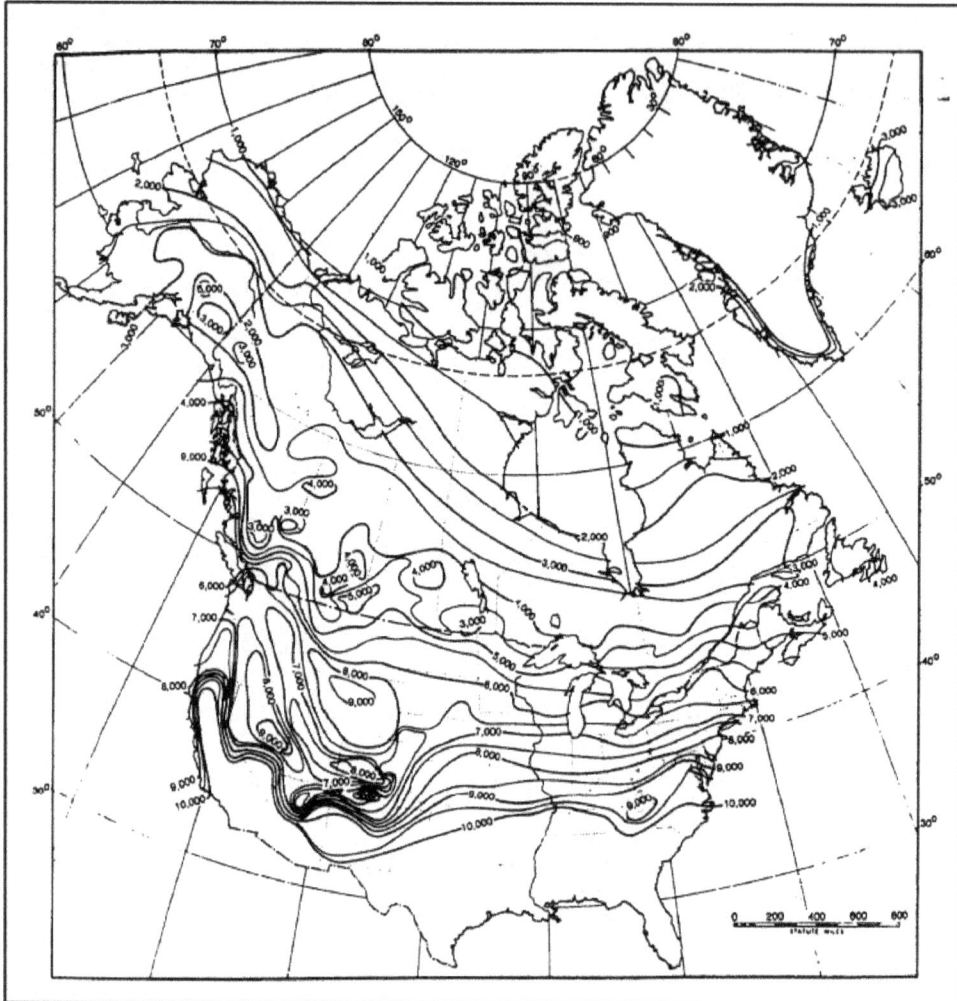

Figure 12-3. Distribution of mean air thawing indexes (°) — North America

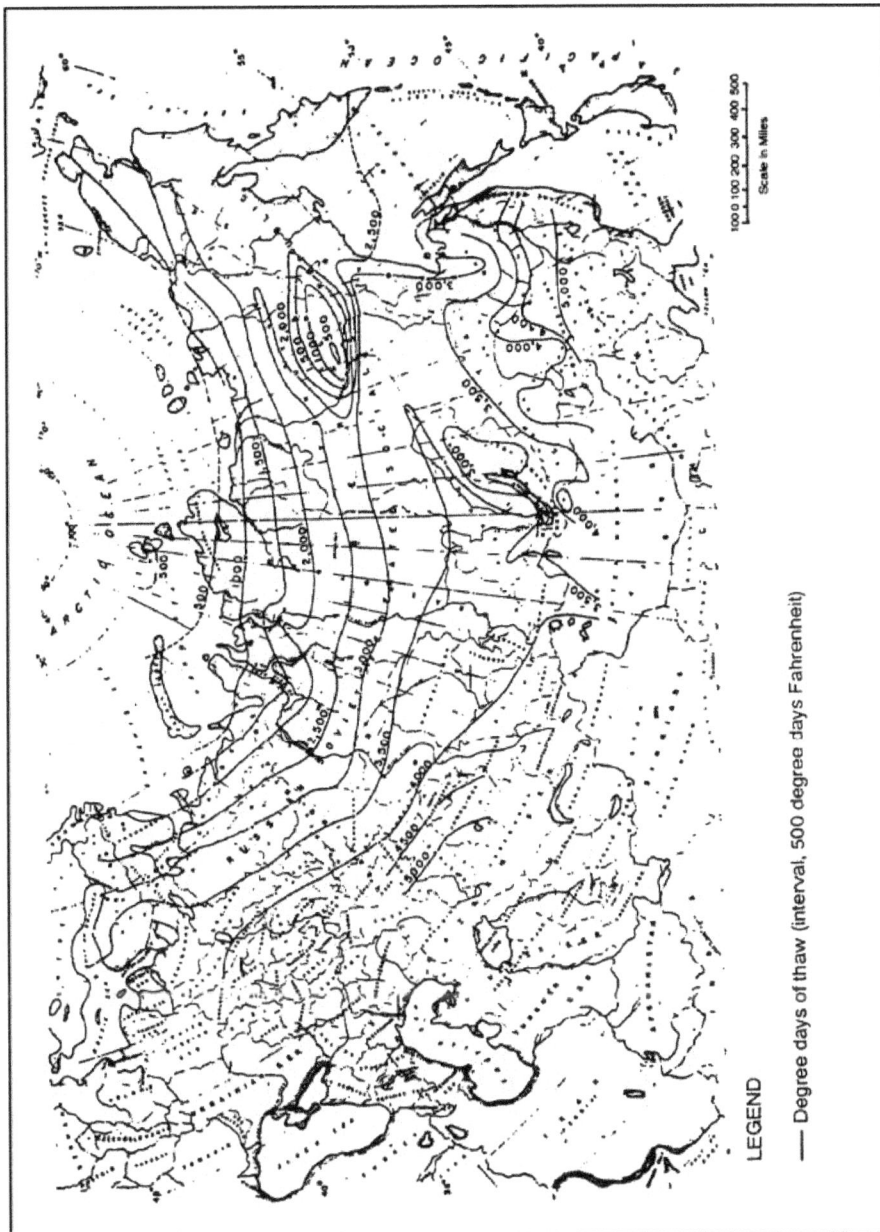

Figure 12-4. Distribution of mean air thawing indexes (°) — Northern Eurasia

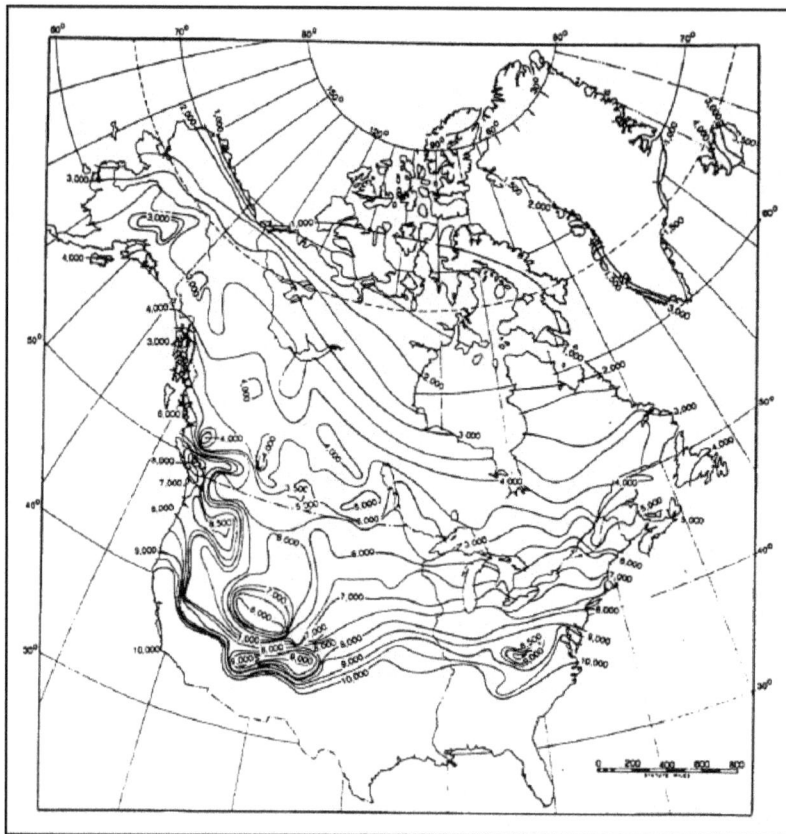

Figure 12-5. Distribution of mean air thawing index values for pavements in North America (°)

SURFACE-THAWING INDEX

12-47. The surface-thawing index may be computed by multiplying the thawing index based on air temperature by a correction factor for the type of surface. For bituminous surfaces, multiply by a factor of 1.65; for concrete, multiply by a factor of 1.5.

12-48. For example, design a road over a permafrost region to preclude thawing of the permafrost. The mean air thawing index is 1,500 degree days (Fahrenheit). Base course material moisture content is 7 percent. Bituminous surface is to be used.

12-49. 1,200 degree days (Fahrenheit) x 1.6 = 1,920 surface-thawing index (air-thawing index) x (bituminous correction factor)

12-50. What is the thickness of the base course material required to prevent thawing of the permafrost? (See figure 12-2, page 12-9.)

12-51. 150 inches of the base at 5 percent moisture; adjust for moisture difference

$$150 \times 0.87 = 130.5 \; inches$$

12-52. Therefore, for this example, permafrost thawing can be prevented by constructing a 130.5-inch base-course layer.

12-53. If the subgrade is thaw-stable, a soil that does not exhibit loss of strength or settlement when thawed may accept some subgrade thawing because little settlement is expected. If the subgrade is thaw-unstable, a granular embankment 4 to 5 feet thick is generally adequate to carry all but the heaviest traffic. Considerable and frequent regrading is required to maintain a relatively smooth surface. Embankments only 2 to 3 feet thick have been used over geotextile layers that prevent the underlying fine-grained subgrade from contaminating the granular fill. Additional fill may be necessary as thawing and settlement progress. It is usually desirable to leave the surface organic layer intact. If small trees and brush cover the site, they should be cut and placed beneath the embankment. They serve as a barrier to prevent mixing of the subgrade with the embankment material. Figure 12-6, page 12-14, and figure 12-7, page 12-15, assist the flexible pavement designer in determining the depth of thaw and freeze, respectively, beneath pavements with gravel bases. The curves are based on the moisture content of the subgrade soil.

ECOLOGICAL IMPACT OF CONSTRUCTION

12-54. The arctic and subarctic ecosystems are fragile; therefore, considerable thought must be given to the impact that construction activities will have on them. Figure 12-8, page 12-16, shows the long-term (26 years) ecological impacts of construction activities in Fairbanks, Alaska. Stripping, compacting, or otherwise changing the existing ground cover alters the thermal balance of the soil.

12-55. Unlike temperate ecosystems that can recover from most construction activities in a relatively short period of time, the arctic and subarctic ecosystems do not recover quickly. In fact, once disturbed, the arctic and subarctic ecosystems can continue to undergo degradation for decades following the disturbance.

12-56. The project engineer must consider these long-term environmental impacts, from the perspective of a steward of the environment, and the effects such environmental degradation may have on the structure's useful life.

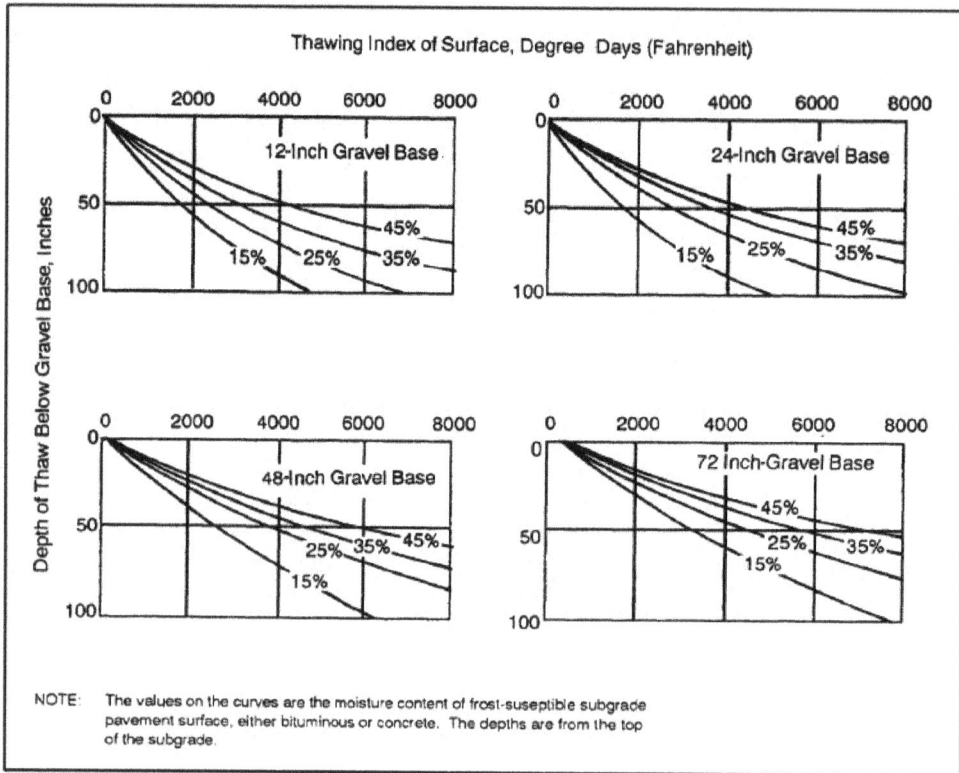

Figure 12-6. Determining the depth factor of thaw beneath pavements with gravel bases

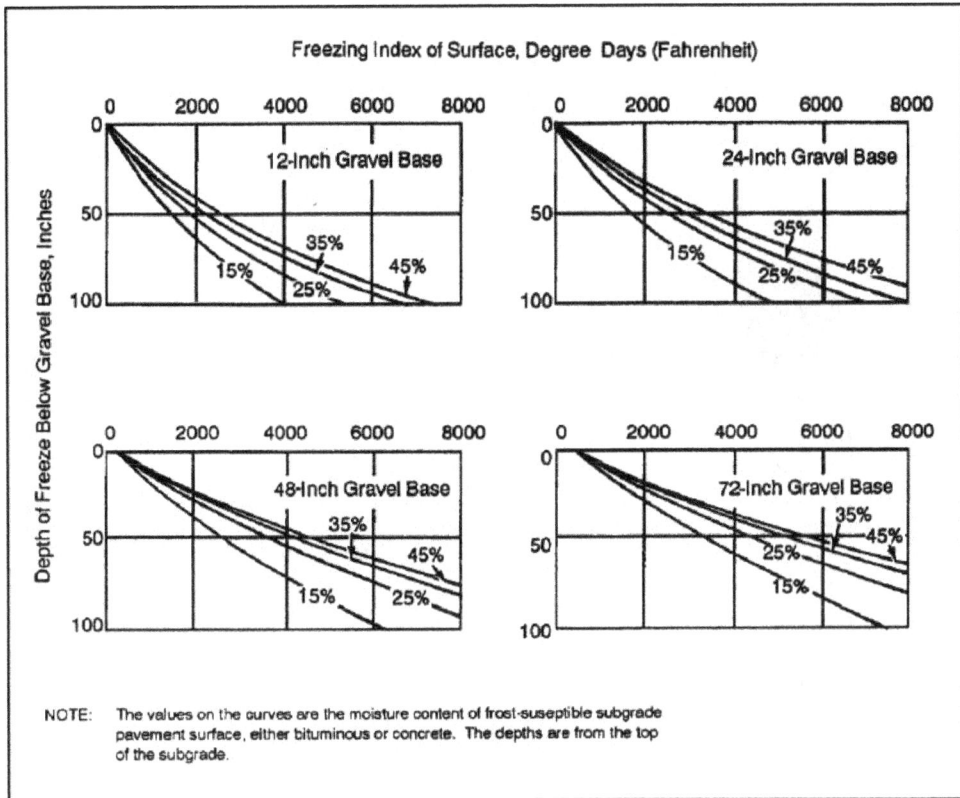

Figure 12-7. Determining the depth of freeze beneath pavements with gravel bases

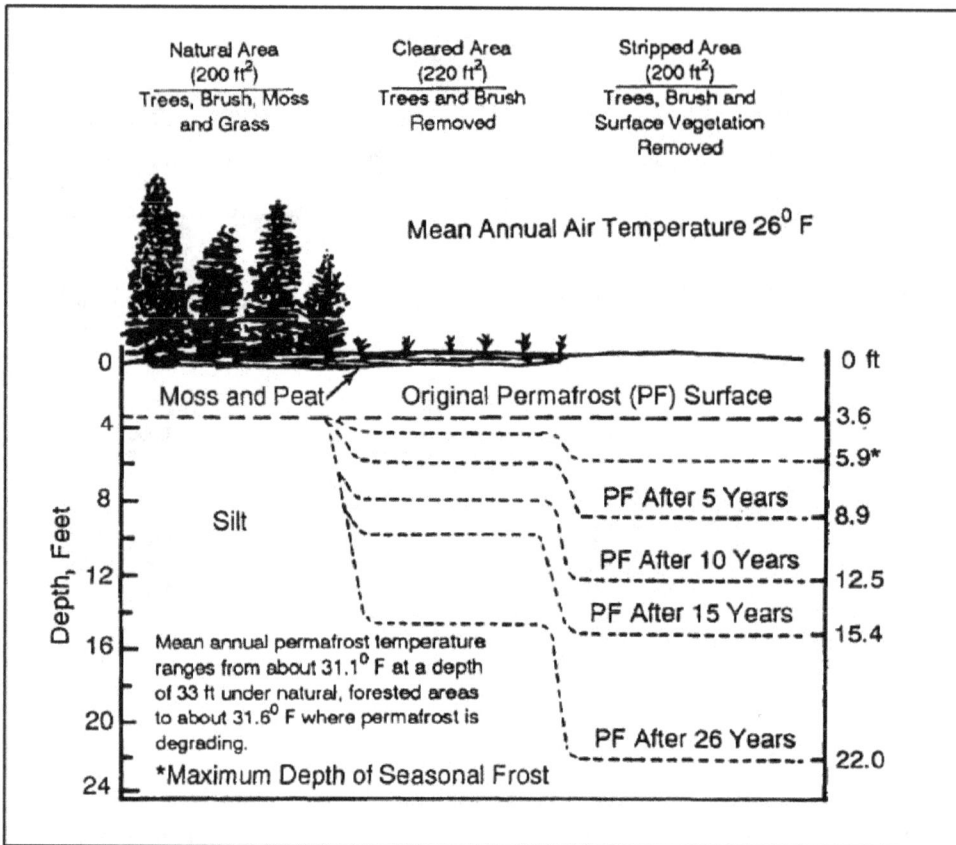

Figure 12-8. Permafrost degradation under different surface treatments

Appendix A

California Bearing Ratio Design Methodology

The CBR is the basis for determining the thickness of soil and aggregate layers used in the design of roads and airfields in the theater of operations. A soil's CBR value is an index of its resistance to shearing under a standard load compared to the shearing resistance of a standard material (crushed limestone) subjected to the same load.

CBR DESIGN METHODOLOGY

A-1. The CBR design methodology has four major parts:

- Evaluate the soil to determine its engineering characteristics (gradation, Atterberg limits, swell potential, Proctor test (CE 55 compaction test) and CBR test). These tests are performed by a Materials Quality Specialist, Military Occupational Speciality (MOS) 51G or a soils testing laboratory according to standard testing procedures (see TM 5-530).
- Evaluate the laboratory data to determine the initial design CBR value and the compactive effort to be applied during construction.
- Evaluate all soil data to determine the final design CBR and construction use of the soil or aggregate.
- Determine the thickness of the soil layer based on the final design CBR and the road or airfield use category (see TM 5-330).

A-2. This appendix focuses on the second and third parts of the methodology.

CBR DESIGN FLOWCHART

A-3. The CBR Design Flowchart (see figure A-1, pages A-2 and A-3) is a useful tool when determining the initial and final design CBR values.

> **Step 1. Look at the Compaction Curves**. For CBR analysis, soils are classified into one of three soil groups:
>
> - Free-draining.
> - Swelling.
> - Nonswelling.

A-4. The compaction curve on page 1 of a Department of Defense (DD) Form 2463 gives an indication as to the group in which a particular soil falls. A U-shaped compaction curve indicates a free-draining soil. A bell-shaped compaction curve indicates that the soil is either a swelling or a nonswelling soil.

> **Step 2. Look at the Swell Curve.** To distinguish between a swelling and nonswelling soil, look at the swell data plotted on a DD Form 1211. If the percent of swell exceeds 3 percent for any soil moisture content, the soil is classified as a swelling soil. If the percent of swell never exceeds 3 percent, the soil is nonswelling.

> **Step 3. Find the Peak of the CE 55 Curve.** The maximum dry density is found at the peak of the CE 55 curve. For free-draining soils, the peak of the curve occurs at the point of the curve where there is no increase in dry density with an increase in moisture.

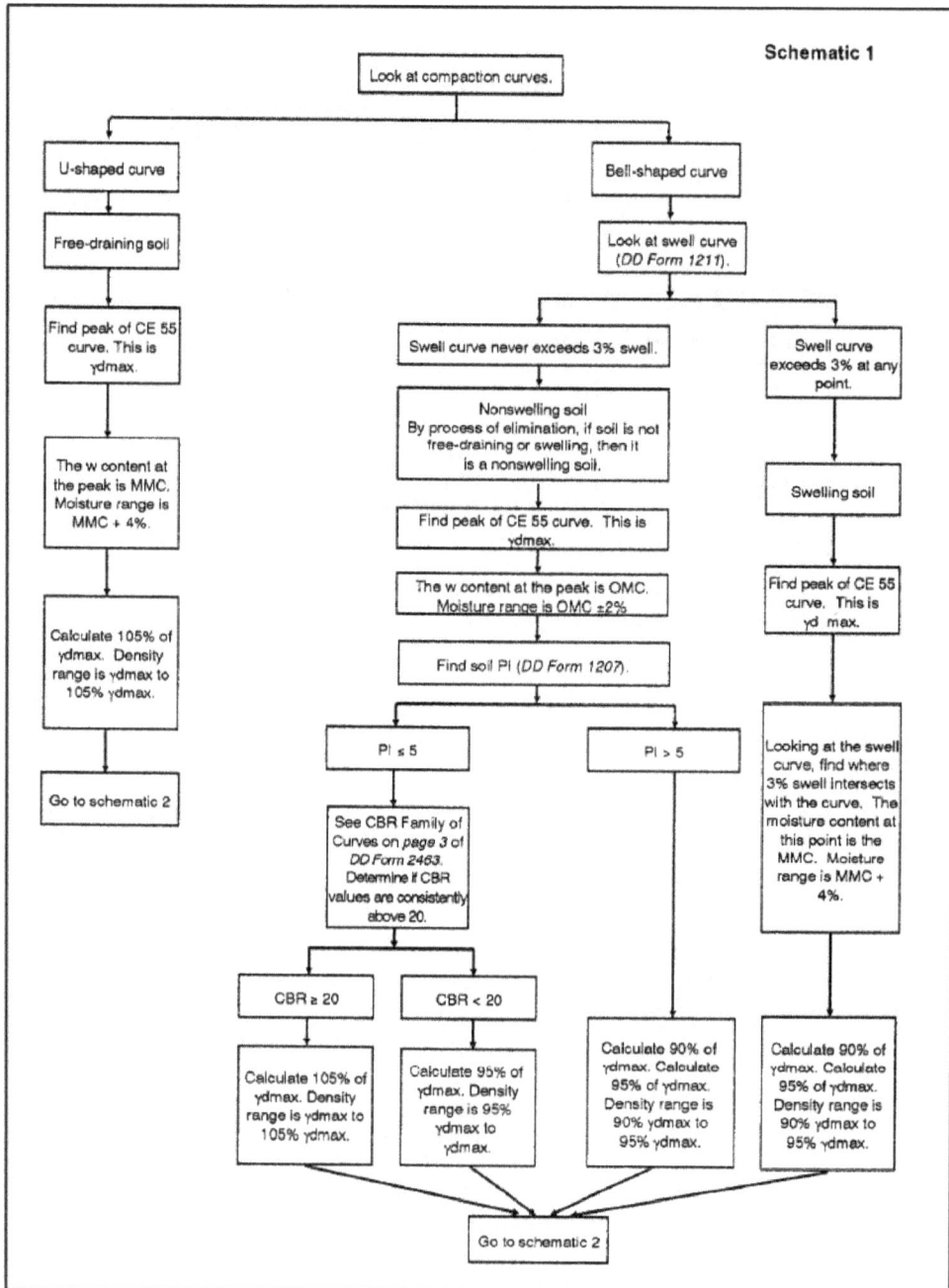

Figure A-1. CBR design flowchart

Schematic 2

Plot the density range as vertical lines on the CBR Family of Curves on *page 3 of DD Form 2463.*

↓

Shade in area between moisture content limits and density limits.

↓

Design CBR is the lowest CBR in shaded area.

↓

Go to Schematic 3

Schematic 3

Gather data on soil:
CBR
LL
PI

↓

Look at the CBR requirements for each soil layer. Is the soil strong enough for the layer in question?

↓

Yes

↓

See TM 5-330, Chapter 7, for base course, subbase, and select specifications. See the requirements for gradation, LL, and PI. Does it meet the specifications?

Yes

No

Place the soil using the design moisture range and the appropriate compactive effort found from the CBR Design.

Either find another soil that has better quality indicators (such as higher CBR, lower PI, lower LL, and better gradation), or consider stabilization to improve the soil, or consider using the soil for a layer lower in the flexible pavement profile. If soil is used in a lower layer, make sure to adjust density range, if needed.

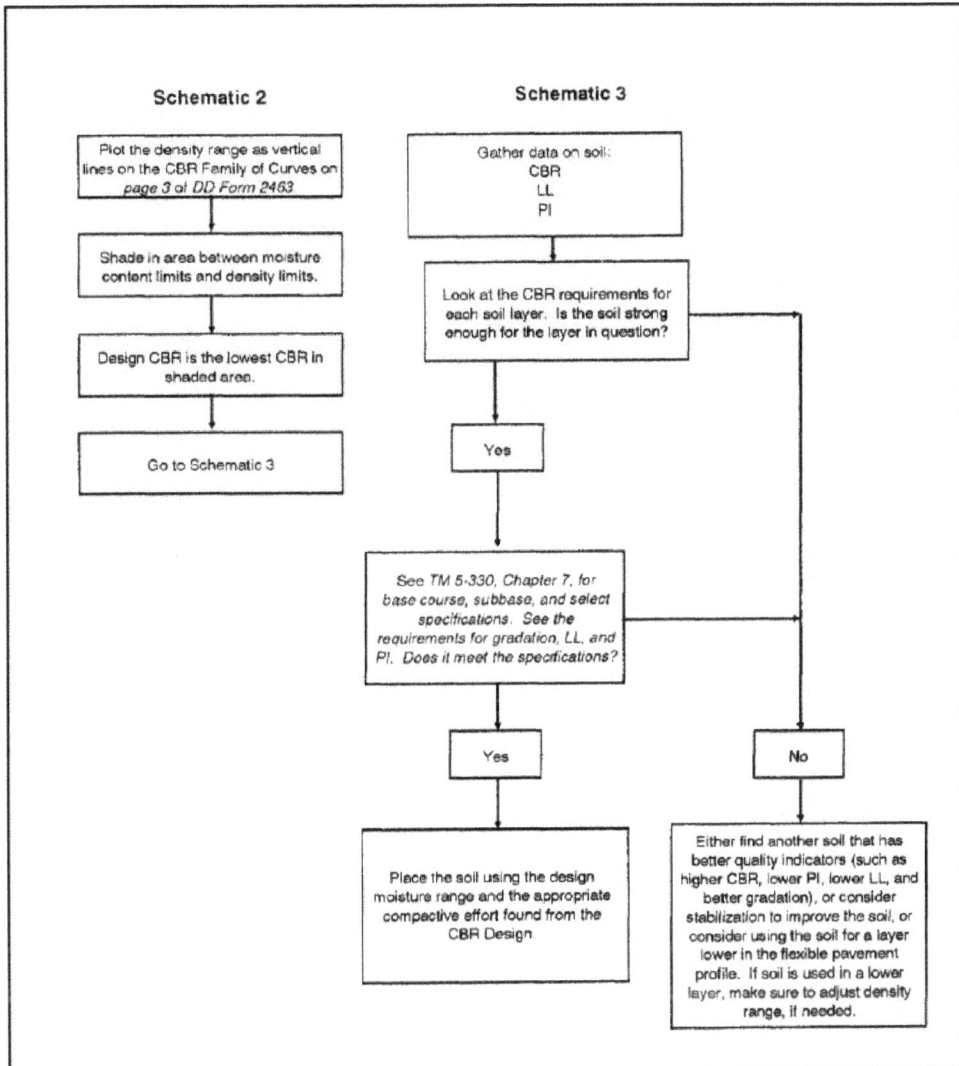

Figure A-1. CBR design flowchart (continued)

Step 4. **Determine the Design Moisture Range.** The design moisture range is influenced by the soil group.

- Free-draining soils. The moisture content that corresponds to the maximum dry density is the MMC. The design moisture range is MMC + 4 percent. For example, if the maximum dry density occurred at 13 percent soil moisture, the MMC would equal 13 percent and the design moisture range would be 13 to 17 percent soil moisture.

- Swelling soils. Army standards permit no more than 3 percent swell to occur after a soil has been placed and compacted. Therefore, swelling soils must be preswelled to a moisture content that will result in 3 percent or less swell. The moisture content at which 3 percent swell occurs is called the MMC. The design moisture range for a swelling soil is MMC + 4 percent. The moisture content corresponding to the maximum dry density is not considered for swelling soils.

- Nonswelling soils. The moisture content that corresponds to the maximum dry density is the OMC. The design moisture range is OMC ± 2 percent. For example, if the OMC is 12 percent, the design moisture range would be 10 to 14 percent.

Step 5. **Find the Soil Plasticity Index.** This step applies to nonswelling soils only. Noncohesive soils (PI 5) can be compacted to greater densities than cohesive soils (PI 5). If the soil is cohesionless, look at the CBR Family of Curves on page 3 of a DD Form 2463 to determine the compactive effort.

Step 6. **Determine the Density Range.** The theater-of-operations, standard compaction range is 5 percent, unless otherwise stated. The maximum dry density is the basis for determining the density range (compactive effort).

- Free-draining soils. They are compacted to between 100 percent and 105 percent maximum dry density.
- Swelling soils. They are compacted to between 90 and 95 percent of maximum dry density.
- Nonswelling soils.
  - PI > 5. Cohesive nonswelling soils are compacted to between 90% and 95% of maximum dry density.
  - PI $\leq$ 15. and CBR < 20. cohesionless, nonswelling soils having CBR values below 20 are compacted to between 95 percent and 100 percent maximum dry density.
  - PI $\leq$ 5. and CBR $\geq$ 20. cohesionless, nonswelling soils having CBR values are compacted to between 100 percent and 105 percent maximum dry density.

Step 7. **Plot the Density Range.** Draw vertical lines on the CBR Family of Curves on page of a DD 2463, corresponding to the density range determined in previous steps. Circle the moisture values that correspond to the moisture range determined in earlier (five moisture values should be circled).

Step 8. **Determine the Initial Design CBR.** Using the CBR Family of Curves, find where each moisture curve, within the moisture range, enters and exits the density-range limit lines drawn in previous step. Find the CBR value corresponding to the lowest entrance or exit point of each curve. This is the initial design CBR.

Step 9. **Gather the Soil Data.** The CBR is not the only criteria used when determining where to place a soil in the road or airfield design. Criteria for the LL, the PI, and the grain-size distribution must also be satisfied. Subbase and select material criteria are found in table A-1. Base-course gradation requirements are found in table A-2.

Step 10. **Determine the Final Design CBR.** Using the initial design CBR value, determine if the soil material is suitable for use as a base, subbase, select, or subgrade layer material. Next, look at the gradation requirements for use in that layer. Finally, look at the LL and the PI criteria. If a soil material meets all the criteria for use in a soil layer, then it can be placed in that layer. If it fails to meet all criteria for its intended use, consider using the material in another layer. The final design CBR is determined by the criteria that the soil material meets. The following examples illustrate this point:

- A soil material has an initial design CBR value of 65. Based on CBR value, this material is suitable for use in a base-course layer for a road. The maximum aggregate size is 1.5 inches. When evaluating the soil against the base-course criteria in table A-2, we find that the percent passing the Number 40 sieve is 6 percent, which is less than the minimum allowable for use as a base course. Therefore, we cannot use the material as a base course, but perhaps it can be used as a subbase. By evaluating the soil against the CBR 50 subbase criteria (see table A-1), we find that the soil material meets all criteria. The final design CBR for this soil would be CBR 50. It would be used as a CBR 50 subbase.

- A soil material has an initial design CBR of 37. We are considering the use of this material as a subbase. There are no criteria for a CBR 37 subbase; therefore, we will use the following rule:
  - If a soil material meets the criteria of the next higher subbase, the final design CBR will be the same as the initial design CBR.
  - If a soil material fails to meet the criteria of the next higher subbase but meets the criteria of a lower subbase or select material, the initial design CBR will be adjusted downward to the maximum CBR value of the layer at which all criteria were met, which becomes the final design CBR.

A-5. In our example, if the soil material met the CBR 40 subbase criteria, the final design CBR would be 37 and the road or airfield would be designed with a CBR 37 subbase layer. If the soil failed to meet the CBR 40 subbase criteria but did meet the CBR 30 subbase criteria, the final design CBR would be 30 and the road or airfield would be designed with a CBR 30 subbase. If the soil failed to meet both the CBR 40 and CBR 30 subbase criteria, the material would be evaluated against the select material criteria. If it met the select criteria, the final design CBR would be 20.

Table A-1. Recommended maximum permissible values of gradation and Atterberg limit requirements in subbases and select materials

Material	Maximum Design CBR		Size, Inches		Gradation Requirements Percent Passing				Atterberg Limits			
					No 10		No 200		LL		PI	
	Air-Fields	Roads	Air-fields	Roads	Air-fields	Roads	Air-fields	Roads	Air-fields	Roads	Air-fields	Roads
Subbase	50	50	3	2	50	50	15	15	25	25	5	5
Subbase	40	40	3	2	80	80	15	15	25	25	5	5
Subbase	30	30	3	2	100	100	15	15	25	25	5	5
Select Material	Below 20	20	3	3	—	—	25	—	35	35	12	12

Table A-2. Desirable gradation for crushed rock, gravel, or slag and uncrushed sand and gravel aggregates for base courses

Sieve Designation	Percent Passing Each Sieve (Square Openings) By Weight				
	Maximum Aggregate Size				
	3-Inch	2-Inch	1½-Inch	1-Inch	1-Inch Sand-Clay
3-inch	100				
2-inch	65-100	100			
1½-inch		70-100	100		
1-inch	45-75	55-85	75-100	100	100
¾-inch		50-80	60-90	70-100	
⅜-inch	30-60	30-60	45-75	50-80	
No 4	25-50	20-50	30-60	35-65	
No 10	20-40	15-40	20-50	20-50	65-90
No 40	10-25	5-25	10-30	15-30	33-70
No 200	3-10	0-10	5-15	5-15	8-25

CBR DESIGN PRACTICAL EXERCISE

A-6. In preparation for an airfield construction project in support of a Marine Corps exercise, a soil sample was obtained from Rio Meta Plain, Venezuela. The soil was tested, and the test results are presented in figures A-2 through A-9, pages A-7 through A-14. Using the USCS, the soil was classified as a (GM-GC) (see figure A-2, and figure A-3, page A-8).

A-7. What are the initial and final design CBRs of this soil? Where will this soil be placed in the airfield design?

A-8. Looking at the density-moisture curve on figure A-4, page A-9, we see that the compaction curve is bell-shaped (Step 1); therefore, the soil is either swelling or nonswelling. The swell data is shown in figure A-5, page A-10. Swelling never exceeds 3 percent at any point on the curve (Step 2). The soil is nonswelling.

A-9. Maximum dry density occurs at 125 pcf (figure A-2), and the OMC is 11.3 percent. The design moisture range is 9.3 to 13.3 percent (Steps 3 and 4).

A-10. The LL and PL were determined to be 23 and 18, respectively. PI = LL - PL (23 - 18 = 5); therefore, the soil is cohesionless. We must look at the CBR Family of Curves to determine the compactive effort (Step 5).

A-11. Figure A-6, page A-11, shows the CBR Family of Curves. Note that while some values exceed 20, most of the data points are clustered below 20. Thus, we will compact this soil to a density between 95 percent and 100 percent maximum dry density or between 118.75 and 125 pcf (Step 6). When plotted on the density-moisture curve with the moisture range (9.3 to 13.3 percent), the specification block is formed (see figure A-7, page A-12).

A-12. The CBR Family of Curves, Figure A-8, page A-13, shows the density range superimposed on the curves and the moisture range values circled (in this case we've circled the moisture values 9-13) (Step 7). Following each moisture curve in our range and marking where it enters and exits the density limits, we find that the lowest CBR value occurs at the point where the 13 percent moisture curve exits the 118.75 pcf density limit line. Interpolating between the 13 percent and 14 percent moisture curves, 13.3 percent moisture results in an assured CBR value of 7.5 (see figure A-9, page A-14) (Step 8).

A-13. The initial design CBR of this soil is 7.5. Because of the low CBR, we can consider this soil material for use as select material if it

A-14. meets the criteria of table A-1, page A-5. Looking at DD Form 1207 (figure A-7), we first check the maximum aggregate size and see that it is 2 inches. The maximum allowable aggregate size (see table A-1 for a select material used in the construction of an airfield is 3 inches; therefore, our soil meets this criteria. Next, we check for the percent passing the Number 200 sieve and find that the percent passing the Number 200 sieve is 23 percent. From table A-1 we find that the select criteria allows up to 25 percent of the material to pass the No 200 sieve; therefore, our soil meets this criteria. Finally, we must evaluate the LL and PI of our soil against the select criteria in table A-1. When we do this, we find that our soil's PI of 23 and LL of 5 meet the requirements.

A-15. The final design CBR of the soil is 7.5, and it is suitable for use as a select material.

A-16. The final portion of the CBR design process is to determine the thickness of the road or airfield structure based on the CBR values of the subgrade, select, subbase, and base- course materials available for use in the construction effort. This is discussed in detail in TM 5-330.

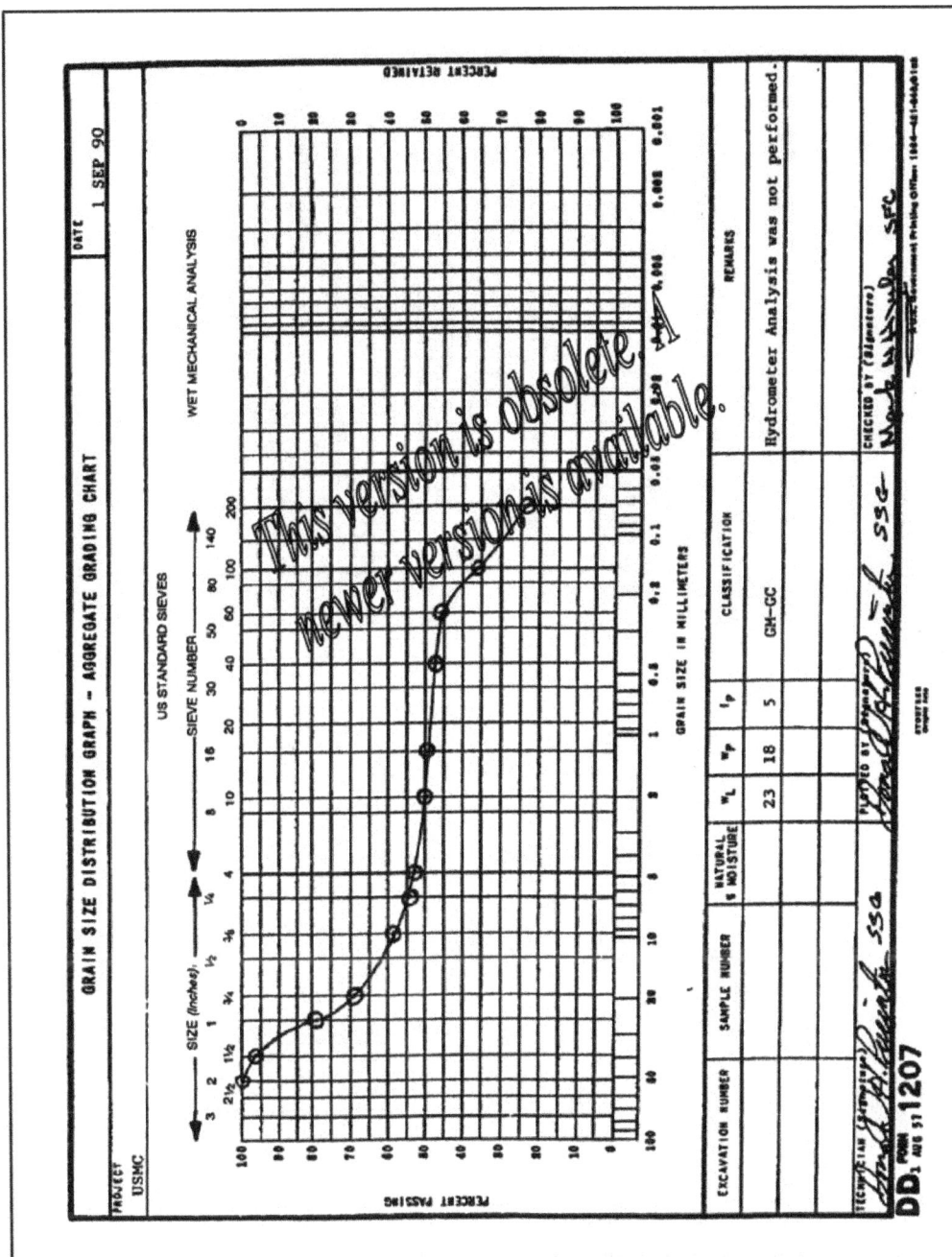

Figure A-2. Grain size distribution of Rio Meta Plain soil

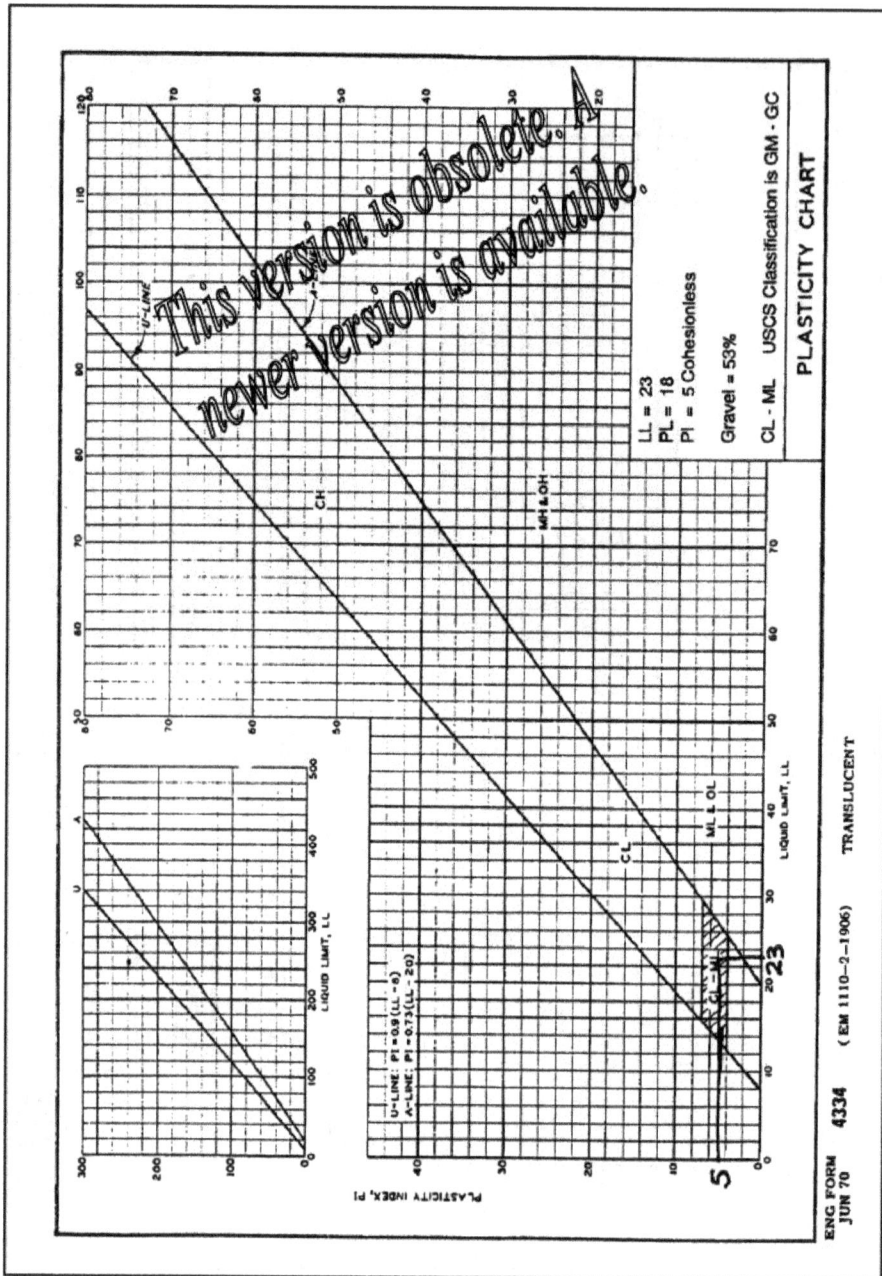

Figure A-3. Plasticity chart plotted with Rio Meta Plain soil data

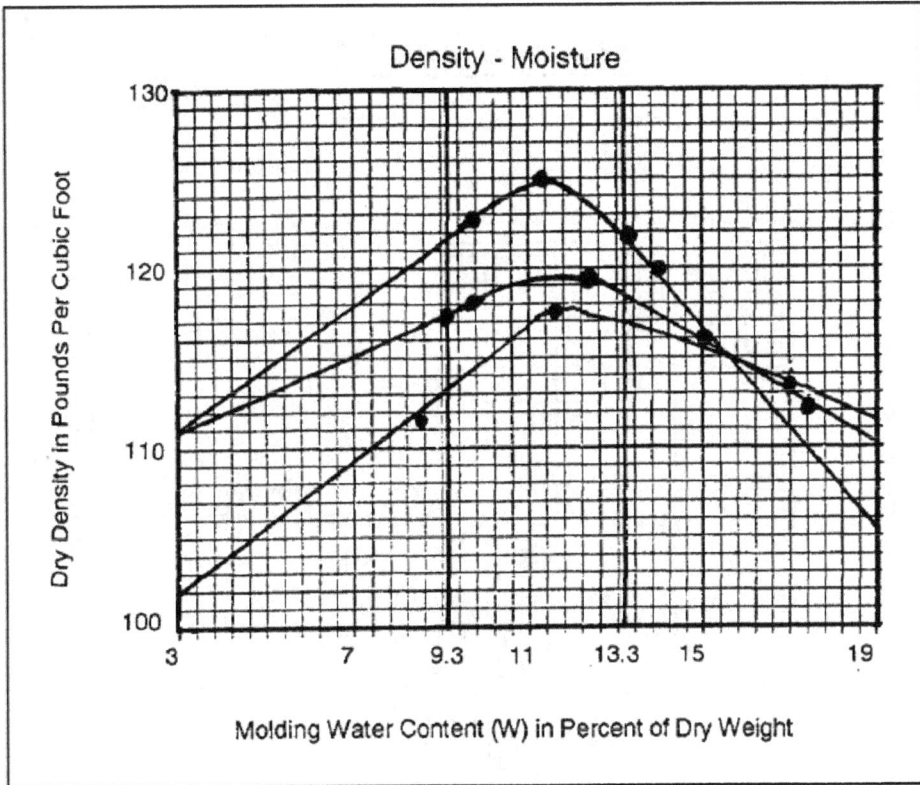

Figure A-4. Density moisture curve for Rio Meta Plain soil

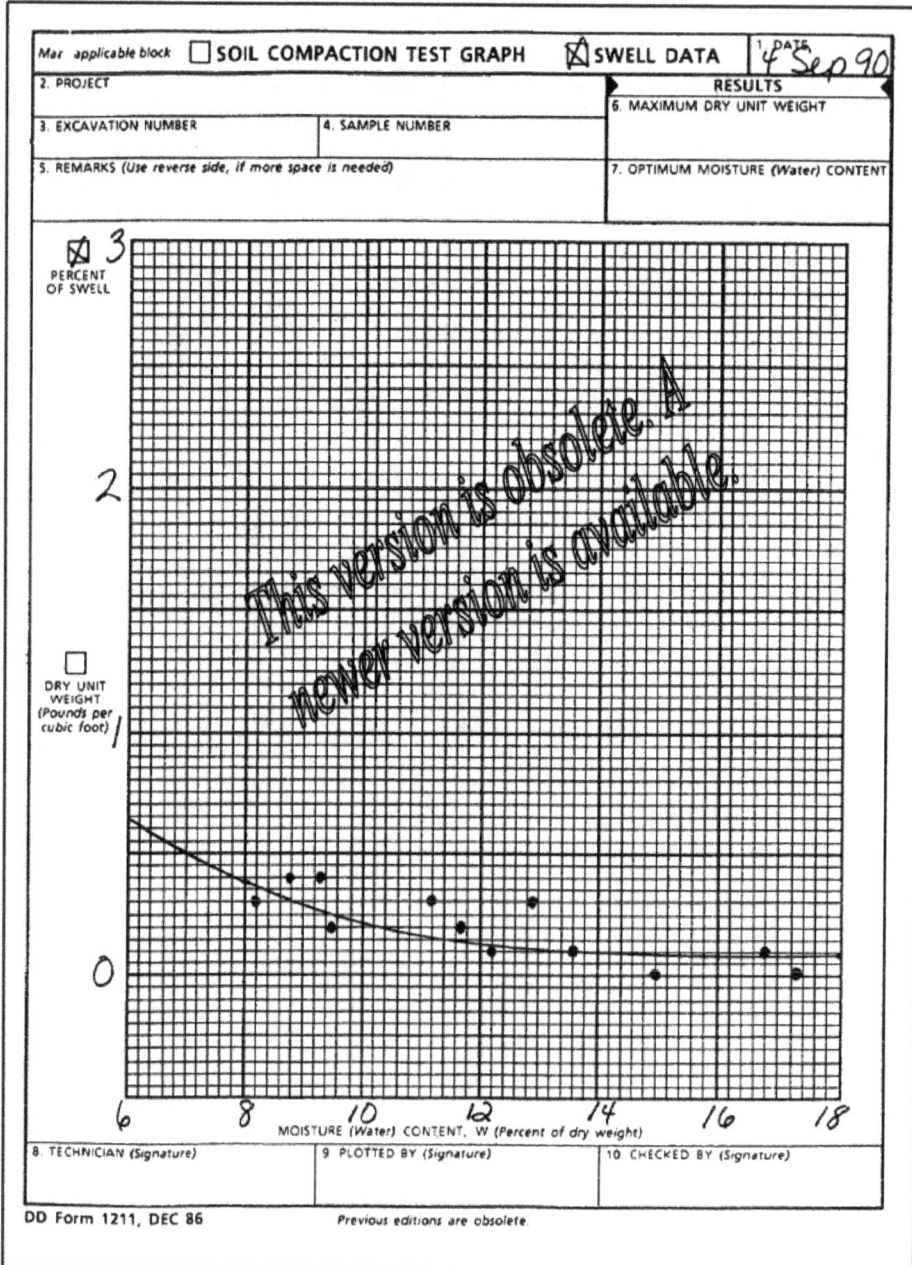

Figure A-5. Swelling curve for Rio Meta Plain soil

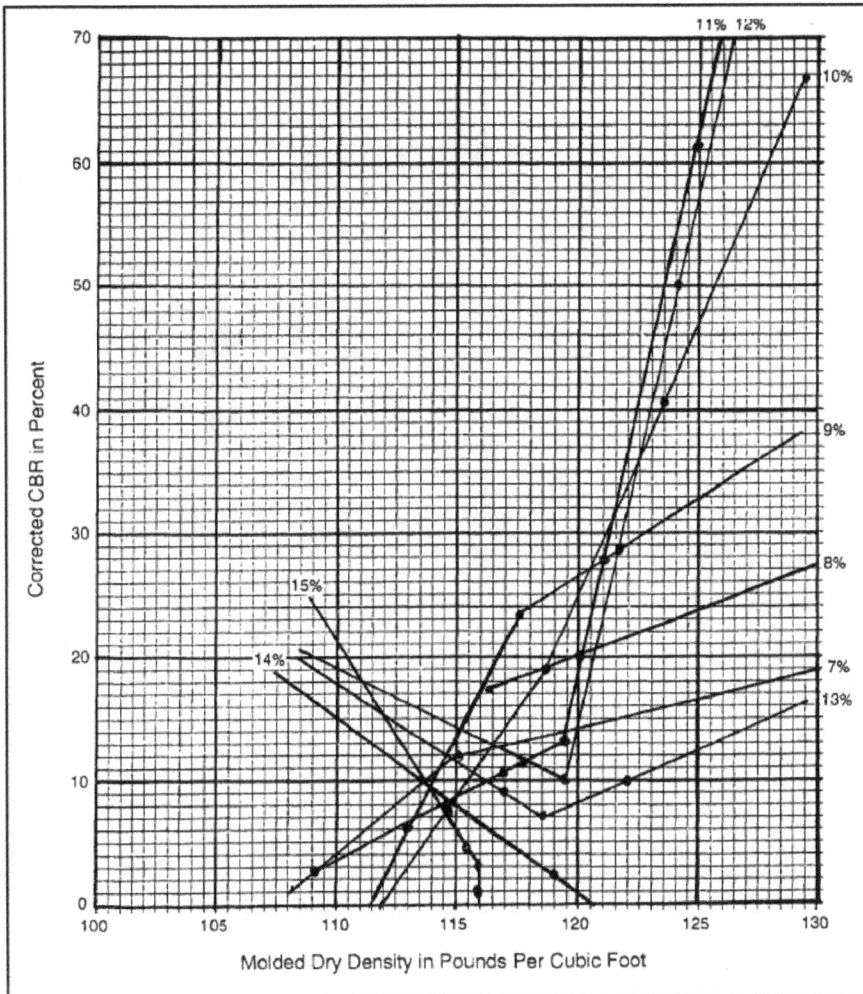

Figure A-6. CBR family of curves for Rio Meta Plain soil

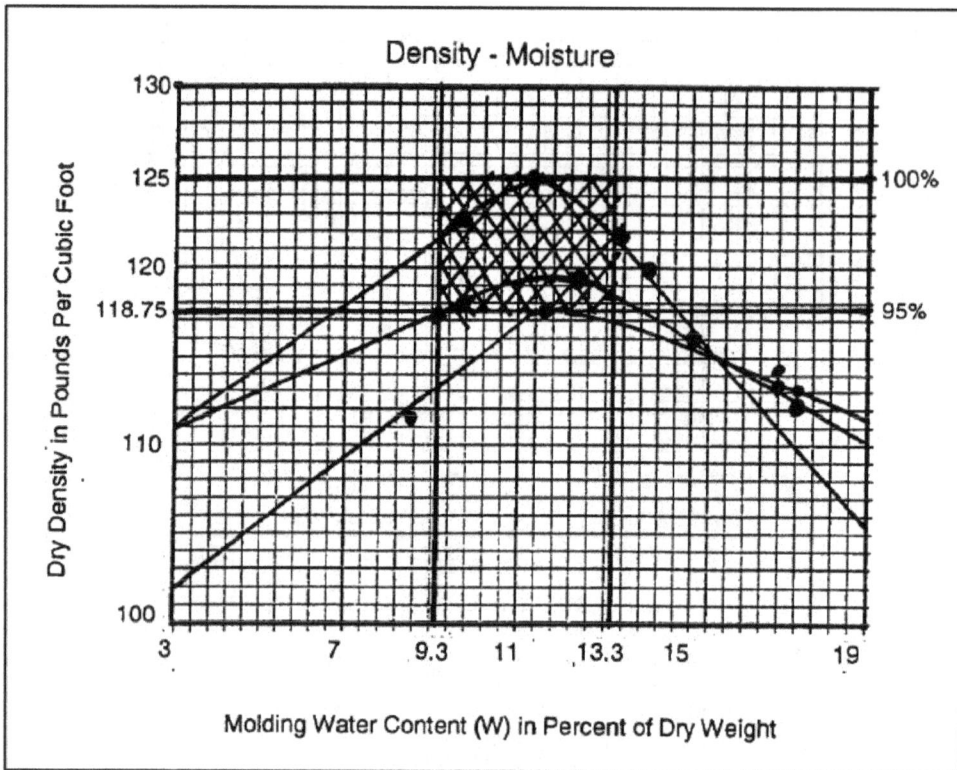

Figure A-7. Density-moisture curve for Rio Meta Plain soil with density and moisture ranges plotted

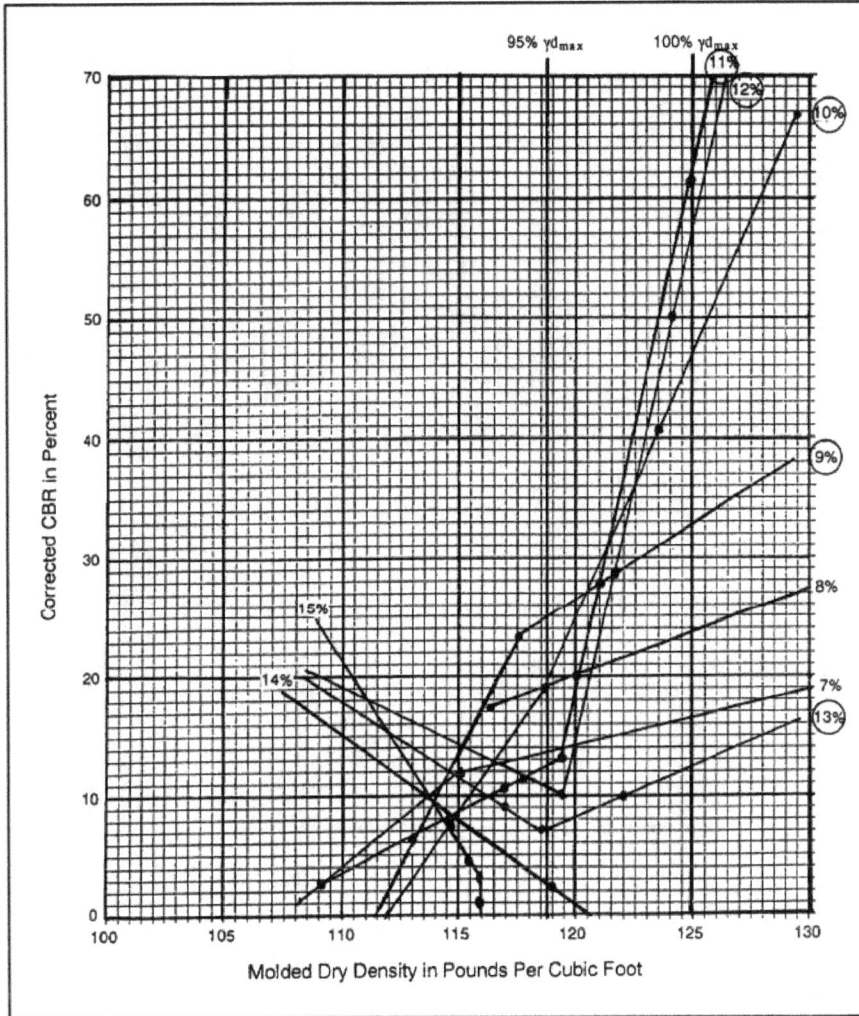

Figure A-8. CBR Family of Curves for Rio Meta Plain soil with density range plotted

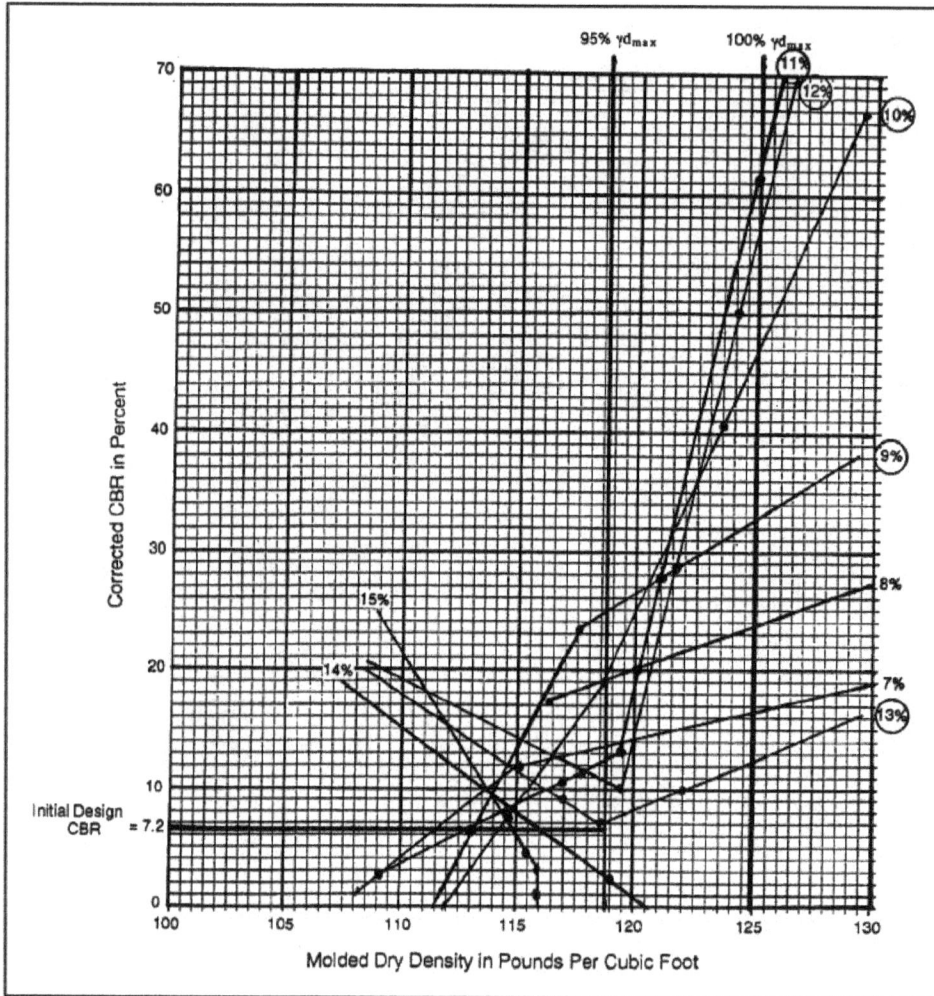

Figure A-9. Design CBR from Rio Meta Plain soil from CBR family of curves

Appendix B

Availability of Fly Ash

B-1. Fly ash is a pozzolanic material that is a by-product of coal-fired, electrical power- generation plants. Because of its pozzolanic properties, it can be used as a—

- Soil stabilizer.
- Liming material.
- Cement component.

B-2. Depending on its calcium oxide (CaO) content, fly ash may be used as a stand-alone product or in combination with other pozzolanic materials.

B-3. Fly ash is divided into two classes, based on their CaO content. They are—

- Class F.
- Class C.

B-4. Class F fly ash has a low CaO content (less than 10 percent) and is not suitable for use as a stand-alone product for engineering purposes. Class C fly ash is often referred to as "high lime" fly ash. Its CaO content must be a minimum of 12 percent and frequently exceeds 25 percent. Class F fly ash originates from hard coal (anthracite and bituminous coal), whereas Class C fly ash originates from brown coal (lignite and subbituminous coal).

B-5. Fly ash occurs throughout the world. In many countries, both classes of fly ash can be found. The likelihood of finding a particular class of fly ash depends on the individual country. Hard coal is more frequently used in industrialized countries in the manufacture of steel. Therefore, in industrialized nations having both hard and brown coal reserves, the amount of brown coals used for power generation is likely to be greater; thus the percentage of Class C fly ash would be greater.

B-6. The major coal-producing countries and the percentages of the hard and brown coal reserves are found in table B-1.

Table B-1. Percentage of hard and brown coal reserves in major coal-producing countries

Country	Hard Coal Reserves (%)	Brown Coal Reserves (%)
South Africa	100	—
India	90	10
Indonesia	2	98
Russia	45	55
Poland	69	31
United Kingdom	100	—
Germany	41	59
United States	48	52
Canada	27	73
Brazil	17	83
Columbia	98	2
China	100	—
Australia	36	64

This page intentionally left blank.

Appendix C

Hazards of Chemical Stabilizers

The major hazards associated with chemical stabilizers and the precautions to take when using them are identified in this appendix. When receiving chemical stabilizers, you should also receive a Material Safety Data Sheet (MSDS) that is prepared by the manufacturer to give you more detailed information on the product. Generally, the hazard of airborne particles is reduced significantly in the open air on the construction site. In this appendix, ventilation is discussed when materials are handled in enclosed areas, such as soils laboratories. Due to differences in manufacturers, the MSDS may list hazards that are slightly different from the ones below.

PORTLAND CEMENT

HEALTH-HAZARD DATA

C-1. Signs and symptoms of overexposure.

- Alkali burns.
- Irritation to the eyes and the respiratory system.
- Allergic dermatitis.

C-2. Emergency and first-aid procedures.

- Irrigate the eyes with water.
- Wash the affected areas of the body with soap and water.

PRECAUTIONS FOR SAFE HANDLING AND USE

C-3. Procedures for cleanup of released or spilled material.

- Use dry cleanup methods that do not disperse the dust into the air.
- Avoid breathing the dust.

CONTROL MEASURES

C-4. Respiratory protection. Use a National Institute of Occupational Safety and Health (NIOSH)-approved respirator.

C-5. Eye protection. Use goggles.

HYDRATED LIME

REACTIVITY DATA

C-6. Avoid acids and fluorine.

HEALTH-HAZARD DATA

C-7. Routes of entry.

- Eyes.
- Skin.
- Inhalation.
- Ingestion.

C-8. Acute reactions.

- Irritation of the eyes.
- Irritation of the skin.
- Irritation of the respiratory tract.
- Irritation and inflammation of the digestive system.

C-9. Chronic reactions.

- Corrosion of the eyes.
- Corrosion of the skin.
- Irritation of the respiratory tract.
- Irritation and inflammation of the digestive system.

C-10. Signs and symptoms of overexposure.

- Irritation of the skin and eyes.
- Irritation of the respiratory tract.
- Irritation of the digestive system.

C-11. Medical conditions aggravated by exposure.

- Respiratory disease.
- Skin conditions.

C-12. Emergency and first-aid procedures.

- If the material gets into the eyes, flush them thoroughly with large amounts of water while holding the eyelids apart.
- If the material gets on the skin, wash thoroughly with soap and water. If the material is inhaled, get medical attention.
- If the material is ingested, do not induce vomiting. Get medical attention.

PRECAUTIONS FOR SAFE HANDLING AND USE

C-13. Procedures for the cleanup of spilled or released materials.

- Wipe the material up and place it in a disposal container (some manufacturers recommend using a vacuum cleaning system).
- Remove the residue with water.

C-14. Procedure for the disposal of waste material. Dispose of the waste according to local, state, and federal regulations.

C-15. Procedure for the handling and storage of material. Store the material in tightly sealed containers away from incompatible materials.

CONTROL MEASURES

C-16. Respiratory protection. Use a dust-filter mask.

C-17. Ventilation. Ensure that the general mechanical ventilation meets the Threshold Limit Value (TLV) and Permissible Exposure Limit (PEL).

C-18. Protective gloves. Use leather or rubber gloves.

C-19. Eye protection. Use well-fitted goggles.

C-20. Other protective equipment. Wear long sleeves and pants.

C-21. Work hygiene. Use good hygiene practices; avoid unnecessary contact with this product.

C-22. Supplemental safety and health data. Use respiratory protection if exposure limits cannot be maintained below the TLV and PEL.

QUICKLIME

FIRE- AND EXPLOSION-HAZARD DATA

C-23. Extinguishing media. Use a dry chemical (CO_2) (water will cause evolution of heat).

C-24. Special fire-fighting procedures. Do not use water on adjacent fires.

C-25. Unusual fire and explosion hazards. The material may burn but does not ignite readily. Some materials may ignite combustibles (wood, paper, and oil).

REACTIVITY DATA

C-26. Conditions to avoid. Moisture.

C-27. Material to avoid. Acids and oxidizing materials.

HEALTH-HAZARD DATA

C-28. Routes of entry.

- Eyes.
- Skin.
- Inhalation.
- Ingestion.

C-29. Signs and symptoms of overexposure.

- Conjunctivitis.
- Corneal ulceration.
- Iritis.
- Skin inflammation. Skin ulceration.
- Lung inflammation.

C-30. Emergency and first-aid procedures.

- Keep unnecessary people away.
- Stay upwind.
- Keep out of low areas. Isolate the hazard and deny entry.
- Wear a self-contained breathing apparatus and full protective clothing. Avoid contact with solid materials. Irrigate the eyes with water.
- Wash the contaminated areas of the body with soap and water.
- Do not induce vomiting if the material is swallowed. If the victim is conscious, have him drink milk or water.

PRECAUTIONS FOR SAFE HANDLING AND USE

C-31. Procedures for the cleanup of released or spilled material.

- Sweep the material into a large bucket.

- Dilute it with water and neutralize it with 6M-HC1.
- Drain into an approved storage container with sufficient water.

C-32. Procedures for the disposal of waste material.

- Put the material into a large vessel containing water. Neutralize with HC1.
- Discharge into an approved disposal area with sufficient water.
- Do not allow calcium oxide (CaO) to enter water intakes; very low concentrations of it are harmful to aquatic life.

C-33. Procedures for the handling and storage of material.

- Protect containers from physical damage.
- Store in a cool, dry place.
- Separate from other storage items and protect from acids and oxidizing materials.

C-34. Precautions to take for personal safety.

- Wear chemical goggles.
- Wear a mechanical filter respirator.
- Wear rubber protective clothing.

CONTROL MEASURES

C-35. Respiratory protection. Use a self-contained breathing apparatus.

C-36. Ventilation. Requires a local exhaust vent.

C-37. Protective gloves. Use rubber gloves.

C-38. Eye protection. Use chemical goggles.

C-39. Other protective equipment. Wear rubber protective clothing.

FLY ASH

HEALTH-HAZARD DATA

C-40. Lethal dose (LD) 50- Lethal concentration (LC) 50 mixture. TLV 10 milligrams/cumulative

C-41. Routes of entry.

- Skin.
- Inhalation.

C-42. Acute reaction. Can cause minor skin irritation.

C-43. Chronic reaction. May cause pulmonary disease if exposed to an excessive concentration of dust for a long period of time.

C-44. Signs and symptoms of overexposure.

- Development of pulmonary disease.
- Irritation of the skin.

PRECAUTIONS FOR SAFE HANDLING AND USE

C-45. Procedures for the cleanup of released or spilled material.

- Vacuum or sweep up the spill.
- Use a dust suppressant agent if sweeping is necessary.
- Wet down large spills and scoop them up.

C-46. Procedure for the disposal of waste material. Dump the material in landfills according to local, state, and federal regulations.

C-47. Procedures for the handling and storage of material.

- Avoid creating dust.
- Maintain good housekeeping practices.

CONTROL MEASURES

C-48. Respiratory protection. Use an NIOSH-approved respiratory for protection against pneumoconiosis-producing dusts.

C-49. Ventilation. Provide local exhaust ventilation to keep below the TLV.

ASPHALT CEMENT

FIRE- AND EXPLOSION-HAZARD DATA

C-50. Flash point. 100 degrees Fahrenheit.

C-51. Extinguishing media.

- CO_2.
- Foam.
- Dry chemicals.

C-52. Special fire-fighting procedures. Wear an NIOSH-approved respirator/full protective gear.

C-53. Unusual fire and explosion hazards. Vapors may ignite explosively.

REACTIVITY DATA

C-54. Conditions to avoid. Heat and flames.

C-55. Hazardous decomposition products.

- CO.
- CO_2.
- Hydrocarbons.

HEALTH-HAZARD DATA

C-56. Signs and symptoms of overexposure. May become dizzy, lightheaded, or have a headache. If this occurs, remove the victim from exposure and provide him with adequate ventilation.

C-57. Emergency and first-aid procedures.

- Avoid contact with the skin.
- Do not ingest. If swallowed, do not induce vomiting. Get medical attention.

PRECAUTIONS FOR SAFE HANDLING AND USE

C-58. Procedures for the cleanup of released or spilled material. Scrape the material into containers and dispose of it according to local, state, and federal regulations.

C-59. Procedures for the handling and storage of material.

- Store below 140 degrees Fahrenheit.
- Do not store near heat or flame.

CONTROL MEASURES

C-60. Respiratory protection. None is required if ventilation is adequate.

C-61. Ventilation. Recommend a local exhaust vent.

C-62. Eye protection. Use normal precautions.

C-63. Other protective measures. Wash the hands and other areas of contact after use.

CUTBACK ASPHALT

FIRE- AND EXPLOSION-HAZARD DATA

C-64. Flash point. 65 degrees Fahrenheit.

C-65. Extinguishing media.

- CO_2.
- Dry chemicals.
- Halogenated agents.
- Foam.
- Steam or water fog.

HEALTH-HAZARD DATA

C-66. Sign of overexposure. Thermal burns to the eyes and skin from the heated material.

C-67. Emergency and first-aid procedures.

- Flush the eyes with cool water for at least 15 minutes.
- Wash the affected areas of the body with soap and water.
- Remove the victim to an uncontaminated area if adverse effects occur from inhalation. Give artificial respiration if the victim is not breathing. Get medical attention.
- Induce vomiting if a large amount was swallowed. Get medical attention.

PRECAUTIONS FOR SAFE HANDLING AND USE

C-68. Procedures for the cleanup of released or spilled material.

- Contain and remove the material by mechanical means.
- Keep the material on an absorbent material.
- Keep the material out of sewers and waterways.

C-69. Procedure for the disposal of waste. Dispose of the material according to local, state, and federal regulations.

C-70. Procedure for the handling and storage of material. Keep the container closed.

CONTROL MEASURES

C-71. Respiratory protection. Use an NIOSH-approved respirator for organic vapors if it is above the TLV.

C-72. Ventilation. Provide local exhaust ventilation to keep it below the TLV.

C-73. Protective gloves. Wear rubber gloves.

C-74. Eye protection. Use safety glasses or goggles.

C-75. Other protective equipment. Wear work clothes and gloves if prolonged or repeated contact is likely.

Glossary

Acronym/Term	Definition
$\geq$	greater than or equal to
$\leq$	less than or equal to
γ	unit weight
γ_m	wet unit weight
γ_d	dry unit weight
A horizon	The upper layer of a typical soil. It contains a zone of accumulation of organic materials in its upper portion and a lower portion of lighter color from which soil colloids and other soluble constituents have been removed.
AASHTO	American Association of State Highway and Transportation Officials
abrasion	Occurs when hard particles are blown against a rock face causing the rocks to break down. As they are broken off, the resulting fragments are carried away by wind. Abrasion also occurs due to water and glacial action.
acid test	Used to determine the presence of carbonates and is performed by placing a few drops of hydrochloric acid on the surface of a rock.
adobe	Calcareous silts and sandy-silty clays, usually high in colloidal clay content, found in the semiarid regions of the southwestern United States and North Africa.
adsorbed water	Thin films of water that adhere to the separate oil particles.
AFM	Air Force manual
AFP	Air Force publication
agronomic dust control	This method consists of establishing, promoting, or preserving vegetative cover to prevent or reduce dust generation from exposed soil surfaces.
AI	Airfield Index
airfield cone penetrometer	Used by engineering personnel to determine an index of soil strengths (airfield index) for various military applications. It serves as an aid in maintaining field control during construction operations.
airfield index	An index used to describe subgrade soil strength, based on data from the airfield cone penetrometer.
alluvial fan	A dry-land counterpart of deltas.
alluvium	Deposits of mud, silt, and other material commonly found on the flatlands along the lower courses of streams.
Alpine glaciation	Takes place in mountainous areas and generally results in the creation of mainly erosional forms. Features are very distinctive and easy to recognize such as a U-shaped profile in contrast to the V-shaped profile produced by fluvial erosion.
amphiboles	Hard, dense, glassy to silky minerals found chiefly in intermediate to dark igneous rocks and gneisses and schists. They generally occur as short to long prismatic crystals with a nearly diamond-shaped cross section.

angular particles	Grains, or particles, with shapes that are characterized by jagged projections, sharp ridges, and flat surfaces.
annular drainage pattern	A ringlike drainage pattern formed in areas where sedimentary rocks are upturned by a dome structure.
anthracite	A hard natural coal of high luster differing from bituminous coal in containing little volatile matter.
anticline fold	Upfolds.
AOS	apparent opening size
APSB	asphalt penetration surface binder
AR	Army regulation
arcuate delta	Arc- or fan-shaped deltas formed when wave action is the primary force acting on the deposited material.
argillaceous	Soils which are predominantly clay or abounding in clays or claylike materials.
asphalt emulsions	A blend of asphalt, water, and an emulsifying agent and is available either as anionic or cationic emulsions.
asphalt penetration surface binder	A special liquid asphalt composed of high penetration-grade asphalt and a solvent blend of kerosene and naphtha.
ASTM	American Society of Testing and Materials
asymmetrical fold	A fold with an inclined axial plane.
augite	The most common of the pyroxenes.
B horizon	The layer containing soluble materials washed out of the A horizon. This layer frequently contains much clay and may be several feet thick.
backfill	To refill (as an excavated area) usually with excavated material.
barchan dune	The simplest and most common of dunes. Usually crescent-shaped. The windward side has a gentle slope rising to a broad dome that is cut off abruptly on the leeward side.
basalt	Very fine-grained, hard, dense, dark-colored extrusive rock which occurs widely in lava flows around the world.
base exchange	The process of replacing cations of one type with cations of another type in the surface of an adsorbed layer
batholith	A great mass of intruded igneous rock that for the most part stopped in its rise to the earth's surface at a considerable distance below the surface.
batter	A receding upward slope of the outer face of a structure.
batter piles	Those piles that are driven at an angle with the vertical. They may be used to support inclined loads or to provide lateral loads.
bearing capacity	The soil's ability to support loads which may be applied to it by an engineering structure
bearing pile	A pile driven into the ground so as to carry a vertical load.
bedded sediments	Parallel layers of sediment lying one on top of the other.
bedding planes	The surface that separates each successive layer of a stratified rock from its preceding layer, a depositional plane, or a plane of stratification.
bed load	Material too heavy to be suspended by erosional agents for great distances at any one period of time. Consists mainly of coarse particles

	that roll along the ground
bentonite	A clay of high plasticity formed by the decomposition of volcanic ash, which has high swelling characteristics.
bird's-foot delta	Resembles a bird's foot from the air and is formed in instances where fluvial processes have a major influence on deposited sediments.
bite test	A quick and useful method of distinguishing among sand, silt, or clay. In this test, a small pinch of the soil material is ground lightly between the teeth and the grittiness determined.
bitumen-lime blend	A system in which small percentages of lime are blended with fine-grained soils to facilitate the penetration and mixing of bitumens into the soil.
blanket cover	Any material that forms a (semi) permanent cover and is immovable by the wind it serves to control dust.
blind drainage	Sections of a stream channel which are blocked at both ends.
boulder clay	Another name, used widely in Canada and England, for glacial till.
braided drainage pattern	A drainage pattern resulting from the dividing and reuniting of stream channels. The pattern commonly forms in arid areas during flash flooding or from the meltwater of a glacier. The stream has attempted to carry more material than it is capable of handling.
breccia	A rock consisting of sharp fragments cemented by finer-grained material.
breaking test	Performed only on material passing the number 40 sieve. Used to measure the cohesive and plastic characteristics of the soil.
buckshot clay	Clay of the southern and southwestern United States which, upon drying, crack into small, hard lumps of more or less uniform size.
bulking of sands	The volumetric increase in a dry or nearly dry sand caused by the introduction of a slight amount of moisture and disturbance of the soil.
buttress	A projecting structure of masonry or wood for supporting or giving stability to a wall or building; a projecting part of a mountain or hill; a broadened base of a tree trunk or a thickened vertical part of it; something that supports or strengthens.
C	celsius; centigrade; clay
C horizon	The weathered parent material.
calcareous	Soils which contain an appreciable amount of calcium carbonate, usually derived from limestone.
calcite	A soft, usually colorless to white mineral distinguished by a rapid bubbling or fizzing reaction with dilute hydrochloric acid. It is the major component of sea shells and coral skeletons and often occurs as well-formed, glassy to dull, blocky crystals.
caliche	The nitrate-bearing gravel or rock of the sodium nitrate deposits of Chile and Peru; a crust of calcium carbonate that forms on the stony soil of arid regions.
California Bearing Ratio	A measure of the shearing resistance of a soil under carefully controlled conditions of density and moisture.
capillary fringe	Capillary action in a soil above the groundwater table.
capillary moisture	When dry soil grains attract moisture in a manner somewhat similar to the way clean glass does.

capillary saturation	When the soil is essentially saturated.
carbonation	The chemical process in which carbon dioxide from the air unites with various minerals to form carbonates.
carbonic acid	A weak dibasic acid H_2CO_3 known only in solution that acts to dissolve carbonate rocks.
cavern	An underground chamber often of large or indefinite extent.
CBR	California Bearing Ratio
C_c	coefficient of curvature
CE	compactive effort
CE 55	compactive effort, 55 blows per layer
cementation	This occurs when precipitates of mineral-rich waters, circulating through the pores of sediments, fill the pores and bind the grains together.
CH	clays, highly compressive (LL>50)
chemical stabilization	Relies on the admixture to alter the chemical properties of the soil to achieve the desired effect, such as using lime to reduce a soil's plasticity.
chemical weathering	The decomposition of rock through chemical processes. Chemical reactions take place between the minerals of the rock and the air, water, or atmosphere.
chert	A rock resembling flint and consisting essentially of cryptocrystalline quartz or amorphous silica.
chlorite	A very soft, grayish-green to dark green mineral with a pearly luster. It occurs most often as crusts, masses, or thin sheets or flakes in metamorphic rocks, particularly schists and greenstone.
cinder	The slag from a metal furnace; a fragment of ash; a partly burned combustible in which fire is extinct; a hot coal without flame; a partly burned coal capable of further burning without flame; a fragment of lava from an erupting volcano.
CL	clay, low compressibility (LL<50)
Class C fly ash	Has a high CaO content (12 percent or more) and originates from sub-bituminous and lignite (soft) coal.
Class F fly ash	Has a low CaO content (less than 12 percent) and originates from anthracite and bituminous coal.
clastic sediments	Deposit of rock particles dropped from suspension in air, water, or ice.
clay minerals	Form soft microscopic flakes which are usually mixed with impurities of various types (particularly quartz, limonite, and calcite). When barely moistened, as by the breath or tongue, clays give off a characteristic somewhat musty "clay" odor.
clay stone	A calcareous concretion formed in a bed of clays; a dull earthy feldspathic rock containing clay.
cleavage	The tendency of a mineral to split or separate along preferred planes when broken.
cm	centimeter(s)
cm^3	cubic centimeter(s)
CMS	concrete-modified cement

coal	An accumulation and conversion of the organic remains of plants and animals under certain environments.
coarse-grained rocks	Those rocks that have either crystals or cemented particles that are large enough to be readily seen with the unaided eye.
coarse-grained soil	Those soils in which half or more of the material is retained on a Number 200 sieve.
compaction	The reduction of volume and increase in density that results from the application of downward stress to a material. The stress moves the particles closer together, reducing the volume of air voids and increasing the unit weight (density) of the material.
competence	The maximum size of particles capable of being moved by a stream.
complex dunes	Dunes which lack a distinct form and develop where wind directions vary, sand is abundant, and vegetation may interfere. They occur locally when other dune types become overcrowded and overlap.
compressibility	That property of a soil which permits it to deform under the action of an external compressive load.
conchoidal fracture	A fracture surface that exhibits concentric, bowl-shaped structures like the inside of a clam shell.
conglomerates	Rocks composed of rounded fragments varying from small pebbles to large boulders held together by a natural cement.
contact moisture	When water is brought into the capillary zone from the water table by evaporation and condensation.
continental glaciation	Occurs on a large, regional scale affecting vast areas. Characterized by the occurrences of more depositional features than erosional features. The glaciers can be of tremendous thickness and extent.
coquina	Consists essentially of marine shells held together by a small amount of calcium carbonate to form a fairly hard rock. Coquina is widely used for the granular stabilization of soils along the Gulf Coast of the United States.
coral	Calcareous, rocklike material formed by secretions of corals and coralline algae. The white type is very hard, while the gray type tends to be soft, brittle, and extremely porous.
coral sand	Consists of decomposed coral which may be combined with washed and sorted beach sands.
crevasse filling	A material contained within a deep crevice, or fissure, in a glacier or the earth.
cross-beds	Individual layers within a bed that lie at an angle to the layers of adjacent beds, typical of sand dune and delta front deposits.
CSS	cationic slow-setting
cu	cubic
C_u	coefficient of uniformity
cutback asphalt	A blend of an asphalt cement and a petroleum solvent.
d	desirable base and subbase material
daily mean temperature	An average of the maximum and minimum temperatures for one day or an average of several temperature readings taken at equal time intervals during the day, generally hourly.
DA	Department of the Army

DBST	double bituminous surface treatment
DCA 70	a polyvinyl acetate emulsion
DD	Department of Defense
debris avalanche	The rapid downslope flowage of masses of incoherent soil, rock, and forest debris with varying water content. They are shallow landslides resulting from frictional failure along a slip surface, essentially parallel to the topographic surface, formed where the accumulated stresses exceed the resistance to shear.
debris flows	Occur when a debris avalanche increases in water content. They are caused most frequently when a sudden influx of water reduces the shear strength of earth material on a steep slope, and they typically follow heavy rainfall.
deflation	Occurs when loose particles are lifted and removed by the wind. This results in a lowering of the land surface as materials are carried away.
degree days	(As used in this FM) the difference between the average daily air temperature and 32 degrees Fahrenheit. The degree days are minus when the average daily temperature is below 32 degrees Fahrenheit (freezing degree days) and plus when above (thawing degree days).
delta	The alluvial deposit at the mouth of a river.
delta kame	Well-sorted glacial deposits made up of gravels and sands.
dendritic drainage pattern	A treelike pattern composed of branching tributaries to a main stream. It is characteristic of essentially flat lying and/or relatively homogeneous rocks.
density	The weight per unit volume.
desert pavement	See "lag deposits."
design freezing index	The freezing index of the average of the three coldest winters in 30 years of record, or of the coldest winter in a 10-year period, if 30-year data are unavailable.
diatomaceous earth	Composed essentially of the siliceous skeletons of diatoms (extremely small unicelled organisms). It is composed principally of silica, is white or light gray in color, and is extremely porous.
diorite	A granular crystalline igneous rock commonly composed of plagioclase and hornblende, pyroxene, or biolite.
dip	The inclination of a bedding plane.
dirty sand	A slightly silty or clayey sand.
DM	draft manual
dolomite	A mineral consisting of a calcium magnesium carbonate found in crystals, as well as in extensive beds, as a compact limestone.
dome	An upfold that plunges in all directions.
D_r	relative density
drag	Folding of rock beds adjacent to a fault.
drumlins	Asymmetrical streamlined hills of gravel till deposited at the base of a glacier and oriented in a direction parallel to ice flow.
dry density	Weight of solid fraction of a soil material divided by the volume of the soil material. Synonymous with dry unit weight.
dry strength test	See "breaking test."

durability	The resistance to slaking or disintegration due to alternating cycles of wetting and drying or freezing and thawing.
dust palliative	Material used to prevent soil particles from becoming airborne.
e	void ratio
E	east
earthflow	A landslide consisting of unconsolidated surface material that moves down a slope when saturated with water.
earth-retaining structure	Used to restrain a mass of earth which will not stand unsupported.
earthquake	A shaking or trembling of the earth that is tectonic in origin; movement along a fault.
elongate delta	Long, relatively narrow delta formed where tidal currents have a major impact on sediment deposition.
EM	engineer manual
e_{max}	void ratio in the loosest condition possible
e_{min}	void ratio in the most dense condition possible
e_n	natural void ratio
end moraines	Ridges of till material pushed to their locations at the limit of the glacier's advance by the forceful action of the ice sheet.
eolian	Descriptive term implying action by wind.
EOS	equivalent opening size
EPA	Environmental Protection Agency
erosion	The transportation of weathered materials by wind or water.
eskers	Winding ridges of irregularly stratified sand and gravel that are found within the area of the ground moraine created by continental glaciation.
evaporites	Sedimentary rocks (as gypsum) that originated by evaporation of seawater in an enclosed basin.
exfoliation	A type of weathering that involves the breaking loose of thin concentric shells, slabs, spalls, or flakes from rock surfaces.
extrusive igneous rock	Those rocks formed by extrusion from the earth in a molten state or as volcanic ash.
F	Fahrenheit; frictional resistance to sliding
FAA	Federal Aviation Administration
fat clay	Fine, colloidal clay of high plasticity; classified as (CH) by the USCS.
faults	Fractures along which there is displacement of the rock parallel to the fracture plane. Once-continuous rock bodies that have been displaced by movement in the earth's crust.
fault scarp	A cliff or escarpment directly resulting from an uplift along one side of a fault.
fault zone	An area in which there are several closely spaced faults.
feldspars	Any of a group of crystalline minerals that consist of aluminum silicates with either potassium, sodium, calcium, or barium; essential constituents of nearly all crystalline rocks.
felsite	A very fine-grained, usually extrusive igneous equivalent of granite.

FHWA	Federal Highway Administration
fill saddles	Narrow saddles used to hold hauled material.
filtration geotextiles	Used when soils may migrate into drainage aggregate or pipe. It prevents the soil migration and thus maintains water flow through the drainage system.
fine-grained soil	Those soils in which more than half the material passes a Number 200 sieve.
fissility	Capable of being split or divided in the direction of the grain or along natural planes of cleavage.
FLS	forward landing strips
fly ash	Fine solid particles of ashes, dust, and soot carried out from burning fuel (as coal or oil) by a draft. This pozzolanic material consists mainly of silicon and aluminum compounds that, when mixed with lime and water, forms a hardened cementitious mass capable of obtaining high compression strengths.
FM	field manual
foliated	Metamorphic rocks that display a pronounced banded structure as a result of the deformational pressures to which they have been subjected.
footwall	The block below a fault plane.
fracture	The way in which a mineral breaks when it does not cleave along cleavage planes.
fragipan	See "hardpan."
free water	When the zone of saturation is under no pressure except from the atmosphere.
freezing index	The number of degree days between the highest and lowest points on a curve of cumulative degree days versus time for one freezing season. It is used as a measure of the combined duration and magnitude of below-freezing temperature occurring during any given freezing season. The index determined for air temperatures at 4.5 ft above the ground is commonly designated as the "air freezing index," while that determined for temperatures immediately below a surface is known as the "surface freezing index."
frost action	Processes which affect the ability of soil to support a structure when accumulated water in the form of ice lenses in the soil is subjected to natural freezing conditions.
frost boil	The breaking up of a localized section of a highway or airfield pavement when subjected to traffic. During the process of thawing, the melted water produces a supersaturated or fluid subgrade condition with very limited or no supporting capacity. The traffic imposes a force on the pavement and thus on the excess water in the subgrade, which in turn exerts an equalizing pressure in all directions. This pressure is relieved through the point of least resistance (up through the pavement surface) and produces a small mound similar in appearance to an oversized boil.
frost heave	An upthrust of ground caused by freezing of moist soil (as under a footing or pavement).
frost-melting period	An interval of the year during which the ice in the foundation materials is returning to a liquid state. It ends when all the ice in the ground is

	melted or when freezing is resumed. Although there is generally only one frost-melting period, beginning during the general rise of air temperature in the spring, one or more significant frost-melting intervals may occur during a winter season.
frost-susceptible soil	Soil in which significant ice segregation will occur when the necessary moisture and freezing conditions are present.
frothy	A sample that appears pitted or spongy.
ft	foot; feet
Fuller's earth	Unusually highly plastic clays of sedimentary origin, white to brown in color. Used commercially to absorb fats or dyes.
g	gram(s)
G	specific gravity; gravel
gabbro	A granular igneous rock composed essentially of calcic plagioclase, a ferromagnesian mineral, and accessory minerals.
gabbro-diorite	A series of dense, coarsely crystalline, hard, dark-colored intrusive rocks composed mainly of one or more dark minerals along with plagioclase feldspar.
gabion	Large, steel wire-mesh baskets usually rectangular in shape and variable in size. They are designed to solve the problem of erosion.
gal	gallon(s)
galena	A bluish gray mineral (PbS) with metallic luster consisting of lead sulfide, showing highly perfect cubic cleavage, and constituting the principal ore of lead.
gap-graded	Soil contains both large and small particles, but the gradation continuity is broken by the absences of some particle sizes.
garnet	A brittle and more or less transparent usually red silicate mineral that has a vitreous luster, occurs mainly in crystals but also in massive form and in grains, is found commonly in gneiss and mica schist, and is used as a semiprecious stone and as an abrasive.
GC	clayey gravel
geogrid	A geotextile that is constructed with relatively large openings that act to lock soil particles in place (confining them) and adding strength to the soil.
glacial lake deposit	Occurs during the melting of the glacier. Many lakes and ponds are created by meltwater in the outwash areas.
glaciofluvial	Relating to, or coming from streams deriving much or all of their water from the melting of a glacier.
glassy	Fine-grained rocks with a shiny smooth texture showing conchoidal fracture. An example is obsidian, a black volcanic glass.
GM	silty gravel
gneiss	A foliated metamorphic rock corresponding in composition to granite or some other feldspathic plutonic rock. It is medium- to coarse-grained and consists of alternating streaks or bands.
gouge	Crushed and altered rock.
GP	poorly graded gravel
graben	A block that is downthrown between two faults to form a depression.
gradation	The distribution of particle sizes in the soil.

grain size	See "particle size."
granite	A coarsely cystalline, hard, massive, light-colored rock composed mainly of feldspar and quartz, usually with mica and/or hornblende.
gravitational water	See "free water."
gravity fault	See "normal fault."
grit test	See "bite test."
ground moraine	Glacial deposits that are laid down as the ice sheet recedes.
groundwater table	The upper limit of the saturated zone of free water.
gumbo	Peculiar, fine-grained, highly plastic, silt-clay soils which become pervious and soapy, or waxy and sticky, when saturated.
GW	well-graded gravel
H	high compressibility
halite	Rock salt.
hanging wall	The block above a fault plane.
hardness	The resistance to scratching or abrasion by other minerals or by an object of known hardness.
hardpan	A general term used to describe a hard, cemented soil layer which does not soften when wet and may be impervious to water.
h_c	height of capillary rise
HCl	hydrochloric acid
headwalls	Bowl-shaped areas with slope gradients often 100 percent or greater. It is usually the junction for several intermittent streams that can cause sharp rises in the groundwater levels in the soil mantle during winter storms.
hematite	A mineral (Fe_2O_3) constituting an important iron ore and occurring in crystals or in a red earthy form.
homocline fold	A rock body that dips uniformly in one direction (at least locally).
hornblende	A mineral that is the common dark variety of aluminous amphibole.
horst	An upthrown block between two faults.
hummocky terrain	Terrain composed of many depressions and uneven ground.
hydration	The chemical union of a compound with water.
hydrolysis	A chemical process of decomposition involving splitting of water molecules and subsequent reaction with various minerals.
hydrophytes	A vascular plant growing wholly or partly in water especially a perennial aquatic plant having its overwintering bulbs underwater.
hygroscopic moisture	The water adsorbed from atmospheric moisture when the soil is in an air-dry condition.
igneous rocks	Those rocks that have solidified from molten material which originated deep within the earth's mantle. This occurred either from magma in the subsurface or from lava extruded onto the earth's surface during volcanic eruptions.
impervious soils	Fine-grained, homogenous, plastic soils, and coarse-grained soils that contain plastic fines. Soils that do not allow for the transmission of significant amounts of water.

intrusive igneous rocks	Those rocks that have cooled from magma beneath the earth's surface.
Ip	plasticity index
joints	Rock fractures along which there has been little or no displacement parallel to the fractured surface.
k	coefficient of permeability; subgrade reaction
kames	Conical hills of sand and gravel that were deposited by heavily laden glacial streams that flowed on top of or off of the glacier.
kame terraces	Roughly linear deposit of sand and gravel associated with alpine glaciation.
kaolin	A fine, usually white clay that is used in ceramics and refractories as a filler or extender.
kettle holes	Pits formed by the melting of ice which had been surrounded by or embedded in the moraine material.
kip	kilopound (1,000 pounds)
L	low compressibility
lag deposits	Gravel and pebbles that are too large to be carried by the wind and so accumulate on the earth's surface in the form of a sheet. They ultimately cover the finer-grained material beneath and protect it from further deflation.
laterites	A residual product that is red in color and has a high content of the oxides of iron and hydroxides of aluminum.
laterite soils	Residual soils which are found in tropic regions. Many different soils are included in this category and they occur in many sections of the world. They are frequently red in color and in their natural state have a granular structure with low plasticity and good drainability. When moistened with water and remolded, they often become plastic and clayey to the depth disturbed.
latosols	A leached red and yellow tropical soil.
lava	A viscous liquid that flows out a volcanic vent or from fissures along the flanks of a volcano. It can flow many miles from the crater vent. Molten material at the earth's surface.
lb	pound(s)
LCF	lime-cement-fly ash
lean clay	Silty clays and clayey silts, generally of low to medium plasticity.
lignin	A by-product of the manufacture of wood pulp.
limbs	The sides of a fold as divided by the axial plane.
lime	A dry white powder consisting essentially of calcium hydroxide.
lime-cement-fly ash	A mixture of lime, cement, and fly ash used as a soil stabilization admixture.
limestone	A soft to moderately hard rock that is formed chiefly by accumulation of organic remains (as shells or coral), consists mainly of calcium carbonate, is extensively used in building, and yields lime when burned.
limonite	A native hydrous ferric oxide of variable composition that is a major ore of iron. Occurs most often as soft, yellowish-brown to reddish-brown fine-grained earthy masses or compact lumps or pellets. It is a

	common and durable cementing agent in sedimentary rocks and the major component of laterite.
LL	liquid limit
LM	light duty mat
LMS	lime-modified soil
loam	A general agricultural term, applied most frequently to sandy-silty topsoils which contain a trace of clay, are easily worked, and support plant life.
loess	Thick accumulations of yellowish-brown material composed primarily of windblown silt. The silty soil is of eolian origin characterized by a loose, porous structure, and vertical slope. It covers extensive areas in North America (especially in the Mississippi Basin), Europe, and Asia (especially north-central Europe, Russia, and China).
longitudinal dune	A dune elongated in the direction of the prevailing winds.
LOTS	logistics-over-the-shore
luster	The appearance of a mineral specimen in reflected light. It is either metallic or nonmetallic.
M	silt
M2	a type of compass
magma	Molten rock material within the earth.
magnetite	A black isometric mineral of the spinal group that is an oxide of iron and an important iron ore.
marble	A soft, fine to coarsely crystalline, massive metamorphic rock which forms from limestone or dolomite. It is distinguished by its softness, acid reaction, lack of fossils, and sugary appearance on freshly broken surfaces.
marl	A soft, calcareous deposit mixed with clays, silts, and sands, often containing shells or organic remains. It is common in the Gulf Coast area of the United States.
mature stream	Has a developed floodplain and, while the stream no longer fills the entire valley floor, it meanders to both edges of the valley.
max	maximum
MB	mechanical blending
MBST	multiple bituminous surface treatment
MC	medium curing
mean freezing index	The freezing index determined on the basis of mean temperatures. Temperatures are usually averaged over a minimum of 10 years and preferably 30 years.
mean temperature	The average temperature for a given time period, usually a day, a month, or a year.
mechanical stabilization	Relies on physical processes to stabilize the soil, either altering the physical composition of the soil (soil blending) or placing a barrier in or on the soil to obtain the desired effect (such as establishing a sod cover to prevent dust generation).
meniscus	Curved upper surface of a water column.
metamorphic rock	Those rocks that have been altered in appearance and physical properties by heat, pressure, or permeation by gases or fluids.

METT-T	mission, enemy, troops, terrain, and time available
MH	silt, highly compressible (LL<50)
mica	Any of various colored or transparent mineral silicates crystallizing in monoclinic forms that readily separate into very thin sheets.
micaceous soil	Soil that contains a sufficient amount of mica to give it distinctive appearance and characteristics.
MIL-STD	military standard
min	minimum
mineral	A naturally occurring, inorganically formed substance having an ordered internal arrangement of atoms. It is a compound and can be expressed by a chemical formula.
ML	silts, low compressibility (LL<50)
mm	millimeter(s)
MM	medium duty mat
MMC	minimum moisture content
Mohs' Hardness Scale	A simple scale used to measure the hardness of a mineral.
MO-MAT	A commercial material used as an expedient surface.
monocline fold	A rock body that exhibits local step-like slopes in otherwise flat or gently inclined rock layers. Common in plateau areas where beds may locally assume dips up to 90 degrees.
mountain glaciation	See "Alpine glaciation."
moraine	An accumulation of earth and stones carried and finally deposited by a glacier.
MSDS	Material Safety Data Sheet
muck (mud)	The very soft, slimy silt or organic silt frequently found on lake or river bottoms.
mud cracks	Polygonal cracks in the surface of dried-out mud flats.
mudstone	An indurated shale produced by the consolidation of mud. Primarily a field term used to temporarily identify fine-grained sedimentary rocks of unknown mineral content.
mulch	A protective covering (as of sawdust, compost, or paper) spread or left on the ground especially to reduce evaporation, maintain even soil temperature, prevent erosion, control weeds, or enrich the soil.
multilayer pavement	A pavement that consists of at least two layers, such as a base and wearing course, or three layers, such as a subbase, base, and wearing course.
muskeg	Peat deposits found in northwestern Canada and Alaska.
N	north; normal force or interlocking force. The fraction of the weight of an object that rests on a surface.
naphtha	Any of various volatile often flammable liquid hydrocarbon mixtures used chiefly as solvents and dilutents.
NAVFAC	Naval Facilities Engineering Command
NE	northeast
NIOSH	National Institute of Occupational Safety and Health

no	number
nonfoliated	Massive metamorphic rocks that exhibit no directional structural features.
nonfrost-susceptible materials	Crushed rock, clean sandy gravel, slag, cinders, or any other cohesionless material in which ice segregation does not occur under natural freezing conditions. Uniformly graded soils, containing less than 10 percent of grains smaller than 0.02 mm are nonfrost-susceptible. Well-graded soils containing less than 3 percent by weight smaller than 0.02 mm are nonfrost-susceptible. In soils with less than 1 percent of grains smaller than 0.02 mm, ice lenses will not form under field conditions.
normal fault	Faults along which the hanging wall has been displaced downward relative to the foot wall. Common where the earth's surface is under tensional stress so the rocks are pulled apart. Characterized by high angle (near vertical) fault planes.
NSN	national stock number
nuclear moisture density tester	An instrument that provides real-time, in-place moisture content and density measurements of a soil.
NW	northwest
o	organic soil
obsidian	A hard, shiny, usually black, brown, or reddish volcanic glass which may contain scattered gas bubbles or visible crystals.
odor test	A test used to determine if a soil is organic. A strong odor indicates an organic soil.
off-site admixing	This is done when in-place admixing is not desirable and/or soil from another source provides a more satisfactory treated surface. This can be accomplished with a stationary mixing plant, or by windrow-mixing with graders in a central working area.
OH	organic soil, highly compressible (LL>50)
oiled earth	An earth-road system made resistant to water absorption and abrasion by means of a sprayed application of slow-or medium-curing liquid asphalts.
OL	organic soil, low compressibility ()
old-age stream	A stream in which the gradient is very gentle, the water velocity is low, there is little downcutting, and lateral meandering produces an extensive floodplain.
olivine	A very hard, dense mineral which forms yellowish-green to dark olive-green or brown glassy grains or granular masses. It is often found in very dark, iron-rich rocks, particularly gabbro and basalt.
OMC	optimum moisture content
organisms	Living creatures, plants, animals, and microorganisms. The acids produced by their life processes hasten the decomposition of rock masses near the surface. The wedging action caused by plant root growth hastens the disintegration process.
outwash plains	Result when melting ice at the edge of the glacier creates a great volume of water that flows through the end moraine as a number of streams rather than a continuous sheet of water.
overthrust fault	Low-angle (near horizontal) reverse faults.

oxidation	The chemical union of a compound with oxygen.
paneling	Solid barrier fences of metal, wood, plastic, or masonry used to stop or divert sand movement.
parabolic-shaped dunes	Crescent-shaped dunes with two tips that point upwind. They typically form along coastlines where the vegetation partially covers the sand or behind a gap in an obstructing ridge.
parallel drainage pattern	Drainage pattern characterized by major streams trending in the same direction. They indicate gently dipping beds or uniformly sloping topography.
particle size	Sizes of individual grains as determined by the use of sieves.
PBS	prefabricated bituminous surfacing
pcf	pound(s) per cubic foot
peat	A term which is frequently applied to fibrous, partially decayed organic matter or a soil which contains a large proportion of such materials. Large and small deposits of peat occur in many areas and present many construction difficulties. Peat is extremely loose and compressible.
pedology	The study of the maturing of soils and the relationship of the soil profile to the parent material and its environment.
pegmatite	A coarse variety of granite occurring in dikes or veins.
PEL	permissible exposure limit
perched water table	A localized zone of saturated soil above the normal groundwater table; created by the localized presence of relatively impervious soil layers.
period of weakening	An interval of the year which starts at the beginning of the frost-melting period and ends when the subgrade strength has returned to normal period values.
peridotite	Any of a group of granitoid igneous rocks composed of ferromagnesian minerals, especially olivine.
permeability	The property of soil which permits water to flow through it.
PF	permafrost
physical weathering	The disintegration of rock. Rock masses are broken into smaller and smaller pieces without altering the chemical composition of the pieces.
PI	plasticity index
pier foundation	A type of support normally used only for very heavy loads.
pile foundation	A load-bearing member which may be made of timber, concrete, or steel. It is generally forced into the ground.
pistol-butted trees	Downslope-tipped trees that are small as a result of sliding soil or debris or as a result of active soil creep. They are a good indicator of slope instability for areas where rain is a major component of winter precipitation.
pit-run coral	Consists of fragmental coral in conjunction with sands and marine shells.
PL	plastic limit
plagioclase	A triclinic feldspar.
plasticity	The ability of a soil to deform without cracking or breaking.
plasticity index	The difference between the liquid and plastic limits.

plastic limit	The lowest moisture content at which a soil can be rolled into a thread 1/8 inch in diameter without crushing or breaking.
platy grains	Extremely thin grains, compared to their lengths and widths. They have the general shape of a flake of mica or a sheet of paper.
plunging fold	Folds that dip back into the ground at one or both ends.
POL	petroleum, oils, and lubricants
precipitates	Salts that have become insoluble, separated from solution, and been deposited.
prefabricated mesh	Heavy woven jute mesh, such as commonly used in conjunction with grass seed operations; can be used for dust control of nontraffic areas.
psi	pounds per square inch
Pt	peat soil
pumice	A volcanic rock full of cavities and very light in weight; used especially in powder form for smoothing and polishing.
pyrite	A common mineral that consists of iron disulfide (FeS_2), has a pale brass-yellow color and metallic luster, and is burned for the manufacture of sulfur dioxide and sulfuric acid.
pyroclastics	Volcanic materials that have been explosively ejected from a vent.
pyroxene	Hard, dense, glassy to resinous minerals found chiefly in dark igneous rocks and, less often, in dark gneisses and schists. They usually occur as well-formed short, stout, columnar crystals that appear almost square in cross section.
quarry	An open excavation made into rock masses by drilling, cutting, or blasting.
quartz	An extremely hard, transparent to translucent mineral with a glassy or waxy luster.
quartzite	A compact granular rock composed of quartz and derived from sandstone by metamorphism. It is an extremely hard, fine- to coarse-grained massive rock that forms from sandstone.
quick silts	Very fine sands and silts that are compacted in the presence of a high water table that pump water to the surface.
radial drainage pattern	A drainage pattern in which streams flow outward from a high central area. Normally found on domes, volcanic cones, or rounded hills.
rakes	Inclined braces used to support wales.
RC	rapid curing
recessional moraines	Sediments deposited when a receding glacier halts for a considerable period of time.
recrystallization	To recrystallize again or repeatedly.
rectangular drainage pattern	A drainage pattern characterized by abrupt 90-degree changes in stream directions. It is caused by faulting or jointing of the underlying bedrock.
recumbent fold	A fold with an axial plane that has been inclined to the point that it is horizontal.
residual soil	Unconsolidated deposits resulting from the weathering of rock material in place.
resins	Dust palliatives used as either surface penetrants or surface blankets; usually lignin based.

retaining walls	Constructed for the purpose of supporting a vertical or nearly vertical earth bank that, in turn, may support vertical loads.
reverse fault	Results when the hanging wall of a fault becomes displaced upward relative to the foot wall. These are frequently associated with compressional forces which accompany folding.
ripple marks	Parallel ridges formed in some sediments. They may indicate the direction of wind or water movement during deposition.
road tar	Viscous liquids obtained by distillation of crude tars extracted from coal.
rockfall	A mass of falling or fallen rocks.
rock flour	Finely ground rock particles, chiefly silt-sized, resulting from glacial abrasion.
rock riprap	A foundation or sustaining wall of stones or chunks of concrete thrown together without order (as in deep water); a layer of this or similar material on an embankment slope to prevent erosion.
rockslide	A usually rapid downward movement of rock fragments that slide over an inclined surface.
rounded particles	Those in which all projections have been removed and few irregularities in shape remain. They approach spheres of varying sizes.
rough tillage	Uses chisel, luster, or turning plows to till strips across nontraffic areas. This method works best with cohesive soils that form clods.
RS	rapid setting
RT	road tar
RTCB	road tar cutback
S	south; degree of saturation; sand; permissible stress
SA	soil-asphalt
saddles	Low points on a ridge or crest line, generally a divide between the heads of streams flowing in opposite directions.
sag pond	A slump block depression that fills with water during the rainy season.
salt solutions	Water saturated with sodium chloride or other salts applied to sand dunes to control dust.
sand dune	A hill or ridge of sand piled up by the wind commonly found along shores, along some river valleys, and generally where there is dry surface sand during some part of the year.
sand grid	A honeycomb shaped geotextile measuring 20 feet by 8 feet by 8 inches deep when fully expanded. It is used to develop a beachhead for logistics-over-the-shore operations. It is also useful in expedient revetment construction.
sandstone	A hard clastic sedimentary rock composed mainly of sand-size (1/26 mm to 2 mm) quartz grains, often with feldspar, calcite, or clay.
SBST	single bituminous surface treatment
SC	slow curing; soil-cement
schist	Metamorphic crystalline rock having a closely foliated structure allowing division along approximately parallel planes. It is a fine- to coarse-grained rock composed of discontinuous thin layers of parallel mica, chlorite, hornblende, or other crystals.

SCIP	scarify/compact in-place
scoria	Rough vesicular cindery lava, common in volcanic regions and generally forms over basaltic lava flows. It is somewhat denser and tougher than pumice, and the gas bubbles which give it its spongy or frothy appearance are generally larger and more widely spaced that those in pumice.
SCS	Soil Conservation Service
SE	southeast
sedimentary rock	Formed of mechanical, chemical, or organic sediment, such as rock formed of fragments transported from their source and deposited elsewhere by water (as sandstone or shale); rock formed by precipitation from solution (as rock salt or gypsum); rock formed from inorganic remains of organisms (as limestone comprised of shells and skeletons).
seepage force	The drag force that moving water exerts on each individual soil particle in its path.
shale	A fissile rock that is formed by the consolidation of clay, mud, or silt, has a finely stratified or laminated structure, and is composed of minerals essentially unaltered since deposition.
shearing	When one portion of a block of uniform soil fails, or slides, past another portion in a parallel direction.
shear planes or plane of failure	The surface along which shearing action takes place.
shear strength	The resistance to shearing.
shelter belts	Barriers formed from hedges, shrubs, or trees which are high and dense enough to significantly reduce wind velocities on the leeward side.
shrinkage	Reduction in volume when the moisture content of a soil is reduced.
shrinkage limit	The moisture content a which a soil occupies the boundary between the semisolid and solid states.
sieve	A device with meshes or perforations through which finer particles of a mixture (as of ashes, flour, or sand) of various sizes are passed to separate them from coarser ones, through which the liquid is drained from liquid-containing material, or through which soft materials are forced for reduction to fine particles.
silicon dioxide	See "quartz."
single-layer pavement	A stabilized soil structure on a natural subgrade.
siltstone	A rock composed chiefly of indurated silt.
sinkhole	A depression in which drainage collects and communicates with a cavern or passage. Normally located in regions underlain by limestone.
SL	soil lime
slaking test	Used to assist in determining the quality of certain shales and other soft rocklike materials. The test is performed by placing the soil in the sun or in an oven to dry and then allowing it to soak in water for a period of at least 24 hours. The strength of the soil is then examined. Certain types of soil will completely disintegrate, losing all strength.
slate	A dense fine-grained metamorphic rock produced by heat and pressure action on shales so as to develop a characteristic cleavage.

slumps	Slope failures where one or more blocks of soil have failed on a hemispherical, or bowl-shaped, slip surface; they may show varying amounts of backward rotation into the hill in addition to downslope movement. They usually occur in deep, moderately fine or fine-textured soils that contain a significant amount of silt or clay.
slurry	A watery mixture of insoluble matter (as mud, lime, or plaster of paris).
SM	silty sands and poorly graded sand-silt mixture
SP	poorly graded sand
soil	The entire unconsolidated material that overlies and is distinguishable from bedrock. It is composed principally of the disintegrated and decomposed articles of rock.
soil creep	A relatively slow-moving type of slope failure.
spall	To break up (ore) with a hammer usually preparatory to crushing; to reduce (as irregular stone blocks approximately to size by chipping with a hammer; to cause to break off in spalls; to break off chips, scales, or slabs from the surface or edge often as the result of a rapid change of temperature; to split off particles as the result of bombardment in such a manner that a large part remains used of a surface, target, or nucleus. A fragment removed from a rock surface by weathering (few exfoliations detach themselves from the parent mass in the form of lenses).
specific gravity	The ratio of a substance weight (or mass) to the weight (or mass) of an equal volume of water.
"speedy" moisture content test	An accurate test providing a very rapid moisture content determination.
S-S	silica-sesquioxide ratio
SS	slow setting
stability	The ability of a soil to support loads.
stoss side	The side from which ice flows.
stratified glacial deposit	A glacial deposit consisting of layered sediments.
stratified rock	See "sedimentary rock."
strike	The trend of the line of intersection formed between a horizontal plane and the bedding plane being measured.
strike slip fault	A fault that is characterized by one block being displaced laterally with respect to the other; there is little or no vertical displacement.
subangular particles	Those particles that have been weathered until the sharper points and ridges of the original angular shape have been worn off.
subrounded particles	Those particles that have undergone considerable weathering so that they are somewhat irregular in shape and have no sharp corners and few flat areas.
surface blanket	A blanket cover over the soil surface to control dust. Materials used to form the blanket include aggregates, prefabricated membranes and mesh, bituminous surface treatments, polyvinyl acetates (with or without fiberglass scrim reinforcements) and polypropylene - asphalt membranes.
suspended load	That portion of material within a transporting medium that is lifted far

	from the earth's surface, is sustained for long periods of time, and is distributed through the entire body of the current.
symmetrical fold	A fold with a vertical or near-vertical axial plane.
synclinal fold	Downfolds.
SW	well-graded sand; southwest
talus	A fan-shaped accumulation of mixed fragments of rock that have fallen because of weathering of a cliff or steep mountainside.
tension cracks	Those cracks that relieve stress in the soil mantle.
terminal moraines	See "end moraines."
texture	The relative size and arrangement of the mineral grains making up a rock.
throw	The vertical displacement along a fault.
thrust fault	Reverse faults that dip at low angles (less than 15 degrees) and have stratigraphic displacements commonly measured in kilometers.
till	Unstratified glacial drift consisting of clay, sand, gravel, and boulders intermingled.
till plains	See "ground moraine."
TLV	threshold limit value
TM	technical manual
TO	theater of operations
topsoil	A general term applied to the top few inches of soil deposits. Topsoils usually contain considerable organic matter and produce plant life.
toughness	A rock's resistance to crushing or breaking.
transported soil	Materials that have been transported and deposited at a new location by glacial ice, water, or wind.
tranverse dune	Wavelike ridges formed perpendicular to prevailing wind direction and separated by troughs. They resemble sea waves during a storm.
trellis drainage pattern	A drainage pattern in which tributaries generally flow parallel to the main streams, eventually joining them at right angles.
trenching	The cutting of a trench either transversely or longitudinally across a dune to destroy its symmetry.
tuff	A term applied to compacted deposits of the fine materials ejected from volcanoes, such as more or less cemented dust and cinders. Tuffs are more or less stratified and in various states of consolidation. They are prevalent in the Mediterranean area.
u	undesirable base and subbase material
uniformly graded soils	Soil consists primarily of particles of nearly the same size.
unloading	A form of physical weathering that results from the relief of pressure on a rock unit due to the removal of overlying materials.
unstratified glacial deposit	Heterogeneous mixtures of different particle types and sizes ranging from clays to boulders that were directly deposited by glacial ice.
US	United States
USCS	Unified Soil Classification System
V	total volume; unit weight

Va	volume of air
varved clay	A sedimentary deposit which consists of alternating thin layers of silt and clay.
vegetative treatment	Method of soil stabilization.
vol	volume
volcanic ash	Uncemented volcanic debris, usually made up of particles less than 4 mm in diameter. Upon weathering, a volcanic clay of high compressibility is frequently formed. Some volcanic clays present unusually difficult construction problems, as do those in the area of Mexico City and along the eastern shores of Hawaii.
$\mathbf{V_m}$	wet unit weight; wet density
$\mathbf{V_s}$	volume of solids
$\mathbf{V_v}$	volume of voids
$\mathbf{V_w}$	volume of water
w	moisture content
W	total weight; west
wales	A horizontal constructional member (as of timber or steel) used for bracing vertical members.
$\mathbf{W_d}$	dry weight
weathering	The physical or chemical breakdown of rock. It is the process by which rock is converted into soil.
well-drained soil	Soils that allow for the significant transmission of water (essentially clean sands and gravels).
well-graded soil	Has a good representation of all particle sizes from the largest to the smallest and the shape of the grain-size distribution curve is considered "smooth."
wet density	The weight of a soil material (including moisture fraction) divided by the volume of the soil material. Synonymous with wet unit weight and field density.
$\mathbf{w_L}$	liquid limit
$\mathbf{w_p}$	plastic limit
$\mathbf{W_s}$	weight of solids
$\mathbf{W_w}$	weight of water
yd	yard(s)
ADM	area defense management
alidade	In plane tabling, a straight edge having a telescopic sight or other means of sighting parallel to it.
aquifer	Any geologic formation containing water.
artesian	Refers to ground water confined under hydrostatic pressure.
ASCS	Agricultural Stabilization and Conservation Service
attitude	The position of a structural surface relative to the horizontal and expressed by the strike and dip.
bedrock	The lowest level of unbroken, solid rock. It is overlaid in most places by soil or rock fragments.

borings	The chips, fragments, or dust produced in drilling or driving a hole into the earth's surface.
bpi	bits per inch
cobble	A rock fragment larger thn a pebble and smaller than a boulder (64 to 256 millimeters in diameter) that is somewhat rounded or otherwise modified by abrasion in th ecourse of transport.
conglomerates	Rocks composed of rounded fragments, varying from small pebbles to large boulders, and held togher by a natural cement.
contour line	A line connecting points of equal elevation above or below a datum plane.
CP	command post
DC	District of Columbia
DIA	Definse Intelligence Agency
EROS	Earth Resources Observation System
ERTS	Early Remote Tracking System
fluvial	Pertaining to rivers and streams.
geology	The science that deals with the physical history of the earth, rocks that compose the earth, and the physical changes that the earth has undergone or is undergoing.
geomorphology	The orgin and development of the topography of the continents.
geophysical exploration	Locating and studying underground deposits of ores, mineral, oil, gas, and water.
geosyncline	A downward trough of the earth's crust where sediment accumulates.
HCI	hydrochloric
hydrogeologic	Refers to geologic features that may indicate the presence of water.
hydrologic cycle	A cycle in which water is evaported from the sea, then precipitated from the atmosphere to the surface of the land, and finally returned to the sea by rivers and streams.
IR	infrared
LANDSAT	land satellite
leaching	The process where the more soluble compounds are removed by percolating groundwater.
lithification	Conversion of unconsolidated sediments into solid rock.
m	meter(s)
MSS	Multispectral Scanner System
photogeology	The art and science of using photo images to determine the geology of an area.
plane table	A drawing board mounted on a tripod. It is used in the filed for obtaining and plotting survey data.
PO	post office
porosity	The state or quality of being porous, expressed as a percentage of the volumn of the pores of a rock to the total volumn of its mass.
rhyolite	A fine-grained igneous rock that is rich in silica. The volcanic equivalent of granite.
schist	Metamorphic srystalline rock having a closely foliated structure that

allows division along approximately parallel planes. A fine- to coarse-grained rock that is composed of discontinuous thin layers of parallel mica, chlorite, hornblende, or other crystals.

SD	South Dakota
SLAR	side-looking radar
stratigraphic sequence	The classification, correlation, and interpretation of stratified rocks.
topography	The relief features or surface configurations of an area.
USGS	United States Geological Survey
UT	Utah

This page intentionally left blank.

References

SOURCES USED

These are the sources quoted or paraphrased in this publication.

JOINT AND MULTISERVICE PUBLICATIONS

FM 5-430-00-1/AFPam 32-8013, *Volume I. Planning and Design of Roads, Airfields, and Heliports in the Theater of Operations: Road Design* 26 August 1994.

FM 5-430-00-2/AFJPam 32-8013, *Volume II. Planning and Design of Roads, Airfields, and Heliports in the Theater of Operations: Airfield and Heliport Design.* To be published within 6 months.

TM 5-530. *Materials Testing.* (NAVFAC MO-330/AFM 89-3) 29 September 1994. (superseded by FM 5-530)

TM 5-803-13. *Landscape Design and Planting Criteria.* (AFM 126-8) 6 August 1988.

TM 5-822-2. *General Provisions and Geometric Design for Roads, Streets, Walks, and Open Storage Areas.* (NAVFAC DM5.5: AFM 88-7, Chapter 5). 14 July 1987.

TM 5-822-4. *Soils Stabilization for Pavements. (AFM 88-7, Chapter 4).* 1 April 1983.

TM 5-822-5. *Pavemen Designs for Roads, Streets, Walks, and Open Storage Areas.* (AFM 88-7, Chapter 3) 12 June 1992.

TM 5-822-8. *Bituminous Pavements - Standard Practice.* (AFM 88-6, Chapter 9) 30 July 1987. (rescinded)

TM 5-830-3. *Dust Control for Roads, Airfields, and Adjacent Areas.* (AFM 88-17, Chapter 3). 30 September 1987.

ARMY PUBLICATIONS

FM 5-446. *Military Nonstandard Fixed Bridging.* 3 June 1991. (superseded by FM 3-34.343)

TM 5-330. *Planning and Design of Roads, Airbases, and Heliports in the Theater of Operations.* 6 September 1968 (To be superseded by FM 5-430). (superseded by FM 5-430-00-1 and FM 5-430-00-2)

TM 5-337. *Paving and Surfacing Operations.* 21 February 1966. (superseded by FM 5-436)

TM 5-331A. *Utilization of Engineer Construction Equipment, Volume A: Earthmoving, Compaction, Grading, and Ditching Equipment.* 18 August 1967. (superseded by FM 5-434)

TM 5-545. *Geology.* 8 July 1971.

AR 420-74. *Natural Resources: Land, Forest, and Wildlife Management.* 1 July 1977. (superseded by AR 200-3)

Mil Std 621A. *Test Methods for Pavement Subgrade, Subbase, and Base Course Materials.* 22 December 1964. (canceled)

EM 1110-2-1901. *Seepage Analysis and Control for Dams.* 30 September 1986.

NAVFAC DM 7.1. *Design.* 1985. (obsolete)

Special Report 83-27. *Revised Procedures for Pavement Design Under Seasonal Frost Conditions*, US Army Corps of Engineers. September 1983. Office of the Chief of Engineers, Washington, DC 20314. (obsolete)

NONMILITARY PUBLICATIONS

American Society of Testing and Materials (ASTM) D1633. *Compression Strength of Molded Soils Cementing Cylinders.* 1990. (superseded by ASTM D1633-00)

Paeth, R.C., M.E. Harward, E.G. Knox, and C.T. Dyrness. *Factors Affecting Mass Movement of Four Soils in the Western Cascades of Western Oregon.* Soil Science Society of America, Vol. 35. 1971: 943-947.

Swanston, D.N. *Mass Wasting in Coastal Alaska.* US Department of Agriculture, Forest Service, Pacific Northwest Forest and Range Experiment Station, Research Paper, PNW-83. 1969:15.

Swanston, D.N. *Mechanics of Debris Avalanching in Shallow Till Soils of Southeast Alaska.* US Department of Agriculture, Forest Service, Pacific Northwest Forest and Range Experiment Station, Research Paper. 1970.

DOCUMENTS NEEDED

These documents must be available to the intended users of this publication.

ARMY PUBLICATIONS

DD 1207. *Grain Size Distribution Graph* - Aggregate Grading Chart. 1 August 1957. (This version is obsolete. A newer version is available.)

DD 1211. *Soil Compaction Test Graph/Swell Data.* December 1986. (This version is obsolete. A newer version is available.)

DD 2463. *California Bearing Ratio (CBR) Analysis.* December 1986. (This version is obsolete. A newer version is available.)

ENG 4334. *Plasticity Chart.* June 1970. (This version is obsolete. A newer version is available.)

Index

TM 3-34.64/MCRP 3-17.7G
25 September 2012

By order of the Secretary of the Army:

RAYMOND T. ODIERNO
General, United States Army
Chief of Staff

Official:

JOYCE E. MORROW
Administrative Assistant to the
Secretary of the Army
1135412

RICHARD P. MILLS
Lieutenant General, USMC
Deputy Commandant for
Combat Development and Integration

DISTRIBUTION:

Active Army, Army National Guard, and United States Army Reserve: Not to be distributed; electronic media only.